The Palgrave Handbook of Methodological Individualism

Nathalie Bulle · Francesco Di Iorio
Editors

The Palgrave Handbook of Methodological Individualism

Volume II

Editors
Nathalie Bulle
Sociology
French National Center for Scientific
Research
Sorbonne University
Paris, France

Francesco Di Iorio
Department of Philosophy
Nankai University
Tianjin, China

ISBN 978-3-031-41507-4   ISBN 978-3-031-41508-1   (eBook)
https://doi.org/10.1007/978-3-031-41508-1

© The Editor(s) (if applicable) and The Author(s), under exclusive license to Springer Nature
Switzerland AG 2023

This work is subject to copyright. All rights are solely and exclusively licensed by the Publisher, whether
the whole or part of the material is concerned, specifically the rights of translation, reprinting, reuse
of illustrations, recitation, broadcasting, reproduction on microfilms or in any other physical way, and
transmission or information storage and retrieval, electronic adaptation, computer software, or by similar
or dissimilar methodology now known or hereafter developed.
The use of general descriptive names, registered names, trademarks, service marks, etc. in this publication
does not imply, even in the absence of a specific statement, that such names are exempt from the relevant
protective laws and regulations and therefore free for general use.
The publisher, the authors, and the editors are safe to assume that the advice and information in this book
are believed to be true and accurate at the date of publication. Neither the publisher nor the authors or
the editors give a warranty, expressed or implied, with respect to the material contained herein or for any
errors or omissions that may have been made. The publisher remains neutral with regard to jurisdictional
claims in published maps and institutional affiliations.

Cover illustration: Gianluca Cavallo, "Pantheon", 2021, oil on canvas 130x130, source: https://gianlucac
avallo.com

This Palgrave Macmillan imprint is published by the registered company Springer Nature Switzerland AG
The registered company address is: Gewerbestrasse 11, 6330 Cham, Switzerland

Paper in this product is recyclable.

Contents

Methodological Individualism and Key Research Fields in the Social Sciences and Humanities

Collective Intentionality and Methodological Individualism   3
Jens Greve

Methodological Individualism and Collective Representations   29
Pierre Demeulenaere

Kelsen: Methodological Individualism in the Social Theory of Law   53
Stephen Turner

Methodological Individualism and the Foundations of the "Law and Economics" Movement   77
Jean-Baptiste Fleury and Alain Marciano

Methodological Individualism and Social Change   103
Michael Schmid

Applying Methodological Individualism in the Analysis of Religious Phenomena   127
Salvatore Abbruzzese

vi   Contents

Immigration, Identity Building and Multiculturalism: A Methodological Individualism Approach   151
Simon Langlois

Individualistic Models in Collective Norm-Based Regulation: The Positive, Normative and Hermeneutic Dimensions   177
Emmanuel Picavet

Analytical Tools and Exemplar Case Studies

Examples of Sociological Explanation in Terms of Methodological Individualism   203
Raymond Boudon

Collective Action   225
Anthony Oberschall

Methodological Individualism in Weber's Sociology of Religion   249
David d'Avray

Unintended Consequences   271
Karras J. Lambert and Christopher J. Coyne

Individual Choice and Collective Identities   297
Günther Schlee

Understanding Religious Radicalization: The Enigma of Beneficial Violence   325
Hans G. Kippenberg

Risk Takers or Rational Conformists: Extending Boudon's Positional Theory to Understand Higher Education Choices in Contemporary China   351
Ye Liu

Methodological Individualism and Formal Models   373
Werner Raub and Arnout van de Rijt

Controversial Issues Surrounding Methodological Individualism

Holistic Bias in Sociology: Contemporary Trends   403
Ieva Zake

Methodological Individualism and Reductionism   423
Francesco Di Iorio

Contents   vii

Methodological Individualism Facing Recent Criticisms from Analytic Philosophy   447
Alban Bouvier

Individualism-Holism Debate in the Social Sciences: Political Implications and Disciplinary Politics   473
Branko Mitrović

Economics: A Methodological Individualism in Search of Its Own Incompleteness   497
Olivier Favereau

Methodological Individualism in Terms of Rational-Choice and Frame-Selection Theory: A Critical Appraisal   525
Richard Muench

Interactionism and Methodological Individualism: Affinities and Critical Issues   551
Natalia Ruiz-Junco, Daniel R. Morrison, and Patrick J. W. McGinty

Explaining Social Action by Embodied Cognition: From *Methodological Cognitivism* to *Embodied Individualism*   573
Riccardo Viale

Methodological Individualism and its Critics: A Roundtable Discussion

Methodological Individualism, Naive Reductionism, and Social Facts: A Discussion with Steven Lukes   605
Steven Lukes, Nathalie Bulle, and Francesco Di Iorio

Methodological Individualism and Institutional Individualism: A Discussion with Joseph Agassi   617
Joseph Agassi, Nathalie Bulle, and Francesco Di Iorio

Methodological Individualism and Methodological Localism: A Discussion with Daniel Little   633
Daniel Little, Nathalie Bulle, and Francesco Di Iorio

viii Contents

**Methodological Individualism and Critical Realism: Questions
for Margaret Archer** 659
Nathalie Bulle and Francesco Di Iorio

Author Index 669

Subject Index 681

Contributors

Salvatore Abbruzzese Department of Sociology, University of Trento, Trento, Italy

Joseph Agassi Tel-Aviv University, Tel Aviv-Yafo, Israel;
York University, Toronto, ON, Canada

Raymond Boudon Department of Sociology, Academy of Moral and Political Sciences, Sorbonne University, Paris, France

Alban Bouvier Philosophy and Social Sciences, Institut Jean Nicod, Sciences & Lettres University, Paris, France;
Aix-Marseille University, Paris, France

Nathalie Bulle Sociology, French National Center for Scientific Research, Sorbonne University, Paris, France

Christopher J. Coyne Department of Economics, George Mason University, Fairfax, VA, USA

David d'Avray Department of History and Sociology, University College London, London, UK

Pierre Demeulenaere Department of Sociology, Sorbonne University, Paris, France

Francesco Di Iorio Department of Philosophy, Nankai University, Tianjin, China

x   Contributors

Olivier Favereau Economics, University of Paris-Nanterre, Nanterre, France

Jean-Baptiste Fleury Economics, Sorbonne University, Paris, France

Jens Greve Sociology, University of Heidelberg, Heidelberg, Germany; University of Bielefeld, Bielefeld, Germany

Hans G. Kippenberg Comparative Religious Studies, University of Bremen, Bremen, Germany

Karras J. Lambert Department of Economics, George Mason University, Fairfax, VA, USA

Simon Langlois Department of Sociology, Laval University, Quebec City, QC, Canada

Daniel Little Philosophy and Sociology, University of Michigan-Dearborn and University of Michigan, Dearborn, Ann Arbor, MI, USA

Ye Liu Department of International Development, King's College London, London, UK

Steven Lukes Politics and Sociology, New York University, New York, NY, USA

Alain Marciano Department of Economics, University of Montpellier, Montpellier, France;
THEMA Cergy Paris Université, Cergy, France;
Karl Mittermaier Centre for Philosophy of Economics, University of Johannesburg, Johannesburg, South Africa

Patrick J. W. McGinty Sociology, Western Illinois University, Macomb, IL, USA

Branko Mitrović Faculty of Architecture and Design, Norwegian University of Science and Technology, Trondheim, Norway

Daniel R. Morrison Sociology, University of Alabama in Huntsville, Huntsville, AL, USA

Richard Muench Zeppelin University, Friedrichshafen, Germany;
Otto Friedrich University, Bamberg, Germany

Anthony Oberschall Department of Sociology, University of North Carolina, Chapel Hill, NC, USA

Emmanuel Picavet Department of Philosophy, Université Paris 1 Panthéon-Sorbonne, Paris, France

Werner Raub Department of Sociology, Utrecht University, Utrecht, The Netherlands

Natalia Ruiz-Junco Sociology, Auburn University, Auburn, AL, USA

Günther Schlee Professor of Social Anthropology, Arba Minch University, Arba Minch, Ethiopia;
Director Emeritus at the Max Planck Institute for Social Anthropology, Arba Minch University, Halle/Saale, Germany

Michael Schmid Sociology, Neubiberg University, Neubiberg, Germany

Stephen Turner Department of Philosophy, University of South Florida, Tampa, FL, USA

Arnout van de Rijt Sociology, European University Institute, Fiesole, Italy

Riccardo Viale Behavioral Sciences and Cognitive Economics, University of Milan Bicocca, Milan, Italy

Ieva Zake Department of Humanities and Social Sciences, Millersville University, Millersville, PA, USA

List of Figures

Examples of Sociological Explanation in Terms of Methodological Individualism

Fig. 1   The structure of the Prisoner's Dilemma   206

Fig. 2   Comparison of model of "chain-reaction" innovation with model of individual innovation: Evolution of cumulative proportion of individuals who have introduced the innovation Months after start of process (*Source* Coleman, Katz & Menzel [1957])   217

Methodological Individualism and Formal Models

Fig. 1   Coleman's diagram   375

Fig. 2   Diekmann's (1985) Volunteer's Dilemma ($U > K > 0$; $N \geq 2$)   378

Fig. 3   Watts-Strogatz graphs (Watts & Strogatz, 1998) with $n = 20$ nodes and $k = 4$ ties per node, and random tie probabilities $p = 1$ (left), $p = 0$ (middle), and $p = 0.075$ (right)   388

Economics: A Methodological Individualism in Search of Its Own Incompleteness

Fig. 1   Map of contemporary economic theories of individualist methodology   511

List of Tables

Risk Takers or Rational Conformists: Extending Boudon's Positional Theory to Understand Higher Education Choices in Contemporary China

Table 1	A sample of the university and field form	358
Table 2	The detailed profile of respondents from different fields of study and types of institutions	360
Table 3	The detailed number and profile of respondents' socioeconomic and demographic characteristics	361

Methodological Individualism and Key Research Fields in the Social Sciences and Humanities

Collective Intentionality and Methodological Individualism

Jens Greve

1 Introduction

While collective intentionality has only recently become a topic of social philosophy, the debate about individualism has shaped social theory since the emergence of the corresponding special sciences toward the end of the nineteenth century. Both debates relate to the other, even if the relation between them has not been often discussed (Donati & Archer, 2015, pp. 36–37).

Traditionally, intentionality has been predominantly considered as a property of individuals. This holds for mainstream philosophy as well as for many approaches within the sociological tradition (even if there is a current of a dissenting perspective, like in Durkheim or in Mead). Collective intentionality designates a form of intentionality which cannot be understood simply in a summative way, i.e., as "simple summation, aggregate, or distributive pattern of individual intentionality" (Schweikard & Schmid, 2021). For example, two persons who make a walk together do not simply intend individually to go their own way. Therefore, the question arises to what extent intentionality has to be understood as a concept which has to be extended

---

J. Greve (✉)
Sociology, University of Heidelberg, Heidelberg, Germany
e-mail: jensgreve@gmx.de

University of Bielefeld, Bielefeld, Germany

© The Author(s), under exclusive license to Springer Nature
Switzerland AG 2023
N. Bulle and F. Di Iorio (eds.), *The Palgrave Handbook of Methodological Individualism*,
https://doi.org/10.1007/978-3-031-41508-1_1

beyond individual mental states. Thus, like in methodological individualism, the concept of collective intentionality prompts the question of how individual and collective phenomena are related to each other.

In the first part of this contribution, I will clarify the context of the debates on collective intentionality by means of a brief recapitulation of the concept of methodological individualism. Subsequently, reductive and non-reductive approaches to collective intentionality are presented, as well as attempts to find alternatives to these (this concerns in particular a relational version of collective intentionality).

2 Methodological Individualism

Since this chapter is on collective intentionality, I will be brief here. Certainly indisputably, the concept of methodological individualism is formulated in contrast to holistic positions. *"The contrary principle of methodological holism states* that the behavior of individuals should be explained by being deduced from (a) macroscopic laws which are *sui generis* and which apply to the system as a whole, and (b) descriptions of the positions (or functions) for individuals within the whole" (Watkins, 1973 [1952], p. 150).

In this sense, for Durkheim, social facts are phenomena that are not based on individual properties, but are external to individuals because they can exert an external constraint on them. Durkheim concluded from this that social phenomena cannot be explained by tracing them back to individual actions, but that social phenomena as entities sui generis have to be explained by other social phenomena (Durkheim, 1982 [1894]). Secondly, holism can be identified with a Parsonian outlook on role expectations as elements of system integration (e.g., Parsons et al., 1953, see esp. Chapter V).

Contrary to holism, the sociology of Max Weber does not start from social wholes, but from individuals and their actions. Methodologically, this implies that explanations of social phenomena are explanations of individual actions (see esp. Weber, 1968 [1922], Chapter 1).

> Interpretative sociology considers the individual [Einzelindividuum] and his action as the basic unit, as its 'atom'—if the disputable comparison for once may be permitted. In this approach, the individual is also the upper limit and the sole carrier of meaningful conduct… In general, for sociology, such concepts as 'state,' 'association,' 'feudalism,' and the like, designate certain categories of human interaction. Hence it is the task of sociology to reduce these concepts to 'understandable' action, that is, without exception, to the actions of participating individual men. (Weber, 1958, p. 55)

How exactly ontological and methodological perspectives on methodological individualism are connected is a controversial issue. Thus, it has been noticed that Watkins's "classical" definition of methodological individualism directly linked ontological and methodological principles (Hodgson, 2007, p. 3; Lohse, 2019, p. 48; Udehn, 2001, p. 212).

According to this principle [methodological individualism; J.G.], the ultimate constituents of the social world are individual people who act more or less appropriately in the light of their dispositions and understandings of their situations. Every complex social situation, institution, or event is the result of a particular configuration of individuals, their dispositions, situations, beliefs, and physical resources and environment. There may be unfinished or half-way explanations of large-scale social phenomena (say, inflation) in terms of other large-scale phenomena (say, full employment); but we shall not have arrived at rock-bottom explanations of such large-scale phenomena until we have deduced an account of them from statements about the dispositions, beliefs, resources, and interrelations of individuals. (Watkins, 1973 [1957], pp. 167–168)

However, it is questionable whether methodological conclusions can be derived from the ontological truism that social structures are made by people.

… it is possible, both to be a methodological individualist without endorsing ontological individualism, and to endorse ontological individualism without accepting methodological individualism. It is possible to believe in the existence of social wholes, but to advocate an individualist social science. It could be argued, for instance, as we have seen some methodological individualists do, that we only have 'direct access' to the actions of individuals, and that social science, therefore, has to be individualistic. It is also possible, and more common, to believe that only individuals exist and yet, to deny the possibility of an individualist social science. It could be argued, for instance, that large-scale social phenomena are too complex for an individualist social science to be possible. (Udehn, 2001, p. 350f.)

Nevertheless, these options do, in my view, not invalidate the existence of an elective affinity between the ontological and methodological dimension. In the first option, it seems difficult to take into account the way that individuals change social wholes. The same point can be expressed in terms of causalities. Given that social causality is due to social wholes, the observation of changing individual actions nevertheless has to be attributed to the social wholes (like in Durkheim).

6   J. Greve

With regard to the second option, the argument from complexity splits social phenomena in two classes, those that are explainable in individualist terms (the less complex ones) and those that need a holistic explanation (the complex ones). This leads to an unstable position at least in cases where these explanations interpenetrate. This can be seen, e.g., in the position taken by Jepperson and Meyer. "Of course", they argue, "all causal social processes work through the behaviors and ideas of individual persons—this 'ontological truism' […] is a basic premise of all post-Hegelian naturalist social science. But this premise (sometimes called 'ontological individualism') in no way necessitates an explanatory (or 'methodological') individualism" (Jepperson & Meyer, 2011, p. 56).

Jepperson and Meyer distinguish three levels of explanation: the individual level (Jepperson & Meyer, 2011, p. 61) and two structural levels. First: "Social-organizational processes refer to the causal influences attributed to (for example) hierarchic, network, market, and ecological formations" (Jepperson & Meyer, 2011, p. 62). Second: "Institutions are chronically reproducing complexes of routines, rules, roles, and meanings" (Jepperson & Meyer, 2011, p. 64).

Jepperson and Meyer argue that these levels can be distributed to different lawyers of organized action.[1]

> Put another way: individual-level causal pathways capture effects produced by relatively unorganized people. Here causation is generated by their subsocial or elementary social characteristics. In contrast, more structural imageries capture causation generated by more and more collective and complexly organized activities. […] These patterns will have distinct microinstantiations, having to do with (for instance) the opportunities or communications available to people (or other units). But the causality lies with the effects of the role networks themselves, not with variations in the unit-level properties. Similarly, in more institutional imagery, the featured causal influences are those generated by broad complexes of organization and meaning. The micro-instantiations might have to do with (for instance) complementary sets of role enactments, or with enactments of broad ideological scripts. But the causality emerges from the complex organization of roles, routines, and meanings that we call institutional structure—and not from variations in role enactments or in the individuals involved. (Jepperson & Meyer, 2011, p. 68)

---

[1] Following Zahle, Jepperson's and Meyer's position may be classified as "temperate holism" which holds that "[e]xplanations in the social sciences must sometimes be strict holist explanations, i.e. refer to social wholes as wholes, their actions, properties, etc. only" (Zahle, 2007, p. 316).

The upshot of this is the following: Combining an individualist ontology and a holistic methodology does not lead to a perspective where the individualist methodology can be entirely dispensed with (at least not in the cases of "unorganized actions"). If we argue that the recourse to individuals also cannot be dispensed with in cases where structures and institutions come into play—something Jepperson and Mayer to a certain extent admit when they argue that the boundaries of the three levels are blurred (Jepperson & Meyer, 2011, p. 62)—, what we arrive at is a structural-individualist methodology, not a structural methodology.

It has often been objected to the methodological individualism that its ontological or conceptual basis is too narrow to use it as a successful explanatory model.

For example, Steven Lukes argued that already in the basic formulation of methodological individualism exists a contradiction: "It is worth adding that since Popper and Watkins allow 'situations' and 'inter-relations between individuals' to enter into explanations, it is difficult to see why they insist on calling their doctrine 'methodological individualism'" (Lukes, 1968, p. 127).

The same point has been made by Mario Bunge and by Geoffrey M. Hodgson. Bunge proposed to distinguish three basic positions, the individualistic, the holistic, and the systemic. While, according to Bunge, holism is untenable because of its dualism, individualism suffers from its atomism. Bunge contrasts this with systemism, which does not start from single individuals, but from individuals in relations (Bunge, 1996, p. 268). If we equate individualism with atomism it is easy to argue that individualism is not a viable concept of sociality. But it is wrong to see methodological individualism as atomistic (Bulle, 2018; Di Iorio, 2023; Lohse, 2019). For example, Weber's methodological individualism is relational from the beginning since social action consists of the orientation to other actors. It is also not atomistic in the sense Heath describes it:

> The atomistic view is based upon the suggestion that it is possible to develop a complete characterization of individual psychology that is fully pre-social, then deduce what will happen when a group of individuals, so characterized, enter into interaction with one another. Methodological individualism, on the other hand, does not involve a commitment to any particular claim about the *content* of the intentional states that motivate individuals. (Heath, 2020)

Hodgson sees two versions of methodological individualism that may be conflated (the individualist and the systemic in Bunges terminology):

As emphasised above, much of the confusion in the debate over methodological individualism stems from whether methodological individualism means one or other of the following: (a) social phenomena should be explained entirely in terms of individuals alone; or (b) social phenomena should be explained in terms of individuals plus relations between individuals. The first of these versions (a) has never been achieved in practice, for reasons given above. It has been also shown above that many advocates of methodological individualism fail to specify this doctrine clearly in such narrow terms. It is just as well, as version (a) is unattainable in practice. By contrast, the problem with the second version (b) is that not that it is wrong but the term 'methodological individualism' is unwarranted. The critique here is brief but no less devastating. In modern social theory, structures are typically defined as sets of interactive relations between individuals. (Hodgson, 2007, p. 8)

As with the critique by Lukes, this conclusion is problematic because even strong individualists like Max Weber did not exclude relations between individuals from their individualism. "The term 'social relationship' will be used to denote the behavior of a plurality of actors insofar as, in its meaningful content, the action of each takes account of that of the others and is oriented in these terms" (Weber, 1968 [1922], pp. 26–27).

Nevertheless, the question is not whether relations are included in the analysis but how their ontological status is conceived of. For Hodgson, methodological individualism fails because, with the inclusion of relations, an element is included which cannot be reduced to individual perceptions and beliefs. But this does not follow necessarily from the inclusion of situations or relations. Individuals necessarily are directed to situations. In these situations, it can be maintained that in contrast to natural objects, these relations are not independent of the perception of individuals.

The same point can be made with regard to institutions. Even a strong individualist like Weber does not exclude institutions (understood broadly as regularities of actions or as normative rules) as part of an explanation. But they do not enter as distinct entities with causal powers of their own. In contrast to Hodgson, in the case of Weber's methodological individualism institutions do not enter explanations as actor-independent factors but as parts of the orientations of actors.

… the subjective meaning of action must take account of a fundamentally important fact. These concepts of collective entities which are found both in common sense and in juristic and other technical forms of thought, have a meaning in the minds of individual persons, partly as of something actually existing, partly as something with normative authority. […] Actors thus in part orient their action to them, and in this role such ideas have a powerful, often a decisive, causal significance on the course of action of real individuals. (Weber, 1968 [1922], p. 14)

Based on this, it is easy to determine what, for Weber, the causal relationship between institutions and action consists of, i.e., in action causality, according to which the reasons of the actor are the causes of his actions (Huff, 1984, p. 71). From this point of view, it is now understandable why it can be doubted that, within the framework of a Weberian sociology, one can speak of a causal autonomy of the institutions vis-à-vis the actor or the action. This does not rule out the possibility that institutions are the result of unintended effects, because institutions can of course owe their existence to the insight into the need for regulation in view of the external effects of action. On the one hand, action can have material effects that were not intended by the actor. On the other hand, the actions of others can lead to the action producing unintended effects (others might react in ways not intended or foreseen by an actor). However, it does not follow from this that there can be an autonomy of action processes vis-à-vis actors within the framework of Weberian sociology, but only an autonomy of the effects of actions measured against the respective intentions of the individual actor.

Thus, the question for methodological individualism is not whether relations, structures, and institutions are included in an analysis, but in which way they enter into the analysis.

Nathalie Bulle argues that the distinguishing trait between methodological individualism and holism consists in the way causality is conceived of. "MI considers rational capacities as causal powers on one hand, and social structures as the situational properties underlying the subjective meaning of/the reasons for individual actions on the other hand" (Bulle, 2018, p. 7).

In sum, the question for individualism is not whether explanations refer to relations, structures, or institutions or not but whether they are treated as part of the situation that can have causal effects independent from the actor's own orientation to the situation or not.

I consider this also to be the criterion whether an explanation will count as reductive or not. Reductive is methodological individualism insofar as the idea of an autonomous causal influence exerted by social entities vis-a-vis individuals is discarded. The notion of reduction is multifaceted. As Bulle and Francesco Di Iorio rightly claim, methodological individualism is opposed to atomistic or psychologistic reductionism (Bulle, 2018; Di Iorio, 2023). For methodological individualism, action explanations are always explanations by reasons as causes (Bulle, 2018, p. 15). This does, of course, not rule out that elements of the situation are not being seen by the actors. These elements can sometimes explain the *effect* of an action. The overlooked branch might let you stumble; unrecognized expectations of others can make your behavior inappropriate—given it is considered as such—; individual preferences can lead to segregation (Schelling, 1978) and so on. Nevertheless, an explanation by reasons as causes does not become questionable due to the existence of unintended effects of action, because the action itself cannot be explained by its effects.

3   Collective Intentionality

Collective intentionality is understood to denote a variety of approaches and phenomena: taking a walk together, painting a house together, dancing a tango, but also the beliefs of groups or organizations (such as university faculties). The focus of the analysis, however, is mostly on simple forms of collective intending and not on those that are inherent in more complex social formations, which are usually based on explicit common agreements (Bratman, 1993, p. 98). It is no coincidence that these simple forms are the focus of the work on collective intentionality because they display where the characteristic of collective intentionality is seen (nevertheless, Kutz argues that the more institutionalized forms of collective intentionality should be given more attention, Kutz, 2000). Thus, it makes a difference whether two people walk next to each other or whether they go for a walk together. Since the movement of their bodies is the same in both cases, the difference between the two cases must lie in the intentional structures realized by those who walk.

Collective intentionality encompasses more than just knowing that actions are related to each other. In this sense, collective intentionality goes beyond strategic interdependencies (in which the participants also coordinate their behavior with regard to the behavior of others). There's also something like committing to a common goal and committing to shared support—as in building a house together, lifting a piano, etc. (Bratman, 1999).

What matters is that not only do both parties want to go for a walk but that they do so in a way that relates to each other's intentions in an appropriate way. Reductionist approaches attempt to clarify these intentional structures without deviating from the assumption that intentionality is individual intentionality.

Approaches that are labeled "reductionist" in the debates about collective intentionality assume that the mental states that are required, e.g., to go for a walk together, can be represented as individual intentions. Non-reductionist approaches, on the other hand, assume that an analysis which understands collective intentionality in this way, misunderstands the essence of collective intentionality.

The question of reducibility concerns three dimensions. The first is the dimension of "content", i.e., the question whether or not the meaning of what it means to do something together can be understood by the individual intentions.

The second dimension is called the dimension of "mode". Proponents of a non-reductionist approach to mode argue that to relate to others in collective intentionality has to be understood in a fundamentally different way from relations to others marked by an individual mode, like in strategic interdependencies not marked by the idea of doing something together.

The third dimension refers to the question of the "subject" of collective intentionality, i.e., the question whether the bearers of it are individuals or some kind of a collective entity. Since the proponents of the "collective" perspective are non-dualists, this dimension is not framed in an ontological way—the idea of an independent collective substance is rejected. Rather the non-reductionist perspective means that the idea of collective intentionality entails the assumption that some kind of collective bearer has to be assumed in order to understand the concept of collective intentionality.

4 The Reductionist Perspective

Perhaps the best-known "reductionist" perspective is that of Michael Bratman. This does not exclude that intentional states are determined by factors external to the individual (Bratman, 1993, p. 112), but that shared intentionality exists beyond individual states.

> Supposing, for example, that you and I have a shared intention to paint the house together, I want to know in what that shared intention consists. On the one hand, it is clearly not enough for a shared intention to paint the house together that each intends to paint the house. Such coincident intentions do not even insure that each knows of the other's intention or that each is appropriately committed to the joint activity itself. On the other hand, a shared intention is not an attitude in the mind of some superagent consisting literally of some fusion of the two agents. [...] My conjecture is that we should, instead, understand shared intention, in the basic case, as a state of affairs consisting primarily of appropriate attitudes of each individual participant and their interrelations. (Bratman, 1993, pp. 98–99)

Bratman's basic analysis reads:

We intend to J if and only if

1. (a) I intend that we J and (b) you intend that we J
2. I intend that we J in accordance with and because of 1a, 1b, and meshing subplans of la and 1b; you intend that we J in accordance with and because of 1a, 1b, and meshing subplans of 1a and 1b.
3. 1 and 2 are common knowledge between us. (Bratman, 1993, p. 106)[2]

In "I intend that we J" (Bratman, 1999) Bratman addresses three main objections to his analysis. What all three critiques have in common is that they assume that Bratman's attempt to explain shared intentionality through the intentions of the people involved presupposes a problematic concept of intentionality.

According to Stoutland (1997), actors can only intend something that they can do themselves. "An agent can intend only to do something herself.[...] We may call this the own action (OA) condition" (Bratman, 1999, p. 148). A weaker version, which Bratman calls a control condition, can be found in Baier (1997). According to this condition, "one cannot intend what one does not take oneself to control" (Bratman, 1999, p. 149). While in the case of the first condition it is not possible to have an intention that, for example, another person should paint a house, under this condition, it would be possible to intend to do so provided that there was sufficient control over the other person's actions (Bratman, 1999, p. 149). In contrast, Velleman (Velleman, 1997) considers the control condition as insufficient. While it is possible for one person to control the action of another if there is sufficient one way or reciprocal control of actors, a problem arises when this control is not yet in place.

Overall, the following problem arises from the objections: there can only be talk of a joint intention if all those involved already have a "participatory intention" (intending the subplans to mesh). However, from the point of view of the single participants, this cannot literally be intended for the others. "Consider [...] roommates who are intending to make dinner together. According to Bratman, each must have the intention that 'we will make dinner'. But neither roommate could bring it about that the group makes dinner together. That would require one person's intentions determining what another would do" (Risjord, 2104, p. 192). The concept of a genuine collective intentionality derived from individual intentions thus seems to become unattainable. Bratman accepts that the problem exists, but

---

[2] Some extensions and explanations are given later (Bratman, 2014). They can be ignored in our context.

points out that these objections do not preclude the existence of such shared participatory intentions (Bratman, 1999, p. 159).

A second criticism, developed mainly by Searle, consists of the argument that the analysis is not sufficiently equipped to exclude counterexamples (see below).

Hence, the general objection to a reductive analysis along the lines proposed by Bratman is that what constitutes a shared intention cannot be described adequately on the basis of individual intentions (Bratman, 1999, p. 145; Schweikard, 2008, p. 22; Schweikard & Schmid, 2021; Tuomela, 2005).

5   Searle's Proposal: Irreducibility and Individualism

The objection to a reductive analysis, that what constitutes a shared intention can only be determined insufficiently, motivates John Searle's thesis, which assumes the irreducibility of the structure of collective intentionality. "Collective Intentionality is a biologically primitive phenomenon that cannot be reduced to or eliminated in favour of something else" (Searle, 1995, p. 24). At the same time, however, Searle assumes that the ability to use the concept of collective intentionality can, in principle, belong to a solitary subject.

Let us look at the reductionist version that Searle criticizes, i.e., the analysis proposed by Raimo Tuomela and Kaarlo Miller. It should be noted that Tuomela and Miller's proposal differs from Bratman's analysis in a number of respects. As already noted before, in the analysis of collective intentionality, three ways in which collectivity is asserted are distinguished: in terms of *content*, *mode*, and *subject* (Schweikard & Schmid, 2021). In the case of content, it is a person's intention to do something together—this is the substance of the intention. This approach is found in Bratman. Approaches that proceed from the mode also tie them to individual intentions, but these form a specific type of intentions that can be distinguished from I-intentions. Tuomela and Miller, who speak of a we-mode, choose this approach: "In all, the we-mode in the case of joint intention amounts to saying that the participants must have collectively accepted 'We together will do X' (or one of its variants) for their group, and they must have collectively committed themselves to doing X" (Tuomela, 2005, p. 333). In contrast to Bratman's analysis, for Tuomela and Miller, collective intentionality thus contains a reference to a group (on the question of how the group exists, see below in Sect. 6).

A member Ai of a collective G we-intends to do X if and only if

(i). Ai intends to do his part of X;

(ii). Ai has a belief to the effect that the joint action opportunities for X will obtain, especially that at least a sufficient number of the full-fledged and adequately informed members of G, as required for the performance of X, will (or at least probably will) do their parts of X;

(iii). Ai believes that there is (or will be) a mutual belief among the participants of G to the effect that the joint action opportunities for X will obtain. (Tuomela & Miller, 1988, p. 375)

In Searle's assessment, the analysis suffers from the fact that it is not sufficient, i.e., there are cases that meet the definition but intuitively do not meet the sense of a common intention. He gives the following example that starts with business people who have studied Adam Smith's doctrine of the invisible hand in a business school and then develop the following intentions and beliefs in terms of Tuomela and Miller's analysis:

1. A intends to pursue his own selfish interests without reference to anybody else, and thus, he intends to do his part toward helping humanity.
2. A believes that the preconditions of success obtain. In particular, he believes that other members of his graduating class will also pursue their own selfish interests, and thus help humanity.
3. As A knows that his classmates were educated in the same selfish ideology that he was, he believes that there is a mutual belief among the members of his group that each will pursue their own selfish interests, and that this will benefit humanity. (Searle, 1990, p. 405)

From Searle's point of view, the conditions given by Tuomela and Miller might be given, but, nevertheless, there is no collective intentionality. Searle's conjecture is that this type of counterexample poses an insurmountable obstacle to reductionist analysis, since a direct move to circumvent this problem leads to circularity. "We are tempted to construe 'doing his part' to mean doing his part toward achieving the *collective* goal. But if we adopt that move, then we have included the notion of a collective intention in the notion of 'doing his part'" (Searle, 1990, p. 405). Hence the analysis would become circular. For Searle, there is a dilemma: either we accept circularity or we end up with inadequate analyses. His supposition is that the notion of cooperation cannot be broken down into individual ego intentions, even when mutual beliefs are added (Searle, 1990, p. 405). Searle's thesis is not that of a universal proof of impossibility—therefore it could be that a new analysis satisfies the claim of being sufficient and non-circular (Searle, 1990, p. 406)—, even if Searle ultimately believes that either an infinite regress or circularity will occur (Searle, 1995, p. 95). Since it might prove impossible to provide such an analysis, he outlines an alternative, saying that we deal with a 'primitive' phenomenon. His second argument against reductionist approaches is that it is not necessary for collective intentions, "we-intentions", to be reduced in content to "I intend statements" (Searle, 2010, p. 46).

According to Searle's analysis, the mode of collective intentionality cannot be reduced to individual intentions and beliefs, but he nonetheless considers the bearers (the subject) of it as individuals only. Searle formulates this by stating two conditions for a concept of collective intentionality:

(1) It must be consistent with the fact that society consists of nothing but individuals. Since society consists entirely of individuals, there cannot be a group mind or group consciousness. All consciousness is in individual minds, in individual brains.

(2) It must be consistent with the fact that the structure of any individual's intentionality has to be independent of the fact of whether or not he is getting things right, whether or not he is radically mistaken about what is actually occurring. [...] One way to put this constraint is to say that the account must be consistent with the fact that all intentionality, whether collective or individual, could be had by a brain in a vat or by a set of brains in vats. (Searle, 1990, pp. 406–407)

Searle's concept of collective intentionality is very influential. Those who view individualistic analyses of collective intentionality as circular refer not least to Searle's critique of reductionist proposals (Schmid, 2005, p. 161). Nevertheless, many have criticized the strong internalism of Searle's analysis. Meijers summed up these criticisms when she formulated: "There has to be somebody 'out there,' so to speak, for collective intentionality to be possible" (Meijers, 2003, p. 179). Radical internalism probably cannot be refuted definitively (Johansson, 2003), but there are a number of problems that arise for Searle.

Thus, the question arises of what the collective aspect of collective intentionality can consist of, if we-intentions can also be located in brains in a vat. The radically individualistic approach leads to a skeptical problem. Why can actors or observers assume that the respective contents of the we-intention are identical at all? Searle himself states: "Collective intentionality in my head can make a purported reference to other members of a collective independently of the question whether or not there actually are such members" (Searle, 1990, p. 407). But then the question arises, what exactly is collective here (Donati & Archer, 2015, p. 40; Gilbert, 2015, p. 841).

Ingvar Johannsen (Johansson, 2003) points out that Searle's radical mental internalism is in tension with his ontological stance. Searle wants to avoid a dualism between the mental and the physical by claiming that they are not separate worlds. However, since physical reality goes beyond the physical realizations in brains, one cannot assume radical internalism with regard to the physical realizations of the brain (Johansson, 2003, p. 249).

Since Searle also assumes that people can be wrong with regard to the existence or non-existence of the collective intention, external criteria are required for the assessment of the existence or non-existence. However, within a radically internalistic conception, these are not accessible at all, regardless of the states in the respective brains. At this point, it can also be argued in the spirit of Ludwig Wittgenstein that there have to be indicators of mental states that are understandable to others: "Suppose that everyone had a box with something in it which we call a 'beetle'. No one can ever look into anyone else's box, and everyone says he knows what a beetle is only by looking at his beetle.—Here it would be quite possible for everyone to have something different in his box. […] … the box might even be empty" (Wittgenstein, 2009 [1953], para 293). And we can add Donald Davidson's thought that intelligibility presupposes that we cannot be completely mistaken about world states (Davidson, 1984, p. 168). But brains in a vat must be necessarily mistaken in a fundamental sense.

Interestingly, there is also a building block in Searle's concept of sociality that speaks against the possibility of a solipsistic conception of individualism. He assumes that intentionality is to be located in a network of intentions and these in turn are dependent on background capacities (the background) (Searle, 2010, p. 30). And he assumes that these background assumptions must be shared for a society to function (Searle, 2010, p. 156).

Searle wants to block the path to a collectivist idea of subjecthood by the strict internalism of his second condition. But is this compelling and are there no alternatives to it? Individualism, as will be argued below, by no means requires internalism. Thus, it is possible to argue individualistically and against solipsism, that is, for Searle's first condition but against the second. In the following, we will first consider another dimension of the debate, namely the question of whether there is an independent subject of collective agency beyond individuals.

6 Collective Subjects

Group Minds

The thesis of an independent group mind is represented by Philip Pettit. In doing so, he not only starts from an assumed failure of the reduction of we-intentions (Pettit, 2003, p. 187) but also from two further observations. Based on a voting dilemma described by Kornhauser and Sager, which consists of the fact that the voting results in collective decisions are not independent of the procedures, Pettit asserts an independent intentionality of groups. One example Pettit gives is as follows.

	p?	q?	r?	p&q&r?
A judges that	not p	q	r	not p&q&r
B judges that	p	not q	r	not p&q&r
C judges that	p	q	not r	not p&q&r
A-B-C judges that	p	q	r	not p&q&r

> Suppose that a group of three people, A, B and C, have to make up their views as a corporate agent on four issues: whether p, whether q, whether r and whether p&q&r. And imagine that the group is member-responsive in a majoritarian way, being disposed on any issue to form the judgment supported by a majority of members. The matrix ... shows that majority voting may lead them to judge as a group that p, that q, that r and—on the basis of a unanimous vote—that non-p&q&r. Thus it shows that if the group is to satisfy rational sensitivity, as the simulation of agency requires, then it must breach majoritarian responsiveness. (Pettit, 2014, p. 93)

Similar to Arrow's theorem or Condorcet's paradox, there is no clear rational way of aggregating individual attitudes. Pettit adds another argument: groups have their own history that shapes them in terms of their current decisions (Pettit, 2003, p. 178). Both arguments justify assigning groups the status as intentional and personal subjects (Pettit, 2003, p. 175). Pettit accepts that we are dealing with different subjects than those represented by ordinary persons, because groups do not have their own faculties of perception and memory and do not form graduated degrees of beliefs and preferences (Pettit, 2003, p. 182). Nevertheless, according to Pettit, their ontological status is independent but not mysterious, since groups supervene over individuals and thus ultimately remain bound to them: "that if we replicate how things are with and between individuals in a collectivity—in particular, replicate their individual judgements and their individual dispositions to accept a certain procedure—then we will replicate all the collective judgements and intentions that the group makes" (Pettit, 2003, p. 184).

Then, however, the fact that no clear aggregation rule for collectively binding votes can be found does not in itself speaks in favor of explaining the choice of a method of aggregation as the achievement of the group as such (it ultimately consists in the "dispositions" of the individuals). In other words, which voting rule is used is itself the result of individual decisions. Since ultimately only natural persons act—"the group can only act through the actions of its members" (Pettit, 2003, p. 183)—, the decisions for the choice of actions characteristic of the group must be located in the beliefs and desires of the natural persons.

The fact that there is no real scope for independent group processes is shown by the fact that in cases of doubt (i.e., a possible contradiction between collective and individual assessments), the individual takes precedence: "I hold that natural persons have the inescapable priority and that in this kind of case it will be up to the natural person to decide whether or not to cede place to the institutional, acting in furtherance of the collective goal and in neglect of his or her priorities" (Pettit, 2003, p. 190).

Finally, Pettit believes that the question of individualism/collectivism and that of atomism/non-atomism (meaning the difference between internalist and externalist positions) is independent of the question of the existence of a group mind (Pettit, 2003, p. 191; 2014). Ultimately, it is a question of the singularity or non-singularity of groups (Pettit & Schweikard, 2006, p. 36). "The discussions of joint action and group agency are all directed, one way or another, toward this question. Those discussions culminate, if we are right, in the claim that, yes, there can be group agents of this relatively novel variety. Singularism, quite simply, is false" (Pettit & Schweikard, 2006, p. 36). While the thesis of non-singularism (group mind) might be combined with collectivism, Schweikard and Pettit hold that in the case of human groups, it is more plausible combined with an individualistic thesis, "because the story we told in arguing for non-singularism certainly involves people coming together on the basis of their individual, intentional attitudes" (Pettit & Schweikard, 2006, p. 36–37).

Plural Subjects

Margaret Gilbert does not share Pettit's view that it is possible to speak of a group's mind. Nevertheless, this is only a terminological difference. Her reason for rejecting Pettit 's view is that the attribution of mind presupposes a capacity for experience and consciousness, but this is not the case with groups (Gilbert, 1992, p. 309). As we have seen, Pettit also rejects the view that groups can be equated with beings capable of sensations.

Nevertheless, Gilbert concurs with Pettit in assuming that there can be beliefs of groups, and she also opposes "summative" notions according to which group beliefs are functions of individual beliefs. According to her, a number of "summative analyses" can be distinguished here: group beliefs could be a function of beliefs shared by all, they could be based on the beliefs of the majority, they could be determined by a representative group or a representative individual. Gilbert rejects all of these analyses. At the heart of her critique is the use of a test question, i.e., that it seems legitimate to use sentences of the type such as "The University's research board believes that women are more creative than men" (Gilbert, 1992, p. 255). In her view, the summative requirements do not satisfy this claim. Let us look at her alternative proposal.

In Gilbert's view, group beliefs can be said to exist when the members of a group have mutually acknowledged, as a group, to hold the belief "that p". Shared acceptance, according to Gilbert, means that there is shared knowledge among all participants that they hold this belief (Gilbert, 1992, p. 306). According to Gilbert, the bearer of a group belief is a plural subject. Now, according to Gilbert, one merit of this analysis is that it becomes possible to separate group beliefs from personal beliefs. A personal deviation from the group belief is therefore conceivable. Thus, statements such as the following are possible without contradiction: "Qua a member of Tom's family I may believe Tom should have got the job; qua department member I may be of the opinion that he was the worst candidate; as for my personal view, I may think he fell somewhere in the middle" (Gilbert, 1992, p. 305).

Thus, group belief is not based on A and B individually believing X in order to speak of a group belief. This is not necessary because Gilbert holds that group belief is based on the intention to accept a belief as a group. From a purely logical point of view, "We believe that P is a suitable candidate" and "I believe that P is a suitable candidate" need not be related (Gilbert, 1992, p. 304). There may be a tendency on the part of group members to conform personal beliefs to group beliefs, and conversely, individual beliefs may also influence group beliefs. It is therefore likely to be observed that group beliefs are shared by at least some group members (Gilbert, 1992, p. 304). Gilbert can leave this relationship between group belief and individual belief open because group belief is not defined on the basis of individual belief.

At the same time, however, the group belief remains bound to the individual intention, in the sense that without the acceptance to share the group belief, the group belief does not come into being at all. An "independent" group conviction does not exist. "I suggest that what is both logically necessary and logically sufficient for the truth of the ascription of group belief here is, roughly, that all or most members of the group have expressed willingness to let a certain view 'stand' as the view of the group" (Gilbert, 1992, p. 289). Now this is not only true for the group belief but already for the group (the plural subject) itself. For Gilbert, plural subjects emerge when there is a mutual willingness to share certain actions (or to contribute to an action goal) and when there is a mutual knowledge of this willingness (Gilbert, 1992, pp. 185, 225).

Beliefs of groups are accordingly based on a two-stage process. First, a group is formed, and then there is a common recognition of holding a particular belief as a group. The two-stage process also explains why Gilbert could accuse Bratman's proposal of allowing a unilateral terminability of collective intentionality (Gilbert, 2015, p. 840). For her approach, at first sight, this problem does not arise because the recognition of commonality has already occurred before the individual opinions may differ. As a second-order problem, however, it also returns in her approach, because if a group member terminates his/her membership, she/he is also no longer subject to the compulsion to recognize the view shared by the group as a group view.

Even though the process of group formation may well be implicit (Gilbert, 1992, p. 197), Gilbert's model can be considered a contractualist conception, which is in no necessary contradiction to an individualist conception, because group persuasion is based on individual agreements to hold a certain view on behalf of the group (Gilbert, 2010). For some critics, such as Hans-Bernhard Schmid, this individualistic foundation goes way too far: "…the ontological basis of plural subject theory is an almost atomistic concept of the individual, which underlies the plural subject with its pre-constituted intentions" (Schmid, 2005, p. 220).

7   The Relational Alternative

Schmid, in contrast, aims for a non-reductionist conception. Against Searle's individualism (which he calls "subjective" individualism—because it refers to agency, as opposed to "formal" individualism, which refers to the content of intention), Schmid invokes Bratman's consideration that shared intentions consist of an interlocking web of intentions. "Shared intentions are intentions of the group. But I argued that what they consist in is a public, interlocking web of the intentions of the individuals" (Bratman, 1999, p. 143).[3]

Schmid assumes that by starting from this intersubjective-relational concept, the individualistic reduction thesis can be avoided as well as the thesis of a collective subject. Schmid's criticism of reductionism starts from the supposed failure of formal individualism (here he follows Searle), but beyond that, Schmid also doubts Searle's individualistic theory of agency. As Schmid argues, following Anthonie Meijers (Meijers, 2003), the assumption that a single individual (a brain in a vat) is capable of possessing collective intentionality is problematic (see above). Meijers' and Schmid's critique demonstrates that the double critique, that of a notion of a collective subject and that of individualism, only leads to a critique of individualism if one equates it with solipsism. Searle's position is that of philosophical internalism, i.e., that a person's mental states are determined solely by his "internal" states. It is precisely this premise that Meijers and Schmid rightly doubt.

In place of an internalist position, they maintain an externalist one, i.e., one according to which the states of consciousness of a person are dependent on states in his environment. Now this assumption does not contradict individualism. The latter, too, can assume that natural and social objects can be found in the environment of individuals to which individuals relate. Natural and social objects occur in situations. Social objects are other actors who are also able to orient themselves meaningfully to the actions of others (Parsons, 1951, p. 4; Parsons et al., 1951, pp. 14–15). What Meijers and Schmid's critique eliminates is thus an analysis according to which a collective intention is present when an isolated individual has a collective intention. What is not ruled out is an analysis of the form according to which collective intention consists of two or more mutual oriented persons. So, we can state that we are still dealing with an individualistic view when we say that shared intentions presuppose that at least two individuals have appropriate intentions.

The criticisms by Meijers and Schmid can be agreed exactly at the point that the thesis of a collective subject is undermined by a relationist conception. What Schmid seems to ignore is the fact that the relational character of social phenomena does not refute the thesis that social phenomena are ultimately based on individuals (Schmid, 2005, p. 237). In Schmid's case, a vacillation can be observed here. He too wants to hold on to the dependence on individual intentions, but nevertheless, he does not want to understand collective intentionality as a product of individual intentions. Schmid's analysis therefore leads to an ambiguous situation. If A and B must intend to walk

---

[3] Public here refers to common knowledge.

together in order for collective walking to occur, but this cannot be traced back to individual intentions, what additional intending must be added in order to speak of collective intending? On the one hand, Schmid wants to hold on to the necessity of the appropriate individual intentions; on the other hand, it does not seem sufficient if these are given.

> It is not the case that what is related here - what individuals think, feel, or intend when they think, feel, or act together - is logically independent of the relation itself, that is, the commonality of this thinking, feeling, or acting. In a sense, the relation of relation and relata here is one of mutual founding. No common intending without individuals intending - but the relevant intentionality of individuals is what it is only in relatedness itself. (Schmid, 2005, p. 239 [my translation])

In a reductive version, collective intending is dependent on there being at least one second person who also has an appropriate intention. In Schmid's formulation, it is not only the adequacy of the intentions but also the presence of the relation itself that is necessary for collective intending to be present. But how can the relation depend on more than the presence of the appropriate intentions of the participants? The alternative is to adhere strictly to the relational analysis as a complete analysis. For example, if we assume that it has not been possible to give a non-circular definition of friendship, it does not follow that a friendship can exist even if the persons who are friends do not have the appropriate attitudes toward each other.

A clear ambivalence can also be found in the proposal of Margaret S. Archer and Pierpaolo Donati. They also consider a relational view as an alternative to individualism and holism. Like Meijers and Schmid, they are not convinced by Bratman's and Searle's individualism. Against a holistic alternative, however, they too hold to the necessary attachment to individuals.

On the one hand, their concept of the relational subject is also directed against the assumption of an ontological priority of relations over consciousness. Instead, they emphasize the starting point with the human subject: "A subject is, first and foremost, an agent and actor apprehended in his or her singularity as a human person" (Donati & Archer, 2015, p. 53). On the other hand, according to the authors, the relation forms its own causally effective reality that cannot be attributed to the actions or exchange relations between individuals. Archer and Donati rightly emphasize that human subjects are shaped by their relations to their environment and to other persons (Donati & Archer, 2015, p. 53). But how does one move from this thesis to the assumption that relations themselves (as opposed to the individuals who bear them) can gain an independent and causally effective existence?

Subject-dependency and subject-independency can hardly be asserted together. The relational alternative here depends on an emergent-theoretical argument, which is in itself extremely controversial.

> A Relational Subject is a subject who exists only in relations and is constituted by the relations that he/she cares for, that is, the subject's concerns. By this we do not claim that the social relation is a subject in and for itself, but rather that the relation has its own (sui generis) reality because it possesses its own properties and causal powers. Such relationality (the relation as a real emergent) is *activity-dependent*, but has its *own structure*, the exercise of whose causal powers acts back upon the constituents (Ego and Alter) of the relation itself. (Donati & Archer, 2015, p. 55)

In contrast to Schmid as well as Donati and Archer, however, I do not see a necessary contradiction to individualism because relational connectedness consists precisely in the way individuals mutually relate to each other. This mutual relating, however, is not a contribution by an independently given relation, but the result of the individual orientations who are in that relation. If these orientations produce the relation in the first place, then the relation as such cannot achieve causal effects of its own.[4]

Following Christopher Kutz, here the difference between the assertion of an irreducibility of content and an irreducibility of common agency can be pointed out: "…the content of agents' intentions can be irreducibly collective so long as the structure of their intentions is straightforwardly individualistic" (Kutz, 2000, p. 13). Even if it is not possible to formulate a non-circular and complete reduction of the content of collective intentionality (the point made by Searle), it does not follow that this collective intentionality is marked by ontological and causal autonomy.

For example, Michael Tomasello et al. assume that human intentionality alone, from a certain stage of development, is characterized by the fact that humans (and only these, in contrast to other animals) are capable of joint intentions. Joint intentions are defined by a mutual understanding that the other person is capable of seeing himself and others as intentional beings directed at the intentions of the other persons (Tomasello et al., 2005,

---

[4] With respect to the question of the causal relevance of collective intentions, the debate on collective intentionality has so far remained silent. However, it is crucial to the question of individualism and holism, as has become clear since Kim's work on emergence (Kim, 2000, 2005). If collective states supervene on individual states, this does not mean that they are independent in an ontological sense. Even if it remains epistemically unclear to which individual states exactly a collective state can be traced (the so-called non-aggregativity), the question remains how collective states can have an independent causal influence if they are ontologically identical to the individual states through which they are realized (Greve, 2012).

p. 681). My point here is not to follow the debate on this research (in particular, the extent to which we are actually dealing with a specific human trait is disputed). What I want to stress is the following. It is plausible to assume that the more complex form of intentional interconnectedness that characterizes joint intentions cannot be traced back to "simpler" forms of intentionality. To understand that the intentions of others are related to one's own intentions is still different from seeing others or oneself as an intentional being. What needs to be added is the idea of a mutual interdependency of intentions. Nevertheless, this more complex intentional structure is realized by individuals since individuals can perceive themselves as relating to each other only if they have the corresponding properties that enable this structure.

8 Conclusion

For methodological individualism, actions are considered to be the result of individual intentions. From the point of view of the concepts of collective intending, this conception might be considered as inappropriate, if there are phenomena of collective intending that cannot be understood as a result of individual intending.

In fact, the corresponding concepts point to some difficulties. In Bratman's conception, the question arises as to whether a shared intention can be produced by individual intentions. For Searle, as we have seen, there is a dilemma here: either we accept circularity (presupposing in advance an understanding of doing something together) or we end up with inadequate analyses of what a collective mode, a we-mode, means. Cooperation, for him, cannot be broken down into individual ego intentions. We-intentions, for him, thus have to be understood as non-reducible. Nevertheless, from the fact that we may be dealing with a "primitive" phenomenon, it does not follow that there is a non-individual carrier. This is what Searle had tried to capture with his individualist thesis. Searle's individualistic thesis, however, led to a solipsistic conception of the individual, close to the notion of atomism rejected by methodological individualism. Nevertheless, solipsism overlooks the fact that people always exist in relation to one another.

The proponents of the group mind thesis seem to assume that there is a separate social reality. Nevertheless, the assertion of dependence on individual intentions is always maintained by the proponents of this view. Thus, even for them, collective intending depends constitutively on the corresponding individual intentions. Gilbert, e.g., considers the processes of group formations even similar to Bratman's idea of shared intentions as intentions to contribute to a common project when she states "that what is both logically necessary and logically sufficient for the truth of the ascription of group belief here is, roughly, that all or most members of the group have expressed willingness to let a certain view 'stand' as the view of the group" (Gilbert, 1992, p. 289).

Finally, in the discussion about relations, it can be maintained that there is a relationist interpretation that is entirely consistent with an individualist interpretation (provided individualism is not equated with solipsism). Collective intentionality according to this view consists in the interrelatedness of individual intentions. Reductive is such a view insofar as it is linked to the thesis that beyond the appropriate relatedness of individual intentions nothing further is required to produce collective phenomena and that these phenomena have no properties that are not determined by the properties of the individuals so related.

References

Baier, A. C. (1997). Doing things with others: The mental commons. In L. Alanen, S. Heinämaa & T. Wallgren (Eds.), *Commonality and particularity in ethics. Swansea studies in philosophy* (pp. 15–44). Palgrave Macmillan.

Bratman, M. E. (1993). Shared intention. *Ethics, 104*(1), 97–113.

Bratman, M. E. (1999). I intend that We J. In M. E. Bratman (Ed.), *Faces of intention: Selected essays on intention and agency* (pp. 142–161). Cambridge University Press.

Bratman, M. E. (2014). *Shared agency: A planning theory of acting together.* Oxford University Press.

Bulle, N. (2018). Methodological individualism as anti-reductionism. *Journal of Classical Sociology*, 1–24.

Bunge, M. (1996). *Finding philosophy in social science.* Yale University Press.

Davidson, D. (1984). *Inquiries into truth and interpretation.* Oxford University Press.

Di Iorio, F. (2023, forthcoming). Methodological individualism and reductionism. In N. Bulle & F. Di Iorio (Eds.), *Palgrave handbook of methodological individualism* (Vol. II). Palgrave.

Donati, P., & Archer, M. S. (2015). *The relational subject.* Cambridge University Press.

Durkheim, E. (1982 [1894]). *The rules of sociological method.* Free Press.

Gilbert, M. (1992). *On social facts.* Princeton University Press.

Gilbert, M. (2010). Culture as collective construction? In G. Albert & S. Sigmund (Eds.), *Soziologische Theorie kontrovers. 50. Sonderheft der Kölner Zeitschrift für Soziologie und Sozialpsychologie* (pp. 383–393). VS Verlag für Sozialwissenschaften.

Gilbert, M. (2015). Joint action. In J. D. Wright (Ed.), *International encyclopedia of the social & behavioral sciences* (Vol. 12, 2, pp. 839–843). Elsevier.

Greve, J. (2012). Emergence in sociology: A critique of non-reductive individualism. *Philosophy of the Social Sciences, 42*(2), 188–223. (First online, 2010). 10.1177/0048393110381770.

Heath, J. (2020). *Methodological individualism—The Stanford encyclopedia of philosophy* (Summer 2020 Edition).

Hodgson, G. M. (2007). *Meanings of methodological individualism.* https://core.ac.uk/display/1638809?utm_source=pdf&utm_medium=banner&utm_campaign=pdf-decoration-v1. Accessed 6 January 2023.

Huff, T. (1984). *Max Weber and the methodology of the social sciences.* Transaction Books.

Jepperson, R., & Meyer, J. W. (2011). Multiple levels of analysis and the limitations of methodological individualisms. *Sociological Theory, 29*(1), 54–73.

Johansson, I. (2003). Searle's monadological construction of social reality. *The American Journal of Economics and Sociology, 62*(1), 233–255.

Kim, J. (2000). Making sense of downward causation. In P. B. Anderson, C. Emmeche, N. O. Finnemann & P. V. Christiansen (Eds.), *Downward Causation* (pp. 305–321). Aarhus University Press.

Kim, J. (2005). *Physicalism, or something near enough.* Princeton University Press.

Kutz, C. (2000). Acting together. *Philosophy and Phenomenological Research, 61*(1), 1–31.

Lohse, S. (2019). *Die Eigenständigkeit des Sozialen. Zur ontologischen Kritik des Individualismus.* Mohr Siebeck.

Lukes, S. (1968). Methodological individualism reconsidered. *The British Journal of Sociology, 19*(2), 119–129.

Meijers, A. W. M. (2003). Can collective intentionality be individualized? *The American Journal of Economics and Sociology, 62*(1), 167–183.

Parsons, T. (1951). *The social system.* Free Press.

Parsons, T., Bales, R., & Shils, E. A. (1953). *Working papers in the theory of action.* Free Press.

Parsons, T., Shils, E. A., Allport, G. W., Kluckhohn, C., Murray, H. A., Sears, R. R., et al. (1951). Some fundamental categories of the theory of action: A general statement. In T. Parsons & E. A. Shils (Eds.), *Toward a General Theory of Action* (pp. 3–29). Havard University Press.

Pettit, P. (2003). Groups with minds of their own. In F. F. Schmitt (Ed.), *Socializing metaphysics. The nature of social reality* (pp. 167–193). Rowman & Littlefield.

Pettit, P. (2014). Three issues in social ontology. In J. Zahle & F. Collin (Eds.), *Rethinking the individualism-holism debate. Essays in the philosophy of social science* (pp. 77–96). Springer.

Pettit, P., & Schweikard, D. (2006). Joint actions and group agents. *Philosophy of the Social Sciences, 36*(1), 18–39.

Risjord, M. (2104). *Philosophy of social science. A contemporary introduction.* Routledge.

Schelling, T. C. (1978). *Micromotives and Macrobehavior.* W.W. Norton & Company.

Schmid, H.-B. (2005). *Wir-Intentionalität: Kritik des Ontologischen Individualismus und Rekonstruktion der Gemeinschaft.* Alber.

Schweikard, D. P. (2008). Limiting reductionism in the theory of collective action. In H. B. Schmid, K. Schulte-Ostermann & N. Psarros (Eds.), *Concepts of sharedness—Essays on collective intentionality* (pp. 89–117). Ontos.

Schweikard, D. P., & Schmid, H.-B. (2021). *Collective intentionality, the Stanford encyclopedia of philosophy* (Fall 2021 Edition). https://plato.stanford.edu/archives/fall2021/entries/collective-intentionality/. Accessed 6 January 2023.

Searle, J. R. (1990). Collective intentions and actions. In P. R. Cohen, J. Morgan, & M. Pollack (Eds.), *Intentions in communication* (pp. 401–415). MIT Press.

Searle, J. R. (1995). *The construction of social reality.* Penguin.

Searle, J. R. (2010). *Making the social world. The structure of human civilization.* Oxford University Press.

Stoutland, F. (1997). Why are philosophers of action so anti-social?" In L. Alanen, S. Heinämaa & T. Wallgren (Eds.), *Commonality and particularity in ethics. Swansea studies in philosophy* (pp. 45–74). Palgrave Macmillan.

Tomasello, M., Carpenter, M., Call, J., Behne, T., & Moll, H. (2005). Understanding and sharing intentions: The origins of cultural cognition. *Behavioral and Brain Sciences, 28*(5), 675–691 and 721–735.

Tuomela, R. (2005). We-intentions revisited. *Philosophical Studies, 125*(3), 327–369.

Tuomela, R., & Miller, K. (1988). We-intentions. *Philosophical Studies: An International Journal for Philosophy in the Analytic Tradition, 53*(3), 367–389.

Udehn, L. (2001). *Methodogical individualism. Background, history and meaning.* Routledge.

Velleman, J. D. (1997). How to share an intention. *Philosophy and Phenomenological Research, 57*(1), 29–50.

Watkins, J. W. N. (1973 [1952]). Ideal types and historical explanation. In J. O'Neill (Ed.), *Modes of individualism and collectivism* (pp. 143–165). Heinemann.

Watkins, J. W. N. (1973 [1957]). Historical explanation in the social sciences. In J. O'Neill (Ed.), *Modes of individualism and collectivism* (pp. 166–178). Heinemann.

Weber, M. (1958). *From Max Weber; Essays in sociology* (Translated, edited and with an Introduction by H. H. Gerth & C. Wright Mills). Oxford University Press.

Weber, M. (1968 [1922]). *Economy and society. An outline of interpretive sociology* (G. Roth & C. Wittich, Eds.). University of California Press.

Wittgenstein, L. (2009 [1953]). *Philosophical investigations.* Wiley-Blackwell.

Zahle, J. (2007). Holism and supervenience. In S. P. Turner & M. W. Risjord (Eds.), *Handbook of the philosophy of science. Philosophy of anthropology and sociology* (pp. 311–341). Elsevier.

Methodological Individualism and Collective Representations

Pierre Demeulenaere

Methodological individualism (MI) is often associated with a model of economic action. As a result, criticism of it tends to equate it with this "individualistic" economic model. However, this kind of reductive critique is in fact irrelevant and ignores the basic arguments of MI, and the misconceptions it introduces should be dispelled. At the same time, MI is often opposed to the "holism" model that is argued to be incompatible with MI because it centers on a "collective" dimension of representations and behaviors (Descombes, 1996). This article seeks to unite these two paradigms by demonstrating that MI is in fact compatible with the existence of norms, institutions, and collective representations. In the first section, I begin by reviewing three basic ideas of the MI tradition. In the second section, I argue that these ideas are entirely compatible with an analysis of institutions, norms, and collective representations, including with in-group identification. This requires, however, that MI not be reduced to an economic or psychological analysis. I devote two sections to discussing and critiquing these two reductive possibilities, where I develop two main arguments simultaneously:

This article was originally published in *L'Année sociologique* 70 (1), 2020, 69–95. Translated and edited by Cadenza Academic Translations. Translator: Adam Lozier, Editor: Faye Winsor, Senior editor: Mark Mellor.

P. Demeulenaere (✉)
Department of Sociology, Sorbonne University, Paris, France
e-mail: pierre.demeulenaere@sorbonne-universite.fr

© The Author(s), under exclusive license to Springer Nature Switzerland AG 2023
N. Bulle and F. Di Iorio (eds.), *The Palgrave Handbook of Methodological Individualism*,
https://doi.org/10.1007/978-3-031-41508-1_2

first, that MI's fundamental ideas need not be reduced to a cost–benefit analysis or to a psychological analysis focused on an unconscious dimension of behavior; and second, that cost–benefit and psychological analyses themselves tend to be inseparable from the collective representations and norms that orient them. The last section deals with rationality in its relationship to collective representations, social norms, and emotions. In principle, rationality is considered to represent, on the one hand, a reasoning effort that seeks to produce "universally" appropriate and adequate responses, and, on the other, a constituent social dimension of individual groups, which may be in conflict with one another. Rationality seeks rational consensus, but it can lead to disagreements as a result of the complexity of the subjects being analyzed, the difficulty of the reasoning involved, the social character of the people engaged in argument, and how desirable the available options are depending on social position. These disagreements also arise in groups that share arguments, based on rationality that is subjective and socially situated, as well as desires that are held in common. This subjective rationality, supported by group emotions, leads to the reinforcement of norms and collective representations until new critical perspectives call them into question.

1   The Three Basic Ideas of Methodological Individualism

The epistemological tradition that produced the definition of "methodological individualism" is founded on three main ideas that I would like to briefly address. They demonstrate both how flexible this tradition is and that it is compatible with the existence of collective representations and the fundamental ideas of what has come to be called holism. I do not want to paint the MI tradition with too broad a brush, nor to equate it with one particular perspective. MI has sought to respond to certain epistemological questions encountered in the social sciences: for example, who acts in social reality? These questions are the subject of myriad debates and disagreements, most of which will not be addressed here. That said, I would like to draw attention to several fundamental ideas that are difficult to avoid and which are in fact associated with the MI tradition.

The first is that social life is not directly subject to overarching social laws[1] that exist at a systemic level. Rather, if such laws exist, or if there are strong regularities at the social level, they must correspond to behavioral regularities

---

[1] This idea is not trivial: Harold Kincaid (1990) still contests it.

of social actors in particular situations, because individuals alone are capable of setting social life in motion. Social life is the result of "individual" actions in the sense that "individuals" are the ones with the power to act in social life. This idea was first posited by John Stuart Mill in a critique addressed to Auguste Comte in which he refutes the existence of laws belonging to a social "system." If such laws were to exist, he argues, they must be based on behavioral regularities: How else would they manifest themselves concretely? Once social life involves actors, observed social regularities must be traceable to the level of these actors (even though it is certainly possible that a single social reality could correspond to several scenarios of individual activities). This principle in no way implies that there are no social "structures" or that these structures have no impact on the behavior of social actors. Mill writes:

> The succession of states of the human mind and of human society can not have an independent law of its own; it must depend on the psychological and ethological laws which govern the action of circumstances on men and of men on circumstances. (Mill, 1882 [1843], 1110)

Of course, Mill does not mention the notion of structure, but of "circumstances." "Structure" is used much later, by George C. Homans, for example, to refer both to the impact of actors on the formation of structures and to the effects structures have on actors' behavior (Homans, 1967). Mill, furthermore, paid close attention to the importance of what he called "ethology," and which, in his definition, referred to the study of the way people's characters develop collectively by adapting to particular circumstances. Subsequent theorizing about a "structural individualism" (Wippler, 1978) amounts to official recognition of what had always been clear: actors do not act in some kind of social vacuum, but in social situations that are defined and "structured," notably by institutions, norms, and collective representations (Demeulenaere, 2011). These structures can of course be related to power relations. The way actors are influenced by all these social aspects, which are conveniently but vaguely translated by the polysemous and relatively ill-defined notion of "structure," deserves to be studied and clarified.

This is why the use of the term "methodological individualism" seems, in retrospect, to be both awkward and a source of confusion: it inevitably evokes an "individualist" aspect. This is likely why Raymond Boudon (1992) spoke of "actionist sociology" and why the literature tends now to employ the terminology of structural individualism. One of the goals of this article is to demonstrate that the concept of the individual is not simple and clear-cut. Furthermore, it does not necessarily lead to an "individualistic" attitude, given, in particular, the existence of collective representations sustained by individuals. That said, it does correspond to a specific ontological dimension that is distinct from that of institutions, as we will see later on. This also means that only individuals are capable of setting social life in motion, and therefore only individuals are, properly speaking, actors. There are certainly such things as collective actors (Coleman, 1990), but in order to discuss their actions, it is necessary to refer to particular individuals and to the institutions that, while they define these collective actors, are themselves sustained by individuals.

The second basic idea of MI relates to the first but is technically distinct. It focuses on the fact that correlations between social variables cannot be directly interpreted in causal terms. In order to establish explicit causality between them, one must analyze the role of the actors responsible for producing the observed situations, actors who are themselves acting within a given set of circumstances, and thus in a set of "structures." Max Weber can be considered one of the first authors to develop this idea more or less explicitly[2] when he refers to observed regularities from a statistical point of view:

> In the absence of such meaningful adequacy we have only an *incomprehensible* (or incompletely understandable) *statistical* probability, even where there is a very significant, precisely quantifiable probability of a regular event occurring (whether overt or psychic). [...]
>
> Only those statistical regularities which correspond to the understandably intended meaning of a social action are in the sense used here understandable types of actions, i.e. "sociological rules." (Weber, 2004, p. 319, emphasis in original)

This proposition gives rise to the entire literature related to "generative mechanisms" (Hedström, 2005) for summarizing causal scenarios that allow observed relationships between variables to be interpreted on the basis of individual actions in a given social context. I should note that this recourse to "individual" actions can in fact refer to a cultural or "social" dimension of these actions that is involved in the correlation analysis. Thus, when James S. Coleman (1990) uses Weber's theorized relationship between capitalism and Protestantism as an example to illustrate the structural relationship between a "micro" level and "macro" level (the components that make up what is now

---

[2] In the passage cited, Weber does not directly discuss statistical correlations, nor the fact of interpreting them by relying on meaningful actions. He only discusses the question of knowing whether observed regularities in social actions can be interpreted in terms of "sociological rules." However, I think we can extrapolate from this passage the importance of interpretive mechanisms founded on social actions when it comes to summarizing statistical correlations. That is why I use the somewhat convoluted phrase "more or less explicitly" here, since the quote from Weber does not correspond exactly to what I go on to discuss.

called the Coleman Boat), he refers to cultural tendencies to characterize the micro level. At the macro level is an observed relationship between Protestant doctrine and capitalism that needs explaining. On one side of the micro level of the diagram are "values"; on the other side are typical "economic behaviors" that flow from these values. In other words, for Coleman, in this diagram, the "micro" level includes actors with collective representations, which in turn exist in relation to other social dimensions (the macro level). The fact that social actors, which make up a "micro" level, must be referred to in order to interpret a correlation does not mean that this micro level (or "individual" level, since it refers to effective actions taken by individuals) necessarily corresponds, or must correspond, to actors who lack collective representations. Consequently, and contrary to other authors (Ylikoski, 2012), I do not think it best to avoid referring to concrete individuals when debating the relationship between the micro and macro levels. The existence of "levels" between them does not correspond to a straightforward, unambiguous system of inter-locking hierarchies. Individuals can act in a given situation (which corresponds to a macro situation) with reference to social norms that they respect and which are therefore present at the micro level in the individual's reaction to the situation.

In order to develop a satisfactory explanation, then, one needs to refer to "intelligible" actions. This brings us to the third fundamental principle of the MI tradition: seek to correlate observed social regularities with the intelligible actions that produce them. But what is an intelligible action? Part of this article attempts to answer this question by focusing on collective representations and their meaning from the perspective of MI, as well as the notion of an "individual" dimension of actions and meaning. The basic idea is that, while collective representations do indeed exist, they rely on "individual" capacities to take shape. This is followed by a brief discussion of the relationships between "meaning" and psychological capacities. All that said, the individual dimension of action does not exclude a social aspect of representation formation. This goes for rationality, too, which also has a social dimension. Here, then, rationality is both an individual capacity and the base upon which form groups founded on shared representations. All these groups claim their respective representations to be the suitable, appropriate ones.

2 Institutions, Collective Representations, and Methodological Individualism

If we take these three ideas into account, there is no fundamental divergence between MI and holism. First, from an ontological perspective: social institutions "exist,"[3] and it would be absurd to deny it. The issue is knowing what their mode of existence is. This has been the subject of an immense debate that will not be discussed in detail here. It suffices to address its origin. Both Weber and Durkheim believe that social institutions are a function of shared collective representations. Therefore, Durkheim argues, society, which corresponds to a "collective consciousness," *must not be hypostatized*. It is not a material thing that can be encountered like an object in the road; rather, it results from the combination of individuals' representations:

> In this sense and for these reasons we can and must speak of a collective consciousness distinct from individual consciousnesses. To justify this distinction *there is no need to hypostatise the collective consciousness* [emphasis mine]; it is something special and must be designated by a special term, simply because the states which constitute it differ specifically from those which make up individual consciousnesses. This specificity arises because they are not formed from the same elements. Individual consciousnesses result from the nature of the organic and psychical being taken in isolation, collective consciousnesses from a plurality of beings of this kind. (Durkheim, 1982 [1895], p. 145, note 17)

What matters is understanding how particular collective representations are established and reinforced, as well as how they evolve (or fail to). From an ontological perspective, there is no opposition between individual and collective: individual actors support the collective representations they share.

In principle (and in observation), individuals support four possible types of representations:

* Representations that are universal because they are natural and linked to the existence of human nature: for example, if it is established that an incest taboo exists universally and comes down to human psychology. These representations are both individual and collective as well as natural

---

[3] I do not offer any precise definitions for the complex notions of collective representations, institutions, norms, or groups in this article. Doing so, while preferable, was not possible given the scope of the text. Following Durkheim, I limit myself to mentioning the existence of these social realities in order to try to connect them to the three fundamental ideas of MI as they have been discussed here. More precise definitions would not change the thrust of the argument.

and universal (the collective here is a function of the universality of the members of the species).[4]

- Particular representations, characteristic of particular groups: the details of norms against incest vary from group to group. They are a function of the particular collective representations of a given group.
- Universal social representations that are in no way natural: for example, anyone who wants to legally travel abroad today must carry a passport. This is a universal norm that is obviously not at all natural.
- Particular representations, individual in a strong sense: for example, if a trend is a collective phenomenon that rests on norms of a certain type, there can still be strong and very particular individual variants. In this case, the individual representation cannot serve to support the collective representation, but if it is imitated, it can become collective among the group of imitators.

There is thus no fundamental opposition between the idea that "individual" actors are responsible for social life (because they alone act within the limits of social structures, which are marked by power relations) and the idea that there are shared collective representations. These representations exist, and they extend to varying degrees into groups of varying sizes. Institutions in particular are a function of a set of representations shared by groups of individuals. From this perspective, there is essentially no difference between Weber and Durkheim.

Individual actions thus occur on the basis of representations that are at times individual and at times collective (the collective dimension itself has several possible characteristics). To obtain more precise analyses, the vocabulary needs tidying. Institutions and norms correspond to shared representations, which obviously affect individual behaviors—such is their purpose. The fact that institutions are shared representations lets us easily explain several of their "collective" properties:

- Institutions affect the behavior of individuals, whether these individuals support those institutions or are subjected to them because they are implicated in them by those who do. Institutions are a strong power, because the other people one deals with in social life support them and give them power over others.

---

[4] I mention the example of incest here because of its popularity in the naturalist literature discussed later on, particularly in Turner (2007) and Haidt (2012). This article makes no claims regarding the origins of incest norms. It later notes the existence of competing explanations without weighing in on the debate.

- Through transmission, institutions sometimes outlast the individuals who bring them into existence: a country's borders are thus handed down from generation to generation as long as they are not called into question and overturned.
- Path dependency can make it difficult to change institutions, even when everyone would like to, because it can make coordinating decisions difficult.
- Institutions give rise to groups (which are defined precisely by the presence of functioning norms and institutions) that tend to function like collective actors in two different senses: on the one hand, they often have "representatives" that act in the name of the group; on the other, the relative stability of institutions leads the groups associated with them to react in a relatively stable way in certain domains, in accordance with this institutional stability.
- In certain cases, institutions (the norms associated with them and activated by the actors who support them) can, depending on the collective representations currently in effect, demand "sacrifices" from individuals for the benefit of other group members linked by shared institutions (these other group members are thereby assimilated into the "group"). Weber, for example, gets at this idea when he writes:

Such intermittent political action may easily develop into the moral duty of all members of tribe or people to support one another in case of a military attack, even if there is no corresponding political association; violators of this solidarity may suffer the fate of the sibs of Segestes and Inguiomer—expulsion from the tribal territory. (Weber, 1978 [1922], p. 394)

- Typically, this leads to a belief in the existence of a group that is "superior" to individuals. Beyond a sense of belonging, such a belief arouses a sense of solidarity founded on the belief in belonging to the group. Weber describes this as well:

The belief in group affinity, regardless of whether it has any objective foundation, can have important consequences especially for the formation of a political community. We shall call "ethnic groups" those human groups that entertain a subjective belief in their common descent because of similarities of physical type or of customs or both, or because of memories of colonization and migration; this belief must be important for the propagation of group formation; conversely, it does not matter whether or not an objective blood relationship exists. [...]

On the other hand, it is primarily the political community, no matter how artificially organized, that inspires the belief in common ethnicity.

[...] almost any association, even the most rational one, creates an overarching communal consciousness; this takes the form of a brotherhood on the basis of the belief in common ethnicity. (Weber, 1978 [1922], p. 389)

It is thus clear that MI, which argues that individuals alone act in a particular social context and are constrained by this context and its prevailing power relations, is entirely compatible with the notion that actors act according to collective representations and are constrained by institutions and social situations.

The question now becomes: Should collective representations and institutions be explained through individual motives in the strong sense—that is, motives inherent to the nature of all individuals? If MI is compatible with the existence of institutions and collective representations, does that mean that these can be reduced to individual dimensions in a stronger, more limited sense? This is how Durkheim, as we saw, characterizes the states of collective consciousness: "This specificity arises because they are not formed from the same elements. Individual consciousnesses result from the nature of the organic and psychical being taken in isolation, collective consciousnesses from a plurality of beings of this kind" (Durkheim, 1982 [1895], p. 145, note 17). If there is a plurality, or combination, which combination of elements is needed? For example, the feeling of solidarity that arises within a group obviously makes sense only because the members of a group interact: feeling solidarity all alone is nonsense. However, in order for the members of a group to feel a collective sense of solidarity, each group member must individually be able to feel solidarity. Collective solidarity, as a collective sentiment, requires the individual ability to feel such collective sentiments—otherwise, they could not exist. In order to re-establish the power of collective representations, then, we must first identify the individual aptitudes that make them possible. To address this question, I would like to rule out two unsatisfactory and unsuitable solutions: economic reductionism and psychological reductionism which focuses on the unconscious dimension of behavior.

3 Economic Reductionism

MI is often associated with a model of economic action. From a theoretical perspective, this association is mistaken, as noted above. It is mistaken despite the fact that many people who claim to adhere to MI (in particular, most professional economists) link it to an economic model, and despite the fact that neoclassical economic theory relies on the principles of MI. Fundamentally, the three main principles of MI theory described above in no way imply that social actions are purely economic in nature. Moreover, the economic model of homo economicus is itself relatively imprecise, making it unusable as an individual basis for the collective representations it often presupposes.

It would be beyond the scope of this article to analyze the various models of economic behavior currently in circulation, as it would be to attempt a global unification of economics and sociology. What I would like to demonstrate more specifically is the implicit presence of "collective" themes in the basic cost–benefit reasoning of economics, which MI is generally (mistakenly) reduced to. The present article broadly describes dimensions of behavior that cannot be reduced to costs and benefits (like the importance of seeking appropriate, adequate solutions through reason). The goal of this subsection is to show that even the most orthodox economic analysis based on a homo economicus model tends to assume certain social norms (Demeulenaere, 1996, 2003). The point here is not only to demonstrate that MI is not reducible to cost–benefit analysis, but also to show that cost–benefit analyses themselves tend to presuppose social norms internally (e.g., the definitions of costs and benefits) or externally (e.g., the legitimacy of pursuing certain advantages and avoiding certain costs as opposed to others).

The economic model of homo economicus is imprecise because, first of all, it inevitably uses individual preferences (e.g., smoking or not smoking) to orient actions. However, these individual preferences can in fact have a collective dimension and be reinforced by norms and collective representations, as Amartya Sen (1977) has shown. This means that notions of cost and benefit are not entirely stable: social norms may be integrated into them. Indeed, social norms affect all desires related to consumption and occupation and thus play an important role in defining what a cost or a benefit is.

Second, if, in order to make it more precise, the homo economicus model is restricted to narrowly defined goals (such as making more money), it becomes clear that economic life (owing to preferences and rules, as well as the institutions that organize it) cannot be explained on the basis of this single narrow motivation, which is itself imprecise. Dimensions of economic action other than these narrow ends must then be included, ends which are themselves subject to social norms. For example, in Gary Becker's analysis of fertility, the costs and benefits are essentially monetary, but they also include, according to him, the "psychological" benefit of having a child, which, as before, comes back to social norms.

Finally, from a normative perspective, there is no reason to believe that only narrowly defined economic motives should be taken into account when analyzing economic behavior. In particular, notions of justice are clearly present in economic behaviors, as the field of contemporary economic analysis has shown (Bowles, 2016). This obviously leads to the question of where these feelings about justice come from and how natural they are.[5] In any event, however, they are related to social norms.

At this point, not only can MI not be reduced to a model of economic behavior, but on a more basic level, and inevitably, it is necessary to employ broader interpretative frameworks to summarize the conditions under which an economic model marked by norms and collective representations can be used, as well as where it is relevant.

4   Psychological Reductionism

I find psychological reductionism unacceptable for a basic, unavoidable reason, one that has also been discussed by Weber: there are historically variable social institutions that cannot be deduced from an ahistorical psychology and that therefore cannot be reduced to such a psychology. Thus, capitalism is a particular economic system that cannot be directly deduced from a universal psychology, precisely because other economic systems have existed throughout history.

Weber's position is precise and nuanced: he acknowledges the existence of psychology and its usefulness in interpreting behavior, but he also insists that the historical variability of social norms cannot be directly deduced from it. Here again there is a certain overlap with Durkheim, even though they do not use the same vocabulary. In any case, the three principles of MI discussed in this article do not at all imply that social actions can be reduced to a psychological dimension, even though psychology can play an important role.

Attempts to directly deduce the social norms that structure society from an evolutionary psychology often do so at the cost of standardizing these norms

---

[5] Interestingly, Thomas Piketty's 2019 book, *Capital et idéologie* (published in English in 2020 as *Capital and Ideology*), constantly refers to sentiments of justice with respect to reducing inequalities, but offers no theoretical reflection about what sentiments of justice are and where they come from, nor how they may relate to economic behavior.

and drastically minimizing their historical variation, as in the work of Pascal Boyer (2018). Trying to pinpoint what is stable and relatively unvarying in social norms is interesting and important work, but it must not be undertaken at the expense of paying attention to the importance of social variations. Those psychologists who take seriously the historical variation of norms emphasize contextual factors (such as the development of the state) rather than psychological evolutions per se. Institutional and contextual variations are the decisive factors, as in Steven Pinker's 2011 analysis of the historical trend of decreasing violence, which cites Norbert Elias in elucidating the logic of this evolution. Finally, and somewhat paradoxically, the psychological literature tends to draw attention to the variation of social norms and their attendant representations (Henrich et al., 2001, 2010), which then requires explaining these variations by referencing varied social contexts.

In light of the remarkable development of cognitive psychology and its spread throughout the social sciences, particularly economics, the sense in which MI (here, the three aforementioned fundamental ideas) does not involve psychological reductionism, but is partially linked to psychology, merits closer attention.

In a very broad sense, all representations, individual and collective, are inevitably related to a "psychology" in that they exist only because we have brains that allow them to exist and develop materially. This brings us to a subject that comes under the heading not of sociology, but of neurology on the one hand and the philosophy of mind on the other: How do we move from neural tissue to the lived character of representations? This incontestable reality (assuming we do not adopt the metaphysical perspective of an immaterial soul radically separated from all materiality) of the existence of a neural process of representations, theorized for example by Antonio Damasio (1994), has three major theoretical consequences for the social sciences.

The first is that all the cultural and social diversity of representations and norms is necessarily permitted by this neural process and is compatible with it. All representations have a neural process, and there is clearly a great degree of cultural and social variation with respect to these representations and norms. In other words, the neural and physiological dimension of human nature clearly allows for cultural variation.

It thus makes no sense to oppose the biological and the social, as is often done. It is in fact the "biology" of the human species as expressed by the functioning of the human brain that makes possible a great degree of diversity in culture, representations, and social situations. But this immense diversity does not mean that we leave the domain of neural processes. Brains are always there, and they clearly allow for this cultural and social variation, which does not occur on a metaphysical, immaterial plane.

To recap, the goal is not to oppose biological and social, nor nature and culture, but to analyze cultural and social variation in the context of a single species with one set of capacities. In this regard, two things need to be identified: structural behavioral constants and major cultural and social variations. Both these constants and variations can be linked to structures in the mind, which are flexible and indeterminate enough to permit this variation but sufficiently formatted and defined (like the human need for coherence noted by Weber) to restrict the mind to limited possibilities.

Between the hard-naturalist and hard-constructivist approaches in the social sciences, there is currently no consensus or even attempt to reach common ground. The hard-naturalist approach seeks to determine the origin of social behaviors through neural analysis and the Darwinian mechanisms that select for them. For example, a study of the DNA of 13,000 Australians (Hatemi et al., 2011, cited in Haidt, 2012) found a significant difference between the genes of conservatives and liberals. These genes are related to the neurotransmitters glutamate and serotonin, which play a role in the brain's responses to threats and fear—responses that themselves are selected for through evolution. I cannot speak to the quality of the study, but it is hard to see how evolution and fluctuations in political behavior generally in a given population could strictly correspond to genetic differences—this would be absurd. Moreover, even though the human species is subject to the evolutionary process and should be situated in it, there is no reason to think that this process alone can explain the details of the historical variations of representations and social norms, which certainly cannot be interpreted solely in terms of species adaptation (Nagel, 2012). Within the framework of traits that adapt to the environment and are thus subject to evolutionary selection, there exists a great variety of historical, social, and cultural mechanisms waiting to be analyzed and explained in their particularities.

On the other hand, that does not mean the correct approach is a radically constructivist one unmoored from any stable point of behavior. Social science's descriptions and explanations refer to relatively stable capacities among the actors being studied—as well as those carrying out the studies. The aforementioned need for coherence can thus be used to interpret both scholarly discourse and the behavior it analyzes. Though it is certainly true that the concept of coherence has evolved over time, and different social conditions favor the desire for coherence to varying degrees, it would be difficult to do without this concept when analyzing a great number of behaviors, beyond cultural diversity.

* * *

On this basis, three approaches are available for interpreting representations and their attendant behaviors: one that relies on an unconscious psychology, one on rationality, and one on particular social norms that actors tacitly admit without directly considering whether they are appropriate and adequate. These three approaches are often in opposition, and indeed they can be legitimately opposed to one another for the localized interpretation of certain social phenomena. However, it is possible to devise a theoretical framework in which these different approaches are interconnected and complementary. In this case, it is the interconnectedness of these approaches that makes a given interpretation complex. Below, I sketch such a framework and highlight three propositions:

1. Emotions and biases play a major role in social life, but they do not act alone. Emotions tend to be associated with representations and social norms that play a specific role and orient emotions in certain directions. Biases themselves presuppose norms whose origins often go overlooked in the psychological literature.
2. Social norms play a central role in social life and affect individual decisions. Individual actors are not constantly trying to reconstruct norms on a rational basis. Rather, they tend to tacitly accept many available norms that, in their view, are self-evident, based on the collective representations in effect in a given group.
3. That said, rationality does play a meaning-making role in the construction of social norms and representations. What is at play here is not a perfect, complete rationality, but a reason-based process that can result in erroneous or inadequate representations. At the same time, these representations are shared and constructed collectively (and are thus group representations). In return, emotions tied to group solidarity or the desirability of various social options play a role in reinforcing social norms. There is thus a rationality dynamic that sometimes stabilizes in shared, relatively uncontested representations and at other times leads to a reassessment of representations and the search for new acceptable ones.

5 How Are Social Norms, Emotions, and Rationality Interconnected?

Contemporary psychology emphasizes an unconscious dimension of behavior that concerns the prerational dimension of moral institutions (Haidt, 2012), the role of biases in reasoning and perception (Kahneman, 2011), and the power of the emotions associated with them (Turner, 2007). For example, Haidt points to the fact that justifications for incest taboos, which tend to be consequentialist, cannot account for the profound sentiments of aversion (among the American students interviewed) about protected (using a condom) sexual relations between a brother and sister. People are disgusted by incest and say it is unacceptable because of the risks of congenital birth defects. This risk is not present when sexual relations are protected, but the aversion (typically) remains. Thus, Haidt argues that judgments are fundamentally based on emotions and not the other way around. In this view, emotions, which are themselves the result of evolutionary selection, undergird social behaviors and social norms, which in turn give rise to ex post facto justifications.

In any case, emotions are a central feature of human and social behaviors, and given their obvious physicality, they cannot be ignored: blushing, for instance, appears to be uniquely human (Boehm, 2012). Emotions have an intensely involuntary dimension: someone whose palms are sweaty because they are nervous did not choose to have sweaty palms (nor to be nervous)—choosing is impossible. That said, the importance of emotions does not mean that social representations can be reduced to them. We might note that emotions are frequently associated with beliefs and representations: being afraid of a plane is linked to the fear that the plane could crash, but if one learns that air transportation is very safe, one should be able to reduce one's fear (though success in reality is not a given), since it is based on a belief that can be changed rationally. Furthermore, emotions are tied to social norms and the representations that validate them. For example, someone who favors the death penalty may be able to enthusiastically defend it and perhaps even stomach the spectacle of an execution with equanimity. In contrast, someone who is against the death penalty might react with indignation and disgust to the same execution. Representations are thus relatively independent of emotions, and collective representations and social norms affect emotions considered to be legitimate in a particular social context. One could argue that emotions such as indignation or the desire for vengeance help determine whether one supports the death penalty, but claiming that emotions alone determine one's position would clearly be overstating the case. Forming opinions involves at least some reflection and relatively autonomous reasoning. For example, whether one feels welcoming or hostile toward immigrants is partly tied to representations indicating whether immigration is economically beneficial or detrimental: reasoning and emotions are relatively independent, though they can reinforce each other.

Emotions are also associated with social norms, as Marcel Mauss (1969 [1921]) argues. They are both partly independent of and dependent on variable social norms, which orient them toward certain attitudes. This illustrates the importance of social norms in people's daily lives; they tacitly refer to them without questioning them. People almost never have a bird's eye view of their situation or possess complete information about it, and they are not apt to consider their objectives perfectly. Social norms play a significant role both in shaping everyday objectives and in one's understanding of the means to achieve them, of representations, and of contemporary values. Social norms are not necessarily established through deliberate decision-making: imitation plays an important role (Henrich, 2016). Thus, to explain why spicy foods are common in places with hot weather, there is no need to presume that people are effectively aware of the fact that spices can kill foodborne pathogens. Norms can be explained simply: young people imitate old people, who are models of wisdom. If people who eat spicy food live somewhat longer than those who do not, then the former will have a better chance of being imitated, since there will tend to be more of them. Over time, and unintentionally, the congruent effects of the combination of natural selection and imitation will lead to a situation in which everyone in a given population eats spicy food (Henrich, 2016). The norm can then be reinforced without anyone being aware of its origin or how it came into effect.

Furthermore, shared social norms can arouse a sense of group solidarity when one group is opposed to others that do not share its norms: "This is the way things should be done, not some other way." Normative conformism is one of the great social universals that applies to all manner of things, such as food (Fiske, 1991). For those who adhere to them, the norms that are conformed to are the correct ones.

Nevertheless, norms are also related to reasons and representations that are more or less precise and developed and that imply a capacity for rationality that is inevitably present in social life. Our idea of rationality must be complete and complex, one that cannot be reduced to satisfying self-interest (Boudon, 2009; Demeulenaere, 2014). On one hand, rationality—the ability to use justified reasoning to come to appropriate, adequate decisions—corresponds to the deliberative aspect of reason. This aspect sets reason apart from thoughtless emotions and thus from what contemporary psychology stresses: emotions, biases, and unconscious attitudes. On the other hand, rationality is in reality a psychological capacity that is itself linked to emotion. This is not a novel theme in the literature on reason; it dates back to Hume if not earlier. Indeed, on a fundamental level, we do not choose to be rational or not. Rationality imposes itself on us: we think and are implicated in acts of reasoning that may be more or less developed and that seek adequate answers, even though a given social sphere may favor deliberative rationality to a greater or lesser extent. Rationality is a human capacity that is related to the functioning of the human mind and is not chosen by it. The human mind does not freely (and certainly not arbitrarily) choose to impose norms associated with rationality on itself. Rather, these norms impose themselves on the mind in a constraining and transcultural manner: this is the very meaning of the concept of rationality, even though rationality does often lead to errors and inadequate judgments (Boudon, 1990). The principles of reason and the self-evident fact of rationality are imposed on us within the limits of our situation, our information, and our unequal capacities (unequal because of our diversity and inequalities of social position). But if we think that two plus two equals four, it is very difficult—in reality, impossible—for us to think that two plus two equals five. As Hume (1982 [1758]) writes, the mind exerts a kind of "pressure" that forces us to recognize what is rationally self-evident. But this pressure also forces us to accept certain errors that appear to be true (Boudon, 1990). The idea of "evidence" Weber refers to harks back to this pressure of the mind, which forces the mind to admit certain things and not others according to its functioning. The effort to persuade others and justify oneself to others described by Hugo Mercier and Dan Sperber (2017) inevitably presupposes rationality norms that make it possible to speak of rationality: the search for appropriate, adequate responses.

It is certainly the case that the more complex the object of analysis, the greater the number of relevant lines of reasoning associated with it. This makes it more difficult to come to a final rational result, and can eventually lead to cacophony, either internal to an individual or among multiple people.

Two other dimensions are involved besides the internal pressure of rationality. First is the fact that objects of analysis, belief, or representation are themselves emotionally charged and arouse desire or aversion (Elster, 1999). Second, although analysis is based on universal (and in this sense individual, natural, and psychological) capacities that constitute the human aptitude for rationality, oriented toward appropriate and universally valid results, this analysis also always occurs within a framework of rationality that is socially inscribed (Durkheim, 1915 [1912]). Beliefs are shared beliefs, and this collective dimension plays a major role in the formation of social representations—to the point that, despite its universalist aspirations, rationality can also give rise to group phenomena.

Fundamentally, representations and collective norms exist in two different orientations with respect to rationality in its interaction with groups. Either norms are directly associated with representations that involve rational deliberation, or they exist beyond the scope of rational judgment. An example of the latter is languages, the preeminent example of norms coordinated within a group.

Any group needs a common language in order to communicate. This common language constitutes a group of shared norms. These norms are, by definition, supra-individual, both because they are shared and because they cannot stem from a single individual. Moreover, a group's common language can also give rise to a sense of in-group belonging. However, language conventions are not the result of rational choices: it is irrelevant whether "the sun" is masculine, as in French, feminine, as in Arabic, or neuter, as in English. What matters is that group members can coordinate among themselves through a shared set of mutually respected conventions. This collective usage cannot be derived from individuals themselves, but it assumes that they can adopt such common usages, respect shared norms, and accept arbitrary conventions. Furthermore, the presence of these shared norms tends to be linked to greater trust between partners to an interaction (Granovetter, 2017), as well as to a sense of solidarity when faced with other groups that share different norms. On the other hand, certain norms directly involve rationality, such as whether the death penalty is legitimate. Contrary to group social norms that do not involve rationality, and which are derived solely from the need for coordination between members of a group (which is defined precisely by these particular conventions), reason has, by nature, a universalist calling that is not limited to local convention. Reason seeks that which is valid in itself and for everyone, as Weber notes. It seeks appropriate, adequate responses. That said, reasoning is also contextual: the death penalty does not mean the same thing in a society without prisons as it does in a society in which prisons have been institutionalized as a form of punishment. There are also beliefs as to whether the death penalty is an effective deterrent to crime. Once it has been agreed on that the role of punishment is not to exact vengeance (even though such sentiments may still be present), debate arises over the legitimacy of sanctions and the reasons to consider the death penalty unacceptable. Yet the complexity of the subject still generates a range of opinions.

And this is precisely the important point: in spite of its universalist aims, rational discussion often produces a variety of opinions. Some of these may be considered mistaken or insufficient in light of the complexity of the subjects involved and the possibility—supposed here for the sake of principle—of an "ultimate" appropriate, adequate solution. However, those who hold an opinion necessarily consider it to be appropriate and adequate for their situation (inevitably one of limited information)—otherwise they would not hold the view in the first place. These representations, whether "adequate" or not, are defined when an uncertain individual relies on the opinions of others to define his or her own. Most often, opinions are shared within groups that are themselves defined by the shared belief in these very opinions.

At this point, common conventions are established, as with a language: "We, the members of such and such group, think this." Furthermore, group emotions are frequently associated with these collective representations, such that individual shared convictions often involve the group identity. Distinct groups can thus oppose each other based on the differences between their representations. The emotions associated with these representations are then reinforced. These emotions have four distinct sources: the cognitive pressure of evidence that leads to the conviction that one is in the right; subjective engagement vis-à-vis this conviction; the desirability of objects of representations; and loyalty to a reference group that supports these representations and that is frequently, for a given individual, at the origin of the shared representation.

Therefore, representations that are rational in principle (in that they involve reasoning that seeks appropriate, adequate solutions) can—because subjects may be complex and representations are often established intersubjectively—lead to conflict between groups that support different representations. At this point, the individual and collective dimensions are not opposed, since they are connected. However, the existence of collective representations inevitably presupposes individual adherence to them that is more or less consciously reflected upon, as well as the logic that comes with it.

This also lets us point out that no natural basis is needed to explain group formation. It is currently common in the naturalist literature to refer to a natural basis for group formation that relies on an evolutionarily developed herd mentality instinct. This is a naturalist reinterpretation of Durkheim and group specificity (Haidt, 2012). It is used to explain numerous experiments that indicate group identification (Akerlof & Kranton, 2010; Sherif, 1966 [1936]) and preference for the members of one's own group. (Weber also theorized this preference in this distinction between internal and external ethics.) The line of reasoning proceeds in several steps: reciprocity norms are essential for cooperation; they occur in groups where these self-same reciprocity norms are respected. Why? The evolutionary perspective holds that united groups are better able to adapt than ones that are not united. By this logic, individuals inclined to belong to united groups outnumber their counterparts, which would explain the fact that we tend to develop group solidarity (Greene, 2013). This may concern the development of the human species generally, which displays more solidarity than other species, notably chimpanzees (Boehm, 2012). But it may also be associated with the history of an evolution within the species: individuals who are capable of group solidarity will outnumber those who are not, since the former are better adapted (Tomasello, 2016). However, these naturalist lines of reasoning ignore the fact that there is no strong ontological basis for groups. Group borders, the perimeters of coalitions, the definitions of identities—all these are essentially variable, so any strong natural dimension of groups is an arbitrary notion. Affiliation with a group depends on representations and lines of reasoning that can considerably change the group's limits and lead to conflicting sentiments of belonging. Moreover, reasoning can be oriented toward the necessity of universal cooperation, even though this is difficult to establish in practice. At this point, the three fundamental principles of MI (only actors act, statistical correlations must produce explanations in terms of mechanisms involving actors, and there is ultimately a reference to an "individual" meaning) are compatible with interpretations of behaviors that focus on psychological dimensions, on the tacit acceptance of social norms, or on the rational capacities of individuals. These three perspectives can give rise to collective representations that often have emotions associated with them. Social norms are thus associated with emotions, which they orient in certain directions. Meanwhile, individual rationality can lead to collective representations that seem acceptable in a given situation for certain groups of actors.

These three perspectives can come into conflict. This is why, fundamentally, MI is not entirely decisive. The use of its principles does not produce a perfectly formed solution to explain behaviors and social situations. MI is not an answer key that solves all problems. Rather, it represents a set of methodological constraints that help determine which types of explanations are acceptable, while leaving open the possibility for multiple explanations to exist in competition. Therefore, the incest taboo could be the result of an evolutionary process of behavioral selection that occurs without our being aware of it, or it could be a culturally variable social norm that is reinforced differently by different groups. Finally, it could be the result of the rational consideration of the consequences of incest that are deemed undesirable based on factual knowledge. These three perspectives are competing, not congruent. Resolving the enigma of why norms against incest are universal requires using both empirical data and knowledge of behavior otherwise acquired (Turner & Maryanski, 2005).

That said, from a certain point of view these three perspectives are compatible. Actors have representations on a rational (though limited) basis. These representations tend to be associated with the groups that support them, and they give rise to emotions that stem from rationality, individual desires, and the desires of other group members. Groups and collective representations are compatible with the principles of MI. Nevertheless, a fundamental principle is to try to take into consideration how actors perceive their collective representations to be relevant and appropriate (even though their perceptions may appear mistaken or senseless and inexperienced from the perspective of an observer). This is why relying on unconscious principles is insufficient, since representations are also connected to logical reasoning.

6 Conclusion

Epistemological discussion in sociology is to a great degree centered on an opposition between MI and what is commonly called holism, which privileges the importance of institutions, norms, and collective representations. This article has sought to demonstrate that the fundamental ideas of MI (once it has been established that reducing them to psychology or economic analysis is unfounded) make possible an integrated analysis of institutions, norms, and collective representations. This in turn makes room for progress toward a unified theory of sociology that integrates the contributions of two traditions that have arbitrarily been deemed irreconcilable. This unified sociology is distinct from psychology and economic analysis, but there is room to more clearly connect it to them. This article has sought in particular to demonstrate that the use of rationality, which usually leads to different points of view based on arguments that are necessarily partial and a function of actors' social positions and limited resources, can also be connected to the establishment of unified groups that share common representations and norms. But it also tends to give rise to perspectives that are critical of these representations and norms. However, this article did not fundamentally address the definition of rationality norms, their effective relationship with behaviors deemed predictable or not in certain circumstances, their relationship with the specifics of norm variation, or their possible interpretation in particular social contexts. All this goes to show that the principles of MI discussed here do not present definitive solutions: I have sought only to lay out a general architecture of reasoning.

Translated and edited by Cadenza Academic Translations

Translator: Adam Lozier, Editor: Faye Winsor,

Senior editor: Mark Mellor

References

Akerlof, G. A., & Kranton, R. E. (2010). *Identity economics. How our identities shape our work, wages and well-being*. Princeton University Press.

Boehm, C. (2012). *Moral origins. The evolution of virtue, altruism, and shame*. Basic Books.

Boudon, R. (1990). *L'art de se persuader. Des idées douteuses, fragiles ou fausses* [The art of persuasion. Questionable, fragile or false ideas]. Fayard.

Boudon, R. (1992). Action. In R. Boudon et al. (Ed.). *Traité de sociologie* [Treatise of sociology] (pp. 21–55). Presses Universitaires de France.

Boudon, R. (2009). *La rationalité* [Rationality]. PUF, "Que sais-je?" series.

Bowles, S. (2016). *The moral economy. Why good incentives are no substitute for good citizens*. Yale University Press.

Boyer, P. (2018). *Minds make societies. How cognition explains the world humans create*. Yale University Press.

Coleman, J. S. (1990). *Foundations of social theory*. The Belknap Press of Harvard University Press.

Damasio, A. (1994). *Descartes' error: Emotion, reason, and the human brain*. Putnam.

Demeulenaere, P. (1996). *Homo oeconomicus. Enquête sur la constitution d'un paradigme* [Homo oeconomicus. Investigation into the constitution of a paradigm]. Presses Universitaires de France.

Demeulenaere, P. (2003). *Les normes sociales. Entre accords et désaccords* [Social norms. Between agreements and disagreements]. PUF.

Demeulenaere, P. (2011). Introduction. In P. Demeulenaere (Ed.), *Analytical sociology and social mechanisms* (pp. 1–30). Cambridge University Press.

Demeulenaere, P. (2014). Are there many types of rationality? *Revista De Sociologia, 99*(4), 515–528.

Descombes, V. (1996). *Les Institutions du sens* [Institutions of meaning]. Éditions de Minuit.

Durkheim, É. (1982 [1895]). *The rules of sociological method* (Translated from the French by W. D. Halls, edited by S. Lukes). The Free Press.

Durkheim, É. (1915 [1912]). *The elementary forms of the religious life* (Translated from the French by J. W. Swain). G. Allen & Unwin.

Elster, J. (1999). *Alchemies of the mind: Rationality and the emotions*. Cambridge University Press.

Fiske, A. P. (1991). *Structures of social life. The four elementary forms of human relations: Communal sharing, authority ranking, equality matching, market pricing*. The Free Press.

Granovetter, M. (2017). *Society and economy. Framework and principles*. The Belknap Press of Harvard University Press.

Greene, J. (2013). *Moral tribes. Emotion, reason, and the gap between us and them*. Penguin Books.

Haidt, J. (2012). *The righteous mind. Why good people are divided by politics and religion*. Vintage Books.

Hatemi, P. K., et al. (2011). A genome-wide analysis of liberal and conservative political attitudes. *Journal of Politics, 73*(1), 271–285.

Hedström, P. (2005). *Dissecting the social. On the principles of analytical sociology*. Cambridge University Press.

Henrich, J. (2016). *The secret of our success. how culture is driving human evolution, domesticating our species and making us smarter*. Princeton University Press.

Henrich, J., Boyd, R., Bowles, S., Camerer, C., Fehr, E., Gintis, H., & McElreath, R. (2001). In search of *Homo economicus*: Behavioral experiments in 15 small-scale societies. *American Economic Review, 91*(2), 73–78.

Henrich, J., Heine, S. J., and Norenzayan, A. (2010). The weirdest people in the world? *Behavioral and Brain Sciences, 33*(2–3), 61–83, discussion 83–135.

Homans, G. C. (1967). *The nature of social science*. Harcourt, Brace & World.

Hume, D. (1982 [1758]). *Enquiries concerning the human understanding and concerning the principles of morals* (L. A. Selby-Bigge & P. H. Nidditch, Eds.). Clarendon Press.

Kahneman, D. (2011). *Thinking, fast and slow*. Farar, Straus & Giroux.

Kincaid, H. (1990). Defending laws in the social sciences. *Philosophy of the Social Sciences, 20*(1), 56–83.

Mauss, M. (1969 [1921]). L'expression obligatoire des sentiments (rituels oraux funéraires australiens) [L'expression obligatoire des sentiments (rituels oraux funéraires australiens)]. In V. Karady (Ed.) *Œuvres: 3. Cohésion sociale et division de la sociologie* [Works: 3. social cohesion and division of sociology] (pp. 269–278). Éditions de Minuit.

Mercier, H., & Sperber, D. (2017). *The enigma of reason. A new theory of human understanding*. Penguin Books-Allen Lane.

Mill, J. S. (1882 [1843]). *A system of logic, ratiocinative and inductive* (8th ed.). Harper & Brothers.

Nagel, T. (2012). *Mind and cosmos: Why the materialist neo-Darwinian conception of nature is almost certainly false*. Oxford University Press.

Piketty, T. (2020 [2019]). *Capital and Ideology*. Translated from the French by Arthur Goldhammer. The Belknap Press of Harvard University Press.

Pinker, S. (2011). *The better angels of our nature: A history of violence and humanity*. Penguin Books.

Sen, A. K. (1977). Rational fools: A critique of the behavioral foundations of economic theory. *Philosophy and Public Affairs, 6*(4), 317–344.

Sherif, M. (1966 [1936]). *The psychology of social norms*. Harper.

Tomasello, M. (2016). *A natural history of human morality*. Harvard University Press.

Turner, J. H. (2007). *Human emotions. A sociological theory*. Routledge.

Turner, J. H., & Maryanski, A. (2005). *Incest. Origins of the Taboo*. Paradigm Publishers.

Weber,M. (2004). *The essential weber: A reader* (S. Whimster, Ed.). Routledge.

Weber, M. (1978 [1922]). *Economy and society: An outline of interpretive sociology* (G. Roth & C. Wittich (Eds.). University of California Press.

Wippler, R. (1978). The structural-individualistic approach in Dutch sociology: Toward and explanatory social science. *The Netherlands Journal of Sociology, 14*(2), 135–155.

Ylikoski, P. (2012). Micro, macro and mechanisms. In H. Kincaid (Ed.), *The Oxford handbook of philosophy of social science* (pp. 21–45). Oxford University Press.

Kelsen: Methodological Individualism in the Social Theory of Law

Stephen Turner

1 Introduction: The Problem of Law and the State

Perhaps the main motivating problem behind the explicit discussion of the issues we now think of as methodological individualism was the problem of the state. The state appears as an agent, with existence and therefore some metaphysical status; in addition it seems to be irreducible to the wills or existence of individuals, however aggregated. In short, the state is a being distinct from individuals, and necessary to explain facts that could not be explained otherwise. Even Max Weber endorsed something like methodological statism, the assumption of the state as a concept, in certain contexts, at least for convenience.

Weber was nevertheless careful to deny metaphysical status to the state. And he also gave an account of how to build up the state as a sociological phenomenon from individual materials. Hence he rejected the "necessity" argument for the "real" existence of the state, and extended this to collective concepts generally. The law, as distinct from the state, presented additional problems. The claim that a legal order existed appeared undeniable. And

S. Turner (✉)
Department of Philosophy, University of South Florida, Tampa, FL, USA
e-mail: turner@usf.edu

© The Author(s), under exclusive license to Springer Nature
Switzerland AG 2023
N. Bulle and F. Di Iorio (eds.), *The Palgrave Handbook of Methodological Individualism*,
https://doi.org/10.1007/978-3-031-41508-1_3

while Weber carefully distinguished the sense in which its existence was necessary to explain actions, such as the actions of a judge, sociologically, from the sense in which it existed as a legal matter (1978 [1968], p. 14); this second sense seemed to elude explanation: it was normative and had a "real" existence at the same time.

The person who devoted the greatest effort to these issues was Hans Kelsen, whose philosophy of law differentiated legal and sociological viewpoints. Kelsen constructed a general theory of law which explicitly focused on what separated the two: the problem of the difference between the individual and the state, and the problem of the normative as distinct from the causal reality of the state. He went beyond this to provide both an extensive critique of collective concepts in sociology and other domains and a general account of the pre-causal primitive mentality that generated the dualism between society and nature that is at the root of the collective concepts he critiqued. For Kelsen, claims about the reality of collective forces were on a par with, and indeed were a form of, the kind of mythological thinking about retribution found universally in primitive societies which persists in various ways in modern society, a point to which we will return.

2 Kelsen on Law and the Individual

The problem that motivated much of the German philosophy of law during the period was related to sovereignty and the specific problem of the ground of law: if the sovereign was the King, and the King's will was the source of law, what grounded the authority of the King? To answer "the law" just produced a circle. And opened the question of whether the King was really sovereign, in the sense of being able to will the law, or act as King apart from any law. These questions were given various answers and the issue of Royal authority came to be transferred to "the state." Did a sovereign state have powers, including the power to command in the form of law, but not limited to that form of command? Did it have, like Kings, a "will?" Was law based on a "will," and if so whose? Was there a collective "will," or a will of parliament, that made a sentence stating a command into a law? If not, what did? (Caldwell, 1997, pp. 13–39).

A will seemed at least to be something objective, though some writers at the time, such as Axel Hägerström, heaped ridicule on the idea of will—pointing out that "the people," or the King, typically weren't even aware that they had willed the law in question, but it was still law (Hägerström, 1953). Kelsen faced this issue directly in terms of what Peter Caldwell has called

Jellinek's paradox: how can the state, conceived as a sovereign, be subject to law? As Jellinek wrote, in a treatise on international law in 1880, "Law [is] possible only on the condition that a directing and coercive force is present" (quoted in Caldwell, 1997, p. 42). But if the force is free to make law, it is also free to ignore or decline to be bound by law. Yet, if the reality of the state consists in the fact that it is a legal person, it depends on these very laws for its status as a being. So if it is not bound by the laws it makes, it is no longer a legal person, or is not acting legally. But what is the coercive force that binds it and makes it conform to those laws?

One can see that this is a rich ground for collective conceptions. The ideology of a royal right to rule by this time had been converted into some notion of kings as representatives of the people. As Ernst Kantorowicz later established (2016 [1957]), this notion was prefigured by the Medieval doctrine of the King's Two Bodies—the actual King and the embodiment of the concept of Royalty and the true Royal will itself, which factions could fight in the name of against the actual king. It was difficult to get away from the association of will and person, so that the idea of sovereignty and the idea of monarchical authority, though it became increasingly abstract, was also difficult to get rid of. But many other aspects of the personalization of authority hung on along with "will," such as the idea that the state was subject to the demands of common morality, and the idea that the state was bound, like a person, by law.

Kelsen's response to the personalization of the state is unequivocal:

> There is no such superman or superhuman organism in society, whose sole reality is the individual human being. What is characterized as a society or a community is either the actual coexistence of individual human beings or a normative system of their reciprocal behavior. Only human beings can have obligations imposed and rights conferred on them to behave in a certain way; only the behavior of human beings can be the content of legal obligations and rights. (1998 [1962], p. 526)

Was this "methodological" individualism or something else: metaphysical individualism, or metaphysical anti-holism about "society?" The initial context is the philosophy of law, and this quotation comes from a paper about the status of the objects of international law, or nations, which poses the problem in a specifically "legal" way. His answer to this is consistent:

> If international law imposes obligations and confers rights on the state to behave in a certain way, this means that it imposes obligations and confers

56   S. Turner

rights on human beings, in their capacity as organs of the state, to behave in this way. (1998 [1962], p. 526)

The "methodology" behind these arguments is that of "legal science." But Kelsen extends this argument to the larger context of thinking about society generally, and in a way that differs from a raw metaphysical claim: instead, he shows that various versions of collective concepts are defective. How he does this, and what makes Kelsen's argument novel, will be the focus of this chapter.

3   From Law to "Society"

The need for a force or will to underwrite the legality of the state leads to notions like the will of the people or "society." The need here is ideological: one needs a placeholder for the sovereign king, and finds it in the sovereign people. But when the problem of the force is solved in this way, the law itself becomes a puzzle: it is obviously different from these things. So how can it regulate and control them? If we stay at the collective level the answer seems to be that the authority of the collective thing is actually exercised by individual people—judges, policemen, officials, and so forth—who act according to the law. This may seem to be a trivial and innocuous point, but Kelsen explains that it is not:

> Let us consider the relatively simple case of a State where one single individual rules as an autocratic or tyrannic way. Even in such a state there are many "tyrants," many people who impose their will on others. But only one is essential to the existence of the State. Who? The one who commands "in the name of the State." How then do we distinguish commands "in the name of the State" and other commands? Hardly otherwise than by means of the legal order which constitutes the State. (2006 [1949], p. 186)

There is nothing "collective" here other than the State in whose name the command is made. But the only meaning of "in the name of the state" is in accordance with the "legal order." The "legal order" is something "whose validity is presupposed by the acting individuals" (2006 [1949], pp. 187–188). But all we need to account for the "State" is this legal order: the legal order is the ingredient that makes the state a state. But the legal order is no more than a system of presupposed concepts, not a metaphysical being with a will. All the "will" is supplied by the individuals involved acting in accordance with this presupposed legal order.

Kelsen goes through a vast number of collective "will" and "community" arguments to make this point again and again: these notions are fictions, fictions that affirm a unity and derive a force from the supposed existence of the unity, but misrepresent a more complex reality of individual variation in "wills" that belie the supposed unity. And he points out the ideological purpose behind these fictions: "In reality, the population of a State is divided into various interest groups which are more or less opposed to each other. The ideology of a collective State-interest is used to conceal this unavoidable conflict of interests" (2006 [1949], p. 185).

The argument that the state's acts are no more than the individual acts authorized by the legal order is an answer to the problem of sovereign power: the presupposed validity of the order is what explains what makes an act a state act, not the intrinsic power of the "sovereign" who performs the act. And this reasoning placed him at odds with Carl Schmitt and many other thinkers who disputed the idea of the "identity" between law and state, and argued for one or another form of the claim that the state had inherent powers beyond the powers given in the law. For Schmitt, the evidence of this was the fact that it was possible to suspend the law—to declare a state of "exception." And whoever had this power was "sovereign." The law applied, for Schmitt, in normal conditions; but deciding what was "normal" was a power reserved to the state, meaning that sovereignty was real and above the law. But in practice the person who could declare a state of exception was authorized to do this by law. Kelsen had a response to this reasoning that preserved both the idea of the identity of state and law and the ultimate dependence of the "state" on individual actions authorized by law.

4 The Idea of Community

The opponents of Kelsen were also opponents of his individualism: typically they argued for the foundation of the state in some sort of collective being, such as "community," or that the law was a part of a "constitution" that included facts other than the law and the fact of authorization. Thus they were committed to the idea of the reality of "society" or of something similar. These arguments could be detached from the question of law, and indeed in Kelsen's time were already well developed in several traditions. But they also had a special role in relation to law: they replaced the sovereign and sovereign authority, the place traditionally occupied by a monarch, with the collective being that the argument appealed to.

Kelsen starts from the observation that the concept of the state already contained an ambiguity between the juridically recognized "state" and a broader concept of the state which included the people and institutions of the larger society. For him, the question becomes twofold. The first is whether this larger sense overcomes the fact of the dependence of the legal "state" on individual action, and is itself not merely a legal person in the sense recognized by the judiciary as an artificial agent represented by authorized individuals, but, in some stronger sense, recognized as real. The second is whether "authorization" and legality itself require some higher sense of collective reality.

Kelsen's reasoning here was complex, in the sense that it is difficult to explain. But his explanations are nevertheless quite clear. They depend, however, on distinctions that at first appear to be entirely separate from the concept of "society" or its conceptual kin. The key distinction is between normative considerations and factual or sociological considerations. We are tempted to confuse the two, in a way that directly relates to the problem of the reality of collective beings. The reason for this confusion, Kelsen argues, is somewhat odd: our usual "sociological" descriptions of the prime collective being, "the state," are dependent on a normative concept. The normative concept is the legal concept of the state. It has juridical significance, and a place in law. But law and the system of legal concepts are normative concepts. They depend not on sociological concepts, such as society, but on other normative concepts. The normative order, however, is itself an empirical fact.

How does this argument for the dependence of factual claims on normative ones work? The key to the argument is the claim that the law is a system of valid norms. As Kelsen puts it, describing his own legal theory, "The statements in which our theory describes its object are … not statements about what is, but statements about what ought to be. In this sense, our theory may also be called a normative theory" (2006 [1949], p. 162). But to be a normative theory is not necessarily to be "non-empirical": "An analytic description of a system of valid norms is no less empirical than natural science restricted to a material given by experience" (2006 [1949], p. 163), that is to say a science without metaphysical elements. A theory of law "becomes metaphysical only if it goes beyond positive law" (2006 [1949], p. 163). Positive law is the law in the statute books: the actual law. It may also include other things recognized juridically, such as customary law, or law established by judicial decisions. But what is recognized juridically as law is open to empirical study in an unproblematic sense. There is nevertheless an important consequence to this reasoning. The laws themselves are normative: they state "oughts."

This seems like an unproblematic distinction. But it points to a problem: a theory of law which says that the foundation of law, which is therefore a part of "law" properly understood, is society, or some other collective authorizer, has to go beyond positive law: it is something outside of the explicit, "positive" law that the law depends on and which is in some sense the source of its validity. If "society" is recognized in the law, as the state is, it becomes part of positive law. So there is a place in a purely legal theory for the state and society. But their place is dependent on law, in the sense of a system of norms that authorizes people, such as judges, to determine what the law is. But their legal status in relation to the individual is that it is a fiction created by an authorized individual, defined by and within law.

The reason appealing to society in a sense other than as a legal fiction for a "foundation" for law is so tempting is that we want to answer the question "what makes law valid?" There is a legal answer to this question: that it is recognized as law by judges sitting in courts. They recognize it as law simply because the law has been produced according to law. If a legislature is authorized, for example, according to a constitution, and in accordance with legal procedures, to make a law, the law would be produced according to law. That fact alone is the source of its "validity." And it is the only fact involved in the judicial recognition of validity. A judicial decision that a law is unconstitutional is one which determines that the law was not produced according to law: that it was produced in violation of the higher authorizing law of the constitution, which is to say that it was not authorized by law. This is the core of Kelsen's "legal positivism."

Is this answer to the question of validity sufficient? Here things get tricky. One problem is this: when we work our way up to the top of the chain of authorizations, to the constitution itself, we are faced with the question "what authorizes that?" We get various non-legal answers to this question in political theory: the "will of the people," for example. But here again we have a concept with a dual meaning: there is a juridically recognized sense of the will of the people, namely the will as expressed through the legal procedures of voting, for example. But constitutions may be founded by revolutions. What then? As Kelsen points out, there is a legal answer to this question as well, but it is an answer that comes from international law, which is in this case a form of customary law (rather than treaty). Under customary international law states are obliged to recognize other states as legal. And the criteria are that the state in question has effective authority, which is to say that its laws, including the law governing its territorial claims, are obeyed. So in this we have a kind of legal surrogate for "the will of the people" which is a fact that can be juridically identified.

60   S. Turner

This nevertheless leads to another puzzle, which Kelsen wrestled with: what is the ultimate source of validity within the legal system itself—the source of legality, so to speak. His answer to this was that legal validity itself required an authorizing norm, which he called the *Grundnorm*, the basic norm, as a matter of logic. Only a norm can ground a norm. By definition, this norm could not itself be "authorized" according to a norm: that would mean it was not basic. It could only be an ungrounded normative assumption of the system of norms itself.

5   The Problem of Description

Kelsen makes an apparently strange claim: that only one description of a given fact may be true. He applies this claim to the Weberian theory of the state as domination. Weber's account, a classification of claims to legitimacy, is a non-collectivist theory. And it is faced with the traditional conundrum of distinguishing the State from a gang of robbers. Kelsen's point here is that even this theory depends on the idea of legitimacy to make this distinction. The participants must believe the domination to be legitimate. But for them to do this and thus to speak of "the state" assumes the validity of the legal order, and forces the sociologist to interpret them in juristic terms:

> There is no sociological concept of the state besides the juristic concept. Such a double concept of the state is logically impossible, if for no other reason because there cannot be more than one concept of the same object. There is only a juristic concept of the State: the State as—centralized—legal order. The sociological concept of a pattern of behavior, oriented to the legal order, is not a concept of the State, it presupposes the concept of the State, which is a juristic concept. (2006 [1949], pp. 188–189)

What is clear for our purposes is that for Kelsen the point of calling it a juristic concept is this: that it is a concept which can be understood in terms of an account of legal science that does not itself appeal to collective entities or to either a theology or metaphysics that appeals to them.

This goes against later analytic philosophy, which, since G. E. M. Anscombe, has taken the view that the same act can have multiple descriptions, for example, intentional and non-intentional ones, without there being any conflict between them (Anscombe, 1979). Correctness of description is a matter of relevance, relative to the purposes of the description. But the basic thought here is characteristic of methodological individualism, properly understood: it is that some descriptions are superfluous from the relevant point of view, but present themselves as essential. They present themselves as essential for the same purposes as those they compete with, which is to say, in the terms of Anscombe, that they are relevant descriptions in the same context.

The "same purposes" are normally the purposes of explanation. The normal claim is that the explanations given in accordance with methodological individualism, meaning explanations that do not appeal to collective objects, are sufficient, and the appeal to collective entities is superfluous for explanatory purposes. They are therefore not "essential." Kelsen, however, takes this argument in a different direction. He does not claim that collective entities are necessary for explanation. Instead, he claims that *normative concepts* are necessary to describe the facts in question. And in addition, he claims that the collective concepts which are alleged to correctly describe these facts are themselves defective. But he also claims that a purely sociological description that does not rely on normative concepts is insufficient. The issue is a complex one, and hinges on the term "unity." Kelsen assumes that collective concepts are assertions about unity, and that unity is what makes a collective object an object. This also requires some disentangling.

The question of what was "purely sociological" is a recurrent theme for him. In *The Pure Theory of Law* (1967 [1960]) he argues the familiar point that the objective validity of the law is a normative matter, to be decided internally within the normative logic of the law. This does not mean that the empirical world conforms to the law or to its logic. Nevertheless there is a connection, because "[e]very by and large effective coercive order can be interpreted as an objectively valid normative order" (1967 [1960], p. 217). We do not need to describe such an effective order in terms of validity: "the relevant human relations can be interpreted … as power relations (i.e., relations between commanding and obeying or disobeying human beings)—in other words, they can be interpreted sociologically, not juristically" (1967 [1960], p. 218). "The sociological interpretation … is a theoretical attitude" (1967 [1960], p. 218) toward the facts it takes to be relevant. The fact that concerns Kelsen is unity: power relations do not explain unity. They are simply interpersonal relations, relations between individuals. But the state, unlike a band of robbers, is normally not understood in this way. It is understood to be a unity. So this is a distinct problem: does understanding the state require it to be understood as a collective entity, as a unity beyond the individuals that act in its name?

A purely sociological description of the normative order would not, he reasons, produce a unity. But what would a purely sociological description be? We can begin with the problem of describing an action of a state official.

We cannot describe the act as authorized, as distinct from "believed to be authorized," without making a normative judgment. But for the act to be a state act, and thus for there to be a state, it must be authorized: believing it to be is enough for a sociological explanation of the action, but not for calling it a state action. The existence of the beliefs cannot tell us whether they are authorized; only the norms themselves can. So a sociological account of a legitimate order is not an account of a "state." The issue that concerns Kelsen, however, is different: is it sufficient or correct to explain the state in terms of norms, or does it require appealing to something collective? The answer to this question is an answer to the question of what makes the state a unity.

The term unity is Kelsen's way of formulating a common issue in relation to collective concepts. John Searle formulates the issue as follows: there are facts that "only exist insofar as they are represented as existing" (Searle, 2010, p. 68). The issue arises for such things as language: are languages collections of more or less convergent idiolects or is there a system? Just because we can represent it as having rules (which we can represent in different ways) is language governed by a system of rules (Davidson, 2005 [1986])? The same issue arises for meaning realism: we can represent meanings as existing, but does the indeterminacy of translation undermine the claim that there are real meanings, i.e., unique facts corresponding to meanings (Roth, 2022)? Similarly for practices: we can represent them as coherent, i.e., as a unity, but is the coherence something that is a part of individual psychology, imposed by the individual practice users on their own activity, or does it inhere in the practice represented as a collective fact (Turner, 2007)?

With law, however, the issue of unity is simply different: the legal order, at least in a mature legal system (but even in the case of tabu, where the tabu person can make something tabu) is hierarchical. States are mature legal orders. Rules are made by people who are authorized by "higher" rules to make the rule. The normative question of the legality of a rule is a question of whether the rule-maker was authorized by the higher rule. This leads to the question of the top rule, the rule authorizing the top authorizer. And this was a source of controversy, to be discussed shortly. Does the hierarchy end in a rule or a top authorizer who is beyond rules: a sovereign? This is a topic we will take up. Kelsen's point about unity is this: the legal order has unity, but as a hierarchical normative system of rules.

This "identity of state and law" theory of the state was the basis of his long-running controversy with Carl Schmitt and others who wished to hold on to a doctrine of sovereignty as the source of law and legal validity. For them, sovereignty was above legal rules, and also the source of legal rules.

Sovereignty, in the sense that was intended, was a supra-individual or collective concept, or seemed to require one. Thus rule in the name of the nation referred to a collective concept that was in some sense prior to law. For Kelsen it was the other way around: the nation was something defined by law, and therefore defined by something normative. His theory of the identity of state and law holds that there is nothing to the state other than that which is defined in the law. Thus the unity of the law as a normative system is the sole source of the unity of the state. And in this we see something crucial: for Kelsen "normative" suffices to do the work that a collective concept purports to do, namely to explain "unity."

Unity is, in effect, Kelsen's test for the reality of collective entities. Weber's approach does not appeal to any such element. Partly this is a result of his narrow definition of sociology as the science of meaningful action. Weber thus does make reference to the legal order, as it appears in the beliefs of people. This remains within the limits of sociology, the explanation of meaningful action, because the beliefs about this order which give it meaning for them are the beliefs of individuals. Beliefs about collective entities would be the beliefs of individuals. But so are beliefs in normative systems, which are not collective entities, as Kelsen conceives such systems, or indeed as Weber does. In short, Weber has a different, and in a sense less ambitious target: he does not attempt to refute collective concepts, but instead to show that the individual actions that need sociological explanation can be explained without reference to them. His point is that collective concepts, understood as real, are superfluous for the explanation of action. Kelsen has a different target: the claim made on behalf of collective concepts, and particularly the idea that they are necessary to account not only for meaningful action, but for the phenomenon of law itself.

Both Weber and Kelsen acknowledge that there is a difference between the law as doctrine and as a judicial practice. As Kelsen says, "we cannot be assured a priori that those patterns of behavior which sociology shows to be actually prevailing among the courts will be identical with those that the legal norms prescribe" (2006 [1949], pp. 169–170). But Kelsen goes beyond this by emphasizing the facticity of law—that law is positive law, the empirically discoverable law as it is found on the books and in judicial decisions, rather than some abstract idea of law to which collective properties, a telos, or an essence can be attributed. This empirically discoverable law has a relation to social reality, but one which is indirect, from the point of view of "normative jurisprudence" itself, which "assumes that an organ ought to execute a sanction only if the norm belongs to an efficacious legal order" (2006 [1949], p. 170). Efficaciousness is the link: belonging to such an efficacious order means that it is probable that the organ—the court—will apply the sanction (2006 [1949], p. 170). This connects the normative order to social reality, but does not identify the two. The unity, however, is in the legal order, the normative fact, not in the non-identical and more heterogenous world of social reality.

6 Unity

Law is a fact, as he repeatedly emphasizes, which can be ascertained empirically by looking at the law books (2006 [1949], p. 163). But as Kelsen later thought, its *validity* is based on a fiction: the ungrounded and ungroundable *Grundnorm*, the norm which validates legal norms as norms. In his earlier writing he simply said "it is presupposed to be valid; it is presupposed to be valid because without this presupposition no human act could be interpreted as a legal, especially as a norm-creating act" (2006 [1949], p. 116). It is presupposed in practice by judges, unconsciously. But there are other kinds of fictions. In his most famous early text he refers to the notions of society that are taken to ground law as "political fictions" (2006 [1949], pp. 184–186). These are fictions in a different sense: they purport to represent factual reality but fail to do so; they are nevertheless instrumental in politics, as justifications. Here is where we find a critical account of collective concepts that goes beyond Weber, and makes key distinctions that are later elaborated in another context that relates to the fundamental question of the nature of society itself. This is a question bound up, for him, with the question of the nature of law, because he considered, as did some of his prominent adversaries at the time, the fact of society to be inseparable from the fact of law. This is what separates him from Hobbes, and from a view of the human individual as an autonomous rational agent, a point to which we will return in the last section.

These arguments rely on a specific contrast: between the unity of the legal system, and the lack of unity in the supposed collective objects which are taken to be the ground of the law, such as "society." Unity is meant here as a criterion of existence: a legal system that if not unified would not be a legal system at all. The question he asks is this: "What is it that makes a system out of a multitude of norms?" (2006 [1949], p. 110). This is a question closely bound up with the problem of validity.

> [i]f one asks why one has to love one's neighbor, perhaps the answer will be found in some still more general norm, let us say the postulate that one must live 'in harmony with the universe.' If that is the most general norm of whose

validity we are convinced, we will consider it the ultimate norm....One would then have reached a norm on which a whole system of morality could be based. (Kelsen, 2006 [1949], p. 170)

This is a model static normative system. To be such a system is to be able to derive its elements from its unquestioned premises. A dynamic system is one in which norms are "created through acts of will by individuals who have been authorized to create norms by some higher norm" (2006 [1949], p. 171). In neither case is there reference to anything other than individuals. But there is unity: the higher rule resolves conflicts between lower rules, or provides a means of doing so. Unity amounts to there being a way in law to resolve conflicts between laws by reference to the hierarchy of norms.

This gives us a model of unity. His question becomes this: does the state have a unity, and therefore a reality, distinct from the unity provided by the unity of law? He provides a way of testing this claim, and concludes that it fails this test:

The assertion that the State is not merely a juristic but a sociological entity, a reality existing independently of its legal order, can be substantiated only by showing that the individuals belonging to the same State form a unity and that this unity is not constituted by the legal order but by an element which has nothing to do with law. However, such an element constituting the "one in the many" cannot be found. (Kelsen, 2006 [1949], p. 183)

Without the legal order, in short, there is no state: it is the element which constitutes the one in the many. The unity of the state derives from the legal facts. To claim otherwise would require identifying an element that unifies but has nothing to do with law.

The significance of this form of argument becomes apparent when he applies it, which he does not only in the context of legal theory, but in a critique of various theories of society and of the crowd. His basic point is that the unity of the thing we call society, its character as a natural object like an organism, is simply assumed. But when we inquire into the basis for this assumption, the idea becomes not only elusive but false. In the *General Theory* he presents his views in a characteristic way: he goes through a series of arguments for the supposed unity of the thing which stands outside of law and grounds it, and refutes each one.

Kelsen considers the claim that the natural basis for the notion of society as a unit that can be the basis of the state is the fact of interaction itself, so that it is an "actual social unit of interaction" (2006 [1949], p. 184). The supposed fact here is that the people in a society have more or more intense interactions with others in the society, and this makes for a natural fact of unity. He observes that this is empirically false: some people in the society have more intense interactions with people outside the society.

> interaction is not limited to people living in the same space. Thanks to present day means of communication, the liveliest exchange of spiritual values is possible between people scattered over the whole earth. In normal times, state borders are no hindrance to close relationships between people. (2006 [1949], p. 184)

Moreover, the contours of "society" as imagined by the theories do not correspond to the contours of "society" assumed by the theory. "If one could exactly measure the intensity of social interactions, one would probably find that mankind is divided into groups in no way coinciding with existing states" (2006 [1949], p. 184). The contours of "society" assumed by theories which take the state to be grounded are in fact taken over from juridical concepts, which are tacitly assumed (2006 [1949], p. 184).

The second argument he considers is that the social body is constituted by a common will or interest, or "a 'collective sentiment', a 'collective conscious-ness', a kind of collective soul, as a fact which constitutes the community of the State" (2006 [1949], p. 184). This would require that actual indi-viduals thought the same things "and are united in this common willing, thinking, feeling" (2006 [1949], p. 184). He dismisses the idea as implau-sible: "a real unity…exists only among those who actually are in an identical state of mind, and it exists only in those moments when this identity actu-ally prevails" (2006 [1949], p. 184). This would happen only under unusual circumstances. "To assert that all citizens of a State permanently will feel or think in the same way is an obvious political fiction, closely similar to that which the interaction theory was supposed to embody" (2006 [1949], p. 185). A third line of argument—the theory of the State as a natural organism found, for example, in Otto von Gierke—Kelsen dismisses as a political fiction of this kind, noting Gierke's use of it for ethical purposes. Gierke says that the need to sacrifice oneself in battle for the state must be a grounded moral obligation which one does not have to equal individuals—only something "higher." For Gierke it is the "whole" with "superior value … as compared to its parts" (2006 [1949], p. 186).

Kelsen returns to this topic in relation to Freud, in the Freudian journal *Imago* (1922, pp. 97–141); in the course of his article he also discusses Émile Durkheim, Gabriel Tarde, Gustave Le Bon, Georg Simmel, William McDougall on group mind, and others. Kelsen's point in these commen-taries is that social interaction is sufficient to explain the facts in question without positing a group mind. The distinction is especially relevant to the later discussion of reciprocity and revenge as a basis for law, to which we will return, which is also his basis for rejecting the idea that law depends on collectives.

The distinction between "social" in the sense of social interaction and "social" in the sense of belonging to a higher entity called "society" was a well-known distinction at the time, but primarily in French and American contexts: in France, it was represented by Tarde's alternative to Durkheim, which was well-received and elaborated on by American sociologists Charles Ellwood (1901), E. A. Ross (1909), MacDougall (1920), and group mind (Borch, 2013). In the *Imago* paper, Kelsen shows his detailed knowledge of some of these writers, notably Durkheim, but he is especially concerned to refute Freud's *Group Psychology and the Analysis of the Ego* (1959 [1921]).

Kelsen's text, a response to Freud's late work on group psychology, broadened the issues to consider in group mind theories generally. Freud commented on Le Bon's "deservedly famous" *The Crowd* (Le Bon, 1895), which Freud used as a starting point for his own analysis of the question of "what… is a 'group'" and "How does it acquire such a decisive influence over the mental life of the individual?" (Freud, 1959 [1922], p. 4). Le Bon's text relied on the notion of unity. It defined the issue of group mind in terms of this specific question: "If the individuals of the group are combined into a unity, there must be something to unite them, and this bond might be precisely the thing that is characteristic of a group" (Freud, 1959 [1922], p. 5). Le Bon, Kelsen says, did not answer this question. Like Le Bon, Freud reasoned that the conscious mind is a small part of mental life, and that the something that was the source of the unity would be found in the unconscious (1959 [1922], p. 5).

Although the starting point of the analysis is the temporary crowd, the analysis was extended to such artificial groups as the Church. Freud adds to Le Bon's account of the possible unconscious processes: not only "the racial unconscious," handed down concealed in tradition and contagion, but "the unconscious repressed" which also has "archaic" sources (1959 [1922], p. 5). The source, and the source of unity, is the shared unconscious and therefore repressed libidinal attachment to a common object, the leader, which is transformed into group identity and transmitted to the heirs.

Kelsen's issue is with the crucial step in this theory of society—the application of reasoning about the group soul character of temporary crowds to institutionalized social life. His comments on Freud are revealing. He notes that Freud accounts for group psychology, but without ever claiming that it is other than the psychology of the individuals involved: "For Freud there are no other souls than individual souls," though they may identify with many mass souls or groups (Kelsen, 1922, p. 119). But Freud's own account of groupness depends on a shared libidinal object, a leader, and extends the notion of attachment to this object to authority generally. This is a reductive account of the nature of authority, paradigmatically the authority of the father, and in *Totem and Taboo* (1918 [1913]), the result of the repressed memory of his murder, which is mysteriously transmitted intergenerationally (Kaye, 1993; Turner, 2020).

Kelsen's critique of this is simple: the psychological processes involved in adherence to a normative system and the psychology of adherence to an ideology are different from the psychology of leadership (1922). These involve their own different psychological processes. In the case of stable institutionalized social life, the fact that life is regulated by a system of norms is decisive. Similarly, the realization of an idea, an ideology, involves its own psychological processes. Thus "what are called 'states' is something completely other than the phenomenon described by Le Bon as 'masses' and explained by Freud psychologically" (1922, p. 119).

Turning a normative order, which involves individuals and their reciprocal relationships, into a person-like collective being is a kind of theologizing. Kelsen illustrates this with a passage from one of Freud's sources, Robertson Smith, writing about the ritual of the shared sacrificial meal (Beidelman, 1974), in which unity comes from the penetration of the ritual substance—which in ancient times was a holy victim—into their bodies, producing a sacred bond.[1] The parallel to the Eucharist is obvious (1959 [1922], pp. 135–136). But the ritual substance is a "real" bond only in a theological sense. Nor, Kelsen thought, is it necessary to explain anything: what is needed is a recognition of the place of norms in creating the "structures" and shared objects that are being theologized about. Kelsen's point is similar to his point about sovereignty and kingship: the theologized things are unnecessary for explanation. They have been wrongly granted the explanatory power that the existence of the norms is sufficient to explain. But the effect of Kelsen's reasoning is not to eliminate the explanatory problem. It is instead to shift the burden of explanation to the problem of norms itself.

---

[1] Smith's theory, and Freud's use of it, are discussed in Beidelman (1974, pp. 53–61).

7 The Individual and Primitive Law

From a certain point of view, Kelsen is backing himself into a corner with these arguments: how can one appeal to the idea of a normative legal order without invoking such notions as consent, or otherwise appealing to supra-individual concepts? The classical legal positivism of John L. Austin understood law as the commands of the sovereign. H. L. A. Hart's updated version of legal positivism (1961) adapted the contemporary notion of speech acts from his Oxford colleague, J. L. Austin, to interpret law as the product of a specific kind of speech act that enacted law (Lacey, 2006, p. 144).

The difference between Kelsen's "legal positivism" and the legal positivism of H. L. A. Hart, which anglicized and transformed it, is shown by the notion of enactment. Hart thought that law was made into law by a verbal formula: a distinctive speech act (1961, p. 36), and that sovereignty, to be a continued fact rather than one which died with the sovereign, itself depended on rules of continuity which could be changed but had general acceptance. Hart regarded customary law as a real but modest part of legal systems. This is a plausible picture of a mature legal system. Hart did not regard international law as law, and focused his account of law on a revision of the idea of law as the commands of a sovereign. Kelsen, in contrast, had a concept of primitive law, and considered international law, which has no law-giver, to be based on customary law, and that even treaties were to be understood as based on a customary principle: *Pacta sunt servanda* ["agreements must be kept"] (Kelsen, 1944, p. 30). For Kelsen, there was not only law before the enactment of the kind Hart refers to, but enacted law was based on prior customary law, including "primitive law," which is to say law before judges, rulers, and states. International law, for him, was a form of primitive law.

Hart's account of law was consistent with methodological individualism in this way. For Hart, the problem of determining what the law is was a matter of having a rule of recognition, and for there to be more or less general consent to the validity of this rule, and acceptance, supported by "social pressure" (1961, p. 92). Each law did not need consent; the system as a whole, based on the rule of recognition, did. Within the whole decisions could be made on traditional legal grounds. The "whole" here is, as with Kelsen, a normative system, not a collective being. The kind of "general consent" needed was not absolute consent by everyone under the law, or by an organic unity, but an effective acquiescence to the system itself, which in practice was carried out by lawyers and professionals who knew the details of the rules (1961, p. 60).

The idea of law as the commands of a sovereign that was the basis of John Austin's legal positivism, and persists in a modified form in Hart, assumes a state. Kelsen was interested in something more fundamental. Since he defined law as a coercive order, he regarded normatively accepted sanctions as the mark of "law." Law allowed actions that were otherwise illegal or violations of the normative order to be imposed legally in the form of sanctions. This was fundamental: law defined the same act, such as killing, as either a violation or as a legally valid sanction. But this distinction predated states and sovereigns and did not require them. It was part of what Kelsen called primitive law, and also international law, which lacked the kinds of courts and enforcers that were characteristic of a state. It was a normatively regulated system, but a self-help system, that made the same distinctions between acts, such as attacks on another country that could be either legal sanctions or acts that were violations of the normative order. And like primitive law, international law was customary law, together with treaties, which were analogous to contracts.

Kelsen, commenting on the problem of the origin of the state, discussed the two basic theories: the contractual one, and the force theory, which treated the state as the result of a sheer assertion of power. He suggested that it was probably a kind of mixture of the two: coercion together with acceptance. Such a state could be a source of law in the sense of the older legal positivism. But law, and the sense of legality that distinguished acts as violations or sanctions, preceded states, and his concern was with the problem of the deep origins of law, in pre-state societies.

Kelsen's argument was a complex, ethnographically based, inversion argument. Not only did he invert the story of the origin of law, so that law came before sovereigns, he inverted the story of the origins of causality in relation to norms and the story of the relation of the individual to society. He developed this account in a book called *Society and Nature* (1946), published in Karl Mannheim's Sociology and Social Reconstruction book series, which contains a vast number of examples from pre-state societies taken from the anthropological literature. One consequence of his account is to eliminate the mysteries of social contract theory, which depended in Schmitt on the mysterious flash of lightning that transforms the fearful mass into a state, and in Talcott Parsons on the equally magical appearance of a normative order as a solution to the problem of Hobbesian state of nature conflict. Each of these accounts, and social contract theory generally, depend on a notion of the individual that Kelsen rejects.

His account inverts the usual way of thinking about the relation of society to nature. This dualism, he says, is alien to primitive man. "Modern Science" accepts this dualism as a problem, which it tries to resolve into a monism by "conceiving society as a part of nature and not nature as a part of society" (1946, p. vii). Kelsen's main target in the book is causality. As he puts it in the first page,

> Causality is not a form of thought with which human consciousness is endowed by natural necessity: causality is not, as Kant calls it, an "innate notion." There were periods in the history of human thought when man did not think causally—that means that man connected the facts perceived by his senses not according to the principle of causality but according to the same principles which regulated his conduct toward other men. (1946, p. vii)

He followed this surprisingly Durkheimian point with the comment that "The law of causality as a principle of scientific thought first appears at a relatively high level of mental development." How do primitive people think about nature, and the connections between facts? They do so "according to social norms, especially according to the *lex talonis*, the principle of retribution" (1946, p. vii), and with a strong element of emotion.

The "primitive" response to illness or death is paradigmatic: it is seen as the result of action by a malign agent, such as a spirit of the dead, or the result of the violation of a tabu. He gives the example of a Kaffir who broke a piece of an anchor, and died shortly afterward. The anchor was regarded as sacred. But the connection between the events was not causal, in the modern sense that it could be generalized and falsified, but a connection between two concrete events modeled on human action: "The anchor, imagined as a personal being, has taken vengeance on the injurer, just as men, because of injuries done to them, take and are entitled, if not obliged, to take vengeance" (1946, p. 4).

This kind of thinking is combined with a different sense of the individual, marked by a lack of ego consciousness of sense of themselves as individuals. In part, this is connected to their relation to the world, which is dominated by fear, such as the fear of the souls of the dead who may take vengeance, as well as fear of the environment, which is understood in terms of animism and as filled with powerful spirits. The result is a sense of self as a part of nature and subordinate to powerful forces beyond ego's control, a "lack of self-confidence, which manifests itself clearly in the magic which occupies a central position among all primitive people's" (1946, p. 11).

The lack of ego consciousness, "the negative side of a mentality completely determined by social life" (1946, p. 11), is crucial to Kelsen's account. He gives many examples of the lack in language: the fact that terms like "my" refer not to the self but the group, or that the "I" of agency does not exist in early Semitic, where the expression is not "I kill," but "here killing"

(1946, p. 11). This applies to responsibility as well: we-ness predominates, and injuries are to be repaid by the group, not the perpetrator. In this and many things, "man does not regard himself as a separate individual, but only as a member of a collectivum" (1946, p. 18). This collectivum includes dead ancestors, who are sometimes thought to exist as animals, and who are feared. Social norms are enforced through this fear. The world at large, nature, is thus interpreted as an extension of the human world. Rather than asking "How did it happen?" they ask "Who did it?" (1946, p. 42). In legal terms, they replace causal thinking with "imputation": a concern with assigning responsibility rather than generalization.

With the problem of assigning responsibility, we come to the problem of "society" itself. If a person is killed, they can't exact vengeance themselves. In any case, because of their reduced individuality, the injury is to the group. "It is the family of the murdered individual, i.e., the oldest society itself, which exercises retribution" (1946, p. 56). The principle of retribution, in turn, especially retribution from superhuman authorities, Kelsen argues, is the foundation of such customs as sacrifice, obligatory gift giving, fetishes, and tabu, and is largely inseparable from magic, which calls on superhuman forces (1946, p. 70). "Superstition" is just social interpretation of nature. And Kelsen argues against the idea that misfortunes arising from the violation of a tabu are "automatic": the idea of automaticity is itself modern; for the primitive, it is a form of retribution.

"The freeing of the interpretation of nature from the principle of retribution" was accomplished by ancient Greek philosophers. And this applied to law itself: Protagoras "taught that the specific technique of the state order, which reacts to a socially harmful deed with a coercive act directed at the wrongdoers, is not justified by the religious idea of retribution, but by the rational intent of prevention" (1946, p. 246). With these changes we go beyond the primitive mentality.

The effect of these arguments is to provide a grounded alternative to the puzzle of the origin of an apparently non-individual, transcendent fact, of law or the normative system, without appeal to collective facts. But it does so by posting a primitive "individual" who is not a modern rational agent, but something quite different: a person living in a world dominated by superhuman agents, ghosts, retribution, and a nature which is "social" in the sense that its objects are themselves agents. More important, this individual is person who thinks as a "we," as a group member, rather than as an evolved "ego," with a strong sense of self and self-interest. For this person, law, or a system of norms, is primary, and they understand themselves as a "we." Law begins with the principle of retribution which obliges the "we" to respond to injury to the "we." This is the principle of international law as well as primitive law.

At no point does Kelsen need to depart from his basic methodological individualism. That individuals think as a "we" does not make them into a real "we" with causal powers. That they subordinate themselves to a normative system is a result of their epistemic situation. They live in fear of the world, which for them is populated with superhuman agents whose powers they both invoke and propitiate. The norms themselves are those taken from their dead ancestors, who exact vengeance for their violation. The norms are thus like law: they are rules with sanctions. But the norms are embedded in nature itself, nature understood on the model of human society. But the understanding they have of this nature comes from the ancestors, and supplies the specific norms.

Thus we have a normative system without either transcendence or the superhuman agency of "society." The only transcendental source is in the imagination of the primitive man, in the epistemic situation Kelsen describes: one in which causality in the modern sense is absent and emotion predominates, in which retribution, which is a natural extension of animal reaction to injury, provides the framework of understanding. Extended to the "we" by the undeveloped ego, it becomes customary law. The sanctions, and the normative expectations themselves, do not come from "society" but from the "we," which includes the dead ancestors inhabiting nature: a "we" of actual and imagined individuals. Thus we have law on an entirely "individual" basis. But we arrive there through a reconstruction of the epistemic world of primitive man.

8   Law and Norms Without a Group Mind

The status of the normative system, either the system of custom or the coercive system of the law, remained a puzzle, however. The temptation to locate it in some sort of real collective—a collective consciousness, the term of Durkheim— rather than individuals, persisted. Kelsen had originally formulated his account of law from within, by looking at actual law and its logically necessary conditions. He then showed that legal terms were necessary for social description in the case of the state understood as a unity, but also that the acts of the state were acts of individuals. This was not a completely satisfactory resolution of the problem of the status of the norms themselves, and became more of an issue when he moved from the German language sphere to an Anglo-American one, and with the rise of Logical Positivism. What the term "logic" had meant in the post-Kantian philosophical world in which he began was "relations of conceptual dependence" rather than "deductive relations that preserved truth by virtue of the form of the sentences that expressed them." In the new environment, it became clear that there was a problem with normative statements, which were not "truths" in the sense of truth statements, but oughts, whose logic was problematic (see Olen & Turner, 2016).

At the same time, there were alternative attempts to solve the question of the objective status of cultural objects which avoided attributing them to some sort of group mind: Popper's World III was the most visible example. But this had a precursor in Rudolph Carnap's rarely discussed concept of *geistige Gegenstände*, or "higher cultural objects" (1967, pp. 231–233; including the State, p. 231), a product of his long association with Freyer (Tuboly, 2022), who had articulated a notion of "objective mind," which he traced from sources like Hegel (Freyer, 1991). Kelsen's posthumously published *General Theory of Norms* (1991 [1979]) returned to these questions with the help of the philosophy of "as if" of Hans Vaihinger (2009 [1925]; see Turner, 2010, p. 89). He concluded that the law was an "as if" system in the sense of Vaihinger: not true but not false either, but something treated as if it were real: a "real fiction" (1991 [1979], p. 256). This avoided a group mind, and preserved the idea that normative systems, like the law, which he called a social technology, or the normative order of custom, which was a matter of overwhelming majority opinion, were facts consistent with methodological individualism. In neither case did they require a group mind, society, or special collective reality.

References

Anscombe, G. E. M. (1979). Under a description. *Noûs, 13*(2), 219–233.
Beidelman, T. O. (1974). *W. Robertson smith and the sociological study of religion*. University of Chicago Press.
Borch, C. (2013). *The politics of crowds: An alternative history of sociology*. Cambridge University Press.
Caldwell, P. C. (1997). *Popular sovereignty and German constitutional law: The theory and practice of Weimar constitutionalism*. Duke University Press.
Davidson, D. (2005 [1986]). A nice derangement of epitaphs. In *Truth, language and history* (pp. 89–108). Clarendon Press.
Ellwood, C. (1901). Theory of imitation in social psychology. *American Journal of Sociology, 6*(May), 721–741.

Freud, S. (1918 [1913]). *Totem and taboo: Resemblances between the psychic lives of savages and neurotics* (A. A. Brill, Trans.). Moffat, Yard And Company. https://en.wikisource.org/wiki/Totem_and_Taboo

Freud, S. (1959 [1922]). *Group psychology and the analysis of the ego.* (J. Strachey, Trans. & Ed.). W. W. Norton.

Freyer, H. (1991). *Theory of objective mind: An introduction to the philosophy of culture* (S. Golsby, Trans.). Clarendon Press.

Hägerström, A. (1953). *Inquiries into the notion of law and morals* (K. Olivecrona (Ed.); C.D. Broad, Trans.). Almqvist & Wiksells.

Hart, H. L. A. (1961). *The concept of law.* Clarendon Press.

Kantorowicz, E. (2016 [1957]). *The king's two bodies a study in medieval political theology.* Princeton University Press.

Kaye, H. L. (1993). Why Freud hated America. *Wilson Quarterly, XVII* (2), 118–125.

Kaye, H. L. (2019). *Freud as a social and cultural theorist: On human nature and the civilizing process.* Routledge.

Kelsen, H. (1922). Der Begriff des Staates und die Sozialpsychologie: Mit besonderer Berücksichtigung von Freuds Theorie der Masse. *Imago, 8,* 97–141.

Kelsen, H. (1944). *Peace through law.* University of North Carolina Press.

Kelsen, H. (1946). *Society and nature: A sociological inquiry.* Kegan Paul, Trench, Trubner & Co.

Kelsen, H. (2006 [1949]). *General theory of law & the state.* Transaction Publishers.

Kelsen, H. (1967 [1960]). *The pure theory of law* (2nd ed.). University of California Press.

Kelsen, H. (1998 [1962]). Sovereignty. In S. L. Paulson & B. Litschewski Paulson (Eds.), *Normativity and norms: Critical perspectives on Kelsenian themes* (pp. 523–536). Oxford University Press.

Kelsen, H. (1991 [1979]). *General theory of norms* (M. Hartney, Trans.). Clarendon Press.

Lacey, N. (2006). *A life of H.L.A. hart: The nightmare and the noble dream.* Oxford University Press.

Le Bon, G. (1895). *The crowd: A study of the popular mind.* T. Fisher Unwin. https://www.files.ethz.ch/isn/125518/1414_LeBon.pdf

McDougall, W. (1920). *The group mind: A sketch of the principles of collective psychology with some attempt to apply them to the interpretation of national life and character.* Cambridge University Press. https://www.gutenberg.org/files/40826/40826-h/40826-h.htm

Olen, P., & Turner, S. (2016). Was sellars an error theorist? *Synthese, 193,* 2053–2075. https://doi.org/10.1007/s11229-015-0829-7

Roth, P. A. (2022). What does translation translate? Quine, Carnap, and the emergence of indeterminacy. In S. Morris (Ed.), *The philosophical project of Carnap and Quine* (pp. 154-176) Cambridge University Press.

Ross, E. A. (1909). *Social psychology: An outline and source book.* The Macmillan Company.

Searle, J. (2010). *Making the social world: The structure of human civilization*. Oxford University Press.

Tuboly, A. T. (2022). The constitution of geistige Gegenstände in Carnap's *Aufbau* and the importance of Hans Freyer. In C. Damböck, G. Sandner, & M. G. Werner (Eds.), *Logischer Empirismus, Lebensreform und die deutsche Jugendbewegung* (Logical Empiricism, Life Reform, and the German Youth Movement) (pp. 181–206). Springer. https://doi.org/10.1007/978-3-030-84887-3

Turner, S. (2007). Practice then and now. *Human Affairs, 17*, 110–125.

Turner, S. (2010). *Explaining the normative*. Polity Press.

Turner, S. (2020). Freud in many contexts. Book review symposium on Howard L. Kaye (2019) *Freud as a social and cultural theorist: On human nature and the civilizing process*, Routledge, New York. *Society, 57*(3), 269–275. https://doi.org/10.1007/s12115-020-00478-3

Vaihinger, H. (2009 [1925]). *The philosophy of "as if": A system of the theoretical, practical and religious fictions of mankind* (C. K. Ogden, Trans.). Harcourt Brace and Company.

Weber, M. (1978 [1968]). *Economy and society: An outline of interpretive sociology* (G. Roth & C. Wittich, Eds.). University of California Press.

Methodological Individualism and the Foundations of the "Law and Economics" Movement

Jean-Baptiste Fleury and Alain Marciano

1 Introduction

Originally, "law and economics" consisted in studying the economy by taking into account the institutions and, more narrowly, the legal rules that affect the individuals in their economic activities. This "law and economics" movement, in its modern form, can be traced back to Henry Simons in the 1930s at the University of Chicago—he was the first economist to be hired at the Chicago Law School there. Subsequently, the field started to take shape and be structured in the second half of the 1940s, emerging from the Free Market Study

---

J.-B. Fleury
Economics, Sorbonne University, Paris, France
e-mail: jean-baptiste.fleury@sorbonne-universite.fr

A. Marciano (✉)
Department of Economics, University of Montpellier, Montpellier, France
e-mail: alain.marciano@umontpellier.fr

THEMA Cergy Paris Université, Cergy, France

Karl Mittermaier Centre for Philosophy of Economics, University of Johannesburg, Johannesburg, South Africa

© The Author(s), under exclusive license to Springer Nature Switzerland AG 2023
N. Bulle and F. Di Iorio (eds.), *The Palgrave Handbook of Methodological Individualism*,
https://doi.org/10.1007/978-3-031-41508-1_4

(1946–1952) and the Antitrust (1952–1957) projects, which aimed at understanding the legal underpinnings of competitive markets.[1] In the 1960s, other seminal contributions moved the field forward, notably Ronald Coase's acclaimed article "The Problem of Social Cost" (1960), Guido Calabresi's "Some Thoughts on Risk Distribution and the Law of Torts" (1961) as well as Gary Becker's Crime and Punishment: "An Economic Approach" (1968)—to mention only the major founders of the field. These contributions suggested important shifts in how to construe the links between economics and law, because they led scholars to develop an economic analysis of law, which no longer focused on economic phenomena but consisted in using economics to analyze any type of legal phenomena. In this "new law and economics movement", Richard Posner stands as the leading figure, after his seminal *Economic Analysis of Law* (1973).

In this paper, these two approaches will be (and still are) labeled "law and economics," although they are substantially different in many respects.[2] Indeed, the biggest difference probably lies in the scope of economics' methods and the boundaries of economics. Early law and economics scholars simply enriched traditional economic analysis with considerations taken from law. "New law and economics" scholars pertained to the broader movement of expansion of economics' scope: microeconomics now encompassed legal problems. In spite of these differences, both kinds of scholars used economics as their main tool of analysis. In other words, both kinds of scholars were methodological individualists. The present chapter describes, consequently, both approaches (and the historical shift from the former to the latter), with a special emphasis on the role and place of individuals within their framework. Indeed, the phenomena that economics, and even more specifically law and economics, studies necessarily have a social dimension or can be viewed as having an aggregate dimension, a sort of social nature that go largely beyond mere individual decisions (Arrow, 1994, p. 3; Basu, 1996, p. 269). Reciprocally, all individual actions are necessarily part of a legal or institutional system in such a way that one can hardly separate individual actions from these rules. How did these scholars handle the analytical relationship between social entities and individual behavior?

Some answered by arguing that it was difficult for economists, impossible indeed, to explain such social phenomena only in individual terms. For instance, Robert Ahdieh wrote that using methodological individualism

---

[1] For an historical perspective and details about the scope of these projects, see, for instance, the works of Robert Van Horn (2009, 2011).

[2] For a discussion of these differences see Medema (1998), Harnay and Marciano (2009), and Marciano (2016).

"offers a poor window into important areas of legal and economic analysis. In these areas, it is essential *to go beyond an individualistic focus*, and engage the place of social and institutional factors as direct causes, and not merely indirect influences, in the explanation of social and economic phenomena." (2011, 56; italics added) This view echoed Joseph Agassi's (1960, 1975) on "institutional individualism," which, to some commentators (Toboso, 2001) could easily be viewed as standing halfway between methodological individualism and methodological holism. Therefore, even those unwilling to embrace methodological holism would be forced to abandon what some have termed narrow and "reductionist" (Rutherford, 1994, p. 27) or "strong (strict)" (Neck, 2021) methodological individualism in favor of a "weak (mild)" (ibid.), degenerated form of methodological individualism. This is not far from saying that methodological individualism cannot really be used in economics and law and economics. Or, as Geoffrey Hodgson said about institutional economics, that the term may be "unwarranted" (2007, p. 220).

Such a distinction between a "strong" and a "weak" form of methodological individualism seems to be a mere philosophical abstraction. Most of the major methodological individualists like Carl Menger, Max Weber, Georg Simmel, Karl Popper, Friedrich Hayek, Raymond Boudon were supportive of a social, systemic, institutional, and anti-reductionist individualistic methodology. For instance, Raymond Boudon defines methodological individualism as being such that

> any social phenomenon must be analyzed as the effect of individual actions obeying reasons and motivations configured by the context.[3]

Indeed, methodological individualism does not imply to conceive human beings as (isolated) atoms living as if they were suspended in a social vacuum. To the contrary, Raymond Boudon insists, the social, political, cultural, and, for that matter, legal, contexts in which individuals are embedded represent parameters of individual behavior, although they should not be counted as causes that determine it (Boudon, 2010).

Thus, from this perspective, law can be viewed as one of the parameters that allow individuals to reach their ends and that affect the social outcome. Eventually, though, these parameters are endogenized, that is, the analyst has to provide an explanation for the origins and functioning of the social context in terms of individual behaviors. Under that perspective, law becomes an

---

[3] The original quotation states that "le paradigme de l'IM" is such that "tout phénomène social doit être analysé comme l'effet d'actions individuelles obéissant à des raisons et à des motivations paramétrées par le contexte" (2010, 16).

object of study in itself. Law and economics developed in two strands that perfectly illustrate that view of methodological individualism. This is what we show in this chapter. Old law and economics studied the effect of legal rules on individual choices and social outcomes, whereas new law and economics studied the rules themselves as the result of individual choices.

2 Law and Economics: The Origins at Chicago

One of the first economists who played a major role in the development of "law and economics" was Aaron Director. A professor of economics at the University of Chicago Law School, as Henry Simons before him, Aaron Director "is often considered the father of Chicago law and economics" (Van Horn, 2011, p. 1527). His interest in law came from his desire to understand how and how far competitive markets should be regulated. Hence, Aaron Director was primarily interested "in the analysis of the abuses of competition or monopoly and the meaningfulness of such devices as tie-in arrangements" (Levi, 1966, p. 4; see also Posner, 1975, p. 758). As George Stigler (1992, p. 455) later stressed, when he reviewed the history of the field, "the first systematic application in America of economics to law was the use of price theory to explain economic phenomena involved antitrust cases. Director made creative use of price theory to phenomena such as tie-in sales and patent licensing and assisted colleagues such as John McGee and Lester Telser in their respective studies of predatory competition and resale price maintenance."

George Stigler was himself another prominent figure of Chicago economics and an indirect actor in the development of law and economics. Like Aaron Director, George Stigler was interested in questioning the regulation of the economy and the control of monopoly. In the late 1930s and early 1940s, George Stigler wrote on monopoly (Stigler, 1938, 1942). Later he discussed regulation (Stigler, 1964a, 1964b, 1971; Stigler & Friedland, 1962) and antitrust (Stigler, 1955, 1963, 1966). George Stigler's concerns with monopoly can also be illustrated by the research program that he set when he was recruited at Chicago in 1957, with the monies of the Walgreen fund. George Stigler aimed at studying the "causes and effects of governmental control over economic life" (Stigler to Walgreen, cited in Nik-Khah & Van Horn, 2016, p. 31). From this perspective, George Stigler has repeatedly claimed that, among the many tasks economists should fulfill, one was to contribute to the formulation of economic policy. In that regard, he found "two pieces of knowledge" to be "essential": first, measuring "the quantitative effects of various policies" (Stigler, 1972, p. 5) and, second, evaluating "the influence of governmental structure on economic policy" (p. 6).

These details illustrate a major feature of the law and economics movement as it emerged at Chicago in the 1940s and 1950s. The economists who pushed it "had no interest in the law or, for that matter, in legal problems" (Priest, 2006, p. 354). They were not interested in the origins of the legal rules that frame economic activities. They viewed economics as defined by its subject matter—economic activities, the production of wealth—and paid attention to rules because of their influence on economic activities. Rules mattered because they formed the context in which these activities took place. Studies such as G. Warren Nutter's (1951) dissertation— an empirical study on "The Extent of Enterprise Monopoly in the United States, 1899–1939"—, Arnold C. Harberger's (1954) article on "Monopoly and Resource Allocation," George Stigler's (1955) "Mergers and Preventive Antitrust Policy"; George Stigler and Claire Friedland's (1962) study of the effects of the regulation of electrical utility rates, stand as good illustrations of what these economists were doing.

Looking at their attempts to study the effect of the institutional context— in particular legal rules—on the functioning and structure of markets, and eventually on economic outcomes, one may think at first sight that such an approach bore resemblance to that of the old institutionalists, who nonetheless were known for having adopted methodological *holism* rather than individualism. It is deceiving. First of all, these economists were—at least from their perspective—using a "theoretical" approach grounded on microeconomics ("price theory"). They were not pragmatic like lawyers (Stigler, 1972, p. 8), nor inductive like old institutionalists.[4] That was one of the major advantages of applying price theory to legal problems: to provide a rigorous theoretical framework to legal scholars who are mainly interested in case studies.[5]

In George Stigler's view, economics as a science should rest on abstract even if unrealistic assumptions, and among them, that the basic units of analysis are the individuals. These individuals behave rationally and maximize their utility or profit. Moreover, firms were viewed as black boxes within which it did not matter to look. As a corollary, power relationships within firms did not matter, whereas they did matter greatly for the old institutionalists. Then,

---

[4] "The lawyer does not have a theory of legal processes comparable to the economist's theory of economic processes, so the lawyer's contribution to policy formulation is entirely pragmatic" (Stigler, 1972, p. 8).

[5] When he referred to the origins of law and economics and the first applications of economics to law, Stigler (1992, p. 455) explained that "professional economic analysis was replacing amateur economics of the lawyer."

economists assess various rules or law (or public policies) in terms of efficiency (rather than in terms of justice).[6] And the ways to assess these inefficiencies rested on individual premises: the costs and benefits, although measured at the aggregate level, were solely based on individual evaluations. Important conclusions stem from these individualistic premises, for instance that "[t]he ineffectiveness of regulation [of the electricity market] lies in two circumstances," one of them being that each individual user "may move" and choose a different provider in George Stigler and Claire Friedland's (1962) study.

Contrary to John Commons's and other old institutionalists, Aaron Director, George Stigler, and the Chicagoan law and economics scholars can be said to have used an individualist methodology in that they did not study the legal foundations of capitalism (as the title of Commons's, 1924 book read), nor did they study how laws and courts underpin or shape the economic system and its evolution. They did not consider, contrary to Commons, that "the role of the legal system" was "the centerpiece of [their] analysis" (Medema, 1994, p. 189). In law and economics grounded in price theory, the legal system, laws, and courts, form the context in which individuals maximize their objective functions. They do not *determine* the choices individuals make.[7] This might be one of the ambiguities of this kind of approach that will also characterize Ronald Coase's and Gary Becker's works. The environment simply shapes the set of constraints, under which individual rational choice is made, but not individuals' very motives and motivations. Moreover, market competition and other abstract features of the economic system were not necessarily shaped by law: they are somewhat a distinct social entity, which legal rules and policy decisions disturb. That was different for the old institutionalists who resorted to arguments closer to methodological holism, in claiming that individuals cannot be seen as rational *because* they were influenced by rules, customs, and other environmental factors. Indeed, the influence of various institutions such as technology or norms and cultural forces were not the prime focus of scholars in price theoretic law and economics.

---

[6] "The basic question of policy formulation for the economist is not whether the goal is desirable but whether the means are efficient. Is the end in view actually achieved, and at the least cost?" (Stigler, 1972, pp. 4–5).

[7] To clarify this point, we could refer to Lars Udehn's (2002, p. 489) point about social institutions that "appear only in the explanandum or, better, the consequent of an explanation, but never in the explanans, or antecedent."

3 Ronald Coase: From Old to New Law and Economics

Although the situation did not really change after the publication of Ronald Coase's article, "The Problem of Social Cost" (1960), the paper is widely acknowledged as a turning point in the history of the field, marking the "origin [of]… the modern law and economics movement" (Hovenkamp, 1990, p. 494) also labeled "new" law and economics (Posner, 1975). Indeed, Ronald Coase's subsequent influence came from his insistence on the centrality of property rights—in his view, exchange dealt primarily with property rights instead of goods and services.

In "The Problem of Social Cost" (1960), Ronald Coase started his analysis of how to deal with the divergence between the social and private costs in the presence of harmful effects by considering that external effects were a reciprocal problem, in which both parties, the tortfeasor and the victim, were involved. Assuming that transaction costs were nil, these parties can easily bargain to decide who will bear the costs of the external effect, irrespective of the initial assignment of liability for the damage. In other words, in a world without transaction costs, institutions and the initial delimitation of legal rights do not matter. The negotiation that then takes place is very similar to a market transaction, in the standard (neoclassical) sense of the word. It is not a surprise if George Stigler characterized the "Coase theorem" without referring to transaction costs, and that he mentioned competition instead: "[t]he Coase theorem thus asserts that under *perfect competition* private and social costs will be equal" (Stigler, 1966, p. 113; italics added). "Perfect competition" evacuates the reference to institutions and to the legal environment in which transactions take place. In that case, the analysis can be said to be individualistic (see Marciano, 2018a; Medema, 2020). This result would prove eventually highly influential on subsequent developments of the economic analysis of law, in Richard Posner's work notably (see below).

However, Ronald Coase did not believe that transaction costs could effectively be equal to zero. He maintained throughout his career that the economist should consider a world of positive transaction costs. Transaction costs were of significant importance in constraining the market exchange of these rights. In doing so, Ronald Coase's work departed slightly from the univocal relationship between law (seen as a constraint) and the workings of competitive markets that characterized Aaron Director's and others' work. Pure market transactions were not to be opposed to constrained ones: exchange was unescapably embedded within an institutional arrangement that implied a specific set of transaction costs and consequences on the allocation of resources. Transaction costs mainly captured the essence of the constraints on agents' choice derived from the institutional-legal context. Some commenters even noted (see for instance Pratten, 2016, p. 121) that Coase collapses the many "different aspects of social reality (those aspects necessary to achieve co-ordination) into a single category," transaction costs. In any case, as Steven Medema (2016) put it, "the legal regime both determines the initial resting places of rights and influences the level of transaction costs," while "transaction costs, in turn, impact the ability to exchange legal rights."

Having acknowledged the existence of positive transaction costs, Coase also insisted on the need to minimize them, "either by the competitive process or by careful regulation" (Pratten, 2016, p. 121). The way Coase approached the relationships between economics and policy was to consider that policy was a form of institutional change (Bertrand, 2016). Where government intervention was called upon, alternative institutional arrangements had to be analyzed to establish that government rules or regulations would be efficient. Ronald Coase insisted in the careful historical and empirical examination of the legal institutional setting to provide the analysis with realistic assumptions and evaluation of the benefits of various institutional changes. In many cases, this provided the starting ground for policy recommendations that aimed at maximizing allocative efficiency with the help of the market mechanism.

Therefore, Ronald Coase's reliance on the benchmark of economic efficiency, which is itself based on individual valuation, has an individualistic dimension. However, efficiency is always viewed within an institutional framework. In a number of famous studies, Ronald Coase pinpointed that inefficiency came from a lack of definition of property rights. Once these defined and enforced by the regulatory institution, exchange between individuals would ensure that resources are allocated to their most productive use. This can be illustrated notably by Ronald Coase's recommendations to change the broadcasting policy in the UK. Ronald Coase offered to break the public monopoly by giving a transferable property right to each frequency and let bargaining set the prices that led to an efficient allocation of resources. Yet, he also considered that individuals' preferences were embedded in a certain set of social conventions. He notably proposed that the system remains public, and criticized a conservative proposal to introduce privately run companies to add to the public supply, as he was well aware that such a proposal would be met with intense skepticism by a population attached to public service (Levy & Peart, 2020, p. 250).

Thus, Ronald Coase's analysis remains individualistic in the sense that no social entity is endowed with agency: only individuals do act. Yet, Ronald Coase departed from the radical individualistic stance characterizing some of the more formal post-World War II economic theories by rejecting reductive assumptions about individuals, refusing notably to assume that they were utility maximizers. Indeed, Ronald Coase (1978a, p. 208) found economic theories of utility "sterile" because they do "not tell us why people choose as they do" (Coase, 1988, p. 5), what are "the purposes which impel people to action" (Coase, 1978b, p. 244) and "for which they engage in economic activity" (Coase, 1978a, p. 208). He insisted on this very point, economics is "the study of man as he is and the economic system as it actually exists" (Coase, 2012). He also criticized economic theories of the firm because they presuppose the existence of firms and focus on how they function (the process of production) without explaining why individuals come to create firms (Coase, 1988, p. 5).

In a way, Ronald Coase's approach reminded of Aaron Director's, George Stigler's, and the economists who had preceded him in that he was endeavoring to enrich mainstream economists' individualistic analyses of exchange and allocation of resources with considerations regarding the legal environment. Contrary to how his "theorem" would be used in subsequent economic analyses of law, his interest in law and economics remained that of "an economist" (see, for instance, Coase in Epstein et al., 1997, p. 1138), as he, for instance, wrote "The Problem of Social Cost" as "an essay in economics" that was "aimed at economists," not legal scholars. What [he] wanted to do was to improve our analysis of the working of the economic system," not of the law (Coase, 1993, p. 250).

Nonetheless, improving the analysis of the economic system required to provide individualistic explanations for the emergence and functioning of institutions. For instance, in "The Nature of the Firm" (1937), he opened the door to an analysis of institutions as emerging out of individual choice. Indeed, the choice of having interactions through a market or to set up a firm to organize interactions is a matter of transaction costs (Coase, 1937). From this perspective, the existence of firms remained explained in individualistic terms. Firms are a nexus of contracts that are devised and signed by individuals. Regarding the regulatory regime, Ronald Coase also sketched ideas about legal change, as he addressed the reinforcing dynamics between individuals and institutions (Medema, 2016). A given property rights distribution implied some benefits and costs, thus, in return, incentivized those who lose to pressure for legal change. Although not the center of his work, "Coase's analysis of the regulatory environment in Britain is replete with illustrations of business interests attempting to shape state action to their own benefit" (Medema, 2016, p. 299).

4 Law as an Incentive Toward Social Optimum: The Birth of an Economic Analysis of Law

Besides "The Problem of Social Cost" (Coase, 1960), another article, published in 1961, played an important role in the development of a new relationship between law and economics: "Some Thoughts on Risk Distribution and the Law of Torts," written by Yale law professor Guido Calabresi. With this article, Guido Calabresi did what Aaron Director, George Stigler, and Ronald Coase had not done: he used economics to analyze a legal problem. Guido Calabresi's article thus contributed to change the nature of the field, and accordingly to change its methodology. Partly, this was due to how Guido Calabresi envisaged the nature and role of law. To Guido Calabresi, the purposes of law, in the specific case of his article, accident law, were both to spread losses and minimize the costs of accidents. Not long after his first paper, Guido Calabresi (1965a) introduced the concept of "deterrence" in his analysis to characterize another goal at which accident law should aim. Guido Calabresi's focus on deterrence implied that law could act as an incentive on individual behaviors. This means, as a corollary, that analyzing how individuals react to laws and other rules becomes crucial to understand how the law can minimize the costs of accidents. Therefore, an analysis of the functioning of the legal system can be made at the individual level. This is exactly what Guido Calabresi did when he discussed how individuals would react to different types of liability rules. Although he argued that individuals were not as rational as economists assume, Guido Calabresi nonetheless also claimed that they were sufficiently rational to react to the incentive effect of legal rules. In other words, although the legal system was depicted as having a collective objective, the analysis proceeds by using an individualistic method—at least, partially, as we will show in the next section. With this approach, Guido Calabresi departed from the standard view about tort law as a mechanism to ascribe liability to faulty individuals, who should consequently pay for the accident they had caused. More broadly, in the 1960s, a "fundamental tension" existed between those who saw tort law as "an instrument for admonishing currently undesirable civil conduct" and those who see it as "a means for compensating injured people" (White, 1980, p. 147).

Besides Guido Calabresi, some economists also endowed law with a social purpose, and made its deterrent effect one of its mechanism.[8] One of them was Simon Rottenberg, whose 1965 article on liability used an individualistic approach while ascribing a collective objective to the law. Rottenberg (1965, p. 108) explained that "the primary economic object of a liability rule applied to activities causing personal injuries or death is the prevention of accidents, and this because either of these occurrences deprives society of the output the injured or dead person may have produced had the accident not occurred." In other words, preventing accidents was meant to "maximize... social welfare" (p. 108). Elsewhere in his article, Simon Rottenberg mentioned the need to reach a "social optimum" (p. 113). Thus, accident law and liability rules have an instrumental role to play and indeed, to use Rottenberg's words, a "social purpose" (p. 108). To reach this goal, and to define a proper liability rule, however, one could use economics and an individualistic method. It was, in Rottenberg's view, a matter of incentives that "may take the form of costs imposed upon those whose behavior causes accidents." (p. 108) For instance, Simon Rottenberg analyzed how air carriers would react to different types of legal rules. He ended up with a liability rule that incorporated "self-insurance." This was the most appropriate rule because it would "cause the maximizing behavior of the carriers to coincide with the requirements of the interests of society in the aggregate" (p. 114).

Another landmark paper in the analysis of law and deterrence is obviously Gary Becker's (1968) "Crime and Punishment: An Economic Approach." Gary Becker's article does not seek to provide an economic analysis of law per se. The object is not to study how laws are defined or devised. Like Guido Calabresi and Simon Rottenberg, Gary Becker wrote a paper about the proper amount of sanction (legally defined) and probability of arrest (determined by the whole law enforcement process) that maximize social utility. In other words, his goal was to study what are the best ways to control crime. Thus, because crime combat is costly, an optimal response to criminal activities imply to allow for a certain amount of crime that is too costly to deter, and answer the question "how many offenses *should* be permitted and how many offenders *should* go unpunished?" (Becker, 1968, p. 170, emphasis in original).

Anticipating his later definition of economics as an approach to human behavior (e.g. Becker, 1971, 1976), Gary Becker used an economic model to analyze crime, thus offering a very different approach from the criminological

---

[8] Harold Demsetz's (1964) "The Exchange and Enforcement of Property Rights" is yet another example.

mainstream of the time, dominated by the influence of the delinquent subcultures approach inspired by sociology (see Fleury, 2021). In Gary Becker's initial model, self-interested and rational individuals maximize an expected utility function. Thus, individual calculations are the basic unit upon which a social welfare function that aggregates individuals' welfare can be built. However, Gary Becker's objective remains to minimize social costs. He indeed moved to the minimization of a social cost function, which, again, balances the costs of producing enforcement as well as the harm generated by crime, all calculated from the aggregation of *individual* costs and harms. Gary Becker (1968, p. 181) notably claimed that such an approach to crime control would enable to "go beyond catchy phrases" and give "due weight to the damages from offenses, the costs of apprehending and convicting offenders, and the social cost of punishment." Subsequently, scholars in the economics of crime modeled social welfare functions to compute the difference between aggregate individual gains and harms of crime measured in monetary units (see, for instance, Polinsky & Shavell, 2000).

As with other above-mentioned approaches, Becker's economics of crime control starts with individuals and focuses on how their responses to incentives influence the level of crime. But here again, individual choice is partly influenced by the institutional and cultural setting for a few different reasons. Firstly, in Gary Becker and George Stigler's 1974 article, the malfeasance of enforcers who accept bribes depends partly on what the authors call "the supply of honesty" (p. 3), although strictly economic considerations such as repeated interactions also play a strong part. Secondly, because, as illustrated by Gary Becker's 1968 paper, rules that lead to a certain level of punishment appear only to be prices or costs that are included in the individual cost–benefit analysis. So these rules are part of the constraints individuals have to take into account to make their decisions. These rules are exogenously given. They are not explained, not even in individual terms. This will also characterize the economic analyses of law enforcement that were developed after Gary Becker's seminal paper. These works explored various types of punishments, from fines to imprisonment, the effects of various liability rules, and discussed their links with individuals' behaviors toward risk and final social outcome in terms of crime control (see Polinsky & Shavell, 2000).

Overall, the economic analysis of optimal enforcement is largely an endeavor at normative economics. An important question, in this respect, explores whether law enforcement should be a strictly public matter or whether there is room for private enforcement (see Polinsky & Shavell, 2000). Gary Becker and George Stigler (1974), notably, argued in favor of perfect competition among private enforcers, suggesting that the public monopoly of enforcement may be inefficient. The assumption that institutions, particularly legal rules, emerge to improve efficiency, provides generally the intellectual link bridging normative analysis of crime control to a more positive analysis of legal rules.

Gary Becker's discussion of fines, although still within a normative outlook, certainly paved the way for the subsequent development of more "positive" economic analyses of law. To Gary Becker, fines are, in numerous cases, an optimal sanction from the collective point of view, because they ensure that individual harm done by criminals or delinquents has been correctly compensated at a very low cost (far lower than imprisonment and other heavier sanctions). This led him to consider that sanctions, as said, mostly defined by law, should be studied from the point of view of economic theory. Of course, the first aspect that comes to mind is the magnitude of sanctions: Becker provides a policy guide to establish optimal criminal sanctions, defined by criminal laws, in order to maximize social wealth.

But that is not all. Gary Becker's analysis of fines addresses the nature of sanctions as well. As he (1968, p. 193) noted, "legislation usually specifies whether an offense is punishable by fines, probation, institutionalization, or some combination." Here, the overarching model is one reminiscent of Arthur Pigou's welfare approach. Crime is defined as uncompensated harm, and fines act like a tax to bring compensation and induce criminals to reduce their production of negative external effects. Gary Becker, then, conjectures that if fines were to be adopted as the primary mode of punishment, then it would change the social purpose of law and enforcement policies, as fines aim primarily at compensating harm, and less at deterring or inflicting vengeance. Under that approach, legal proceedings would become an activity that seeks to assess the amount of harm done and to devise optimal fines. "Much of traditional criminal law would become a branch of the law of torts, say, 'social torts'... A criminal action would be defined not by the nature of the action but by the inability of a person to compensate for the 'harm' that he caused" (Becker 1968, p. 198).

5 William Landes and Richard Posner: Toward an Economic Analysis of Law

Discussing the nature of sanctions and their social effect opened, therefore, the door to a positive analysis of how law enforcement and sanctions emerge and evolve in society, in both the private and public realms. The intellectual trajectory of William Landes illustrates nicely the shift from law and economics to an economic analysis of law, thus anticipating Richard Posner's work.

William Landes initially studied, from a traditional law and economics perspective, the effects of fair employment laws on economic outcomes and unemployment in the mid-1960s. More precisely, William Landes (1967) had used expected utility theory to model the decision of firms to comply or violate anti-discrimination laws. Then, after having defended his PhD in 1966, William Landes used the same framework to analyze a phenomenon that was troubling him: that less than 10 percent of criminal cases went to trial. To solve this puzzle, William Landes analyzed "the conditions under which a pretrial settlement or trial will take place" (Landes, in Downs et al., 1969, p. 505). In his model, pretrial arrangements were the results of individual optimization. Indeed, William Landes assumed that not only the suspect (1969) or the defendant (1971) but also the prosecutors were utility maximizers—"the basic assumption of the model is that both the prosecutor and the defendant maximize their utility, appropriately defined, subject to a constraint on their resources" (Landes, 1971, p. 61). More precisely, the utility of prosecutors is assumed to depend on "the expected number of convictions weighted by their respective S[entences]" (Landes, 1971, p. 63). Therefore, he argued, "the prosecutor and defendant would reach a plea bargain on a sentence if both could be made better off compared to risking an uncertain trial outcome" (Landes, 1997, p. 34). That was the Coase theorem applied to prosecution. Landes, therefore, shifted emphasis from an analysis of the effects of specific laws on economic behavior, to an economic analysis of legal procedures, here, pretrial arrangements. Moreover, the shift in emphasis exemplified the individualistic analysis of a social entity (the judicial process).

William Landes's work also foreshadowed important conclusions in the economic analysis of law: that the judicial process tended to move toward efficiency. This idea had already been developed by Guido Calabresi about some areas of the common law in the 1960s. In William Landes's work, the optimal level of conviction, which depended on the amount of resources prosecutors devoted to cases, was reached as the result of prosecutors' individual rational decisions. In other words, prosecutors' individual behavior contributed to the efficiency of the legal system. Yet, William Landes does not clarify the premises on which such an individualistic outlook rest. Indeed, on the one hand, the only consideration taken into account in prosecutors' calculus is the expected number of convictions while, on the other hand, prosecutors make their decisions by maximizing "the community's welfare for a given resource level" (Landes, 1971, p. 63). How does the individual decision rule coincide with a social optimum? William Landes does not answer the question. He thus does not explain the articulation between individual choice and social ouctomes.

A similar difficulty characterized Richard Posner's work on the economic analysis of law, that he developed alongside (and sometimes in collaboration with) William Landes.[9] Richard Posner followed both paths of Ronald Coase and Guido Calabresi by discussing the efficiency of the entire Anglo-American legal system, using, like Guido Calabresi (1961, 1965a, 1965b, 1967, 1970), a framework based on the tools of economics (see Posner 1971, p. 202; 1973b, p. 399). This resulted in Posner framing the common law as a method "to allocate responsibilities between people engaged in interacting activities in such a way as to maximize the joint value, or, what amounts to the same thing, minimize the joint cost of the activities" (Posner, 1973a, p. 98). Common law was understood by Richard Posner as "a pricing mechanism designed to bring about an efficient allocation of resources" (Posner, 1987, p. 5). In cases when high transaction costs prevent markets from working properly, the common law replaces the market: "[T]he common law prices behavior in such a way as to mimic the market" (Posner, 1992, p. 252). As such, judicial decisions can be said to be made "as if" they intended to replicate the functioning of a perfectly competitive, efficient, market (Backhaus, 2017 [1978]; Marciano, 2018b).

Richard Posner's defense of the efficiency of the common law was nonetheless individualistic, partly because the social efficiency of markets, as the result of exchange among maximizing individuals, constituted the benchmark. To Richard Posner, wealth maximization is a superior criterion of justice, firstly because it derives from consent—as it necessarily comes out of market transactions. Moreover, wealth maximization avoids the pitfalls of utilitarianism, in particular interpersonal comparisons. Indeed, Richard Posner claimed, "a dollar is worth the same to everyone" (2000, p. 1,170). Precisely because "it treats a dollar as worth the same to everyone" (2000, p. 1,154), wealth maximisation, or the Kaldor-Hicks criterion, "leaves out of normative consideration...distributive justice" (ibid.; see also Posner, 1985, p. 104). What matters only is allocative efficiency, in other words "welfare is increased when a policy inflicts a dollar loss on the losers from it and confers a dollar and five cents gain on the winners even though the losers are not compensated" (Posner, 2000, p. 1,170). To Richard Posner, questions of distributive justice

---

[9] Richard Posner is a key scholar in the development of the economic analysis of law. The field, ever since its creation in the early 1970s, addressed topics ranging from judicial decisions and the formation of rules within the legal system, to informal rules and norms in primitive societies.

concern only the legislatures—public policy, not courts. Hence, methodological individualism grounds the judge's calculus of wealth maximization, as it is based solely on the aggregation of individual evaluations in money estimates.

Another aspect of the individualistic dimension of Richard Posner's work comes from the links between the efficiency of the system and the behavior of judges. Richard Posner insisted on that the Common Law is an efficient legal system and this was because it was a judge-made law. He indeed believed (and claimed) that judges "are guided by concern with economic efficiency" (Posner, 1971, p. 223) and "think in economic terms" (p. 224), so that one can "assume that judges make their decision in accordance with the criterion of efficiency" (Posner, 1973a, p. 325). Or, one could say, *each* judge decides in accordance with the criterion of efficiency by comparing the costs and benefits associated with the action under scrutiny. Thus, the criterion for judging the results of judicial decision was strictly grounded on individual welfare. Yet, as with William Landes's 1971 weakly justified decision rules for prosecutors, precise reasons for why individual decision-making and social efficiency coincide were not provided. In other words, Richard Posner jumped from an individualistic analysis to a global, aggregate conclusion. Thus, scholars such as Rutherford (1994) have noted that Posner's approach actually leaned closer to functionalism. Indeed, what Posner initially did was to give purpose (efficiency) to a social entity (law).[10]

Thus, in a way that was very similar to Gary Becker's and William Landes's works, Richard Posner was not interested in judicial behavior in itself. From this perspective, Richard Posner's (1993) later economic analyses of judicial behavior —in which judges are assumed to be "rational" (Posner, 1993, p. 3) and to "respond rationally to ordinary incentives" (p. 1)— were a by-product of his primary interest in the efficiency of the legal system (Marciano et al., 2020).

During the 1970s and 1980s, the rapidly growing subfield of the economic analysis of law, occasionally bolstered by some contributions in the fields of public choice, developed analyses of rules, constitutions, norms, and customs, which relied on strong individualistic premises. In these cases, social institutions are shown to emerge from repeated games. Although these approaches rested on a narrower conception of methodological individualism, where agents are not influenced by external and environmental factors (such as ethical views, norms, etc.)—they are only driven by the maximization of a very limited set of variables, such as income—these models are not devoid of considerations for the social context. One may mention, for instance, that

---

[10] In functionalist approach, different social institutions perform a specific function that contributes to the stabilization and survival of societies.

the origins of the rules of the games to be repeated are not specified, and ultimately imply some sort of reliance to an institutional environment, or that there is common knowledge among agents. Such description extends to the larger movement labeled "New Institutional Economics," comprised of the Property Rights approach developed by Armen Alchian and Harold Demsetz (1973, for instance), Oliver Williamson's works in the theory of organizations, as well as works in history, such as those of Douglass North. Like Richard Posner's work, that movement also sought to endogenize institutions in an economic framework based on individual choice (Rutherford, 1994).

6 Heterodox Economic Analysis of Law: Calabresi

To conclude our overview of the use of an individualist methodology in the law and economics movement, it is important to come back to the views of one of the founders of the field, already mentioned in a previous section, Guido Calabresi. He stood as one of the most important representatives, if not the leader, of the so-called "New Haven School of Law" (Medema & Mercuro, 2006), a school that studied the place of courts in the regulatory regime, the latter aiming at solving market imperfections. Although he was the first to claim that law had deterrent effects on individuals (whether or not the latter are rational), and the first to formulate the intuition for the Coase theorem, Guido Calabresi's work deserves a section of its own because of his use of methodological individualism that differs slightly from what the other scholars above-mentioned did.

Guido Calabresi (1961, p. 506) was one of the first to claim with his "pure loss distribution theory" that, from the perspective of economic analysis, the assignment of the burden of costs, the distribution of losses, and the assignment of liability to one party or the other have no impact on the final allocation of resources: "[t]here are naturally, some situations where… it actually does not matter who bears the loss initially" (1961, p. 506; see also Calabresi, 1965a, pp. 725–726; 1968, p. 67). Guido Calabresi, to put it in clearer terms, reached the same result as Ronald Coase and[11] scholars eventually talked about the Coase-Calabresi theorem.[12] The similarity between the findings of Ronald Coase and Guido Calabresi seems even clearer when the latter insisted on the importance of bargaining: "situations in which it will not matter which of two activities initially bears the cost of an accident are all the situations in which the two or more possible accident-causing activities are related by bargaining" (1965a, pp. 725–726). Indeed, through bargaining, "the least expensive way to minimize the loss will be sought out and used whichever of the two is initially liable" (p. 726). From this perspective, Guido Calabresi's analysis is as individualistic as Ronald Coase's.

Furthermore, like Ronald Coase, Guido Calabresi doubted the practical validity of that result because it could hold only if the economy was perfectly competitive. But Calabresi also departed from economists in claiming that a bargaining process would work only if individuals were perfectly rational. None of these two conditions are actually fulfilled in the real world. Monopolistic firms benefit from a strong market power in a potential bargain while human beings have little bargain power because of the several cognitive biases that affect them.[13] Monopolies and other forms of deficiencies in competition exert a strong influence on the choice set of individuals. Therefore, Coasean bargains—so dear to Richard Posner and George Stigler—could not work in the real world. It was indeed unlikely that the optimal amount of activities could be reached by "voluntary" arrangements among members of society. What such arrangements could bring was necessarily influenced by the power firms benefited from. For instance, Guido Calabresi (1970, p. 50) wrote that, in the case of accidents, "people individually do not or cannot voluntarily insure against accident risks to the degree they collectively deem desirable," which hinted at the need for accident laws to take into account that market failure.

This had implications on how Guido Calabresi approached methodological individualism. Guido Calabresi claimed that automobile manufacturers

---

[11] The story has been recounted many times, including by Guido Calabresi himself: the first version of his 1961 article, written in 1957, included a reference to causation, and accordingly a pre-Coase version of the Coase theorem. Although convinced that he was right, Guido Calabresi removed this passage after Ward Bowman had told him that he was wrong because of what Pigou had written. And he eventually regretted it (see Kalman, 2014). Had he not followed Bowman's suggestion, a "liberal version" of the Coase theorem "would have been available at the creation," and those who "tried to make of economic analysis of law a basis for blindly supporting the status quo would have found their path more difficult" (Calabresi, in Shapiro, 1991, p. 1484).

[12] For an analysis of the similarities and differences between Ronald Coase and Guido Calabresi, see Alain Marciano (2012), and Steven Medema (2014).

[13] This is why it could be argued that "Many of the ideas of behavioural law and economics were hence already implicit in Calabresi's writings" (Faure, 2008, p. 75).

were most of the time in the best position to determine how to minimize the costs of accidents. From this perspective, it seems that firms (or workers or drivers and pedestrians) are treated as categories. It is unclear however whether Guido Calabresi treated these categories as individual entities or, to the contrary, if reasoning with categories meant that individuals disappeared behind the categories to which they belonged. It is likely that here, drivers or pedestrians are not considered as individuals but, rather, as the expression of a specific group, acting only as members of a group. The treatment of the individuals by Calabresi is methodologically ambiguous. It certainly alters the individualistic dimension that characterizes the Coasean bargaining process.

Consequently, the claim that it did not matter who bears the loss initially was no longer valid. To Guido Calabresi, starting points, that is the rights initially granted to individuals, do matter in the bargaining process. In the case of accidents, because workers or pedestrians were not able to assess the risks of accident as well as employers or car manufacturers were, whether the liability was put on the firms than on the drivers, or on the employers rather than on the workers was crucial to minimize the costs of accidents.

Pushing the analysis further, Guido Calabresi also argued that, since starting points and liability rules matter, distributional problems should be taken into account in the analysis. This meant that Calabresi opposed Posner and many economists, for whom economics should focus on the allocation of resources, while distributional questions should be rejected as outside the realm of economic analysis of law. Now, such questions play a significant part in conditioning individual choice. In standard economic theory, individuals are assumed to agree with the condition of choice, and with the initial distribution of wealth. Guido Calabresi rejects that starting point, as we said above. To Calabresi, the traditional stance that wealth distribution is outside of the scope of neoclassical economics does not apply for legal issues: legal decisions and rules determine conditions of choice and starting points, hence, they have to be considered by the legal scholar.

Therefore, Guido Calabresi departed from Richard Posner's work by insisting that criterions of fairness and justice be used to judge the various methods to reduce costs suggested by an economic analysis of law. What mattered, ultimately, was that legal rules "comply with our general sense of fairness" (Calabresi, 1970, p. 26). Contrary to Richard Posner, efficient solutions that violate a general sense of fairness are to be abandoned, as when a decision would, for instance, overly favor rich people against poor people. Law and economics are conceived as a discussion that goes back and forth between distribution problems and choice. On the one hand, there are general goals and collective values decided at the collective level. On the other hand, there are concrete everyday questions that are too complex to be dealt with collectively. In the latter case, economics brings interesting analytical tools, although, again, economics does not offer ultimate goals. This is why Guido Calabresi argued for a certain level of loss spreading in accident law, partly to compensate for individuals' lack of rationality, partly to avoid "social dislocation" (see Hackney, 2007, pp. 115–120).

Perhaps one recurrent difficulty in Guido Calabresi's approach was that he seemed to have failed to offer a precise theory that would explain how these goals and criterions of justice would be determined (Hackney, 2007, p. 138). Efficiency is one important goal; however, Guido Calabresi's own take on contemporary society's preferences over the distribution of wealth led him to advocate a system that also protects the poor and the aged (Hackney, 2007, p. 139). This is the source of Guido Calabresi's "middle theorizing": using a quite individualistic scientific framework alongside considerations about collective value judgments where one relies also on "indications of society's distribution preferences" (ibid. p. 139). Here, Guido Calabresi's position clearly departed from the standard neoclassical analysis of law, according to which, notably after the clarifications of Steven Shavell (1981, see also Kaplow & Shavell, 1994), distributional concerns should only be tackled through the tax system, leaving law dealing only with efficiency (see also Hackney, 2007, p. 147). But although economists have consistently striven to produce narrowly individualistic accounts of both legal problems and the tax system within the frameworks of economics and game theory, Guido Calabresi's attempts to tame economics with justice and ethical considerations have resulted in a less individualistic and perhaps fuzzier methodological approach that was regularly criticized by, among others, Richard Posner.

7   Conclusion

In this chapter, without trying to be exhaustive, we have reviewed the theoretical approaches and results adopted by the main contributors—the founders, indeed—to the law and economics movement. Our goal was to reflect on their use of methodological individualism. *All* of them considered that the analysis of legal phenomena had to start from individual behavior, even as these very behaviors were embedded, to various degrees, though not determined, in legal and institutional frameworks. In that, they followed on the path opened by the authors mentioned in the introduction, of this paper. With this method, economic approaches have led to original and influential outcomes, both in the study of how legal rules shape economic outcomes (the "old" law and economics) and in the economic analysis of law. It allowed analysts to provide a clearer, and, to many, a more rigorous way of evaluating the outcomes of rules and decisions, based on individual welfare.

References

Agassi, J. (1960). Methodological individualism. *The British Journal of Sociology, 11*(3), 244–270.

Agassi, J. (1975). Institutional individualism. *British Journal of Sociology, 26*(2), 144–155.

Ahdieh, R. (2011). Beyond individualism in law and economics. *Boston University Law Review, 91*(1), 43–86.

Alchian, A. A., & Demsetz, H. (1973). The property right paradigm. *Journal of Economic History, 33*(1), 16–27.

Arrow, K. (1994). Methodological individualism and social knowledge. *American Economic Review, 84*(2), 1–9.

Backhaus, J. (2017 [1978]). Lawyers' economics vs. economic analysis of law. A critique of Professor Posner's economic approach to law by reference to a case concerning damages for loss of earning capacity. *Munich Social Science Review, 3,* 57–80.

Basu, K. (1996, February 3). Methodological individualism. Resurrecting controversy. *Economic and Political Weekly,* 269–270.

Becker, G. S. (1968). Crime and punishment: An economic approach. *Journal of Political Economy, 76*(2), 169–217.

Becker, G. S. (1971). *Economic theory.* Alfred Knopf.

Becker, G. S. (1976). *The economic approach to human behavior.* The Chicago University Press.

Becker, G. S., & Stigler, G. J. (1974). Law enforcement, malfeasance, and compensation of enforcers. *Journal of Legal Studies, 3*(1), 1–18.

Bertrand, E. (2016). Coase's empirical studies: The case of the lighthouse. In C. Ménard & E. Bertrand (Ed.), *The Elgar companion to Ronald H. Coase* (pp. 320–332). Edward Elgar.

Boudon, R. (2010). *La sociologie comme science.* La découverte.

Calabresi, G. (1961). Some thoughts on risk distribution and the law of torts. *Yale Law Journal, 70*(4), 499–553.

Calabresi, G. (1965a). The decision for accidents: An approach to non-fault allocation of costs. *Harvard Law Review, 78*(4), 713–745.

Calabresi, G. (1965b). Fault, accidents, and the wonderful world of Blum and Kalven. *Yale Law Journal, 75*(2), 216–238.

Calabresi, G. (1968). Transaction costs, resource allocation and liability rules—A comment. *Journal of Law and Economics, 11*(1), 67–73.

Calabresi, G. (1970). *The costs of accidents—A legal and economic analysis*. Yale University Press.

Coase, R. H. (1937). The nature of the firm. *Economica, 4*(16), 386–405.

Coase, R. H. (1960). The problem of social cost. *Journal of Law and Economics, 3*(October), 1–44.

Coasse, R. A. (1978a). Economics and contiguous discipline. *Journal of Legal Studies, 7*(2), 201–211.

Coasse, R. A. (1978b). Discussion. *American Economic Review, 68*(2), 244–245.

Coase, R. H. (1988). The nature of the firm: Influence. *Journal of Law, Economics, and Organization, 4*(1), 33–47.

Coase, R. H. (1993). Law and economics at Chicago. *Journal of Law and Economics*, 36 (1), Part 2, John M. Olin Centennial Conference in Law and Economics at the University of Chicago, pp. 239–254.

Coase, R. H. (2012). *Saving economics from the economists*. http://hbr.org/2012/12/saving-economics-from-the-economists

Commons, J. R. (1924). *Legal foundations of capitalism*. University of Wisconsin Press.

Demsetz, H. (1964). The exchange and enforcement of property rights. *Journal of Law and Economics, 7*(October), 11–26.

Downs, A., Landes, W. M., Rottenberg, S., & Hoffman, R. B. (1969). Round table on allocation of resources in law enforcement. *American Economic Review, 59*(2), 504–512.

Epstein, R., Becker, G. S., Coase, R. H., Miller, M. H., & Posner, R. A. (1997). The roundtable discussion. *University of Chicago Law Review, 64*(4), 1132–1165.

Faure, M. G. (2008). Calabresi and behavioural tort law and economics. *Erasmus Law Review, 1*(4), 75–102.

Fleury, J. B. (2021). Crime. In P. Fontaine & J. Pooley (Eds.), *Society on the Edge: Social Sciences and Public Policy in the Postwar United States*. Cambridge University Press.

Hackney, J. R., Jr. (2007). *Under cover of science: American legal-economic theory and the quest for objectivity*. Duke University Press.

Harberger, A. C. (1954). Monopoly and resource allocation. *American Economic Review, 44*(2), 77–87.

Harnay, S., & Marciano, A. (2009). Posner, economics and the law: From law and economics to an economic analysis of law. *Journal of the History of Economic Thought, 31*(2), 215–232.

Hodgson, G. M. (2007). Meanings of methodological individualism. *Journal of Economic Methodology, 14*(2), 211–226.

Hovenkamp, H. (1990). The First Great Law & Economics Movement, *Stanford Law Review, 42*(4), 993–1058.

Kalman, L. (2014). Some thoughts on Yale and Guido. *Law and Contemporary Problems, 77*(2), 15–43.

Kaplow, L., & Shavell, S. (1994). Why the legal system is less efficient than the income tax in redistributing income. *Journal of Legal Studies, 23*(2), 667–681.

Landes, W. M. (1967). The effect of state fair employment laws on the economic position of nonwhites. *American Economic Review, 57*(2), 578–590.

Landes, W. M. (1971). An economic analysis of the courts. *Journal of Law and Economics, 14*(1), 61–107.

Landes, W. M. (1997). The art of law and economics: An autobiographical essay. *The American Economist, 41*(1), 31–42.

Levi, E. H. (1966). Aaron Director and the study of law and economics. *Journal of Law and Economics, 9*(October), 3–4.

Levy, D., & Peart, S. (2020). *Towards an economics of natural equals*. Cambridge University Press.

Marciano, A. (2012). Guido Calabresi's economic analysis of law, Coase and the Coase theorem. *International Review of Law and Economics, 32*(1), 110–118.

Marciano, A. (2016). Economic analysis of law. In A. Marciano & G. B. Ramello (Eds.), *Encyclopedia of law and economics*, Springer. https://doi.org/10.1007/978-1-4614-7883-6_598-1

Marciano, A. (2018a). Why is "stigler's coase theorem" stiglerian. A methodological explanation. *Research in the History of Economic Thought and Methodology, 36A*, 127–155

Marciano, A. (2018b). An introduction to Juergen Backhaus's "Lawyers' economics vs. economic analysis of law". *Munich Social Science Review* (New Series, vol. 2).

Marciano, A., Melcarne, A., & Ramello, G. (2020). Justice without romance: The history of the economic analyses of judges' behavior, 1960–1993. *Journal of the History of Economic Thought, 42*(2), 261–282.

Medema, S. G. (1998). Wandering the road from pluralism to Posner: The transformation of law and economics, 1920s–1970s. In *The transformation of American economics: From interwar pluralism to postwar neoclassicism, history of political economy* (Annual Supplement 30, pp. 202–224).

Medema, S. G. (2011). A case of mistaken identity: George Stigler, 'the problem of social cost', and the coase theorem. *European Journal of Law and Economics, 31*(1), 11–38.

Medema, S. G. (2014). Juris prudence: Calabresi's uneasy relationship with the Coase theorem. *Law and Contemporary Problems, 77*(2), 65–95.

Medema, S. G. (2016). Ronald Coase and the Legal-Economic Nexus. In C. Menard & E. Bertrand (Eds.), *The Elgar Companion to Ronald Coase* (pp. 291–304). Edward Elgar Publishing.

Medema, S. G. (2020). The Coase theorem at sixty. *Journal of Economic Literature, 58*(4), 1045–1128.

Medema, S. G., & Mercuro, N. (2006). *Economics and the law: From Posner to post-modernism and beyond* (2nd ed.). Princeton University Press.

Neck, R. (2021). Methodological individualism: Still a useful methodology for the social sciences? *Atlantic Journal of Economics, 49*, 349–361.

Nik-Khah, E., & Van Horn, R. (2016). The ascendancy of Chicago neoliberalism, in S. Springer, Birch, K., & MacLeavy, J. (Eds.), *The Handbook of Neoliberalism*. Routledge.

Nutter, G. W. (1951). *The extent of enterprise monopoly in the United States.* University of Chicago Press.

Polinsky, A. M., & Shavell, S. (2000). The economic theory of public enforcement of law. *Journal of Economic Literature, 38*(1), 45–76.

Posner, R. A. (1971). Killing or wounding to protect a property interest. *Journal of Law and Economics, 14*(1), 201–232.

Posner, R. A. (1973a). *Economic analysis of law.* Little, Brown and Company.

Posner, R. A. (1973b). An economic approach to legal procedure and judicial administration. *Journal of Legal Studies, 2*(2), 399–458.

Posner, R. A. (1975). The economic approach to law. *Texas Law Review, 53*(4), 757–782.

Posner, R. A. (1985). Wealth Maximization Revisited, Notre Dame Journal of Law, *Ethics and Public Policy*, 2, 285–105

Posner, R. A. (1987). The law and economics movement. *American Economic Review, 77*(2), 1–13.

Posner, R. A. (1993). What do judges and justices maximize? (The same thing everybody else does). *The Supreme Court Economic Review, 3,* 1–41.

Posner, R .A. (1992). Economic Analysis of Law, Fourth edition. Little, Brown and Co.

Posner, R. A. (2000). Cost-benefit analysis: Definition, justification, and comment on conference papers. *Journal of Legal Studies, 29*(S2), 1153–1177.

Pratten, S. (2016). Coase on the nature and assessment of social institutions. In C. Ménard & E. Bertrand (Ed.), *The Elgar Companion to Ronald H. Coase*, Cheltenham, Edward Elgar, pp. 110–128.

Priest, G. (2006). The rise of law and economics: A memoir of the early years. In F. Parisi & C. K. Rowley (Eds.), *The origins of law and economics: Essays by the founding fathers* (pp. 350–382). Edward Elgar.

Rottenberg, S. (1965). Liability in law and economics. *American Economic Review, 55*(1/2), 107–114.

Rutherford, M. (1994). *Institutions in economics The old and the new institutionalism.* Cambridge University Press.

Shapiro, F. R. (1991). The Most-Cited Articles from The Yale Law Journal, *Yale Law Journal*, *100*(5), 1449–1514.

Shavell, S. (1981). A Note on efficiency vs distribution equity in legal rulemaking: Should distributional equity matter given optimal income taxation. *American Economic Review, 71*(2), 414–418.

Stigler, G. J. (1938). Social welfare and differential prices. *Journal of Farm Economics, 20*(3), 573–586.

Stigler, G. J. (1942). The extent and bases of monopoly. *American Economic Review, 32*(2), Papers Relating to the Temporary National Economic Committee, 1–22.

Stigler, G. J. (1955). Mergers and preventive antitrust policy. *University of Pennsylvania Law Review, 104*(2), 176–184.

Stigler, G. J. (1963). United States v. Loew's Inc.: A note on block-booking. *The Supreme Court Review, 1963,* 152–157.

Stigler, G. J. (1964a). Competition and concentration. *Challenge, 12*(4), 18–21.

Stigler, G. J. (1964b). Public regulation of the securities markets. *Journal of Business, 37*(2), 117–142.

Stigler, G. J. (1966, October). The economic effects of the antitrust laws. *Journal of Law & Economics, 9*, 225–258.

Stigler, G. J. 1971. The Theory of Economic Regulation, *The Bell Journal of Economics and Management Science, 2*(1), 3-21.

Stigler, G. J. (1972). The law and economics of public policy: A plea to the scholars. *Journal of Legal Studies, 1*(1), 1–12.

Stigler, G. J. (1992). Law or economics? *Journal of Law & Economics, 35*(2), 455–468.

Stigler, G. J., & Friedland, C. (1962, October). What can regulators regulate? The case of electricity. *Journal of Law & Economics, 5*, 1–16.

Toboso, F. (2001). Institutional individualism and institutional change: The search for a *middle way* mode of explanation. *Cambridge Journal of Economics, 25*, 765–783.

Udehn, L. (2002). The changing face of methodological individualism. *Annual Review of Sociology, 28*, 479–507.

Van Horn, R. (2009). Reinventing monopoly and the role of corporations. In P. Mirowski & D. Plehwe (Eds.), *The road from Mont Pélerin: The making of the neoliberal thought collective* (pp. 204–237). Harvard University Press.

Van Horn, R. (2011). Chicago's shifting attitude toward concentrations of business power (1934–1962). *Seattle University Law Review, 34*(4), 1527–1544.

White, E. G. (1980). *Tort Law in America. An intellectual history*. Oxford University Press.

Methodological Individualism and Social Change

Michael Schmid

1 The Problem

The following chapter deals with the question of whether sociological theories of social change—as many of its representatives contend (Galt & Smith, 1976, p. 7; Lauer, 1973, pp. 21–22; Strasser & Randall, 1979; Wiswede & Kutsch, 1978, p. 203)—are able to *explain* the development or change of societies in accordance with the corresponding scientific procedural postulates that can be derived from Methodological Individualism. I will begin by clarifying the logic of scientific explanation in anticipation of the answer to this question. Afterward I will compile different (paradigmatic) versions of the theory of social change, and finally I will examine in each case whether they actually do justice to the commonly declared aim of explaining social changes.

Methodological Individualism and Its Logic of Explanation

Within the context of the problem specified above, the following consideration can be seen as an (in principle approving) comment on Karl Popper's repeatedly quoted request that "the task of social theory is to construct and to

---

M. Schmid (✉)
Sociology, Neubiberg University, Neubiberg, Germany

© The Author(s), under exclusive license to Springer Nature
Switzerland AG 2023
N. Bulle and F. Di Iorio (eds.), *The Palgrave Handbook of Methodological Individualism*,
https://doi.org/10.1007/978-3-031-41508-1_5

analyse our sociological models carefully in descriptive or nominalist terms, that is to say, in terms of individuals, of their attitudes, expectations, relations etc.—a postulate which may be called 'methodological individualism'" (Popper, 1961, p. 136). In order to explicate this passage, I shall, as a first step, present an *explanatory model* which respects the preconditions and limits outlined by Popper and which can serve to clarify some of the misunderstandings associated with both its rejection and its defense.[1] Clearly, Popper's mandate to the social sciences is to develop (and empirically test) "models" in order to explain the actions of actors (as "intentional" or "rational") who must take care of solving their action problems on the basis of their value beliefs and their expectations about the kind of relationship in which they find themselves. To the extent that such explanations can only be given by using (general and substantial) laws, the question arises as to what kind of laws are to be considered. Since theories should consist of laws that serve to grasp the (causally effective) "capacities", "potential" or "production function" (Cartwright, 2001) of an event to be investigated and i.e., in the present case the dynamics of human problem-solving actions, we have to search for a *theory of individual action.*[2]

Insofar as the terms of individual theories of action are restricted to those factors that (mentally or psychologically) generate or organize actions, the descriptions of the peculiarities of action situations in the face of which the actors have to act cannot be logically derived from them. Social science model explanations, as Popper demands, cannot represent an "explanation by subsumption" (Fetzer 2022) or—as William Dray called it—"covering law explanation"[3] which is sufficiently characterized by the possibility of logically deriving a (social science) explanandum if the antecedence-condition of the respective (action-theoretical) law hypotheses have been realized. Rather, we need additional and law-independent information about the actor

---

[1] Following Udehn (2001), methodological individualists defend at least three different kinds of "social explanation": (a) an anti-holistic but "reductive" version, (b) Coleman's "bathtub-model" of explanation and (c) Popper's "situational logic". In contrast to Udehn, who seems to prefer Poppers "logic of the situation", I reject this explanatory model as it deviates from Popper's criticism (cf. Schmid, 1979, pp. 16–27). And as, equally, "logical reductionism" is untenable (cf. Schmid, 2017) and the logic of "Coleman' boat" rather unclear, I argue in favour of a *micro-foundational conception of explanation* (Little 1998); defending the same idea Reinhard Wippler (1978) speaks of "structural-individualistic" explanations. For (historical, logical and theoretical) details of this approach see Schmid (2006), Maurer 2017), Maurer and Schmid (2010, pp. 57–145).

[2] As there are several candidates (Etzrodt, 2003; Miebach, 2006) we have to compare these different theoretical offers with regard to their logical form, their content and their explanatory scope; to the degree that the various proposals can be in principle united under one theoretical roof—contrary to the very widespread Kuhnian view (Ritzer, 1975)—it cannot be true that social theory is characterized by an (open) set of logically independent "paradigms".

[3] As such Dray (1957) had criticized the Hempel-Oppenheim model of explanation as either trivial or incomplete.

constellations, i.e., their number and social relations, which together "constitute" the action situation (Popper, 1994, pp. 162f., 166ff.). Regarding this condition, social explanations can be characterized as "context explanations" (Eberlein, 1971). Care must be taken to ensure that the respective situation-characteristics can be interpreted "in the light" of the action theory used and thus in such a way that it becomes visible to what extent these situational attributes are (empirically) "relevant" for the goals and expectations of the actors. A theory of action that is not able to do this cannot be used to explain social phenomena.

The resulting task therefore requires the creation of an appropriate explanatory argument by means of a sequence of successive steps. As a first step, we must be able to explain how, with what objectives and on the basis of what assumptions about what they can achieve on the basis of their abilities, the actors plan, organize and thus determine or fix their actions. Afterward one can turn to the (further) problem of whether and in what way the expected success of each actor depends on the fact that he relies on co-actors who have to act in the face of similar circumstances as he does himself. In the most general sense, this opens up the necessity for each actor to "take into account" the (expected and evaluated) actions of others in order to align his "plans" and "performances" and thus to "orient"[4] his own actions to those of his co-actors. The respective actors can do this in various ways (Schimank, 2000): on one hand, they may organize their actions by "observing" each other alternately or one-sidedly; on the other, "orientation" may indicate that actors can "influence" their respective "opportunities" and "restrictions" in a mutual or one-sided, albeit indirect way, and finally "orientation" can mean that the actors take up direct "interactions" with each other. In all three cases, we need to find out what "social" situation the actors think they are in and what accommodation problems they face *as a result*. The nature and quantity of these problems determine, on the one hand, the considerations on the basis of which the actors *strategically* align their actions with each other, and, on the other, by help of which institutionally regulated relationships they can arrive at *solutions to their accommodation problems*, and whether they maintain their institutions, want to change or give them up. Explanatory modelling serves to answer the associated questions.

Of course, it is not certain that the actors will have succeeded in the course of their relationship management insofar as actions sometimes have ill-considered or undesirable consequences and all too often fail. Accordingly, the "success" of their action-exchange, or the amount and security

---

[4] At this point, I will take over the term proposal of Thomas Scheff (1967).

of the associated returns, depends on which collective consequences can be set and identified due to the fact that their actions—along the three "orientation dimensions" mentioned above—are "interdependent".[5] Not only recently has social theory become aware that, on the one hand, dependencies can lead to the obstruction of one's own action success, but on the other, that it must also be true that without the participation and involvement of like-minded or at least tolerant, but sometimes also will-less and powerless co-actors, an action cannot be realized. Accordingly, a further explanatory step should subsequently clarify how the collective consequences of their interdependent actions, which can be aggregated under these conditions,[6] "have an effect" on the further actions of each individual actor and how, as a result, the respective "definition" of their action situation[7]—and thus their "perception" and "evaluation" of its problematic character—changes or does not change. From these socially mediated influencing variables, one can distinguish "natural" factors, which an explanatory model may treat as "external" parameters to the extent that they are not to be understood as a consequence of the joint problem-solving action of the actors, but make themselves largely felt as uninfluenceable "selective environment(s)".

Strictly speaking, social science modelling, insofar as it wants to follow the guidelines of methodological individualism, should therefore distinguish four explanatory steps: the explanation (1) of the socially or situationally "embedded" individual action (by help of a theory of individual action), (2) the explanation of the structure (and course) of the mutual exchanges or interactions which selectively allow for the solution of different action problems, (3) the explanation of their collective or distribution consequences (for participants as well as uninvolved persons) and (4) the explanation of the recursions of these consequences of action to the interpretations of the situation and motivations for action of the respective actors. Regarding the requirement that social structures have to be necessarily explained by considering individual actions, one may say that social-scientific explanatory arguments have the (logical) form of "multi-level" or "micro-foundational explanations".

This view has three implications that are important for the subsequent assessment of the logical and substantive performance of theories of social change.

First, micro-foundational explanations cannot be defended as an *alternative* to the deductive-nomological explanatory approach proposed by Popper,

---

[5] I hereby adopt a terminological proposal of James Coleman (1990).

[6] On the logic of "aggregation" see Schmid (2009).

[7] On the explication of what is meant by a "definition of the social situation" compare Esser (2004, pp. 109–150).

Hempel and Carnap. They, too, proceed deductively and consider a set of laws. But the nomological knowledge that finds its way into micro-founding explanations is exclusively embodied in the action theory used in each case and deals with the mental prerequisites of the action design of single agents. This suggests the thesis that laws of this kind are to be regarded as *psychological laws of action*, the validity and explanatory relevance of which does not depend on the fact that there should be "sociological" laws. Instead, many social scientists interested in micro-foundations suspect that there are *no* laws about situational feature distribution, about the process dynamics of interactions and their distributional structural collective consequences.[8] This does not mean that action-exchanges and structures do not exist but it means that the *potential* that drives social dynamics and that represents the "animating part of every social model" (Popper, 1985, p. 361) can *only* be derived from the theories of action that micro-founding explanations have to presuppose. Thus, social relations and distribution structures resulting from the constellations of such individual actions do not act themselves; at most, they define the scope (or limitations) for action of the agents that result from perverse or desired "collective effects" of joint action (Boudon, 1977). Of course, the ambiguity of the concept of causality alone does not prohibit anyone from assuming a "causal force" to these "opportunity limits" as well as to the "action potentials" of the actors; however, one should then terminologically distinguish the restraining "structural causality" from the energizing "action causality", which can only be inherent in actors (Lloyd, 1986), in order to state that *only* these actors can generate, maintain and change their social relations (Goldthorpe, 2007, p. 169); talk of the "social self-dynamics", "Eigenlogik" or "systemic logics" and so forth does not capture the creativity of human action and is in this sense without content (cf. Greve, 2015).

If, however, there are no "systemic" or—as Popper called them—"sociological" laws (Popper, 1974, pp. 22, 62, 67), then social scientists—secondly—cannot hope to be able to develop macro-theories suitable for explanation; methodological individualists do not know any "structural" laws. Conspicuous macroscopic "regularities" and repeatedly observable structural relationships, which convinced empiricists like to consider laws, therefore represent at most contingent descriptions or correlations (Esser, 2004, pp. 19–46). The same applies to the observable and possibly formally describable dynamics of social processes and corresponding interactional mechanisms (Page, 2018). As a consequence of this state of the explanatory art it is nonsensical to distinguish between macro- and micro-theories, to treat them—as is done in many

---

[8] For evidence see Schmid (2006) and Schmid (2017).

textbooks—as self-sustaining "multiple theoretical traditions" (Turner, 2006) and in the end to search for compatibility relationships between them.

Thirdly, the (logical) relationship between structural macro- and psychological micro-theories finally cannot be misunderstood as "reductionist". Non-existent theories cannot be "logically reduced" in the sense of Nagel's proposal (Schmid, 2017); neither should micro-foundational explanations be labelled as "reductive explanations". Rather, they propose "structural models" (Esser, 2002, p. 149) which combine individualistic theories of action with (singular or local) structural or situational descriptions, which a modeler must create with a view to the possible answer to the question of how the existing structural or situational "boundary conditions" affect the "application conditions" of his respective action theory. There is therefore no "explanatory diversity" in the social sciences, but only *one* form of explanation that applies psycho-nomological theories in social contexts to explain dynamic interaction processes and their recursive structural effects on (further) individual actions.

The Concept of "Social Change"

In the following, I reconstruct the content and explanatory claims of theories of social change. The first step, with a view to the narrowly defined topic of this chapter, is to clarify the question of what the social sciences in general and sociology in particular understand by the term "social change". Only then can it be made visible what weight is to be attached to the objections of methodological-individualistic authors against the production of theories of social change.

The meaning and importance attributed to the study of social change fluctuates. On the one hand, we find references to the long-standing venerability of the topic (cf. Dreitzel, 1967; Martindale, 1976) as we come across the thesis that sociology owes its emergence as a scientific-academic discipline to the necessity to cope with the menace associated with the development of (modern) industrial society (McLeish, 1969; Scheuch, 2003). In consequence and still today, many experts regard "the analysis of social change as (the) touchstone (of their explanatory competences)" (Wiswede & Kutsch, 1978, p. vii). On the other hand, one cannot deny that, after the topic of "social change" repeatedly came onto the agenda of social analysis in the wake of the Ogburn study in 1922 and peaked between 1950 and 1980, there is currently hardly any social science change research that wants to meet the far-reaching pretentions the "theory of social change" initially evoked. Instead, critiques of such a research program have increasingly accumulated which show that its claim to explanation—which, as already described, remains basically undisputed—is difficult to fulfill. In consequence, the suspicion that theories of social change could be used to creatively plan and intentionally shape social developments appears increasingly unfounded?[9] And last but not least, the performance assessment of theories of social change is complicated by the fact that there is no agreement among those authors who want to deal with "different models and theories of change" (Smith, 1976, p. 14) as to which theories should play a role in this or which societal process dynamics they intend to deal with (Harper, 1989, pp. 53–97; Vago, 1989, pp. 75–79; Weymann, 1998, p. 17). Moreover, the fact that no consensus can be reached on what logical-semantic character such theories must have in order to serve as explanatory instances,[10] or even on whether a uniform theory of change can be held out at all,[11] does not facilitate the answer to the question of what role methodological individualism must play in the explanation of social change events; since almost every author proposes his own list of theory drafts (Lauer, 1973; Mandelbaum, 1987), it is not even clear which of them are relevant. I take this as a license to limit myself to examining three different theory programs, orienting their compilation to the possibility of showing the different ways in which they miss the standards of microfounding explanatory practice developed above. I start with the *evolutionist* theory program, then go to the structuralist or *functionalist* program variant and finally consider *individualistic* attempts to explain social change. The present treatise ends in a summary assessment of the explanatory merits of sociological theories of change.

Evolutionary Theories of Social Change

The history of the theory of evolution as it has lived through biology has taken as equally a complex course (Mayr, 1982) as its social science reception (Sanderson, 1990). This widespread adoption found its roots in historical-philosophical considerations, as they arose in the eighteenth century and were elaborated in the nineteenth century (Appelbaum, 1970; Mandelbaum,

---

[9] See Etzioni (1968), Garner (1977), and Warren (1977) for such "political" programs.

[10] At least there are two "theory-conceptions" that are difficult to reconcile. On the one hand, one reads that the task of a theory of social change is to present "a conceptual version of social change" (Jäger & Meyer, 2003, p. 14); on the other hand, many authors—and I would like to adhere to this—understand "theory" to mean systems of nomic assumptions (Boudon, 1983, 1984). In using such nomic laws, confusing them with their applications in the form of (situational) "models" should be avoided (as do Galt & Smith, 1976, p. 21).

[11] Galt and Smith (1976) are in favour of this position, while Vago (1980, p. 59) argues against it.

1971; Schneider, 1976). Reflections on the peculiarities and tasks of theories of evolution, which go back to this founding period and in which names such as Adam Smith, Condorcet, Comte, Marx, Durkheim and *above all* Herbert Spencer—and far less Charles Darwin—played a prominent role up to the present time. (Luhmann, 1997; Parsons, 1977; Sahlins & Service, 1973; Steward, 1972; White, 1969).[12]

I will try—with the help of Robert Nisbet (1969), Richard Appelbaum (1970, pp. 15–64) and Anthony Smith (1976, pp. 45–69)—to compile the most important attributes of the social-scientific theory of evolution. Following these authors, it should first be noted that the subject area of this form of theory is "whole societies" or "total systems" (of different scopes and different compositions) while the concept of "system" also might include a civilization "in its *ensemble*" (Comte, 1973, p. 18); in this respect, the search has to look out for a "holistic scheme of social evolution" (Schneider, 1976, p. 16). This scheme deals with both the "statics" of societies and the "social dynamics" of its combined processes (Comte, 1907, pp. 391–534) and it emphasizes that social development is a "*natural*" or "*normal*" and not an extraordinary event that can only be expected in times of crisis and in the wake of catastrophic events.

Of particular importance for the assessment of the explanatory power of evolutionism is the idea that every evolutionary process follows unchangeable and—insofar as they work toward nameable effects of increase or amelioration—"final laws of development" (Jäger & Meyer, 2003, p. 21) or proceeds in accordance with "necessary directional laws" (Mandelbaum, 1971, pp. 114–225). The directionality and the inevitability of social development dynamics coincide. The unambiguity of the evolutionary direction is formulated as "growth" (Nisbet, 1969), "increase in complexity" (Spencer, 1898), as "adaptive improvements" (Sahlins & Service, 1973, p. 15) or in a moral-valued language as "progress" (Spencer, 1897, p. 609). Various authors identify different goals for the course of society, be it the progress of the mind in the consciousness of its freedom (Hegel, 1961), increase in social differentiation (Spencer, 1880, pp. 48–70), the development of recognizing reason (Comte, 1967), the progression from "status" to "contract" (Maine, 1977) or the transition from "savagery" to "barbarism" and finally to "civilization" (Morgan, 1877). The assumed necessity of social development, on the other hand, manifests itself in a threefold characteristic: on the one hand, it is realized by socially anchored social forces the effectiveness of

---

[12] I skip the controversial question of who may count as a "genuine" evolutionist as there are "rather different sorts of evolutionary theories" (Sanderson, 1990, p. 11).

Methodological Individualism and Social Change   111

which cannot be disturbed by external factors; the thesis "change is immanent" applies (Nisbet, 1969, p. 170). On the other hand, the necessity of social development also arises from the fact that the causes of change do not exhaust and therefore always have the same effects; social change therefore proceeds *continuously* and *uniformly*. And finally, the determination of social change documents itself insofar as it retains its accumulative progressiveness even where nonlinear forms such as cyclical, exponential or stage-like can be observed. A famous implication of this thesis emphasizes that all societies will take the necessary steps of development, which is why it is to be expected that all "primitive" present-time societies will "modernize" themselves in the course of their further history.[13]

In what way can this evolutionist tradition of theory be classified from the point of view of methodological individualism? I suspect that a detailed critique is superfluous if one focuses on the most obvious mistake that this school of thought can be accused of. Their main flaw lies in the premise that "whole mankind" or "total" societies are the "entities" that are subject to lawful evolutionary processes *and* that, in the last instance, their expected course remains unaffected by the actions and omissions of the actors. It might be conceded that conceptions such as "ascent", "advancement", "diffusion", "increase", "cumulation", "exponential growth", etc. *describe* these dynamics; however, whether the descriptions thus generated can be inferred from the indication that the pronounced events arise lawfully may represent a thesis that is as bold as it is false. In any case, classical evolutionism has no way of *logically deriving* the developmental trajectories it addresses from the situational problem-solving actions of the actors; micro-founding explanations, thus, remain outside its area of competence. Moreover, it could not be denied that the "social" laws the evolutionists chose as the basis for explanation do not apply factually in all cases. The corresponding failures, therefore, can only be registered as an "anomaly" which—contrary to expectations—does not allow empirically accurate predictions and classifications, and which—as Thomas Kuhn had outlined—can only be got rid of by a paradigmatic revolution. Indeed, the due "antievolutionary reaction" (Sanderson, 1990, pp. 36–49) was not long in coming.[14] The main attack was ridden by functionalism.

---

[13] Therefore, the widely discussed "modernization theory" can be understood as heir to the theory of evolution (Eisenstadt, 1973, pp. 3–46).

[14] For details see Voget (1975, pp. 311–538).

The Functionalist Theory of Social Change

The (modern) functionalist theory of social change starts with the (epistemological) thesis that social change processes can only be observed against a "background of non-change" (Parsons, 1973, p. 72). The logical consequence of this premise is that the sociological theorist must first address the question of the conditions under which a social system can preserve its basal structures (Parsons, 1973, p. 73). In the light of this objective, the advice is to model structure conservation as a consequence of a "stable" or "dynamic equilibrium" (Parsons, 1951, p. 480; 1961a, pp. 60–70; 1973, pp. 73–76).[15] The former stabilization equilibrium is considered to be given if the sociological theory can identify "mechanisms" (Parsons, 1951, Chapters VI and VII, Parsons, 1961a, p. 38, pp. 66–70) which, by "neutralizing" all external and internal disturbances, can secure the preconditions under which the considered "units" are able to reproduce their "patterned relations" (Parsons, 1973, p. 74). A dynamic equilibrium, on the other hand, exists when a system of action has to leave the zone within which the unproblematic reproduction of its structures remains possible, but it succeeds in controlling or erasing the "strains" (Parsons, 1961a), which had built up the pressure of change, by establishing a renewed and variate state of structural equilibrium.

The process of restoring social structures takes place in a lawful manner and is aimed at fulfilling the "functional requirements" of an action system if the respective system or its external borders are to remain in place (Parsons & Bales, 1953, pp. 99–103).[16] In a declared contrast to classical evolutionism,[17] the functional theory of social equilibrium strives to show that the aforementioned laws of system preservation do not take place "behind the backs" of the actors, but that they can be used to give an explanation of how a "pattern of behavior contributes to the maintenance of social stability" (Guessous, 1976, p. 24). In the eyes of Talcott Parsons, such a "functional" explanation can succeed in two steps. First of all, he states that all actors are to be understood as "voluntarist" agents, who can choose between different objectives and are able to acquire knowledge about the action situation in which they have to act (Parsons 1968, 2010). If they have to act in social situations, they are faced with what Parsons called "Hobbes' problem of order" (Parsons, 1968,

---

[15] Neil Smelser (1968, pp. 193–121, pp. 265–268) has endeavored to clarify the concept of equilibrium and to insert Parson's equilibrium models into an extended typology of different kinds of equilibria.

[16] In the present context it may remain undiscussed which "functional prerequisites" the sociological system theory should consider (see for details Parsons [1951, pp. 26–36] and Aberle et al. [1967]).

[17] Once Parsons rhetorically asked "Who now reads Spencer?" and laconically added "Spencer is dead" (Parsons, 1968, p. 3).

pp. 89–94), which consists in the fact that the actors cannot really be sure that their co-actors will support their own plans and intentions. Since each actor is in this situation, they must all try together to adapt their actions mutually and in such a way that the actors can reward and punish themselves alternately for what they do, and thus warrant that in the course of their interactions binding norms and values emerge whose joint observance ensures the stability of their interrelationships (Parson, 1973, p. 75).

How can an "analysis of processes of structure change in social systems" (Parsons, 1961a, p. 70) be carried out within such a framework? Quite early in his career Parsons had accepted the suggestion that the collapse of social control coincides with the need for "structural change" when he distinguished between "change in the system" and "change of the system" (Parsons, 1951, pp. 480–481; 1973, p. 74). In the first case, control-processes organized within the system work toward the "integration" of system cohesion as well as the successful regulation of the "interchanges" that must be maintained between a system and its "environment". In the other case, however, such protective mechanisms do not work or stop to do so, and a system must be able to establish "a new and different" state in order to regain equilibrium, "which must be described in terms of an alteration of its previous structure" (Parsons, 1973, p. 74). A system of action is thus faced with the problem of restructuring its regulatory-relevant institutions (Parsons, 1961a, p. 71). Moreover, since the theory of social systems assumes that the social order ultimately depends on the existence of a system of commonly registered expectations and values that guides joint action (Parsons, 1961b), any transformation of a social relationship becomes a redesign of its respective "cultural system".[18]

However, notwithstanding and unlike classical evolutionism, Parsons is uncertain whether his envisaged "theory of social systems" would find and fully formulate the laws that give shape and direction to social change (Parsons, 1951, pp. 486–490), although he later maintains that the study of the evolution of "total societies" (Parsons, 1966, p. 1) requires knowing them. This uncertainty finds its expression in the fact that Parsons does not speak of "causes of change", but of the "sources" and "forces" of change (Parsons, 1961a, pp. 71–74). In both cases, the pressure to change results from unavoidable "problems of structural inconsistency" (Giesen, 1975),

---

[18] Other functionalist theorists see it differently. Thus, Wilbert Moore (1970, pp. 128–139) declares less the preservation of the "functional prerequisites" of a system of action as the target of social change than the "resolution of human problems", and Bronislaw Malinowski (1977) considers the fulfilment of biological needs of people to be the "reference point" of "functional theory". As Merton (1964, pp. 55–60) showed a long time ago, *all* aspects of "social action" can be analyzed in a functionalistic way. The failures are the same in all cases.

114   **M. Schmid**

which above all arise when the actors can no longer organize their forms of relationship without conflict and without bringing about negative externalities. If it is not possible to avoid or suppress these incompatible or disintegrating lines of action, social dynamics can lead to the abandonment of the old order or finally to a "radical dissolution" of the respective system (Parsons, 1961a, p. 75). As the laws of social change are unknown and therefore can neither be used for deductive explanations nor for predicting the expected change events, the "disturbances in the individual's motivation" that can be identified in this context (Parsons, 1961a, p. 74) must be sufficient to guide research about the possible ways in which the expected process of change must proceed in order to approach a new state of equilibrium.

In order to subject the functionalist theory of social change to a methodological-individualistic evaluation I shall confine myself to a relatively loose list of the following points.

First of all, it can be noticed that Parson's proposal to develop the analysis of social change against the background of a functionalist theory by understanding social change as a "byproduct in the malfunctioning of social control or order" (Hield, 1967, p. 252), while at the same time requiring that sociology should observe an explanatory program in the nomological-deductive sense which has been described above misleads theoretical sociology in a double way. I am not (only) thinking of the conjecture that Parson's "theory" at best provides "the dictionary of a language that possesses no sentences" (Homans, 1987, p. 27) or rather a muddled collection of conceptual "boxes than an ordered axiomatics of hypotheses suitable for explanation" (Homans, 1987, p. 14). Nor am I concerned with reminding the reader that Parson's functionalist analysis of equilibrium, on the basis of its assumption that social change is oriented toward the fulfillment of a nameable list of "functional prerequisites", proceeds teleologically, overlooking the fact that the reference to whatever kind of consequences of a social event neither explains its genesis nor its preservation (Hempel, 1965, pp. 297–330; Nagel, 1961, pp. 520–535).[19] Rather, I have in mind the fact that equilibrium analyses do not allow any explanations at all, but serve exclusively to (formally) prove that there are condition constellations in view of which a (previously defined) state of equilibrium is to be expected or sustained. Such proofs of existence, even if they succeed, contain no indication that or whether these conditions can be realized, which is why corresponding analyses can at best lead to "idealized" and, i.e., factually *wrong* models (Faia, 1986; Turner & Maryanski, 1979). If the functionalist reacts to this fact by modelling the conditions

---

[19] Of course, the "educated" functionalists certainly know this (cf. Merton, 1964, pp. 19–84). However, they do not therefore see any necessity to revise their explanatory scheme.

under which the initially focused equilibrium should be established, from an explanatory logical perspective, he falls into an infinite regress.[20] A second aberration becomes visible when one realizes that it is *logically impossible* to derive substantive assumptions about conditions and processes of change from functional equilibrium analyses (Homans, 1987, pp. 10–14); in fact, Parsons admits several times that one can at best read societal loss of control or social tensions, but also motivational "symptoms of disturbance" (Smelser, 1959, p. 286) as indications that the actors should look out for *new* amelioration policies. Whether this thesis is correct, however, the functionalist theory of action cannot know. Moreover, it may be logically impossible to know this as it does not contain any information about the circumstances by which the actors can anticipate what they will only know later on (Popper, 1961, pp. v–vii).[21] Parallel restrictions obtain with regard to functionalist model-building insofar as the knowledge of the conditions, under which a moral breakdown may be expected, does not specify to the possibilities of a *truly new* equilibrium state.

Sociological functional theory would therefore have reason to reflect on the power of its explanatory concept. In order to judge whether or to what extent functional explanations sufficiently correspond to the above-described principles of a micro-founding explanation of social processes, two critical points have to be considered. First, functionalists should be able to identify the necessary laws for this purpose, and second, they should be engaged in clarifying the logical structure of the corresponding explanatory arguments.

In my opinion, Parsons advocates a description of "law" that is justifiable when he speaks of "analytical laws" and treats them as universal propositions that link the "abstract" elements of a system in a logically labelable way (Parsons, 1968, pp. 622–623). In this context, it is also not completely absurd, as far the theoretically relevant properties of these elements can be interpreted as "variables", to understand the variations that are observable between different factors as evidence of the existence of a causal relationship (Parsons, 1968, p. 750).[22] It should also be true that these nomological

---

[20] This is demonstrated by the attempts of Lewis Coser (1956) and Parsons (1969, pp. 352–404) to analyze conflicts or power processes functionalistically, instead of asking whether the emergence of social conflicts (of various kinds) falsifies the thesis that "entire societies" can always succeed in regaining their equilibrium, and whether and in what way this fact speaks in favour of dropping the identification of "functional analysis" and "empirical theory".

[21] Cancian (1960) had already noted that the success of functionalist change explanations depends on the (non-probable) condition that the analyzed societal system is (factually) closed.

[22] Personally, I do not share John Stuart Mill's understanding of causality which Parsons seems to accept (Schmid, 2015). This implies that I do not see any sense in substituting impossible macro-sociological explanations by help of "historical" or "factorial" explanations as both Nisbet (1969) and Smith (1973) propose.

propositions cannot be taken from mere empirical generalizations (Parsons, 1968, pp. 33–34) and that they are not to be equated with "mechanisms", by which Parsons understands less isolated causal hypotheses than a system-relative combination of processes whose functionality is of theoretical relevance for the attainability of certain target states (Parsons, 1951, pp. 201–202). Moreover, that the logically ordered compilation of such analytical laws is called "theory" is even accepted by his opponents (Homans, 1967). However, this understanding of the law does not answer the question of whether it is correct that the same laws work at all levels of a hierarchically organized action system, nor why the thesis should be accepted that these laws, wherever they work, must aim toward the preservation or recovery of a state of equilibrium. I think that Parsons' inability to remedy this failure is most easily explained when we acknowledge that—as his theory of voluntary action could have told him—there are no laws at the level of interactions and their structural consequences (Homans, 1967, p. 104). This applies not only to "laws of social change" (Parsons, 1951, 486) but also to the laws of equilibrium that Parsons believes he has found, claiming that social systems follow a "law of inertia" (Parsons, 1951, p. 482). In his eyes, such a law ensures that a once established expectation-equilibrium is maintained by deviance-reducing institutions, that counter-forces are provoked by compensating for the severity of the deviation by equally weighted reactions, and by allowing any effort to regain a balanced state of expectation to be publically applauded (Parsons & Bales, 1953). If one considers these assertions in the light of an individualistic theory of action, then they do not formulate universal and valid necessities, but at most they describe special cases of system dynamics, whose actional foundations and structural conditions both have yet to be clarified.

At this point, a further problem may arise, which is why the representatives of a functionalist theory of social change cannot entrust themselves to a micro-founding explanatory practice, although, in well-defined contrast to classical evolutionism, they certainly recognize that every action is based on an (idiosyncratic) decision and that the actors pursue personal intentions and interests (Parsons, 1951, pp. 491–493, pp. 506–507), whose smooth institutionalization does not seem to be self-evident (Parsons, 1951, pp. 36–45). It would have been all too easy to derive from these insights the question of under which circumstances it is advisable, even rational, for the actors to adhere to action strategies of which they can know that the establishment of a consensus of interests, the establishment of communities of values and certainties of expectations, the guarantee of social relations free of conflict and aversions, that the success of communicative understandings, balancing of power and security of supply, the granting of freedoms and property rights and much more *cannot* be counted among the conditions for success of one's own actions. It also remains open why functional social analysis cannot see that the exchange of rewards and punishments does not have to lead to the elimination of the just listed "dysfunctions" without any problems or why socialization efforts and deviance control must be free of charge and why they should ensure a peaceful balance of interests. I think that this short-sightedness is not due to the fact that the functionalist theorists could not see that "strains" and "imbalances" are *unavoidable*. Rather, the inability to describe critical social conditions other than on the basis of the possible functionality of their institutional solutions and the refusal to explain (positively) the difficulties associated with the emergence of such solutions,[23] in my view, results from the fact that sociological functionalism neither has at its disposal a convenient theory of individual action, that explains *why* an actor selects a specific action (Raub & Voss, 1981), nor are we supplied with empirically reliable action-based models of collective action (Mouzelis, 1994, pp. 17, 18).[24] This has lasting consequences insofar as functionalist modelling only reveals in borderline cases—for example, when Parsons speaks of the "Hobbesian problem of order", behind which a Prisoner's Dilemma is buried (Ullmann-Margalit, 1977, pp. 62–64)—that the model-builders really understand the "logic of the situation" in which their actors have to act, and thus the heterogeneity and variate costliness of the *strategic* possibilities the latter are confronted with. A further reason why functionalism cannot adequately explain social processes is that equilibrium as well as transformation models see no need to turn to the *aggregation problem* in order to ask and clarify on the basis of which procedural rules which actions lead to which collective distribution consequences (cf. Lindenberg, 1977). This failure mainly affects structural distribution analyses, which would have to assume that actors with very heterogeneous intentions and resource endowments collide.

---

[23] Moore (1973) contains a list of relevant difficulties and grievances.

[24] Some functionalist change theorists have noticed this omission and tried to develop models of revolutionary action or mass mobilization (Eisenstadt, 1963, 1978; Smelser, 1963). Of course, as one commentator rightly notes, these attempts, as far as they are correct, owe almost nothing to their functionalist premises (Homans, 1987, pp. 10–14). I additionally suspect that these models only in part do justice to the theoretical relevance of strategic action.

Individualistic Theories of Social Change

There are several individualistic explanations of social change phenomena and, as immediately will become evident, their proximity to the methodology of micro-based explanations is obvious. Thus, we find psychoanalytic approaches (McLeish, 1969, pp. 29–51) and status-inconsistency models (Hagen 1962), models based on motivational theories (McClelland, 1961) or on behavioral or learning theories (Kunkel, 1970), frustration theory explanations (Davis, 1962), or explanations which refer to utility and rational theory (Boudon, 1984; Hernes, 1976) or attempts at explanation based on a theory of choice (Rössel, 2005) and finally some syncretic approaches (Wiswede & Kutsch, 1978, pp. 176–203).

Of course, the success proposals of these different modeling offerings are not identical, but they all share the conviction that—as functionalism had already accepted—the distinction between models of societal persistence and models of social change or the corresponding separation of equilibrium and imbalance analyses cannot have any methodological significance; at the same time all these approaches share the opinion that models of change are well advised to separate internally initiated transformations from transformation processes caused by external shocks and that model-building should not hope to identify non-reduceable macro- or system-laws. Therefore, nearly every representative of the individualistic camp accepts that the indisputable possibility of modeling social or interaction-based dynamics *mathematically* or *formally* does not call for the renunciation of their microfoundation, and finally all agree that there is—quite differently than the classical theory of evolution demanded—*no* theoretical reason to favor accumulation and diffusion dynamics, growth and differentiation, adaptation, improvements and accumulations instead of the study of breakdowns, decay, declines, collapses and extinctions or to understand these "degenerative" social dynamics as a logical residual class or as a negligible "byproduct" of equilibrium models or modernization and progress "theories".

In terms of research practice or content, the common focus of such individualistic explanations is to explain the establishment of social structures and their distributional properties as well as their change from the combined, aggregated or even just summed up actions of a plurality of actors, whereby the "analytical primacy", and thus the explanandum of interest to the theoretician, is situated on the system or structural level, while the theoretical, or the "explanatory primacy", lies on the individualistic level (Lindenberg, 1992, p. 7). Individualistic theories of action must be able to make plausible why the actors choose or accept activities or forms of interaction that

Methodological Individualism and Social Change   119

produce the structural and distribution effects to be explained. However, the individualistic modeling proposals mentioned differ not only on the basis of the chosen foundation theory, but above all in *how* they think of the "mediation" between action and structure (Hernes, 1976, p. 536). Sometimes, it seems sufficient for actors—for whatever reason—to have common experiences or joint orientations for action to encourage the theorist to infer the existence of structural effects (Davis, 1962; Hagen, 1962; McLeish, 1969), whereas other explanatory models emphasize the wide and variable range of structural and parametric specifications under which these structural effects arise and continue to have an effect (Boudon, 1984), or the multitude of interaction mechanisms that align all kinds of actions involved (Maurer & Schmid, 2010). At the same time, hardly any representative of an individualistic explanatory practice shares the tendency, often met in evolution theory but also in functionalist circles, to make "whole societies" and their "functional prerequisites" the subject of analysis, or the view that for this purpose the "idea of advancement" (Granovetter, 1979) can be used in a meaningful sense, but they are content with the investigation of structural effects that they can identify in different fields of action where specific coordination requirements of self-interested actors play a recognizable role and which can only be discovered and explained in the light of a theory of action that specifies the social and personal conditions in face of which actors *select* and *determine* their actions (Bates et al., 1998; Esser, 2002; Granovetter, 1978; Wittek et al., 2013); structure change, in consequence, is explained by reference to *any* possible modification of these action conditions.

Consequently, these attempts to explain social structural change, which are guided by theories of individual action, seem to be in accordance with methodological individualism. In the end, however, they regularly contribute to a regrettable misunderstanding in two ways. On the one hand, they overlook the fact that there is only one theory of action (Coleman, 1992, p. 129), which is why the obvious refusal to examine the compatibility and integrability of the different aspects of action that the respective "approaches" use for explanation contributes to the promotion of the inadmissible impression that sociology is a multi-paradigmatic science. And it has the same effects if the various explanatory offers understand themselves as mutually independent and self-sufficient "theories" instead of seeing that they use different aspects of a substantially united theory of action to produce highly situation- and problem-specific models; that the so-called "conflict theory of social change" counts as a self-sustaining approach (Appelbaum, 1970, pp. 81–98; Schneider, 1976, pp. 83–86; Wiswede & Kutsch, 1978, pp. 153–176) is a case in point. In these cases, the defenders of individualistic transformation analyses confuse "theory" and "model" and therefore do not adequately understand the *logic* of their attempts at explanation and systematically underestimate the possible extension of a unified individualistic research heuristics.

2 Result

My thesis is that the theories of social change proposed in various ways cannot meet the demands of methodological individualism to explain the existence and change of "social phenomena" in a logically appropriate way. On the one hand, neither evolutionists nor functionalists have identified valid macro laws that can be entrusted with the explanation of transformation events. On the other hand, by understanding themselves as a "theory" in their own right each individualistic explanatory model also misses two important points. First of all, they regularly fail to question how and to what extent different theoretical proposals can be logically related and thereby connected to a unified approach; and secondly, it remains unnoticed that it may be methodologically unfortunate to restrict one's explanatory claims to just one field of application. Instead, and to remedy all the mentioned insufficiencies, it would be more suitable to suggest explanations of social structures as well as of their change by help of an integrated theory of action that can be applied to a principally indefinite variety of more or less "problematic situations" (Raub & Voss, 1986). In this case it would be possible to align the explanatory practice of Methodological Individualism with Lakatos' "Methodology of scientific research programs" (Schmid, 2023) as well as with the program of a micro-foundational "explanatory sociology" (Maurer, 2017).

References

Aberle, D. F., Cohen, A. K., Davis, A. K., Levy, M. J., & Sutton, F. X. (1967). The functional prerequisites of a society. In N. J. Demerath III & R. A. Peterson (Eds.), *System, change, and conflict. A reader on contemporary sociological theory and the debate over functionalism* (pp. 317–331). The Free Press/Collier-Macmillan Limited.

Applebaum, R. P. (1970). *Theories of social change*. Markham Publishing Company.

Bates, R. H., Greif, A., Levi, M., Rosenthal, J.-L., & Weingast, B. L. (1998). *Analytical narratives*. Princeton University Press.

Boudon, R. (1977). *Effect pervers et ordre social*. Presses Universitaires de France.

Boudon, R. (1983). Why theories of social change fail: Some methodological thoughts. *Public Opinion Quarterly, 47*, 143–160.

Boudon, R. (1986/1984). *Theories of social change. A critical appraisal.* Polity Press.

Cancian, F. (1960). Functional analysis of change. *American Sociological Review, 25*, 818–826.

Cartwright, N. (2001). *The dappled world. A study of the boundaries of science.* Cambridge University Press.

Coleman, J. S. (1990). *Foundations of social theory.* The Belknap Press.

Coleman, J. S. (1992). The vision of foundations of social theory. *Analyse & Kritik, 14*, 117–128.

Comte, A. (1907/1830). *Soziologie.* Aus dem französischen Original ins Deutsche übertragen von Valentin Dorn und eingeleitet von Professor Dr. Heinrich Waenting. Erster Band: Der dogmatische Teil der Sozialphilosophie. Verlag von Gustav Fischer.

Comte, A. (1967/1844). Das Drei-Stadien-Gesetz. In H.-P. Dreitzel (Ed.), *Sozialer Wandel. Zivilisation und Fortschritt als Kategorien der soziologischen Theorie* (pp. 111–120). Luchterhand Verlag.

Comte, A. (1973/1840). The progress of civilization through three stages. In A. Etzioni & E. Etzioni-Halevy (Eds.), *Social change. Sources, patterns, and consequences* (2nd ed., pp. 14–19). Basic Books, Inc. Publishers.

Coser, L. A. (1956). *The functions of social conflict. An examination of the concept of social conflict and its use in empirical sociological research.* The Free Press of Glencoe.

Davis, J. C. (1962). Towards a theory of revolution. *American Sociological Review, 27*, 5–19.

Draw, W. (1957). *Laws and Explanation in History.* Oxford University Press.

Dreitzel, H. P. (1967). Problemgeschichtliche Einleitung. In H. P. Dreitzel (Ed.), *Sozialer Wandel. Zivilisation und Fortschritt als Kategorien der soziologischen Theorie* (pp. 21–91). Luchterhand Verlag.

Eberlein, G. (1971). *Soziologische Theorie heute.* Ferdinand Enke Verlag.

Eisenstadt, S. N. (1963). *The political systems of empires. The rise and fall of the historical bureaucratic societies.* The Free Press/Collier Macmillan Publishers.

Eisenstadt, S. N. (1973). *Tradition, change, and modernity.* Wiley.

Eisenstadt, S. N. (1978). *Revolutions and the transformation of societies. A comparative study of civilizations.* The Free Press/Collier Macmillan Publishers.

Esser, H. (2002). Was könnte man (heute) unter einer 'Theorie mittlerer Reichweite' verstehen? In R. Mayntz (Ed.). *Akteure – Mechanismen – Modelle. Zur Theoriefähigkeit makro-sozialer Analysen* (pp. 128–150). Campus Verlag.

Esser, H. (2004). *Soziologische Anstösse.* Campus Verlag.

Etzioni, A. (1968). *The active society. A theory of societal and political process.* Collier-Macmillan Limited/The Free Press.

Etzrodt, C. (2003). *Sozialwissenschaftliche Handlungstheorien.* UKV Verlagsgesellschaft mbH.

Faia, M. A. (1986). *Dynamic functionalism. Strategies and tactics.* University of Cambridge Press.

Fetzer, J. (2022). Carl Hempel, *Stanford Encyclopedia of Philosophy.* https://plato.stanford.edu/entries/Hempel

Galt, A., & Smith, L. J. (1976). *Models and the study of social change.* Wiley.

Garner, R. A. (1977). *Social change.* Rand McNally Publishing Company.

Giesen, B. (1975). *Probleme einer Theorie struktureller Inkonsistenz. Ein Beitrag zur systemtheoretischen Interpretation sozialen Wandels.* Maro Verlag.

Goldthorpe, J. H. (2007). *On sociology. Vol. 1: Critique and program.* Stanford University Press.

Granovetter, M. (1978). Threshold models of collective behavior. *American Journal of Sociology, 83,* 1420–1443.

Granovetter, M. (1979). The idea of 'advancement' in theories of social evolution. *American Journal of Sociology, 85,* 489–515.

Greve, J. (2015). *Reduktiver Individualismus. Zum Programm und zur Rechtfertigung einer sozialtheoretischen Grundposition.* Springer VS.

Guessous, M. (1967). A general critique of equilibrium theory. In W. E. Moore & R. Cook (Eds.), *Readings in social change* (pp. 23–35). Prentice Hall Inc.

Hagen, E. E. (1962). *On the theory of social change. How economic growth begins.* Dorsey Press.

Hegel, G. W. F. (1961). *Philosophie der Geschichte.* Stuttgart.

Harper, C. L. (1989). *Exploring social change.* Englewood Cliffs.

Hempel, C. G. (1965). *Aspects of scientific explanations. And other essays in the philosophy of science.* The Free Press et al.

Hernes, G. (1976). Structural change and social process. *American Journal of Sociology, 82,* 513–547.

Hield, W. (1967). The study of change in social science. In N. J. Demerath, III & R. A. Peterson (Eds.), *System, change, and conflict. A reader on contemporary sociological theory and the debate over functionalism* (pp. 251–259). The Free Press/Collier-Macmillan Limited.

Homans, G. C. (1967). *The nature of social science.* Harcourt, Brace & World Inc.

Homans, G. C. (1987). *Certainties and doubts. Collected papers, 1962–1985.* Transaction Book.

Jäger, W., & Meyer, H.-J. (2003). *Sozialer Wandel in soziologischen Theorien der Gegenwart.* Westdeutscher Verlag.

Kunkel, J. H. (1970). *Society and economic growth. A behavioral perspective on social change.* Oxford University Press.

Lauer, R. H. (1973). *Perspectives on social change.* Allyn and Bacon.

Lindenberg, S. (1977). Individuelle Effekte, kollektive Phänomene und das Problem der Transformation. In K. Eichner & W. Habermehl (Eds.), *Probleme der Erklärung sozialen Verhaltens* (pp. 46–84). Hain Verlag.

Lindenberg, S. (1992). The method of decreasing abstraction. In J. S. Coleman & T. J. Fararo (Eds.), *Rational choice theory. Advocacy and critique* (pp. 3–20). Sage.

Little, D. (1998). *Microfoundations, method, and causation.*

Lloyd, C. (1986). *Explanation in social history*. Basil Blackwell.

Luhmann, N. (1997). *Die Gesellschaft der Gesellschaft*. Zwei Bände. Suhrkamp Verlag.

Nagel, E. (1961). *The structure of science. Problems in the logic of scientific explanation*. Routledge & Kegan Paul.

Maine, Sir H. (1977). *Ancient law* (Introduction by J. H. Morgan). Dent/Dutton (Everyman's Library).

Malinowski, B. (1977). *A scientific theory of culture and other essays* (With a Preface of Huntington Cairns. Sixth Printing). The University of North Carolina Press.

Mandelbaum, M. (1971). *History, man and reason. A study in nineteenth-century thought*. The Johns Hopkins Press.

Mandelbaum, M. (1987). *Purpose and necessity in social theory*. The Johns Hopkins University Press.

Martindale, D. (1976). Introduction. In G. K. Zollschan & W. Hirsch (Eds.), *Social change. Explorations, diagnosis, and conjectures* (pp. ix–xxv). Wiley.

Maurer, A. (2017). *Erklären in der Soziologie. Geschichte und Anspruch eines Forschungsprogramms*. Springer VS.

Maurer, A., & Schmid M. (2010). *Erklärende Soziologie. Grundlagen, Vertreter und Anwendungsfelder eines soziologischen Forschungsprogramms*. VS Verlag für Sozialwissenschaften.

Mayr, E. (1982). *The growth of biological thought. Diversity, evolution, and inheritance*. The Belknap Press of Harvard University Press.

McLeish, J. (1969). *The theory of social change. Four views considered*. Routledge & Kegan Paul.

McClelland, D. C. (1961). *The achieving society*. Princeton.

Merton, R. K. (1964). *Social theory and social structure* (Ninth Printing. Revised and Enlarged Edition). The Free Press of Glencoe/Collier Macmillan.

Miebach, B. (2006). *Soziologischen Handlungstheorien. Eine Einführung*. Zweite, grundlegend überarbeitete und aktualisierte Auflage. Springer VS.

Moore, W. E. (1970). A reconsideration of theories of social change. In S. N. Eisenstadt (Ed.), *Readings in social evolution and development* (pp. 123–139). Pergamon Press.

Moore, W. E. (1973/1963). *Strukturwandel der Gesellschaft*. Dritte Auflage. Juventa Verlag.

Morgan, L. H. (1877). *Ancient society. Research in the lines of human progress from savagery through barbarism to civilization*. Charles Kerr & Company.

Mouzelis, N. P. (1994). *Back to sociological theory. The construction of social orders*. The Macmillan Press.

Nisbet, R. A. (1969). *Social change and history. Aspects of western theory of development*. Oxford University Press.

Ogburn, W. F. (1966/1922). *Social change with respect to cultural and original nature*. With a New Introduction by Hendrik M. Ruitenbeek. Dell Publishing Company.

Page, S. E. (2018). *The model thinker. What you need to know to make data work for you*. Basic Books.

Parsons, T. (1951). *The social system*. The Free Press/Collier Macmillan.

Parsons, T. (1961a). An outline of the social system. In T. Parsons, E. Shils, K. Naegele & J. Pitts (Eds.), *Theories of society. Foundations of modern sociological theory* (pp. 30–79). The Free Press/Collier Macmillan.

Parsons, T. (1961b). Introduction to Part Four: Culture and the social system. In T. Parsons, E. Shils, K. Naegele, & J. Pitts (Eds.), *Theories of society. Foundations of modern sociological theory* (pp. 963–993). The Free Press/Collier Macmillan.

Parsons, T. (1966). *Societies. Evolutionary and comparative perspectives*. Prentice Hall.

Parsons, T. (1968). *The structure of social action. A study in social theory with special reference to a group of recent European thinkers* (with a New Introduction. 2nd ed.). The Free Press/Collier-Macmillan.

Parsons, T. (1969). *Politics and social structure*. The Free Press/Collier-Macmillan Limited.

Parsons, T. (1973). A functional theory of change. In A. Etzioni & E. Etzioni-Halevy (Eds.), *Social change. Sources, patterns, and consequences* (pp. 72–86). Basic Books, Inc. Publishers.

Parsons, T. (1977). *The evolution of societies* (Edited and with an Introduction by J. Toby). Prentice-Hall.

Parsons, T. (2010). *Actor, situation and normative patterns: An essay in the theory of social action.* Studies in the theory of action (Vol. 2., V. Lidz & H. Straubmann, Eds.). LIT Verlag.

Parsons, T., & Bales, R. F. (1953). The dimensions of action space. In T. Parsons, R. F. Bales & E. Shils (Eds.), *Working papers in the theory of action* (pp. 63–110). The Free Press/Collier Macmillan.

Popper, K. R. (1961). *The poverty of historicism*. Routledge & Kegan Paul.

Popper, K. R. (1974). *The open society and its enemies. Vol. 1: The spell of Plato* (5th ed. [Revised]). Routledge & Kegan Paul.

Popper, K. R. (1985). The rationality principle (1967). In D. Miller (Ed.), *Popper selection* (pp. 357–365). Princeton University Press.

Popper, K. R. (1994). Models, instruments, and truth. In *The myth of the framework. In Defence of science and rationality* (pp. 154–184, M. A. Notturono, Ed.). Routledge.

Raub, W., & Voss, T. (1981). *Individuelles Handeln und gesellschaftliche Folgen. Das individualistische Programm in den Sozialwissenschaften*. Luchterhand Verlag.

Raub, W., & Voss T. (1986). Conditions for cooperation in problematic situations. In A. Diekmann & P. Mitterer (Eds.), *Paradoxical effects of social behavior. Essays in honor of Anatol Rapoport* (pp. 85–103). Physica Verlag.

Ritzer, G. (1975). *Sociology. A multiple paradigm science*. Allyn & Bacon.

Rössel, J. (2005). *Plurale Sozialstrukturanalyse. Eine handlungstheoretische Rekonstruktion der Grundbegriffe des Sozialstrukturanalyse*. VS Verlag für Sozialwissenschaften.

Sahlins, M. D., & Service, E. R. (1973). *Evolution and culture* (With a Foreword by L. A. White). The University of Michigan Press.

Sanderson, S. K. (1990). *Social evolutionism. A critical history*. Blackwell.

Scheff, T. J. (1967). Toward a sociological model of consensus. *American Sociological Review, 32,* 32–46.

Scheuch, E. K. (2003). *Sozialer Wandel. Band 1: Theorien des sozialen Wandels.* Wiesbaden: Westdeutscher Verlag.

Schimank, U. (2000). *Handeln und struckturen: Einführung in die akteurtheoretische sozilologie.* Juventa Velag.

Schmid, M. (1979). *Handlungsrationalität. Kritik einer dogmatischen Handlungswissenschaft.* Wilhelm Fink Verlag.

Schmid, M. (2006). *Die Logik mechanismischer Erklärungen.* VS Verlag für Sozialwissenschaften.

Schmid, M. (2009). Das Aggregationsproblem. Versuch einer methodologischen Analyse. In P. Hill, J. Kopp, C. Kroneberg, & R. Schnell (Eds.), *Hartmut Essers Erklärende Soziologie. Kontroversen und Perspektiven* (pp. 135–166). Campus Verlag.

Schmid, M. (2015). *Soziale Kausalität, Handeln und Struktur. Einige versuchsweise Überlegungen zur Reichweite sozialwissenschaftlicher Kausalitätsvorstellungen* (Script).

Schmid, M. (2017). Theorie, Reduktion und mikrofundierende Erklärung. In G. Wagner (Ed.), *Die Provokation der Reduktion. Beiträge zur Wissenschaftstheorie der Soziologie* (pp. 5–67). Harrasowitz Verlag.

Schmid, M. (2023). *Theorie- und Modellbildung. Mikrofundierende Erklärungen und die Methodologie sozialwissenschaftlicher Forschungsprogramme.* Wiesbaden: Springer SV.

Schneider, L. (1976). *Classical theories of social change.* General Learning Press.

Smelser, N. J. (1959). *Social change in the industrial revolution. An application of theory to Lancaster cotton industry 1770–1840.* Routledge & Kegan Paul.

Smelser, N. J. (1963). *Theory of collective behavior.* The Free Press of Glencoe.

Smelser, N. J. (1968). *Essays in sociological explanation.* Prentice Hall.

Smith, A. D. (1973). *The concept of social change. A critique of the functionalist theory of social change.* Routledge & Kegan Paul.

Smith, A. D. (1976). *Social change. Social theory and historical process.* Longman.

Spencer, H. (1880). *The study of sociology.* Williams & Norgate.

Spencer, H. (1897). *The principles of sociology* (Vol. III). D. Appleton and Company.

Spencer, H. (1898). *First principles* (5th ed.). Williams & Norgate.

Steward, J. H. (1972). *Theory of culture change. The methodology of multilinear evolution.* University of Illinois Press.

Strasser, H., & Randall, S. C. (1979). *Einführung in die Theorien des sozialen Wandels.* Mit Spezialbeiträgen von Karl Gabriel, Hans-Jürgen Krysmanski und Karl Hermann Tjaden Darmstadt and Neuwied. Luchterhand Verlag GmbH.

Turner, J., & Maryanski, A. (1979). *Functionalism.* Benjamin/Cummings Publishing Company.

Turner, J. H. (Ed.). (2006). *Handbook of sociological theory.* Springer Science + Business Media LLC.

Udehn, L. (2001). *Methodological individualism. Background, History and Meaning.* London/New York: Routledge.

Ullmann-Margalit, E. (1977). *The emergence of norms.* Oxford at the Clarendon Press.

Vago, S. (1980). *Social change* (2nd ed.). Englewood Cliffs.

Voget, F. W. (1975). *A history of ethnology.* Holt, Rinehart, and Winston.

Warren, R. L. (1977). *Social change and human purpose: Toward understanding and action.* Rand McNally College Publishing Company.

Weymann, A. (1998). *Sozialer Wandel. Theorien zur Dynamik der modernen Gesellschaft.* Juventa Verlag.

White, L. A. (1969). *The science of culture. A study of man and civilization* (Revised Edition with a New Preface by the Author). Farrar, Straus and Giroux.

Wippler, R. (1978). The structural-individualistic approach in Dutch sociology. Towards an explanatory social science. *The Netherland's Journal of Sociology, 14*, 135–155.

Wiswede, G., & Kutsch, T. (1978). *Sozialer Wandel. Zur Erklärungskraft neuerer Entwicklungs- und Modernisierungstheorien.* Wissenschaftliche Buchgesellschaft.

Wittek, R., Snijders, T. A. B., & Nee, V. (2013). Introduction: Rational choice social research. In R. Wittek & T. A. B. Snijders (Eds.), *The handbook of rational choice social research* (pp. 1–30). Stanford University Press.

Applying Methodological Individualism in the Analysis of Religious Phenomena

Salvatore Abbruzzese

1 Introduction

Few phenomena have been trapped for so long in what Raymond Boudon called *sociological determinism* as those arising from the religious dimension (Boudon, 1979). Even beyond the context of premodern societies, religious phenomena or religious-oriented modes of action have continued to be regarded as the manifest results of social and cultural conditioning. Such interpretation was transversal to the great theoretical approaches of the social sciences developed in the twentieth century: Freudism, culturalism, structural-functionalism, structuralism, neo-Marxism, etc. The conventional wisdom that this influence has been directly exerted by community pressures or by the persistency of the influence of tradition on socially fragile fringes, marked by social poverty and cultural marginalization, led to consider the religious dimension as one of the *social facts* most replete with consequences. On these bases, the subject of religion was invested, including the imaginary it carries given its detachment from the physical experience, being exposed to cultural biases and deprived of the required axiological neutrality.

And yet, the *Methodological Individualism* perspective (henceforth MI) opposes such conceptions, or methodological shortcuts, which assume that

---

S. Abbruzzese (✉)
Department of Sociology, University of Trento, Trento, Italy
e-mail: salvatore.abbruzzese@unitn.it

© The Author(s), under exclusive license to Springer Nature
Switzerland AG 2023
N. Bulle and F. Di Iorio (eds.), *The Palgrave Handbook of Methodological Individualism*,
https://doi.org/10.1007/978-3-031-41508-1_6

the goals and beliefs of others and, generally, the social contexts, exercise forms of mechanistic forces on individuals. By taking the individuals' motivations seriously, MI has achieved its best results in the analysis of the religious phenomenon, highlighting the limits of the deterministic approach, and overcoming them with a more exhaustive description, able to reveal unexpected results so as to considerably renew other branches of sociology, such as political sociology or the sociology of cultural processes. Thus, while the deterministic perspective, according to which the religious dimension is the result of group conditioning, regards the subject as a passive receiver and a mere reproducer of practices related to the sphere of the *sacred*, the individualistic perspective considers her/him an active receiver, for whom the religious dimension is the basis of acting going beyond—as it actually happened—the mere ritual sphere.

In this paper, besides mentioning the most notable achievements in the classics of sociology deriving from an MI-based approach, the advantages for a general theory of social relation deriving from the analysis of beliefs (the general ones and the religious ones specifically) from the MI perspective will be presented.

2 The Supremacy of Sociological Determinism in the Analysis of Beliefs

The immediacy of the deterministic approach in the early stages of non-theological studies of religious phenomena is not obscure and, in the sociological field, this is closely linked to the specific nature of beliefs, and more specifically that of religious beliefs.

If we agree to use the term *belief* to describe the act of "accepting as true" something lacking the certainty deriving from the existence of a proof (Boudon, 1995), we instantly set in place a dimension of individual life where the social environment plays an essential, if not determining, role. Actually, only beliefs can be endorsed from the outset by individuals. This involves trust in the community in which they live, and their perception of the world and of life they have thus produced and acquired. Besides, only beliefs, representations and values make commitment an unavoidable condition for membership, since they clearly define the specific boundary of a shared worldview the (religious) community identifies with. Subscribing to the beliefs emerging in one's "moral environment" (Durkheim, 1895, 1900) is the first condition posed to the individual in order to be acknowledged as a member of the community.

In the realm of beliefs, religious ones hold a prominent position. Beliefs concerning the religious dimension, unlike general beliefs, are not merely opinions but, since they are straightaway prescriptive, they impose specific conduct rules, with deep consequences on the communities (Weber, 1920). In particular, if a deity or a whole pantheon are established in order to ensure the protection of the community, the incomplete acceptance or even the refusal of one single belief by one of its members could nullify the god's protection of the entire community. In such cases, both adherence to the religious belief and scrupulous observance of all the derived conducts, from rituals to everyday life minor actions, represent the basic condition for the individual's permanence in the community (Weber, 1920).

Besides the pressure exerted by the community on the single believer, and unlike any other kind of belief, pressures pertaining to the religious sphere are also the ones where the subject's commitment manifests itself as a pure act of *faith*. Its deep impact on personality development has always been considered a *consequence* of doctrines that claim and obtain acknowledgment not through proofs but merely through an act of faith. Worshipping the deities or a single god leads to modifying one's behavior to the point of creating a unified ethical personality that is completely different from the original one, as it can be recalled exemplarily by such concepts as *rebirth* or *new man*, which can be found in the religious narrative.

The prescriptive nature of voluntary commitment, pledged through faith, turns the commitment to religious beliefs into total submission to the deity, in direct correspondence with the community acknowledging it. This commitment is based on inscrutable inner motives and its prescription has no *reason* deriving from anything other than membership of the community the subject identifies with and by which s/he is recognized. Even the process of rationalization begun at the dawn of the modern world with the protestant dogma of *sola gratia* leaves the believer with no chance of a (rational) choice, except for acknowledgment of the received gift from the deity.

In conclusion, religious beliefs are the first ones that seem to satisfy a deterministic kind of explanation, according to which the subject accepts and subscribes to certain beliefs with, sometimes, crucial consequences for her/his own existence solely by virtue of her/his faith toward the authority that proclaimed them and the community that accepted them. As far as this aspect is concerned, the tradition of the community and the pressure it exerts on its members are assumed to play a crucial role.

Thus, the supremacy of the deterministic approach for religious phenomena seems to be unquestionable. The religious dimension, considered as a set of beliefs concerning the existence of transcendental beings, mediated by a community and systematized by a priestly caste, seems to only be established under the pressure of the milieu within which the subject places her/himself. The very nature of the commitment as an act of faith, together with the explicit condemnation of the non-believer up to her/his expulsion from a *sui generis* community known as Church (Durkheim, 1912), make the commitment decided upon an (irrational) inner motion, supported by tradition and strengthened by the subject's community, the only possible reason—and sufficient in itself—for the commitment itself.

Such interpretation has limited the perspective of MI for a long time, and its establishment in the analysis of religious phenomena has only been possible through the development of a completely different anthropological premise.

Actually, a commitment to religious belief that is only due to the context's conditioning places it in an utterly insignificant position in relation to its genuine development. The acceptance of myths and beliefs based solely on the community's wish for gods' protection and on the mere *adaptation to* the community's narrative, although theologically organized, makes the religious commitment a mere accommodation that cannot originate any transformation or innovation process. It is indeed no coincidence that, in such a perspective, the religious dimension is totally disregarded when analyzing the processes of political, economic and cultural transformation. And even when acknowledged, this only served functional purposes, while on the contrary the actual political processes were the ones deemed crucial.

Now, it is precisely the opposite that has occurred in all the cases examined by Weber. The religious dimension has proved itself pivotal both in shaping the political process and in triggering significant transformations in the economic one. What was meant to be a plain and passive acceptance of certain beliefs concerning the transcendental dimension, fostered by the community and guarded by the "professional guardians of the sacred" (Martin, 1978), proved to be an unexpected source of action, able to initiate a significant number of transformations in the society's structure itself, thus becoming a main agent in the processes of social change.

Determinism thus appears to be inadequate for accounting for the transformation processes generated by religious doctrines. Reducing the religious belief itself to its mere function of community cohesion does not explain its driving force. In order to explain such a potential, the limits set by the anthropology underlying the analysis of the religious dimension had to be overcome. It was necessary to identify through it not only a narrative of the sacred depending on the community bond and an arrangement of rites to celebrate it, but also and above all the formulation of cognitive categories according to which a reorganization of reality (Durkheim, 1912) and an intrinsically consistent way to be part of it is achieved.

Although the commitment to religious beliefs is based on a more or less coercive group conditioning in the course of the learning process, this does not prevent religious knowledge from producing an overarching interpretation of reality, moving from which the believer begins to act on all the different levels of social reality. The consequences go far beyond the sphere of the sacred, and instead deeply modify "the face of the earth" (Psalm 103), namely the whole of social relations and, notably, the relations with the political power (Gauchet, 1985).

Such an interpretative and motivational potential resulting from a single religious doctrine can in the end radically modify the nature of the original commitment which is at first generated and influenced by the environment proposing it (and most times imposing it). This commitment is later consciously endorsed by the individual and consistently practiced for *reasons* considered crucial by her/him. The individual keeps faithful to the religious tradition and continues to accept its participating-emotional variables only when—and for as long as—the cognitive mapping through which every religion understands the world, as well as human life, seems valid and effective to her/him. On its part, religion, during the establishment of the salvation doctrines following the prophetic revelation, continues to be refined and make the proclaimed principles more and more consistent with desired behaviors. In this regard, every religion is a cognitive map of existence, an unceasing reasoning and a steady perfectioning of its internal coherence.

The individual's consciously active commitment not only accounts for the cultural flourishing of religious communities and their ability to adapt and renew in different, if not radically antithetical, contexts, but also for the massive use of energy and cultural resources throughout different ages, which still continues, perhaps even more now than in the past (Berger, 1999).

3 MI and Explanation of the Religious Phenomenon

Moving from subjects almost automatically inscribing themselves in a tradition because of the pressure of social influence to subjects who actively invest themselves, in a significant way and with remarkable consequences on their own as well as on their community's life, unquestionably involves considering their actions as conscious and voluntary.

132   S. Abbruzzese

This is exactly the modality through which the religious phenomenon appeared in the analysis of the classics, an appearance neither Tocqueville, nor Durkheim nor Weber had predicted. Since they were interested in specific themes (the evolution of democracy, the weakness of social cohesion, the extension of the rationalization processes), the three of them were basically uninterested in any religious problematics. Their encounter with the religious phenomenon was the unexpected result of research heading in very different directions. It was actually a notable instance of *serendipity*, the best result of any scientific research. It was by considering any commitment to the religious creed as a conscious commitment based on subjective *reasons*, rather than the consequence of any *causes*, that they implemented, more or less implicitly, the methodology later identified as MI, which proved fundamental for the whole of sociology.

A short overview of the path followed by these authors will clarify the sense and the significance of such an approach.

Religion as a Moral Universe: Alexis De Tocqueville

The comparison between Tocqueville (1805–1859) and his contemporary Auguste Comte (1798–1857) is more illuminating the more radically their personalities, dispositions and choices differ from one another in relation to the same historical-political context (Aron, 1967, 194–279). Although they both consider religion as a reading key of the world and an overarching explanatory system, according to Comte this is limited to the childhood of humankind and, this feature specifically, is deemed to disappear with the development of scientific thinking and positive law. On the contrary, for Tocqueville, religion is an essential element of the social and especially political life of society in modern ages. Far from being a mere relic of tradition, religion is actually more precious the more it enables democratic societies to avoid the drifts that might emerge, especially within the liberal and materialistic society that characterizes the modern world (Manent, 1993).

Such a thought is not derived from theoretical considerations but from empirical research. For the young Enlightenment thinker Tocqueville, who had adopted an agnostic stance since his teens, religion is not one of his main concerns, as he first investigates political phenomena, and in particular the evolution of the current democratic process. During his analysis of democracy in the United States, the role of religion surprises him and, at first, is impossible for him to understand.

In the United States, when the seventh day of each week arrives, the commercial and industrial life of the nation seems suspended; all noise ceases.

A deep repose, or rather a sort of solemn meditation, follows; the soul finally comes back into possession of itself and contemplates itself. During this day, places devoted to commerce are deserted; each citizen, surrounded by his children, goes to a church; there strange discourses are held for him that seem hardly made for his ears. He is informed of the innumerable evils caused by pride and covetousness. He is told of the necessity of regulating his desires, of the delicate enjoyments attached to virtue alone, and of the true happiness that accompanies it. Once back in his dwelling, one does not see him run to his business accounts. He opens the book of the Holy Scriptures; in it he finds sublime or moving depictions of the greatness and the goodness of the Creator, of the infinite magnificence of the works of God, of the lofty destiny reserved for men, of their duties, and of their rights to immortality. Thus at times the American in a way steals away from himself, and as he is torn away for a moment from the small passions that agitate his life and the passing interests that fill it, he at once enters into an ideal world in which all is great, pure, eternal. (Tocqueville, [1835] 2000, 512).

A world of merchants and craftsmen, devoted to work and the search for well-being, who every Sunday stop working to attend sanctuaries where they are told "strange discourses" about the damage caused by greed, about a sublime God loving them and the joys for those who have faith. This seems at first to be a trivial cultural peculiarity, almost folkloric. This is the realm of what Pareto defined as non-logical actions at the beginning of the twentieth century: One subscribes to beliefs and reasonings detached by any means-ends logic, thus entering an ideal world "in which all is great, pure, eternal" and, presumably, completely different from the real world. It is just explaining the *reasons* of the commitment to such universe, so vague and far from the reality, that Tocqueville finds explanations so diverse from the sheer commitment to the religious tradition existing in the foundation of the United States itself.

Rather than the commitment to a tradition, which actually does not surprise him, what impressed Tocqueville was the correspondence the religion establishes with everyday life, with what occurs outside and beyond the boundaries of liturgical rites. Except for some extravagancies of the cults, which upset him, Tocqueville observes the deep, and apparently surprising, continuum existing between the life rules proclaimed from the churches' pulpits and the ordinary life of the liberal, market-based and profit-oriented society. The correspondence is such that he is unable to understand if Sunday sermons, before leading to salvation in the afterlife, might already be useful in this life.

134   S. Abbruzzese

This brings him to refuse at once any anthropology of the suffering subject, oppressed by existence, for whom religion would serve as a feeble consolation, or trapped in the premodern logic of the theological stage where religion dominates through its myths and representations. Moreover, according to Tocqueville, the subject is not induced to enter the religious cosmos through a sense of existential loss in the face of the transience of life and the perspective of death either, but through interest.

> [...] I think that interest is the principal means religions themselves make use of to guide men, and I do not doubt that it is only from this side that they take hold of the crowd and become popular. I therefore do not see clearly why the doctrine of self-interest well understood would turn men away from religious beliefs, and it seems to me, on the contrary, that I am sorting out how it brings them near to them. I suppose that to attain happiness in this world, a man resists instinct in all encounters and reasons coldly about all the acts of his life, that instead of blindly yielding to the enthusiasm of his first desires, he has learned the art of combating them, and that he has been habituated to sacrificing without effort the pleasure of the moment to the permanent interest of his whole life. If such a man has faith in the religion that he professes, it will scarcely cost him to submit himself to the hindrances that it imposes. Reason itself counsels him to do it [...] Not only do Americans follow their religion out of interest, but they often place in this world the interest that one can have in following it. [...] To touch their listeners better, they [the American preachers] make them see daily how religious beliefs favor freedom and public order, and it is often difficult to know when listening to them if the principal object of religion is to procure eternal felicity in the other world or well-being in this one. (Tocqueville, [1835] 2000, 489–9)

In other words, religious beliefs and the derived acts of worship are not at all the results of a persisting religious tradition, memorized and practiced out of habit. Speculating on the subject's *reasons* to subscribe to them, Tocqueville interprets them as the work of a genuine "self-interest rightly understood". Preachers adopt an absolutely rational strategy (without having to interpret rationality in the essentially utilitarian sense), aimed at an actual education of the subject, who consciously refines his character by practicing, in compliance with the divine law, the same virtues that will help him succeed in his life.

But, according to Tocqueville, religions do not just serve success in daily life. Besides educating to self-control, they help to shape a worldview and a view of life that, restraining the conduct and raising the soul to spiritual pleasures, lead it toward the sublime. In doing so, religions eventually inspire the moral side of democratic society, mitigating the undesirable effects produced by practical materialism. This would otherwise make it fall into

Applying Methodological Individualism ... **135**

mere delight for material goods and sheer personal wellness, or at best that of the individual's family. The main risk run by democratic and liberal societies is precisely the pursuit of material well-being alone:

> But while man takes pleasure in this honest and legitimate search for well-being, it is to be feared that he will finally lose the use of his most sublime faculties, and that by wishing to improve everything around him, he will finally degrade himself. The peril is there, not elsewhere. Legislators of democracies and all honest and enlightened men who live in them must therefore apply themselves relentlessly to raising up souls and keeping them turned toward Heaven. It is necessary for all those who are interested in the future of democratic societies to unite, and for all in concert to make continuous efforts to spread within these societies a taste for the infinite, a sentiment of greatness, and a love of immaterial pleasures. If one encounters among the opinions of a democratic people some of those harmful theories that tend to make it believed that everything perishes with the body, consider the men who profess them as the natural enemies of this people. (Tocqueville, [1835] 2000, 513)

Religion as a process of self-education the subject deliberately undertakes, not only aims for the rational control of passions, but also represents an element of moral elevation. This is even more important when the path of a people is not limited to the achievement of well-being and peace but aims for the establishment of an actual civilization. The sole search for material well-being can only cause an increasing loss of social bonds and civic values. According to Tocqueville, a society cannot flourish without an authoritative interpretation of its aims, that is to say without guiding values allowing the subjects to define themselves through the dignity and the specificity of the society they belong to.

> This explains why religious peoples have often accomplished such lasting things. In occupying themselves with the other world they encountered the great secret of succeeding in this one. Religions supply the general habit of behaving with a view to the future. In this they are no less useful to happiness in this life than to felicity in the other. It is one of their greatest political aspects. (Tocqueville, [1835] 2000, 517)

Thus, because religious adhesion is *not* a simple conformism to myths and norms traditionally inherited but is motivated by *reasons* the individual perceives as valid, it leads the believers toward sublime values and provides them with the moral rigor necessary to democratic governments even more than to despotic regimes. Therefore, Tocqueville considers religions as the essential element for helping democracies avoid the fall into practical materialism, which would as a result deprive them of the main principles of attention toward the greater good that are crucial to them.

The American experience led Tocqueville to reorganize his own anthropology, placing a strictly individualistic perspective, purified of any atomistic[1] reductionism, at its core. The social context is far from useless. The subject does not choose the explicative principles and her/his references in order to act morally, shaping her/his own views, but resumes them from the ideas crossing and shaping her/his age. The difference with the determinist perspective is that such explanation does not simply refer to the individual's subscription to the ideas of her/his own age. Beliefs and principles would not be passed from one generation to the next if each of them had not recognized a reasonable advantage in subscribing to them. Religion itself, the most delicate factor of human acting, since it reorganizes its principles in relation to the essential problems of existence, is the result of rational thought the social actor encounters in her/his community and is interested in accepting whenever it does not collide with her/his own material interests or her/his own moral principles. Of all the ideas concerning humankind and its destiny, those deriving from religions are, from a moral perspective, more advantageous (or less costly) to adopt rather than undertaking a personal quest that the subject risks to pursue for her/his whole life, uncertain of achieving any result.

> "Some fixed ideas about God and human nature are indispensable to the daily practice of their lives, and that practice keeps them from being able to acquire them. [...] Among the sciences there are some that are useful to the crowd and are within its reach; others are accessible only to a few persons and are not cultivated by the majority, who need only their most remote applications; but the daily use of this [science] is indispensable to all, though its study is inaccessible to most. General ideas relative to God and human nature are therefore, among all ideas, the ones it is most fitting to shield from the habitual action of individual reason and for which there is most to gain and least to lose in recognizing an authority. The first object and one of the principal advantages of religions is to furnish a solution for each of these primordial questions that is clear, precise, intelligible to the crowd, and very lasting. (Tocqueville, [1835] 2000, 412–3)

---

[1] The difference between methodological individualism and atomistic individualism is the same as that existing between dog as an animal and dog as an astrological figure (Boudon, Bourricaud, *A Critical Dictionary of Sociology*).

Thus, commitment to religious values, even when it is a mere identification with the religion practiced in one's region, does not mark the individual's resignation from the reason but, on the contrary, represents a conscious choice. Tocqueville does not identify anything illogical in the commitment to religious beliefs, instead, contrary to the then-prevailing opinion in France, he considers religion a precious ally for democracy. Only the religious dimension is able to protect citizens from the civic indifference eventually established after succumbing to the fascination of material goods alone. This is a more than dangerous attitude in which democracy sees its own decline.

> When authority in the matter of religion no longer exists, nor in the matter of politics, men are soon frightened at the aspect of this limitless independence. This perpetual agitation of all things makes them restive and fatigues them. As everything is moving in the world of the intellect, they want at least that all be firm and stable in the material order; and as they are no longer able to recapture their former beliefs, they give themselves a master. (Ibid.)

The educational function of religions and their ability to limit the individualistic and materialistic drift the subject inevitably undergoes in a democratic society eventually also explain the subject's reasons for subscribing to a single religious creed. The individual accepts and recognizes the single religion proposed not only by virtue of its cognitive role, as this allows her/him to obtain systematized notions of the world and life that could hardly be formulated on one's own, but also by virtue of moral satisfaction. This derives from the voluntary subscription to an exceptionally positive state of affairs, from which s/he may enjoy the benefits of ordinary life. Therefore, the subject has every *good reason* to adhere to religious beliefs and practices precisely for the results s/he can obtain with such a choice.[2]

The search for the *reasons* of believing rather than the *causes* that drive the subject's commitment enables Tocqueville to push the sociological analysis of commitment to religious beliefs further than he would have been able to if he had just considered it as the result of the direct influence of the community, passed on by tradition and supported by institutions.

---

[2] The notion of "good reasons" was developed by Boudon in his various works on ordinary rationality (see Jean-Michel Morin "Ordinary Rationality Theory (ORT) According to Raymond Boudon", this volume). This notion is based on the Weberian approach to understanding sociology according to which beliefs, preferences and values are based on reasons in the mind of social actors (which are not limited to instrumental goals and do not respond to maximization objectives, as is the case in Rational Choice Theory).

Beyond Communitarianism: Emile Durkheim

The refusal to interpret the religious dimension as a consequence of mere pressure from the community can also be identified in Emile Durkheim's works. Only apparently, as Comte's successor and Fustel de Coulange's disciple, Durkheim considers religion as an expression of the primitive stage of humankind. Actually, his study of the elementary forms of religious life, rather than supporting the passion for the customs and beliefs of primitive societies typical of his age (McLennan 1865, Tylor 1871), complied with his interpretation of religion as a structural element, consubstantial with the society itself since its origins; an element that, by virtue of such coexistence with the society, cannot be confined to its historical dimension but involves its essence itself. It proves to play such a crucial role that its disappearance means the disintegration of community life itself. Religions grant the group identity consistency, and make it the focus of the great gatherings of the community (Mauss, 1905/2013). Similarly, it appears that religions always underlie societies' organization, not only by categorizing time and space, but also by separating the sacred and the profane, namely protecting everything at the foundation of community life as a moral reality, separating it from everyday life and individual interests.

Given such essentiality, the religious dimension, like any other collective phenomenon, does not assert its authority without the individual's consent. According to Durkheim, "The concept which was first held as true because it was collective tends to be no longer collective except on condition of being held as true: we demand its credentials of it before according it our confidence" (Durkheim, 1912/1976, p. 438). Although a religion is proposed by a context the individual lives in and with which s/he identifies, s/he will not grant it credence unless s/he has acknowledged its contents and experienced its substance. While the contents are transmitted during the process of socialization and refer to more or less historically defined events, which are rearranged in the shared narrative that presents them, the substance of the religious dimension is experienced through the realization of the collective rite.

Unlike theological knowledge, the religious dimension cannot be separated from the experience of a real contact, a tangible encounter developed basically, according to Durkheim, in the ritual dimension and specifically in the religious feast. This does not mean giving religion an inferior cognitive status. In fact, Durkheim questions Lucien Lévi-Bruhl's concept of "primitive mentality" (Lévi-Bruhl, 1922/2019), referring to a pre-logic universe where the magic-religious dimension would have found its own breeding ground, thus making it unacceptable in the modern age.

Durkheim does not consider religion as an illogical drift, resulting from a "mentality" surely different from the scientific one. As we have already seen, his methodology, evidenced by his empirical analyses, is in practice similar to Weber's, contrary to the conventional wisdom that opposes them. The case of the rain dance is exemplary in this regard:

> The rites which he [the individual who is called primitive] employs to assure the fertility of the soil or the fecundity of the animal species on which he is nourished do not appear more irrational to his eyes than the technical processes of which our agriculturists make use, for the same object, do to ours. The powers which he puts into play by these diverse means do not seem to him to have anything especially mysterious about them. Undoubtedly these forces are different from those which the modern scientist thinks of, and whose use he teaches us; they have a different way of acting, and do not allow themselves to be directed in the same manner; but for those who believe in them, they are no more unintelligible than are gravitation and electricity for the physicist of to-day. Moreover, we shall see, in the course of this work, that the idea of physical forces is very probably derived from that of religious forces; then there cannot exist between the two the abyss which separates the rational from the irrational. Even the fact that religious forces are frequently conceived under the form of spiritual beings or conscious wills, is no proof of their irrationality. (Durkheim, 1912/1976, p. 26)

Lacking adequate knowledge concerning the atmosphere, the rain dance does not entail a pre-logic mentality but simply explains the rainfall phenomenon according to the only interpretative code available, namely the connections occurring through magic. Moreover, since rain rites were practiced in those periods when rain was expected, the chances for success were highly favorable, thus recording an evident confirmation of the rite's effectiveness.

Basically, even though Durkheim agrees with Comte as far as the existence of a theological stage at the infancy of humankind is concerned, he does not recognize any form of irrationality. The underlying logic is exactly the same as the contemporary one, though lacking the physical-natural knowledge that developed only later on. Thus, the peoples belonging to elementary societies, when they needed it to rain, did not resort to the summoning of transcendental forces by tradition, nor out of pure adherence to a common belief, but because they had "good reasons" to do so, reasons confirmed by the success the rituals often achieved (Boudon, 2010).

Besides recognizing magical forces as underpinning the one possible interpretation of natural phenomena, the adhesion to the expressive forms of religion represented by ritual acts confirms the existence of such forces in the shared experience of the collective ritual. The ceremony itself represents a suspension of ordinary time and marks the beginning of a collective moment during which any ordinary relation is reassessed and re-established according to the evoked foundational event. Consequently, the religious act proves to be at the basis of an experience lived by the whole tribe, while being rationally connected to the practical experience. In so doing, the subject lives an experience that actually brings her/him beyond ordinary material life, lived according to instrumental rationality, in order to direct her/him toward the prime principles in which s/he lives and strengthens her/his own community identity, as well as the moral principles that guide it.

Broadly speaking, dealing with a religion does not mean dealing with a mere commitment to a worldview, originating from a founding myth and influenced by the community context passing it from generation to generation. Although religion is regarded as a mainly social phenomenon, thus exterior and coercive against the individual, the room for recomposition from which the subject carves out her/his own personal position is more and more evident. Actually, though the religious principles appear as a self-consistent whole, the extent to which one subscribes to them varies considerably from principle to principle within the same group of believers. So, one does not subscribe to all beliefs with the same conviction. Those not colliding with the principles of reason (God, soul, sin) are accepted more convincingly; on the contrary the eschatological beliefs, those concerning the "ultimate realities" (judgment after death, hell, paradise), are endorsed with far less intensity (Boudon, 2002). Moreover, the fact that people generally believe more in the existence of heaven than in hell is very revealing. In this regard, the same figure of the Devil often clashes with the evident perplexities of most of the interviewees (Abbruzzese, 1985).

Therefore, if religions are a social phenomenon, exterior and coercive, nevertheless the subject not only tends to subscribe to the various dogmas of faith differently, but also actually strains the values offered by her/his religion through her/his own set of values, making a personal choice through which s/he expresses her/his own priorities. This is nothing new, but only nowadays inquiry techniques, as well as greater attention to the subject's choices and the underlying reasons, have made this visible.

Questioning the individual's reasons for subscribing to a religion, rather than the community's influence in imposing it, allows a shift of the focus onto the cognitive map each religion offers, according to which a religion can be accepted, fully or partly. Following such a perspective, moving from the conditioning *causes* to the explaining *reasons*, it is also possible to reconsider the ritual aspects of single religions. The religious feast itself, in the same way as the Sunday liturgies observed by Tocqueville, does not represent an escape into the irrational. It produces a rational rearranging in the individual's perceptions of the world and life, according to which s/he can build her/his existence in the best way (Isambert, 1982). Thus, the persistency of participation in religious rites and feasts in a fully secularized age appears more and more clearly as the conscious manifestation of a search for meaning that has not been answered in the ordinary world any more than it was in the past (Abbruzzese, 2010).

In conclusion, considering the dominance of the secularization of institutions and the laicization of customs, as well as the surge in free-time opportunities, partaking in religious rites cannot be explained as the result of mere habit or social conditioning. Once this conditioning logic is excluded, any deterministic perspective becomes inapplicable, and the way is open for a sociological understanding of the reasons for any religious action.

Religion as a Reorganization of the World and as an Inner Experience: Max Weber

Max Weber (1864–1920), the author who explicitly adopts the MI paradigm, refuses to define the concept of religion and prefers that of religion- (or magic-)oriented action. Such an action does not differ from social action in general and, similarly, is subject to the analysis of the results by a conscious subject, able to evaluate them. For the people belonging to the elementary society at the dawn of civilization, there is no difference between the magician lighting fire and the one making the rain fall since both actions are considered "magical" acts, that is to say acts provoked by supernatural forces which the magician alone, as charismatically qualified, is able to evoke (Boudon, 2012, 61–65).

According to Max Weber, religious beliefs undergo a real process of rationalization on which they ultimately depend. After that not only the beliefs, but also the whole configuration of the religious dimension is completely transformed. Behind Weber's renunciation of defining religion, there is not only a methodological caution, but also the awareness that religion-oriented action regularly and substantially modifies the dimension of the sacred, making any reduction to unity impossible.

For instance, the definition of the concept of *soul* alone will lead to the birth of a belief in spirits, to the subsequent worship rituals and the ensuing establishment of an order of priests able to arrange, transmit and control these rituals. So, the pre-animistic religions and the following ones will differ substantially, and it will not be possible to reconnect them to one another. Similarly, only the bond between cities and the need to give each city's god the right space will originate a pantheon, moving from which the urgency to characterize each god according to her/his qualities or—as in the case of Rome—according to their abilities will take shape. Only the belief in a unique and omnipotent God raises the issue of God's justification before the evil he paradoxically tolerates, thus opening up the problem of God's possible justifications, the solution of which implies the development of radically new, and otherwise inconceivable, religious situations. More precisely, from such a problem, will be established the concept of life after death, conceived of as a reward for the just and a condemnation for the evil.

Not only does Weber's religious sociology not admit a unique concept to describe religion and include different modes of thought and action from which radically different effects can result, but it also identifies the religious space as a place where, as the process of rationalization advances, new figures develop.

Next to the magician, the figure of the priest appears, as a result of the establishment of the concept of soul. Such an appearance was unimaginable before the transcendental forces remained broadly internal to the object and could therefore be influenced other than by magical coercion. The uncertainty of the magic act, completely entrusted to the skills of the magician and possibly unsuccessful, is followed by the cult of deities, managed by a specific group of charismatically qualified, but also properly trained, people in compliance with the office to which they have been devoted.

Similarly, with the prefiguration of an ethical God, the figure of the prophet, the one who asserts and spreads the salvation doctrine required by this God, is established. Only after the establishment of the prophet and, subsequently, of the group of disciples, the unprecedented dimension of the lay community, a focus of new attentions to everyday life, emerges. Moreover, only after the prophet's revelation, will the issue of evangelization and of different forms of religious presence in the world be raised, introducing the theme of enculturation.

Magician, priest, prophet and lay community lead to configurations of the religious universe from which will be derived religious dimensions for which any integrated definition is no more than a logical exercise.

The dimension of the religion-oriented action postulated by Weber focuses the attention on the subject, his rationalizations and resolutions. The religion moves toward more and more rigorous rationality levels where, besides the increasingly clear definition of deity, also specific expectations of those acknowledging it are developed. The accuracy of ritual acts is followed by the strictness of life's ethic, oriented toward a foundation of unity, and the commitment to follow it. Subsequently, religion works as a rationalizing mechanism for conduct, both toward deities and others. Thus, the subject moves from purely ritual acting to a constant reappraisal of her/his inner life, which results in the development of a unitary ethical personality. Such rationalization of life conduct has significant effects in the political as well as in the economic field.

What appears crucial concerning MI is that not only the commitment to the truths of faith is not explained as a consequence of any community pressure, except in its early phases, but that this commitment undergoes assessments of consistency with the individual's life experience:

> Many more varieties of belief have, of course, existed. Behind them always lies a stand towards something in the actual world which is experienced as specifically 'senseless.' Thus, the demand has been implied: that the world order in its totality is, could, and should somehow be a meaningful 'cosmos.' This quest [is] the core of genuine religious rationalism (...). The avenues, the results, and the efficacy of this metaphysical need for a meaningful cosmos have varied widely. (Max Weber, 1920/1946, p. 281)

This does not simply result in an adaptation to the single salvation doctrine, but can be modified or even refused if said consistency is judged unsatisfactory. For instance, farmers tended to refuse monotheism, since it clashed with their experience according to which work in the fields was subject to uncertainty regarding the weather, which could hardly be considered to be under the will of a sole almighty God, without ascribing to him the trait of supreme indifference. Similarly, the military tended to avoid Christianity, since it preached love for the enemies, and preferred to worship the god Mithras, retaining consistency with military values and its hierarchy. And again, Christianity was avoided by the working class at the end of the nineteenth century, which would rather follow mundane salvation doctrines, since they perceived their living conditions as a consequence of the material conditions of employment relations, rather than of the inevitability of evil in the human condition.

While Tocqueville considers the benefits of religion in relation to the development and protection of democracy, and Durkheim identifies the sacred as the foundation of the community and religion as its custodian, they both identify an active subject who does not passively adhere as a consequence of

144   S. Abbruzzese

the community's pressure, but comes to a willing choice about her/his salvation doctrine or, if s/he does not have any possibility to choose, accepts it, acknowledging the worldviews s/he considers reasonable and experiencing the derived advantages. Therefore, there are various "good reasons" to approach the religious dimension, which range from the advantage of the proposed principles in order to favorably lead one's existence (Tocqueville), to experiencing the gratification of belonging to a moral community (Durkheim), to the willing and aware commitment to the single salvation doctrine since firmly convinced of its content (Luckmann, 1967), to the meaningful role it plays in the individual's belief system (Weber). In all these circumstances, searching for the *reasons* is far more effective and produces far more results than the deterministic approach of seeking the *causes*.

More specifically, through Max Weber, the perspective of MI leads to the possibility of understanding the dimension of religious action even beyond any functional character or any material or moral advantage. Moving from the analysis of religions as external and coercive factors for the individual that present themselves as a whole, to that of religion-oriented action leads the entire sociology to the realm of social action and the subject's reasons for acting. With Max Weber, the reasons for believing are understood more for the internal logic, thoroughly and continuously traced in every single doctrine, rather than for the advantages that might follow for the believer. The Weberian long treatment about the choices of the Protestant Churches' synods, from which Churches' new denominations spring, definitely confirms this (Weber, 1922/2002). Nothing would have emerged from such a dynamic if the analysis of the religion-oriented action had been understood as the mere effect of cultural conditioning.

The Relational Dimension in the Individualistic Approach of the Religious

This short excursus about the classics of sociological thought, meant to explain the effectiveness of research into the *reasons* for believing rather than the *causes* that lead individuals to believe, also has an added value. The analysis of the reasons for the commitment to a single salvation doctrine, whose credibility and legitimacy are acknowledged, allows the opening of a breach in the realm of social relation that has been quite disregarded up until the present day (Donati, 2011).

The adhesion to a belief, precisely because lacking in certainties, is very responsive to the intersubjective dimension. This intersubjective dimension refers to a common world, based on experience, cultural constructs and communication, which allows individuals to develop shared interpretations. Since the subject often perceives the reasons s/he has to subscribe to a certain belief as self-evident or, according to her/him, as evident to all, the sharing dimension becomes crucial (Boudon, 2008). Subsequently, "good reasons" underlie the individual's commitment to the beliefs in question, and lead her/him to see in the intersubjective confirmation the crucial evidence s/he misses. Others are implicitly involved in supporting and sharing those same reasons the subject takes for granted. The "good reasons" logic is thus here willingly and consciously intersubjective.

Such a need to confirm one's reasons through the intersubjective dimension opens a further level of relational motives. The social bond that fuels groups and is founded on a number of motives, which span from material interest to the sharing of values, to simply affection for neighbors, experiences a brand-new motive that keeps it present even in cases lacking all other motives. Apart from material interest, mere habit or a shared passion, the relation with others can actually be steadily reinforced by the continuous need to see the individual's beliefs strengthened through the confirmation of her/his "good reasons" on the level of the intersubjective relation.

The need to see one's reasons confirmed at an intersubjective level leads to a specific mode of relation between individuals whose advantage is by no means material, and cannot be reduced to the emotional dimension. Such a need is not to be mistaken for adhesion to an advocacy group—whether aiming to obtain public good or to support and promote certain values—nor does it derive from the very existence of shared religious or political beliefs, or simply from shared passions. And even less than this can be reduced to the mere gratification derived from the enjoyment of interacting with people because of an affinity of likings and behaviors. This need for an intersubjective confirmation of the individual's beliefs originates a relation that is more significant for the contracting parties, the more the "accepting-as-true" dimension lacks any kind of certainty, and the only reason to subscribe to it is its "manifest plainness", namely its being for all to see. This is why this plainness has to be continuously and repeatedly shared by the equals most relevant to her/him.

Values such as peace, new rights and preservation of the environment can be displayed and, precisely as *values*, can be the object of public statements, become the framework from which a group develops its own mission, represent the identity of a political party and be elevated to the level of shared morality—shared by the institutions as well. It is not the same for beliefs: the realm of what is "accepted as true" does not have such opportunities. Even the semantic label that marks them, that of "I believe in" rather than that of "I believe that", places those who subscribe to any belief in a position of discretion rather than bold assertion. If values can be followed by statements, public demonstration and the formation of associations, beliefs do not have such possibilities. Believing that teaching Greek and Latin in high school helps creating a deep and solid cultural background or, under different circumstances, considering the knowledge of Mathematics as a crucial element for an education in line with new technologies does not originate any kind of public display and cannot but be confined to the level of personal opinions. Believing that autonomy and imagination on the one hand or learning good manners and altruism on the other hand are qualities a child should be encouraged to cultivate does not originate any parents' association, though these kinds of beliefs tend to oppose non-observant laypersons and observant believers (Abbruzzese, 2010).

Precisely because of this inability to be ultimately founded or demonstrated, beliefs come to be the connective tissue of relation networks. The consequence is that they are much more essential than publicly supported values because they involve the very criteria for orientation of political values, which on the contrary are constantly subject to public mobilization.

This has a significant impact on the religious field. As it has been known for seventy years at least (Acquaviva, 1961), the secularization theory is employed to explain the most powerful and evident cultural phenomenon of advanced modernity, namely the decline of religious sensitivity (Dobbelaere, 1981). And yet, with the non-dissolution of the religious presence, an analysis on the decline of the religious dimension, never ceasing to produce new forms of recomposition, has developed (Hervieu-Léger, 2003; Gauchet, 2004; Tylor, 2007). Besides this analysis, some nevertheless identify the end of an entire cultural universe (Cuchet, 2018, e 2021; Delsol, 2021), while others investigate the various forms of religious revival (Berger, 1999, e 2014). As a matter of fact, the secularization theory has reached an impasse. The reasons for this are its undeniable lack of ability to interpret what remains of the religious heritage (Manent, 2019), as well as the clear laicization of the prescriptive horizon on the one hand, and the ever-resurfacing sensitivity of the devotional universe, together with the revival of religious references, on the other hand. Basically, a whole cultural universe left aside by modernity does not cease to exist and reveal itself. This process may possibly involve the transition that brought the various religious doctrines to present themselves—and thus be acknowledged and legitimated—only if they accept to show themselves as *beliefs*. In this regard, the development of secularization, depriving religions of any moral supremacy, has moved them—at least in western culture—to the private level of beliefs. As such, for decades they have been missing from the public sphere, accessing it only when the principles at stake went beyond the realm of beliefs and were set as shared *values* (Beraud, Portier 2015), absolutely necessary for a laicity that tends to stay faithful to its own principles (Marion, 2019).

Since the early twentieth century—according to Weber—the religious sphere has undergone the "pianissimo" of individual relations. There is no more room for prophets (Weber, 1919), and the acting processes they support. Religion's passage from the public to the private sphere, from *principles* guiding the believers' being and acting to beliefs, believed and experienced in the inner forum of the individual's conscience, accounts for both the loss in visibility and the growth of a network of relations that can still be found among religious communities, despite the secularization of institutions and the laicization of customs. The need to confirm one's beliefs, which inevitably grows through relations and ensures the intersubjectivity of believing that stands as a confirmation, could represent the "good reason" at the basis of the religious groups' growth, not *in spite of* secularization, but precisely *due to* it. That is to say, exactly because of the growing inadmissibility of religions as a prescriptive imperative (at least in the Judeo-Christian framework) and their revival as personal *beliefs*, this need continuously reintroduces various kinds of movements and groups, which nevertheless, since they all concern the inner level of belief, are far from establishing an associating tissue able to oppose the principles governing the mundane universe.

4   Conclusion

Current adherence to various religions, meant as a totality of beliefs and practices originating a specific kind of moral community, which includes those adhering to it, cannot be explained as the result of any kind of conditioning. The mundane nature of institutions, and subsequently of the acting inside them, the plainly lay character of the customs characterizing the different fields of life, and the significant exclusion of the religious dimension from the ordinary universe and its express manifestation in places of worship only make any explanation in terms of conditioning absolutely irrelevant.

Today as never before, those who choose to believe individually define their own "reasons to believe" and today as never before, sociological analysis cannot avoid investigating such reasons in order to understand them. Methodological individualism, which led to significant progress in the understanding of the phenomenon of religion-oriented action, also represents the one convenient methodology for analysis of the new reasons underlying not only religious commitment but also the transformation of practices, the rationalization of beliefs and their consequences on the life of those identifying with them.

Religions represent the greatest harvest of behaviors completely abstracted from a pure instrumental logic. The deeply differentiated universe of believers, broadly meant, lives and activates according to logics that originate valid reasons only for them and are absolutely illogical if considered regardless of the contents of the single doctrines. And precisely the specific nature of the "good reasons", to which the believers of any faith subscribe, provides a further reason for them to meet beyond the rites. The nature of belief itself, and religious belief specifically, leads them to seek confirmations that are only possible through the intersubjective dimension. This makes the relational dimension central and accounts for the contrary tendency that eventually represents a specific element of any religious universe, whatever reading map is adopted for reality and the specific salvation doctrine to which one subscribes. The need to see the reasons for one's believing reciprocally confirmed through the intersubjective dimension the relation inevitably activates represents an essential element in the area of shared ceremonies. But this need fuels the sense of "genuine brotherhood" that eventually characterizes the community of believers.

It seems quite clear that it is precisely on the path of MI that the analysis of the religious phenomenon has produced its best results. Only through MI, has it been possible to cross the border of the purely descriptive dimension, where the area of religious phenomena, mainly reduced to cult rituals that could be observed by ethnologists, has been confined. MI alone has made possible an analysis of the religious phenomenon that allowed its understanding beyond the purely emotional dimension and brought it to the level of the cognitive dimension, thus moving from an analysis of the *causes* provoking the believing to that of the *reasons* explaining it. Only through MI is it possible to understand how and why the religious dimension has passed from being a relic of past ages to become an unavoidable presence in western modern society, despite the disappearance of all tradition and despite the decline of the operational capacity of believers in our civilization.

References

Abbruzzese, S., & Calvaruso, C. (1985). *Indagine sui valori in Italia* [Survey of values in Italy]. SEI.

Abbruzzese, S. (2010). *Un moderno desiderio di Dio* [A modern desire for God]. Rubbettino.

Acquaviva, S. S. (1961). *L'eclissi del sacro nella società industriale*. Comunità. Transl. (1979) *The decline of the sacred in industrial society*. Harper & Row.

Aron, R. (1967). *Les étapes de la pensée sociologique*. Gallimard. Transl. (2019). *Main currents in sociological thought: Volume 1. Montesquieu, Comte, Marx, De Tocqueville: The Sociologists and the Revolution of 1848*. Routledge.

Béraud C., & Portier, P. (2015). *Métamorphoses catholiques* [Catholic Metamorphoses]. Editions de la Maison des Sciences de l'Homme.

Berger, P. L. (Ed.). (1999). *The desecularization of the world*, Resurgent Religion and World Politics. Eerdmans.

Berger, P. L. (2014). *The many altars of modernity: Toward a paradigm for religion in a pluralist age*. Walter de Gruyter, Inc.

Boudon, R. (1979). *Effets pervers et ordre social*. PUF. Transl. (1982) *The unintended consequences of social action*. Palgrave Macmillan.

Boudon, R. (1995). *Le juste et le vrai : études sur l'objectivité des valeurs et de la connaissance*. [The just and the true: studies on the objectivity of values and knowledge]. Fayard.

Boudon, R. (2002). *Déclin de la morale ? Déclin des valeurs* [Decline of morals ? Decline of values]. PUF.

Boudon, R. (2008). *Le relativisme* [The Relativism]. PUF.

Boudon, R. (2010). *La Sociologie comme science* [Sociology as a Science]. La Découverte.

Boudon, R., & Bourricaud, F. (1983). *Dictionnaire critique de la sociologie*. PUF. Transl. (1989). *A critical dictionary of sociology*. Routledge.

Boudon, R. (2012). *Croire et savoir. Penser le politique, le moral et le religieux*. [Belief and knowledge. Thinking the political, the moral and the religious]. UF.

Cuchet, G. (2018). *Comment notre monde a cessé d'être chrétien. Anatomie d'un effondrement* [How our world stopped being Christian. Anatomy of a collapse]. Seuil.

Cuchet, G. (2021). *Le catholicisme a-t-il encore de l'avenir en France?* [Does Catholicism still have a future in France?]. Seuil.

Delsol, C. (2021). *La fin de la chrétienté* [The end of Christianity]. Cerf.

Dobbelare, K. (1981). Secularization: A multi-dimensional concept. *Current Sociology, 29*(2), 3–53.

Donati, P. (2011). *Relational sociology. A new paradigm for the social sciences*. Routledge.

Durkheim, E. (1895). *Les règles de la méthode* sociologique. PUF. Transl. (1982). *The rules of sociological method*. The Free Press.

Durkheim, E. (1900). La sociologia ed il suo dominio scientifico [Sociology and its scientific domain]. *Rivista Italiana di Sociologia, 4*(2), 127–148.

Durkheim, E. (1912). *Les formes élémentaires de la vie religieuse*. Transl. (1976). *The Elementary Forms of Religious Life*, George Allen & Unwin Ltd.

Hervieu-Léger, D. (2003). *Christianisme, la fin d'un monde* [Christianity, the end of a world]. Bayard.

Gauchet, M. (1985). *Le désenchantement du monde. Une histoire politique de la religion.* Folio. Transl. (1997). *The disenchantment of the world: A political history of religion.* Princeton University Press.

Gauchet, M. (2004). *Un monde désenchanté* [A disenchanted world]. Les éditions de l'atelier.

Isambert, F.-A. (1982). *Le sens du sacré. Fête et religion populaire* [The sense of the sacred. Celebration and popular religion]. Minuit.

Lévy-Bruhl, L. (1922). *La Mentalité primitive.* PUF. Transl. (2019). *Primitive mentality.* Routledge.

Luckmann, T. (1967/1991). *The invisible religion: The problem of religion in modern society.* Mcmillan.

Manent, P. (1993). *Tocqueville et la nature de la démocratie.* Gallimard. Transl. (1995). *Tocqueville and the nature of democracy.* Rowman & Littlefield Publishers.

Manent, P. (2019). Il cristianesimo e la storia politica dell'Europa [Christianity and the political history of Europe]. In Abbruzzese S., De Ligio G (Eds.), *Cattolicesimo: l'impossibile rinuncia. Riflessioni intorno alle analisi di Pierre Manent* [Catholicism: the impossible renunciation. Reflections around Pierre Manent's analyses]. Rubbettino.

Marion, J.-L. (2017). *Brève apologie pour un moment catholique.* Grasset & Fasquelle. Transl. (2021). *A brief apology for a catholic moment.* University of Chicago Press.

Martin, D. (1978). *A general theory of secularization.* Basic Blackwell.

Mauss, M. (1905). Essai sur les variations saisonnières des sociétés eskimos. Étude de morphologie sociale. *L'Année Sociologique, IX*, 39–132. Transl. (2013). *Seasonal variations of the Eskimo. A study in social morphology.* Routledge.

Marx, K. (1847/1970). Critique of Hegel's 'Philosophy of Right'. Cambridge University Press.

McLennan, J. F. (1865/1970). *Primitive marriage.* University of Chicago Press.

Tocqueville, A. de (1835–1840) *De la démocratie en Amérique.* Transl. (2000). *Democracy in America.* University of Chicago Press.

Taylor, C. (2007). *A secular age.* Press of Harvard University Press.

Tylor, E. B. (1871/2010). *Primitive culture. Researches into the development of mythology, philosophy, religion, art, and custom.* Cambridge University Press.

Weber, M. (1919). *Politik als Beruf, J. C. B. Mohr (Paul Siebeck) Tübingen.* Transl. Science as a Vocation, Hackett Publishing Co, Inc, 2004.

Weber, M. (1920). *Gesammelte Aufsätze zur Religionssoziologie.* Mohr. Transl. (1993). *Sociology of Religion.* Beacon Press.

Weber, M. (1920/1946). The social psychology of the world religions. In H. H. Gerth and C. Wright Mills (Eds.), *From max weber: Essays in sociology* (pp. 267–301). Oxford University Press. [Originally: pp. 237–268 in Weber (1920), vol. 1].

Weber M. (1922). *Die protestantische Ethik und der Geist des Kapitalismus.* Tübingen. Transl. (2002). *The Protestant Ethic and the Spirit of Capitalism: and Other Writings.* Penguin.

Immigration, Identity Building and Multiculturalism: A Methodological Individualism Approach

Simon Langlois

International immigration has given rise to identity and cultural groupings seeking recognition in developed societies. At the same time, many social groups historically ostracized—think of First Nations and Indigenous Tribes in America and Oceania—are demanding respect for their Aboriginal rights and challenging the forced marginalization they have suffered. Mechanical solidarity—based on similarities and on the community of feelings and beliefs—has not disappeared with the advent of organic solidarity resulting from the division of labor in the modern era, well analyzed by Émile Durkheim in *De la division du travail social* (1893). How can we explain the attachment to identity of origin and the development of such ethnocultural groups in developed countries open to international immigration? How can we explain the diversity of social integration policies from one country to another? The search for answers to these two questions has received less attention.

The behaviors of individuals who expatriate are diversified and several scenarios unfold. Some are at odds with their environment of origin and choose to assimilate purely to the majority within their host society. Others want to maintain as much of their national culture as possible—their mother

---

S. Langlois (✉)
Department of Sociology, Laval University, Quebec City, QC, Canada
e-mail: Simon.Langlois@soc.ulaval.ca

© The Author(s), under exclusive license to Springer Nature
Switzerland AG 2023
N. Bulle and F. Di Iorio (eds.), *The Palgrave Handbook of Methodological Individualism*,
https://doi.org/10.1007/978-3-031-41508-1_7

tongue, religion, etc.—and pass it on to their children in their new environment. Others go even further and identify with an ethnocultural minority that promotes difference in the host society. The perspective of methodological individualism (MI) opens up more fruitful avenues for the explanation of these different scenarios and, more broadly, to explain the social processes at work in international migrations and in the emergence of identity groupings at the origin of multiculturalism. The conceptions of methodological individualism being diverse (N. Bulle, 2020), we will distinguish three complementary perspectives of analysis that explain how ethnocultural groupings are formed from the action of individuals and what the implications are that their development brings on life in society. The three perspectives from the MI focus on the cognition of individuals, their membership in social networks and their rationality.

The first perspective comes from cognitive sociology and it explains why individuals who expatriate tend, under the effect of a cognitive attractor, to regroup with other individuals who resemble them by various traits (religion, ethnicity, culture, national traits, etc.) in their new living environment. Far from being isolated, the individual fits into a structured set of relationships whose properties—the second perspective that we will examine—are determined in the analysis of social networks. But considering the position of individuals in such a structure is not enough. It is also important to examine—this will be the third perspective—the rationality or the strong reasons for action of individuals.

Social action must deal with the context in which it takes place and with the constraints that sometimes weigh on them. A number of them are typical of the immigration process. The MI recognizes the existence of systemic and institutional constraints but argues that they are *sui generis* entities, existing independently of individuals (Bulle, 2018). Moreover, the MI enables one to go from the micro to the macrosociological level to explain the diversity of collective identities that result from the migratory flows, the differences of the policies of integration of migrants in the developed societies, the emergence of cultural conflicts and the process of national consolidation in States open to immigration. We will assert that the causal relationship between the microsociological and macrosociological levels is circular in the longitudinal perspective, referring to the Canadian example.

Following the MI perspective, we propose elaborated models of individual behavior that allow to explain the immigration processes and to properly understand the behavioral consequences of different social structures, and we insist on the necessity to establish links between contextual constraints and individual action.

1 Social Circles and Cognitive Attractor

The individuals interact with others to ensure their well-being. They belong to different social circles, a concept put forward by Célestin Bouglé (1910), from the nuclear and the extended family, the village, the city, the tribe, the region or the nation. This belonging makes sense to everyone, as evidenced by the narratives specific to each of the circles, from the family story told by the grandparents to the official national history. How does this shared sense develop? Cognitive sociology shows that there are mental invariants specific to the human species and that a cognitive attractor emerges in social circles. Consider an observation made by Claude Lévi-Strauss on some ancient tribes for whom humanity ceases beyond the immediate group: "But for vast fractions of the human species … humanity ceases at the borders of the tribe, of the linguistic group, sometimes even of the village; to such an extent that a large number of so-called primitive populations refer to themselves by a name that means 'men' … implying that the other tribes, groups or villages do not participate in the virtues—or even of the human-nature—but are at most composed of villains, monkeys or lice eggs" (Lévi-Strauss, 1973). Neuroscience explains how this cognitive mechanism is formed by which individuals give meaning to their behavior within various social circles. Affiliation with groups creates boundaries of meaning that push individuals to value themselves and to devalue those who are not part of them. " (…) such oppositions are observed in all cultures and at all times. How can we explain this? They are a consequence of the grading operator. They result from a cognitive mechanism. Collectives build up by differentiation in a control strategy and they strengthen their internal identity" (Degenne, 2020, 109).

Is the cognitive approach tainted with psychological reductionism, reducing the social to the mental? This objection is consistent with Raymond Boudon's criticism of the work of D. Kahneman and A. Tversky on cognitive biases. "But the origin of these biases remains mysterious. We know that they appear in certain circumstances, but we do not have a clear vision of the conditions that favors or disadvantage their appearance. The notion of bias *recognizes* them; it *designates* them; but it does not *explain* them" (R. Boudon, 2003, 10 underlined by RB). This criticism is too radical. Contemporary research in neuroscience, cognitive psychology and social sciences (anthropology, sociology, experimental economics) has paved the way for the explanations Boudon wants. It is not a question of biologizing the social but rather of considering the biological basis of human behavior. Cognitive mechanisms refer to typical modes of action and thinking, far from psychological reductionism (Di Iorio, 2015, 2020; Bronner, 2006, 2011).[1] More generally, the desired explanation lies in the consideration of the long term. "Today we are talking about the co-evolution of the brain and the natural, and social environment of man. Sociology compatible with this position of principle is necessarily dynamic, centered on processes and transformations induced in feedback by social facts on representations and ultimately on the brain" (Degenne, 2020, 106).

To illustrate the importance of the longitudinal perspective, let us take the prisoner's dilemma. In this classic game, the structure of the interaction system induces players to defect and the pursuit of personal interest plays against collective action. Mancur Olson (1965) argues that the way to obtain cooperation between actors and the way to counter the phenomenon of the stowaway is either to offer personal benefits to participants, or to compel them to cooperate for their own well-being as seen in the unionization of workers in a closed shop in North America. Similarly, the small size of a group of actors favors collective action because they can interact with each other and work together. However, these individualistic strategies stand out from a transversal perspective. But things are changing radically in the longitudinal perspective, because cooperation can indeed be established as a model of action because social actors end up perceiving the disadvantages of defection associated with repeated gestures in time. The defection strategy does not hold when the game is played several times, either because a leader eventually emerges and asserts the advantage of cooperation, or because, with the repetition of the game, the individuals note the deadlock of defection and therefore lean toward cooperation, which thus becomes a cognitive attractor. Time favors the emergence of a structure based on the give-and-take model and coordination appears as an emerging phenomenon if the series of interactions is sufficiently long. The idea of sharing develops through communication between the members of the group (Terence Deacon, 1997). In other words, they develop strong reasons to cooperate and a new rationality emerges. The benefits of cooperation over time give rise to social norms and ethics among the members of a group (Axelrod, 2006; Bowles and Gintis, 2011) and the obligation acquires a social character that becomes necessary for the cohesion

---

[1] Neuroscience research has shown that the brain has an exacerbated sensitivity to conflict. Conflicts get attention for strong reasons (defence of democracy, etc.)—especially in the context of war in Eastern Europe, at the time of writing these lines—but this heightened sensitivity also helps to explain the place given to clashes in the media and to antagonisms of all kinds without social or political importance which nevertheless occupy a significant part of our attention, as shown by the compilation of metadata on the Internet.

of the group, an idea already developed by Célestin Bouglé: "From the association of men emerges a force endowed with a power of pressure as well as of attraction" (1922, pp. 29–30).

A more complex social configuration emerges under a coordinating effect, as seen in a social movement which after a certain time leads to an organized form of collective action (creation of a party, a union, etc.). Belonging to a group then takes on a meaning for individuals and they establish boundaries between them and others that involve entry into a control regime and the elaboration of rules of conduct. Thus, some political parties—for example the Communist Parties—have pushed very far the establishment of borders and control over their members. Social forms established over the years from privileged social bonds (family, tribal, ethnic, national, etc.) facilitate the elaboration of a shared sense that defines the collective situation of individuals and is likely to lead to perverse effects of exclusion, of rejection of others, up to the assertion of differences (us/them) or even a conflictual opposition (us/you).

For individuals who expatriate, the cognitive attractor gives meaning to their group membership and encourages them to join significant others that resemble them in order to face the challenges of setting up in a new environment. Similarly, members of groups already established (village or regional communities, nations, etc.) often welcome newcomers with reluctance, especially if they distinguish themselves by significant differences (languages, religion, socio-economic status, etc.). Let us give an example. Between 1880 and 1929, more than 800,000 French Canadians migrated to the New England States to escape poverty and overcrowding in rural Quebec. They were called "Eastern Chinese" because they accepted lower wages and were looking to build *Small Canadas* (Yves Roby, 2000). The concept of common humanity being, at best, abstract, such oppositions can notably be countered by sociological processes. "The other one is disturbing because it calls into question the meaning behind the identity group. The only thing that can prevent the crystallization of oppositions between identity groups is social mixing, interconnection, overlapping between communities and the emergence of shared interests (…)" (Degenne, 2020, p. 172).[2]

However, the cognitive attractor is not enough to explain the gatherings and concentrations of migrants sharing a unity of meaning or the crystallization of oppositions. Other complementary models from the perspective of

---

[2] According to this interpretation, the hypothesis of a cognitive attractor deserves to be explored further to explain the polarization of contemporary debates on the web and in the public domain, for example. Research has shown that social network users tend to associate with their peers and focus on media and information that reinforces their own social representations and beliefs.

MI must be considered in order to arrive at a valid explanation. The mechanisms of cognition form the biological basis of the behavior of individuals, but a sociological explanation is necessary, considering that individuals fit into structures and systems of action, starting with the web of social relationships that connect them to others. Because individuals are reflective, building relationships with others—after social or geographical mobility, intermarriage, etc.—successfully counteracts the exclusionary effects of the cognitive attractor, just as the repeated experience in the prisoner's dilemma game manages to bring out cooperation instead of defection.

2 Social Networks: Potential for Informal Social Relations

As the life of individuals expands, diverse social relationships are established. These are essential to the quality of life, as the Covid-19 crisis in the years 2020–2022 showed, by forcing them to significantly reduce their social contacts. Information, ideas and opinions circulate through social relations, but also goods and services that contribute to well-being. Social links place individuals in a typical structure: a social network. Such networks are formed on the basis of the connections forged within the family, in educational institutions, in workplaces, in associative life, in the participation of leisure activities, in religious practice, in political commitment, in short, in all spheres of life. Establishing connections with others is the product of each individual's activity. Rather than focusing on social categories such as age, sex or occupation, the sociological analysis of networks emphasizes the study of their structural properties (density, structural hole, centrality, etc.) in order to explain a large number of individual behaviors, but also the emergence of ethnocultural groupings.

We will briefly recall the contours of this particular social structure[3]. In his pioneering work, Vincent Lemieux (1982) defines the social network as an unofficial structure of relationships and connections between individuals as opposed to the official structure of an organization or apparatus. Social network analyses distinguish between the personal network and the total network. The personal network is built around an actor (ego), made up of all the direct relationships that the latter has with other individuals, but also—importantly—it is made up of all the relationships that they have with others.

---

[3] On the paradigm of social network analysis, see notably B. Wellman and Berkowitz (1998); A. Degenne and M. Forsé (2004). [See also Valerie A. Haines in this Handbook].

Anthropologist Robin Dunbar (1996) believes that the brain can physiologically interact with a limited number of other human beings: kinship relationships, friendly and neighborhood relationships, relationships with fellow students, relations with co-workers, etc. The scope of our social world is limited and, beyond a certain number of people estimated by Dunbar to about 150 people, it is not possible to maintain lasting relationships without institutional support. Beyond this number, the social cohesion of a group requires that the authority be institutionalized, which paves the way for the development of collective norms and the establishment of powers that officially regulate social relations. However, as each person is connected to more than a hundred others, a large constellation of direct and indirect connections is built around each ego. The total network is the overall configuration of all the connections that exist between a certain number of individuals. The triad and the clique are examples in graph theory.

Several traits characterize a social network as a structure and distinguish it from an organization. We propose five: (1) the social network has an *unofficial* character; (2) the network offers a *potential* of social relations and connections; (3) the network does not *regulate* its border; (4) the actors in a network are not *specialized*; (5) the connections between the actors become relevant when they have an *influence* on ego. The social network is part of the MI paradigm because its structure and its properties emerge from individual action as illustrated by the five selected traits. We will briefly explain each of them.

Social relations in a network have an unofficial character, as opposed to official relations which take place in an apparatus or an institution. Of course, a network and an organization can overlap, but it is the non-institutionalized nature of relationships that distinguishes them. An informal social network can also be found within an institution (such as a university) and designate a set of unofficial connections between different members pursuing the same objective. Everyone will easily find examples of such networks in the organization in which he/she works. In other words, the social network has no borders, unlike an organization or an institution (nuclear family, school, company, church, etc.).

The social network is made up of potential connections, and many of them are dormant. This is the case for relationships that exist between graduates of an educational institution that can be reactivated after several years. But above all, ego is potentially connected to the relationships of the people with whom he/she was connected. The friends and acquaintances of the latter give rise to the potential of connections which can bring information on job vacancies, on a house, etc. Irish people who left their island in the nineteenth century were potentially connected by a chain of intermediaries more or less long to other compatriots established in New York, Chicago or Montreal. Co-workers of distant cousins could give them information about the possibilities of work when they got off the boat. The structural features of a network such as density or extent are crucial to belonging to their new environment. Stanley Milgram's (1967) pioneering research on the "small world" illustrates this network property as a possibility for relationships. Milgram showed that the number of intermediaries required to connect a given person to any other on the planet was on average six. However, the Internet and the Web today would have reduced this number.

The social network does not regulate its borders while the organization seeks to control its components and its public notably through relations of authority. In a network, on the contrary, there is good "neighborhood" coordination of actors and their connections, but no one is officially or unofficially responsible for coordinating this coordination. While connections are not neutral and power relationships can occur in a network—such as a prostitution network or a network of illegal immigrant smugglers—they are not formalized in established roles.

Actors in a social network are non-specialists. They play several roles depending on the substance of the relations (exchanges of goods, services, information, etc.) and the circumstances of their activation. Ego will be able to find a new apartment or even get information about a position available through a former study colleague who found it in his/her own network.

The connections between actors become relevant when they have an influence on ego; otherwise, they remain dormant. The Italian from Sicily, who has a compatriot from the same village living in Montreal, is in potential contact with the individuals this compatriot knows in Canada; he will be able to obtain useful information and services to settle there as immigrant. Other Italian immigrants will gradually follow by taking advantage of their own networks and thus give birth to a community of meaning (as defined above) in an urban area called "Little Italy." Such neighborhoods have been established in many large immigrant-receiving cities. In his famous article on the strength of weak ties, Mark Granovetter (1973) empirically analyzed the effect of redundancy in social relationships.[4] On the one hand, network will be redundant when the various connections between its members overlap, as the information of some may also be known and shared by others. On the other hand, longer connections will be less redundant and more distant,

---

[4] "The strength of weak ties" is the scientific article that has been most often cited in sociology.

creating a weaker link with Ego. Granovetter thus showed that weak ties proved to be effective in job search for highly skilled workers.

Social networks have formal properties that depend on the substrate (or object) of social relationships—whether one is interested in the flow of information, people or goods—such as extent, density, centrality, articulation, etc. Thus, density is measured by the ratio of actual to potential links in a network. The reference to this property of a network is not new in sociology. Émile Durkheim made an explicit reference to the density of social relations in his explanation of anomic suicide, in advance proof of the importance of this structural property to explain social action.

Having said that, let us go back to immigration. The cognitive attractor encourages the individual who intends to immigrate to turn to social relationships that have a meaning for him, either before migration or once arrived in a new environment. The cognitive attractor gives meaning to a social relationship by establishing a stronger bond of trust with some individuals than with others[5]. The explanation of social action requires going further than the consideration of the sense of belonging to a social circle and the consideration of being part of a structured set of connections between individuals. This is not sufficient. We must also examine the reasons that lead ideal–typical individuals to act. Why immigrate? How to explain the choice of one place rather than another? Answering these questions also calls for considering the rationality of the actors.

3   Immigrants' Strong Reasons

The general sociological theory proposed by Raymond Boudon (2003, 2007) states that rationality takes three forms: utilitarian, cognitive and axiological.[6] Action is rational if the social actor considers it useful or beneficial. But instrumental rationality is not the only type to consider. The actor also relies on beliefs and knowledge (cognitive rationality) or he can be inspired by values and prescriptive beliefs not consequentialist (axiological rationality).[7] The reasons of individuals are not reducible to strictly subjective opinions but they go back to typical ways of thinking and acting, to a shared meaning.

---

[5] Refugees who fled the war in Ukraine in 2022 had no choice of destination, but numerous daily reports from journalists highlighted the importance of social networks activated by many of them.

[6] See Jean-Michel Morin, *Ordinary Rationality Theory (ORT) according to Raymond Boudon* in this Handbook.

[7] For details on axiological rationality see Sylvie Mesure in this Handbook.

Space being limited, we take it for granted that this theory of rationality is well known.

A mixture of strong reasons explains why immigrants establish identity groupings and their various combinations explain the different evolution taken by ethnic communities here and there. The closer the affiliation relationship is, the stronger the identity markers and the more demanding identity control becomes. This is the case, for example, in China Towns, which are maintained over several generations, or in urban areas where traditional migrants of Jewish origin gather. On the other hand, other groups of immigrants will dissolve more quickly and their members will integrate or even assimilate into the host society by geographical mobility, social mobility or as a result of inter-ethnic marriages. The consideration of temporality is essential in the empirical analysis of immigrant collectives. In the absence of strong identity markers, linked for example to language or religion, the passage of time leads to a thorough integration of the majority in the host society. For example, many immigrants of Polish, Italian or Portuguese origin have gradually merged over the generations in the majority in the United States or France.

The explanation of the different scenarios that immigration takes—resulting from a cognitive attractor, inscribed in a reticular logic and based on reasons—must also place the actions of individuals in context. "The general theory of rationality is not disembodied. It assumes *localized* rationality" (Boudon, 2011, p. 70, emphasized by RB).

4 The Social Action Located

In his book *De la démocratie en Amérique I*, Alexis de Tocqueville made two observations on immigration. "It is not the rich who immigrate," he wrote, adding this relevant remark to our statement, "immigrants are in a similar situation." The similar situation includes shared constraints and the MI, taking them into account, does not locate the action of individuals in an abstract social space. It is rare that an individual is able to act only at one's whim. But this does not imply that social constraints determine individual action. These constraints delimit the field of the possible, not the field of the real (Boudon & Bourricaud, 1990, p. 307). The rationality of the social actors takes place in a context that parameters the decisions but does not determine them. Such constraints may change over time and be negotiable depending on the situation. For example, immigration policies regulate the selection of candidates for entry in all countries according to different approaches.

We will identify some contextual elements likely to explain why and how individuals who expatriate come to create ethnocultural communities—more strongly established in certain countries, but not in others—and why such groupings evolve differently in the long run. Six elements are likely to characterize the analogous situation evoked by Tocqueville: the shared causes of migration, the weight of the numbers, the diversity of origins, the proximity of the territory, the channels of communication and the historical links. MI models must take into account these constraints in order to explain immigration.

First, some shared causes—such as hunger, poverty, violence, war, lack of freedom, etc.—are likely to lead to the migration of a large number of people with a common origin and to regroup in the country of entry. This was the case for Irish immigrants who migrated to Canada or the United States in the nineteenth century or for boat people during the Vietnam War.

A second feature of the context to be considered is the weight of numbers. Migrants do not disperse randomly on the planet and do not disperse uniformly in a host country. Tens of thousands of Ukrainians migrated to the plains of Western Canada in the early twentieth century, and the country has the largest Ukrainian diaspora outside of Russia. Because of the many physical and geographical similarities with their country of origin, Canadian territory was a real cognitive attractor for them, an element put forward by the recruitment policy of the Canadian government of the time. In 2022, millions of Ukrainians were forced to flee Russian tanks, and the cognitive attractor and social networks have encouraged many to settle in Canada to rebuild their lives. Other cases illustrate the weight of the numbers. As its size grew, the Haitian community in Montreal attracted even more new immigrants, and so did immigrants from Turkey to Germany.

The diversity of immigrant origins is a third element that has grown in importance over time. The United States, Australia and Canada were first opened to immigrants primarily of British origin (English/Irish/Scottish/Welsh), then to immigrants from the European continent. Canada imposed a special tax on Asian immigrants in the nineteenth and early twentieth centuries and severely restricted the entry of Black people until the 1960s (under the guise of climate mismatch). Subsequently, immigration has become very diverse, with hundreds of different nationalities living side by side in these countries. On the other hand, this diversification of origins is more or less pronounced. For example, France has a high concentration of immigrants from Magreb and its immigrant population is much less diverse than it is in Canada. Britain, for its part, has welcomed large contingents from the Indian continent. We will see later that the diversity of origins plays a role in the direction that multiculturalism has taken in the various states in the twenty-first century.

A fourth element emerges that gives rise to the formation of ethnocultural communities: the proximity of territories that promotes the maintenance of links with the citizens of the country of origin. Millions of citizens of Mexico have settled in the southern United States as well as in California, and Spanish has become ubiquitous in the public space of these parts of the country. Proximity favors people coming and going from both sides of a border, etc.

When the number of applicants for immigration increases, channels of communication are being set up and channels through which migrants pass facilitate the concentration of individuals from the same geographical area and the increase of their number on the soil that welcomes them. This is the case for immigrants to Britain from India and Pakistan in the twentieth century, for example.

Finally, historical links between countries must be taken into account. France has welcomed many nationals from its former colonies in Southeast Asia and Francophone Africa, not to mention immigrants of Maghreb origin. Fleeing starvation, Irish migrants turned mainly to the United States and Canada in the nineteenth century.

These context elements, which vary from one country to another and from one epoch to another, illustrate the importance of considering the social processes that fall under Methodological Individualism in their context. Finally, it should not be forgotten that the typical immigrant must also solve practical problems that he shares with others, linked to a context, that will affect his action. "Many decisions in ordinary life are therefore not based on a calculation of marginal utility or on a call to values, but on a process of practical resolution of the different kinds of constraints that are exerted on him [the subject in situation]" (Pharo, 2011, p. 306).

Once immigrants arrive in their host country, a double social process is at work: they must adapt to their new environment and integrate into it, but by their presence, by their number and by their diversity, they will in turn help change the society in which they settle. A process of circular causality—typical of Methodological Individualism and not found in other theoretical perspectives—then takes shape, which it is important to specify at greater length.[8]

---

[8] On the relation of circularity between the microsociological and macrosociological dimensions see J. Coleman (1990) and A. Bouvier (2011). Let us recall that such a relationship characterizes Max Weber's analysis of Protestant ethics and the spirit of capitalism. Capitalism emerges as an aggregate effect of the action of Calvinist entrepreneurs and in return capitalism modifies the lifestyles and values of individuals, as pointed out by Daniel Bell (1976).

5 Circular Causality Between Microsociological and Macrosociological Dimensions

From the epistemological point of view, the analyses according to the perspective of MI range from the individual to the collective or from the sociable to the societal. "The societal is all that concerns the construction of a society, while the sociable is the society from bottom to top that remains at the level of neighbors, of reciprocal acceptances" (Lemieux, 1976). Sociability characterizes relationships established by individuals within the family, in workplaces, and in living environments and connections rooted in their social networks. The societal refers, on the macrosociological level, to the groupings that result from the aggregation of individual behaviors and to the collective traits of a society such as the formation of ethnocultural enclaves, the advent of multiculturalism and even the rebuilding of national identity. That said, how are the links established between the two levels? By adopting a longitudinal perspective, the circular causal link appears clearly.

Let us first consider the immigration process at the microsociological level from the perspective of MI. The cognitive attractor guides immigrants to places that make sense to them. Second, taking personal networks into account is essential because migrants are not isolated atoms and they are connected to other individuals in a structured set of social relationships, whether it is the expanded kinship network, the social network of their knowledge or the community network of people with whom they share cultural or national traits. Finally, there are strong reasons for some individuals to regroup with significant others and with people who resemble them. These reasons can be instrumental (access to resources that facilitate the installation in a new country), cognitive (knowledge and information about the host country) and axiological (promotion of language or religion).

Microsociological processes lead to a more pronounced community involvement when arriving in the host country. It is legitimate to continue to speak the mother tongue with loved ones, find rewarding to use it in the immediate living environment—for example, in ethnic stores in large cities—and to attend a place of worship where people have the same religious or cultural affiliation. Modern means of communication—Internet, websites, cable TV, low-cost travel, etc.—facilitate the maintenance of links with the country of origin, its language and culture. This accentuates the tendency of the first generations of migrants to retain a justified attachment to their culture of origin. This community inclination, however, is not only a matter of elective affinity. Individuals are more rooted in a community when they find an interest in this attachment, when they have reasons to develop it. Immigrants get the resources to deal with the challenges of a fresh start. Some services are delivered on the basis of ethnic affiliation by companies, religious institutions and even state (social services for the elderly in Chinatown, for example). Immigrants may also turn to their communities if they face various social problems or if they are discriminated against, as is the case for some visible minorities. The community conceals signs of distinction from the majority that can have a value for the individual, be immediately useful or even give the recognition sought. "Each of the actors participates in the formation of these signs, *Nolens volens*, because they feel that the pseudo-institution they constitute fulfils an interesting identification function for the community to which they belongs; it increases their visibility and thus their power; it is therefore useful for each one. This is referred to as an unintended "coordination effect"" (Boudon, 1999, p. 391). This coordination effect is typical of MI.

Normally, strong community involvement fades over the years, when immigrants are better integrated, especially in the case of the second and subsequent generations. Over time, the social and geographical mobility of immigrants, the education of their children or inter-ethnic marriages lead to their integration into a broader whole, promoting the removal of the culture of origin and the ethnocultural collective. This process of integration and even assimilation spans several years and generations and it will be more or less pronounced depending on whether the ethnocultural collective creates or not boundaries of meaning that push their members to differentiate themselves by a strategy of control as seen above. That said, ethnicity remains an important marker for many immigrants—especially those belonging to visible minorities—that is likely to lead to unequal opportunities in many fields (education, housing, jobs, etc.).

At the macrosociological level, international migration is gradually transforming the social fabric of countries open to immigration. Thousands of individual decisions lead to a radical change in the collective. Newcomers tend to settle in cities where they find jobs, services (ethnic shops, etc.) and access to cultural institutions familiar to them (church, mosque, school, etc.). Ethnic neighborhoods are emerging where immigrants of the same origin are in large numbers. Their presence helps to accentuate the differences between large urban centers—places of ethnic and cultural diversity—and small towns and the rural world, which have remained more homogeneous.

Mutations resulting from immigration are not only visible morphologically. They can go so far as to influence the refoundation of the nation as well as the narrative in which it is expressed, and consequently provoke sometimes lively debates. The nation is a group of individuals united by history, shared values, traditions and institutions that impose themselves on them but do not determine them. It is a totality of active individuals whose traits change over time. "Institutions and traditions are the result of human actions and decisions, and can therefore be modified by human actions and decisions," wrote Karl Popper (1945, 14). Analyzed from the perspective of the MI, the nation is changing under the influence of strong international migrations for decades and the national refoundation is intelligible by the presence of newcomers. The case of the United States is an example of this process. Michael Lind (1994) distinguishes four phases of identity in the history of this country over five centuries following the arrival of millions of immigrants: the Anglo-America, the Euro-America, the multicultural America and, finally, the contemporary trans-racial America that he presents as a new "American ethnic identity" without a hyphen.[9] In this perspective, national identity is not based on the internalization of a specific cultural model supposed to characterize the "basic personality" or on collective concepts such as the "soul of peoples" (André Siegfried) or the "spirit of the people" (Volksgeist).

Canada provides an example of several moments of national refoundation (Langlois, 2018). The country was first populated by a large number of Aboriginal nations, scattered over a vast territory, who were marginalized during the colonization by the Europeans from the seventeenth century. New France gave to Canada a francophone identity for more than 200 years until the colony's surrender to England in 1763. From then on, Canada opened up to New England Loyalists fleeing the American Revolution, and subsequently to immigrants from Britain. By 1840 there were 1.8 million English Canadians and 1.4 million French Canadians concentrated in Quebec. National duality and biculturalism became a reality and a major conflict took place between the two nations—English and French Canada—defined linguistically. In the twentieth century, however, international immigration radically changed the landscape of Canada and Quebec with the arrival of several million immigrants who became part of the majority in the English-speaking provinces. Canada has gradually redefined itself as a multicultural state, on the one hand, but also as a multinational state, on the other, recognizing the Quebec nation as well as the aboriginal nations. For its part, the French-Canadian nation had broken out to give birth to the Quebec nation and to Franco-Canadian communities in the English-speaking provinces.

---

[9] These four phases do not take into account the fact that the American continent has been inhabited by a large number of indigenous nations whose claims and desire for recognition resurface in the new millennium, such as the return of the repressed.

The Canadian case provides an example of circularity between the microsociological and macrosociological dimensions from a longitudinal perspective, the process of which can be characterized as follows: immigrants settle in a State which already has a common culture and a historically constituted identity and, by their presence in considerable numbers, they modify over time the morphology of the country and cause it to review the representation it has of itself. By their number, they also force the long-established majority to review discriminatory practices and combat inequalities resulting, among other causes, from differences due to the phenotype. But in return, immigrants must adapt to a new environment and integrate into a society that will lead them—and their offspring—to move away from their national culture of origin in most cases in the long term. This circular process is not the same everywhere. However, states that are open to immigration have faced, and are now facing, to varying degrees, the challenges of reconciling common citizenship and the diversity of their populations, hence the emergence of multiculturalism.

6 Multiculturalism, Universalism and Citizenship

Multiculturalism takes on two complementary meanings.[10] In the first sense, it refers to the coexistence of several cultures in the same country. Multiculturalism thus characterizes the diverse social morphology of a country open to international immigration or the morphology of multi-ethnic countries as in West Africa. The second sense identifies a socio-political system of citizenship.

According to philosopher Will Kymlicka (1995, 2001), multiculturalism promotes a "new culture of society," such as public awareness of racism and cultural diversity, the establishment of positive action programs aimed at increasing the presence of visible minorities in institutions, the modification of school curricula in order to give a place to the contribution of immigrants in the national history, etc. The inclusion of certain cultural characteristics is intended to promote integration into common institutions. According to this argument, it is precisely the failure to adopt these accommodations that would cause immigrants to withdraw into their community. In 1985, the Supreme Court of Canada circumscribed the concept of reasonable accommodation, which is now well known around the world. This concept is based on certain principles for managing situations that involve conflicts such as

---

[10] For a discussion of multiculturalism, see Charles Taylor (1992), Elke Winter (2011b) or Meidad Bénichou (2017). For a critical review, Steven Vertocec and Susanne Wessendorf (2010).

respecting the laws of the country and human rights and seeking to balance the rights and duties of the various parties involved through negotiations.

Multiculturalism has been the object of many criticisms, the main one being that it locks ethnocultural groups in their difference, paving the way for communitarianism. The unexpected result of the excesses of multiculturalism would thus lead to a weak social integration. Canadian writer Neill Bissoondath (himself an immigrant) was concerned about this in his book *Selling Illusions* (1995), arguing that multiculturalism overdeveloped differences rather than promoted integration into the host society. Considered from the perspective of Methodological Individualism, the policy of reasonable accommodations, if it can facilitate the integration of individuals, may as well lead to unintended coordination of identification effects and facilitate the emergence of the dark side of communities. It is not excluded that accommodation in the name of social integration will have the effect of reinforcing the feeling of belonging to particular cultural subgroups, thus creating the opposite effect of the one expected. Such challenges to multiculturalism were acute at the end of the twentieth century in several countries, notably Great Britain and France, but also in Canada (C. Joppke, 2004; S. Vertosec & S. Wessendorf, 2010).

Multiculturalism raises an important question: how to reconcile citizenship and identity in states characterized by the diversity of their population? Citizenship is a common political status that goes beyond cultural, ethnic or socio-economic differences. Open to all, it is a political framework that ensures solidarity transcending any link of identity. In principle, it is blind to ethnic, cultural and socio-economic differences. But citizenship must also deal with identity in a society with a diverse population, as well as with inequalities that specifically affect immigrant populations. However, each State offers a specific answer to the question of reconciliation between citizenship and identity and several models of social and national cohesion are possible. To illustrate that, we will briefly examine the American approach, the French approach and the Canadian approach.

In the United States, respect for the Constitution, consideration for the American flag and attachment to the founding values of the nation mentioned in the preamble of the Constitution—freedom, unity, justice, defense, tranquility and the pursuit of happiness—are the foundation of American citizenship. These elements even form a kind of civil religion, according to Robert Bellah (1967). The American national sentiment reconciles the common national identity and the identity of citizens who remain attached to their ethnicity. "New immigrants identified strongly with the country and were proud to become citizens precisely because it did not require them to fully assimilate culturally. A broader concept of citizenship has absorbed rather than excluded ethnicity" (Lilla, 2017/2018, 77–78). The American motto—E Pluribus Unum—reflects the desire to build a common identity based on the diversity of origins.

Respect for the rights of each person is, however, a difficult ideal to achieve in view of the effects of domination, historic conflicts, reminiscence of racial discrimination, etc. Inequalities persist and discrimination affects certain ethnic groups. As victims of racism, Black Americans have had to— and still have to—fight for respect for their civil rights and to "force America to live up to its principles," (Lilla 2017/2018, p. 79). More broadly, many individuals have good reasons to be critical of the inequalities they are victims of in the United States. The question of respect for the rights of minorities belonging to different social groups—defined by ethnicity, gender, sexual orientation, etc., and not only by phenotype—developed in the wake of the Black people's struggle for civil rights. Such claims based on identity traits extended in other democratic states, especially on the European continent, giving rise to lively debates.

Universal rights are abstract and even utopian in the eyes of people facing injustices and inequalities that persist (access to prestigious schools, good jobs, fair wages, quality housing, etc.). From the point of view of the MI, the Community is closer than the universal in the eyes of individuals who are victims of inequalities because it establishes rights on specific differences. The solutions envisaged—such as quotas for admission to schools or positive discrimination—then reach the immediate interests of individuals.

For its part, France is the motherland of republican universalism which should in principle benefit all citizens regardless of their various affiliations. Many immigrants to its territory, especially those from the Maghreb, have to deal with different forms of inequality of opportunity. French society is also confronted with the inequalities faced by minorities, hence the debates on specific political measures to counter them. Unlike the United States, France refuses to classify its citizens in censuses according to their ethnicity or religion. However, in practice, French society has become increasingly concerned about cultural diversity in its territory, as shown by the interventions of the Secretary of State in charge of immigrants, the obsolescence of the concept of assimilation mentioned in Article 69 of the Nationality Code, the creation of the Agency for the Development of Intercultural Relations and the High Council for Integration, the establishment of a Diversity Observatory in 2007, etc. On the other hand, criticisms of multiculturalism are sharp, as illustrated by expressions like "malaise dans la francité," or "religion du multiculturalisme," while intellectuals conclude that the social world is "fractionalized" (Dubreuil, 2018) and are worried about "identitarism" (Heinich, 2021).

Canada is an example that has pushed forward the definition of multiculturalism and the (re) definition of citizenship. But over the years, the changing nature of international immigration to Canada has contributed to changing the vision of multiculturalism because the origin of its immigrant population is increasingly diverse. For example, Canadian government assistance programs favor funding multi-ethnic coalitions over organized minorities as groups. In addition, many immigrant organizations reject ethnocentrism and express their desire to integrate into a multicultural but non-assimilationist Canada. A new Canadian nationalism is emerging. "While multiculturalism remains an important dimension of Canadian nationalism, this assertion no longer necessarily leads to the accommodation of ethnic groups and the demand for collective rights" (Winter 2011a, p. 52). According to Winter, "a National Community with a large C" would henceforth characterize English Canada as evolving toward a kind of liberal culturalism and "decommunautarization." A new vision of multiculturalism is being developed in English Canada that aims at equality in the context of the growing diversity of its population in order to promote the civic participation of individuals, particularly those defined as "racialized" (Helly, 2002). It is defined less in terms of rights for categorical groups and more in individualistic and liberal terms.

However, Quebec has a different approach than English-speaking Canada closer to many European countries. Open to diversity, Québec society favors interculturalism, which recognizes the contribution of newcomers while protecting the collective rights of the francophone majority (Bouchard, 2012). The Quebec government has adopted legislative measures (language laws and the obligation for immigrants to send their children to French primary and secondary schools, for example) in order to make French the common language of society. The model favored by the Quebec government fosters a concept of secularism that is close to the French model and is different in many respects from Canadian multiculturalism. Significant national refoundation, parallel to that observed in English Canada, is also observed in Quebec.

Several conceptions of multiculturalism now coexist in developed societies. The MI perspective identifies what separates multiculturalism (in its different versions) and universalism and what limits the convergence between the two approaches. Identity movements and communitarian philosophy consider

170   S. Langlois

that values are interwoven in particular cultures that need to be recognized. Community norms easily converge with the immediate interests of social actors participating in an identity movement. However, for Raymond Boudon, values are not arbitrary and it is possible to understand why some of them took place at a given time when they have no basis in contemporary times. He criticizes the idea that it is impossible to reject certain practices or values in the name of respect for minorities. "Indeed, 'communitarianism' is based on a solipsist conception of the group, which in turn is legitimized by the relativistic principle of 'to each his truth,' to each group his truth and his values" (Boudon, 1999, p. 308).

Universalism is erroneously criticized by a sophism that confuses the ideal with its realization. Universalism is indeed a value that cannot be reduced to its degree of completion, leaving differences free to exist and focusing on respect for rights and the attainment of well-being for all. Universal standards are, in principle, weak insofar as they do not immediately serve the interests of individuals. These take time to be admitted and internalized. This is the case for universal suffrage, which has been slow to take root in Western societies; the same applies to the principle of gender equality and the abolition of the death penalty for criminals (still in force in some US states and in some countries.) To what signs do we recognize the rightness and merit of a value? The response of philosophers and social scientists varies. The Anglo-American liberal tradition asserts that special measures are warranted to ensure the full participation of disadvantaged minorities in global society. The criticisms of this philosophical posture are known: individuals cannot be apprehended first as members of a culture, their well-being or freedom is not necessarily linked to the development of all the traits of their culture, etc. To this, the proponents of multiculturalism reply that individual autonomy requires specific measures in certain cases and that we can no longer rely on versions of liberalism that refer to outdated times. Contemporary ethnocultural pluralism needs, according to this liberal philosophical tradition, new answers.

But what do these divergent responses imply in societies? How can diversity and social cohesion be reconciled? The MI provides at least two possible answers.

The first is in the appreciation of an audience not immediately concerned—the impartial spectator in the sense of Adam Smith—who plays a referee role in the appreciation of a value. "It is not consensus that is the basis of correctness, but on the contrary, correctness that sustainably underpins consensus," writes Raymond Boudon (1999, p. 397). The latter further clarified his thinking by showing that cultural diversity was reconcilable with the objectivity of values. "The principle of respect for the individual implies respect for cultural diversity. This diversity is the result of history; it derives from the crucial fact that values are normally expressed in a symbolic way, and thus mobilize effectively 'arbitrary' signs. The principle of respect for the individual therefore implies the absence of any cultural discrimination. On the other hand, it does not imply a relativistic theory of values or a tribalist conception of societies" (Boudon, 2000, p. 337).

The second perspective was proposed by Michel Forsé and Maxime Parodi (2004). In their survey in France on the social representation of justice, they showed that the effort to get along and live together, essential to social cohesion, takes precedence over the imposition of private convictions in cases where self-interest and the desire to impose one's conception of the good weigh little. Individuals give priority to the just over the good as they are in the situation of a fair spectator. The principle favored by the respondents not immediately interested in a conflictual issue is that of justice as an agreement on impartial rules aimed at resolving a conflict and based on the "ethics of discussion" essential to democratic life. Pluralism and diversity in societies preclude the imposition of a single vision of good life.

7   Conclusion

Although there are many descriptive studies on international immigration, not to mention political or normative essays, sociological explanations of immigration are rarer. The theoretical and methodological approaches inspired by Methodological Individualism fill the limits of the description more satisfactorily than the works based on the idea of internalization of cultural models or on socialization by bringing back this social phenomenon from immigration to the actions that comprise it. By identifying the modes of thought typical of social actors, the MI explains why individuals who go abroad choose a particular country, how they go there and how they settle there. Emerging structures are the result of human actions and their influence depends on the actors' ability to interpret. Cognitive attractor, social networks and the rationality of individuals offer fruitful avenues to characterize the various situations observed empirically in countries open to immigration. Thus, the aggregation effects produced by the actors' decisions and their collective beliefs illuminate the emergence of ethnic neighborhoods more satisfactorily than habitus in the sense of culturalism or economic structure determination. Similarly, because they settle in large numbers in urban centers, immigrants transform the social morphology of many societies and the differentiation between cities and the rural world is accentuated.

The increase in the number of immigrants questions the national identity in a more or less marked way according to the societies, a trend that can even go as far as the national refoundation and the development of new narrative discourses. Considering migratory phenomena from a longitudinal perspective highlights the causal relationship of circularity between the microsociological and macrosociological levels, either "a double bottom-up and bottom-down causality that allows the adaptation of the global level to the individual level and vice versa" (Di Iorio, 2020, 32). Thus, international immigration more or less strongly transforms the host societies and, in return, migrants must adapt to the host society. As can be seen from this causal circularity, Methodological Individualism is not reductionist and it recognizes the systemic nature of the social phenomena it intends to explain, and this applies to international immigration in developed societies.

References

Axelrod, R. (2006). *The evolution of cooperation*. Basic Books.

Bell, D. (1976). *The cultural contradictions of capitalism*. Basic Books.

Bellah, R. N. (1967). Civil religion in America. *Journal of the American Academy of Arts and Science, 96*(1), 1–21.

Bénichou, M. (2017). *Le multiculturalisme* [Multiculturalism]. Bréal.

Bissoondath, N. (1994). *Selling illusions. The cult of multiculturalism in Canada*. Penguin.

Bouchard, G. (2012). *L'interculturalisme. Un point de vue québécois* [*Interculturalism. A Quebec Perspective*]. Boréal.

Boudon, R. (1999). *Le sens des valeurs* [The meaning of values]. Presses universitaires de France.

Boudon, R. (2000). Pluralité culturelle et relativisme [Cultural Plurality and Relativism]. In Will Kymlicka et Sylvie Mesure (Eds.), *Comprendre* (pp. 311–339). Number 1. *Les identités culturelles*.

Boudon, R. (2003). *Raison. Bonnes raisons* [*Reason. Good reasons*]. Presses universitaires de France.

Boudon, R. (2007). *Essais sur la théorie générale de la rationalité. Action sociale et sens commun* [Essays on the general theory of rationality. Social action and common sense]. Presses universitaires de France.

Boudon, R. (2008). *Le relativisme* [*Relativism*]. Presses universitaires de France.

Boudon, R. (2011). La théorie générale de la rationalité, base de la sociologie cognitive [The general theory of rationality, basis of cognitive sociology]. In Fabrice Clément & Laurence Kaufmann (Eds.) *La sociologie cognitive* [Cognitive sociology] (pp. 43–74). Éditions de la Maison des sciences de l'homme.

Boudon, R., & Bourricaud, F. (1990). *Dictionnaire critique de sociologie* [A Critical Dictionary of Sociology]. Presses Universitaires de France, 3rd édition.

Bouglé, C. (1910) *Qu'est-ce que la sociologie ?* [*What is sociology?*]. Félix Alcan

Bouglé, C. (1922). *Leçons de sociologie sur l'évolution des valeurs.* [*Sociological lessons on the evolution of values*]. Félix Alcan.

Bouvier, A. (2011). Individualism, collective agency and the 'micro-macro' relation. In I. C. Jarvie and J. Samora (Eds.), *Handbook of philosophy of social science* (198–215). Sage Publications.

Bowles, S., & Gintis, H. (2011). *A cooperative species.* Princeton University Press.

Bronner, G. (2006). L'acteur social est-il (déjà) soluble dans les neurosciences? [Is the social actor (already) soluble in neuroscience?]. *L'Année sociologique, 56*(2), 331–352.

Bronner, G. (2011). Invariants mentaux et variables sociales [Mental invariants and social variables]. In Fabrice Clément & Laurence Kaufmann (Ed.), *La sociologie cognitive* (pp. 43–74). Paris : Éditions de la Maison des sciences de l'homme.

Bulle, N. (2020). Trois versions de l'individualisme méthodologique à l'aune de l'épistémologie [Three versions of methodological individualism in the light of epistemology]. *L'Année sociologique, 70*(1), 97–128.

Bulle, N. (2018). Methodological individualism as anti-reductionism. *Journal of Classical Sociology, 19*(2), 161–184.

Coleman, J. (1990). *The foundation of social theory.* Harvard University Press.

Deacon, T. (1997). *Symbolic species. The co-evolution of language and brain.* Norton.

Degenne, A. (2020). *Dynamiques des structures sociales.* [*Social Structure Dynamics*]. L'Harmattan.

Degenne, A., & Forsé, M. (2004). *Les réseaux sociaux. Une approche structurale en sociologie* [*Social networks. A structural approach in sociology*] 2e édition, Armand Colin.

Di Iorio, F. (2015). *Cognitive autonomy and methodological individualism. The interpretative foundations of social life.* Springer.

Di Iorio, F. (2020). Individualisme méthodologique et réductionnisme [Methodological individualism and reductionism]. *L'Année sociologique, 70*(1), 19–44. Tranl. This volume.

Dubreuil, L. (2018). *La dictature des identités* [The dictatorship of identities]. Gallimard.

Dunbar, R. I. M. (1996). *Grooming, gossip and the evolution of language.* Cambridge University Press.

Durkheim, É. (1893). *De la division du travail social.* [On the division of social labor]. Felix Alcan.

Forsé, M., & Parodi, M. (2004). *La priorité du juste. Éléments pour une sociologie des choix moraux.* [The priority of the just. Elements for a sociology of moral choices]. Presses universitaires de France.

Granovetter, M. (1973). The strength of weak ties. *American Journal of Sociology, 78*, 1360–1380.

Heinich, N. (2021). *Oser l'universalisme. Contre le communautarisme* [Daring to be universal. Against communitarianism]. Le bord de l'eau.

Helly, D. (2002). Minorités ethniques et nationales : les débats sur le pluralisme culturel [Ethnic and national minorities: debates on cultural pluralism]. *L'Année sociologique*, *52*(1), 147–181.

Joppke, C. (2004). The retreat of multiculturalism in the liberal state: Theory and policy. *British Journal of Sociology*, *55*(2), 237–257.

Kymlicka, W. (1995). *Multicultural citizenship. A liberal theory of minority rights*. Oxford University Press.

Kymlicka, W. (2001). *Politics in the Vernacular: Nationalism, multiculturalism and citizenship*. Oxford University Press.

Langlois, S. (2018). *Refondations nationales au Canada et au Québec* [*National Refoundations in Canada and Quebec*]. Septentrion.

Lemieux, V. (1982). *Réseaux et appareils. Logique de systèmes et langage des graphes*. [*Networks and devices. System logic and graph language*]. Edisem.

Lemieux, V. (1976). Un homme et une œuvre: Paul Mus. [A man and a work: Paul Mus] *Cahiers internationaux de sociologie*, *60*, 129–154.

Lévi-Strauss, C. (1973). *Race et histoire* [*Race and History*]. UNESCO.

Lilla, M. (2017/2018). *The once and future liberal after identity politics*. Trad. Fr. *La gauche identitaire. L'Amérique en miettes*. Stock.

Lind, M. (1994). *The next American nation. The new nationalism and the fourth American revolution*. Free Press.

Milgram, S. (1967). The small world problem. *Psychology Today*, *1*, 61–67.

Olson, M. (1965). *Logic of collective action*. Cambridge (Mass.), Harvard University Press.

Pharo, P. (2011). Réalisme cognitif et dépendance pratique [Cognitive realism and practical dependence]. In Fabrice Clément & Laurence Kaufmann, (Eds.) *La sociologie cognitive* [Cognitive sociology]. Éditions de la Maison des sciences de l'homme, 299–324.

Popper, K. (1945). *The open society and its enemies*. Routledge.

Roby, Y. (2000). *Les Franco-Américains de la Nouvelle-Angleterre : Rêves et réalités*. [Franco-Americans in New England: Dreams and Realities]. Sillery: Éditions du Septentrion.

Taylor, C. (1992). *Multiculturalism and the the politics of recognition*. Princeton University Press.

Tocqueville, A. de. *De la démocratie en Amérique I et II*. [*Democracy in America*].

Vertocec, S., & Wessendorf, S. (Eds.). (2010). *The multiculturalism backlash: European discourses, policies and practices*. Routledge.

Wellman, B., & Berkowitz, S. D. (Eds.). (1998). *Social structures. A network approach*. Cambridge Univesity Press.

Winter, E. (2011a). L'identité multiculturelle au Canada depuis les années 1990: de la consolidation à la mise en question? [Multicultural identity in Canada since the 1990s: From consolidation to questioning?] *Canadian Ethnic Studies/Études ethniques au Canada*, 35–57.

Winter, E. (2011b). *Us, them and others: Pluralism and national identities in diverse societies*. University of Toronto Press.

Individualistic Models in Collective Norm-Based Regulation: The Positive, Normative and Hermeneutic Dimensions

Emmanuel Picavet

1 Introduction

Methodological individualism is concerned with letting individual reasons play an important part—perhaps the most fundamental one—in the explanation of collective phenomena or in the study of the justification of collective choices which impact social organization (Boudon, 1979; Coleman, 1990). Empirical explanation and norm-based justification play distinct and complementary roles. Although the reliance on good reasons offers a common ground, methodological individualism as an explanatory and predictive enterprise with respect to social phenomena, or as a means of understanding and interpretating social processes, should be distinguished from a privilege given to individual-related notions in systematic (philosophical) explanations of the connections between values, beliefs, principles, etc., (that is to say, sources of good reasons) and actions, norms or institutions. We shall not be concerned here with normative individualism in the sense of a personal commitment to defend or criticize social arrangements on the basis of what happens to individuals, with a view to the values and beliefs of the individuals themselves.

Social functionings and processes are infused with norms and institutions which are given a regulatory role. This kind of collective, norm-based

---

E. Picavet (✉)

Department of Philosophy, Université Paris 1 Panthéon-Sorbonne, Paris, France
e-mail: Emmanuel.Picavet@univ-paris1.fr

© The Author(s), under exclusive license to Springer Nature Switzerland AG 2023

N. Bulle and F. Di Iorio (eds.), *The Palgrave Handbook of Methodological Individualism*, https://doi.org/10.1007/978-3-031-41508-1_8

regulation involves positive dimensions (which have to do with explanation, prediction and modeling), as well as normative and hermeneutic ones. The purpose of this contribution is to highlight some key features of the integration of these dimensions. Methodological individualism offers perspectives for understanding the challenges of this integration, for different modes of political regulation.

Collective choices that are deemed "to be selected" according to norms of the preferable are best understood from a norm-based (or "normative") perspective as enlightened by correct explanations of how and why social agents behave and how their behaviors aggregate into social facts or patterns. Conversely, matters of legitimacy or acceptability have a structuring role in individual choice; for example, it is quite usual to consider that the better the rules are accepted as legitimate, the more confident we can be in assuming that they will be followed (engaging in normative argument in support of social rules would ordinarily be regarded as pointless in ordinary life if this linkage were broken).

Thus, a consistent development of methodological individualism in empirical social science and in social philosophy is the investigation of the status of individual reasons as they are hypothesized, modeled or theoretically framed for the purposes of organization, policy or collective action. The justificatory schemes in policy, and social life more generally, are usually based on the use of models, theories or general representations of social functionings and social mechanisms, with a role for individual reasons. The interplay of the positive, normative and hermeneutic dimensions is quite complex, however, and it raises issues of consistency and relevance.

Norm-based regulatory decision-making consists in using norms (such as those social norms which can be impacted in a voluntary manner, strict legal rules or freely endorsed charters) in collective or public decision-making. Classically, the latter aims at avoiding an unruly state of affairs, or an unruly kind of development in time. "Unruly", here, means that the social world is distant from a model social world in which a normative appreciation of the state of affairs is appropriate (in the generic format "things are as they should be in this or that respect", with "should" being interpreted in the light of a substantial norm or set of norms).

In such a setting, individuals are under focus, quite inevitably, because their conducts have a causal impact on the state of the social world. Their own reasons—which are, in principle, accessible to existing justifications—are involved in the achievements or failures of collective action. Justification and explanation are thus correlated in social practice. Individual conduct provides the linkage and, as we shall see, the hermeneutic capacities of individuals are quite important in this respect.

Focusing on the norm-based variety of regulatory decision-making, the articulation of the positive, normative and hermeneutic dimensions in the status and role of hypothesized individual reasons will be examined in the perspective of methodological individualism. Even though there are time-honored good reasons to differentiate these dimensions from one another, trying to get a synthetic picture of their mutual articulation may be useful, owing to the practical and theoretical significance of their mutual support.

First of all, it will be asked whether methodological individualism is attuned to a correct understanding of the connection between collective guidance and individual reasons (Sect. 2). The challenges of responsibility in the use of models and representations with regulatory or organizational purposes (a kind of responsibility that can be called "epistemic") will then be addressed and the institutional side of the relationship between beliefs, individual ways of life and collective attitudes will be addressed in connection with methodological individualism (Sect. 3).

2 Norm-Based Order and the Challenge of Individual Reasons

The Individual and the Collective in Norm-Based Regulatory Action

Organized collective action—e.g. in political or administrative action, or in the strategies of major corporate actors—relies on norms, which are supposed to be applied to a sufficient degree, if the granted transformative powers are to be effective at all. What does "applied" mean? As argued by Wittgenstein, rule-following isn't a model to be taken literally and this has consequences for social analysis (Reynaud, 2002). Norms are often formulated as rules and relying on the rules, Livet and Conein (2020, pp. 197–198) explain, consists either in (1) using them with pragmatic goals, especially with coordination purposes, (2) making normative or axiological statements about states of affairs in a way which explicits or modifies the presentation of them or (3) applying norms or rules which are in fact "principles", whose use in the real world is problematic and unspecified.

Explaining the role or situation of individual actors has a reflexive dimension for individuals because collective-level action (in public policies for instance) often relies on models or representations of the individual itself.

When justification is possible, the models and representations are supposed to be true, acceptable or, at the very least, relevant in some important respects. They lead to artificially designed norms (the object of "normative design" nowadays). These norms, in turn, allow collective action of policies to be based on accepted collective goals or principles (such as efficiency or social-justice goals and principles). Norms are guidelines which are passed onto the individuals (who are supposed to follow them and to convert them into guidelines for their own conduct). Thus, it makes sense to say that the collectively hypothesized nature and qualities of the individuals matter for the individuals themselves, as they indirectly impact their lives. This is highlighted in those approaches of regulation, especially in the Marxist tradition, which stress the importance of historically evolving social conventions in the constitution of general architectures of regulation which condition the concrete forms of human life.

Insofar as we care about the common good, we are potentially concerned with norm-based regulatory action whenever we intend to impact society through collective decision-making, making use of norms. More precisely, whenever regulating human affairs is in view, this sort of norm-based ambition consists in seeing to it that, through the use of norms, an essential order in human affairs is *preserved* (as an equilibrium between the existing forces or as a dynamic path which exhibits properties of stability or resilience) or *enhanced* (for example, making progress in the direction of increased efficiency, social justice or environmental viability).

This is an important aspect of political action of course, maybe the central one, as argued in theories of the "regulatory State" which frame State action as regulatory power in market economies. Similarly, it is present in private or public–private partenarial ventures which aim at implementing systems of norms which can be freely endorsed, such as norms of Corporate social responsibility, "green finance" standards in the choice of investments, etc. It is also involved in efforts to harness the complexity of large sets of teams, activities and problems—for instance, in setting rules for big, multinational corporations. This proximity between different contexts accounts for the rise of the notion of "governance" to a large extent.

On the one hand, from Plato to Castel de Saint-Pierre and authors in later times, the ambition of letting a norm-based order emerge through theory and well-chosen assumptions has fostered utopianism and hope. Today, regulation is still a theme in thinking about general social transformations (Klein, 2022). On the other hand, an attitude of skepticism is frequent because individual motives are not necessarily what they "should" be or what they are supposed to be. Certainly, individual motives are not alien to social discourse about them: they can be influenced by dominant representations about how they can or should be thought of. This is not always sufficient, however, to warrant the alignment of collective assumptions about individual motivations with individual motivations as they are. This tension is perceptible in normative and positive explanations alike.

On the normative side, the individual's conception of "what should be done" at the individual level isn't necessarily harmonious with the predominant collective view about what the individual should do. In addition, if the conception of "should" is determined by enforcible rules (such as legal rules or otherwise socially coercive rules), the individual' determination of one's own correct rule-following individual behavior might differ from the collectively predominant and authoritative determination thereof, due to the individual's lack of information, or maybe to the over-simplified character of benchmark models which underpin collective rules or, alternatively, because of interpretative divergence.

The very generality of collective guidelines might also be at odds with the sheer singularity or novelty of the contexts in which individual decision-making occurs, causing *ex post facto* discrepancy between actually revealed motivations in individual behavior on some occasions and predominant views about what individuals should have done, following a predominant view about the way to deal with new situations on the basis of benchmark interpretations of rules and benchmark views about what it takes to apply the rules, or follow them. At the same time, what might be called an "interpretative regulation" plays an important part in the management of ambiguity and it interacts with the kind of "strategic regulation" which relies on expectations and the evaluation of risks (Padioleau, 2000).

On the positive side of the matter, collectively predominant assumptions about what individual motivations are can simply be false. Owing to the fragility of models or theories, the resulting discrepancy between observation and prediction hardly comes as a surprise. The policy implications, however, are numerous and, more often than not, they amount to serious collective challenges. Thus, epistemic fragility in financial modeling poses a threat to the credibility and effectiveness of financial regulation (Walter, 2010).

All in all, both hope (in reform for instance) and skepticism about letting a norm-based order emerge draw attention to a family of lasting problems, to do with the compatibility, interaction or progressive integration of the representations of individual motivation and action on the one hand, and actual individual motives and behavior on the other hand. At root, we find a generic difficulty which follows from the collective necessity to rely on simplified and partly conventional views about individual reasons and the correlative motives in action. Such views have no probability of giving a true picture of real-life individual concerns. Thus, methodological individualism might have difficulties to account for collective arrangements and guidance which rely on simplified and potentially false assumptions about individual reasons. Is it possible, however, to accommodate an "individualistic" perspective to the necessities of collective guidance in the face of complexity, hardly predictable novelty and interpretative subtleties and difficulties in real-world social life?

Hermeneutic Capacities, Problem–solution Pairs and Methodological Individualism

Interpretation is another important ingredient in the collective use of models of the individual, especially if we are interested in norm-based regulatory decision-making. In this case, indeed, the relevant "order" or adequate social model is intended somehow to materialize in the actual functionings or states of the social world, and in the rules which are deemed to lead to them. This involves problems of meaning and interpretation: in order to grasp the nature and value of the "order" which is longed for, individuals need information and they need useful normative shared references (which enable them to give significance and legitimacy to the enforced or proposed norms). For example, they need information about the reasons for the unsatisfactory state of society or social routines. The kind of normative shared references they need may comprise relevant principles and values which make a certain kind of social order desirable. The emergence and use of such normative references are hardly reducible to a logic of self-interest maximization because the associate kind of legitimacy, and the expectations that legitimacy will be recognized by others as well, are usually backed by impersonal reasons, not just personal reasons like self-interest reasons. All this plays a role in normative justification, hence also potentially in individual motivation and empirical behavior. From the point of view of methodological individualism, this gives reason to deal with a fairly rich picture of the individual, at a good distance from the simplified models which deal with probability assessments and pre-determined interests only.

Furthermore, hermeneutic tasks are involved in the connection between collective goals or models on the one hand, and individual action on the other hand. Ideally, the proposed rules should be understood as adequately correlated with collective aspirations, both *causally* (how are they conducive to the desired states of affairs?) and *hermeneutically* (are the rules followed in an adequate way and are the rules and their expected effect expressive of correctly interpreted justificatory principles or values in the first place?). In addition, the resulting situation should be identified as belonging to the looked-for kind of situation. This further identification problem calls for guidance in information and interpretation all the same.

These elements enable collective guidance to make sense for individuals and they play a role in the legitimacy—or at least the acceptability—of constraints. Collective guidance isn't always based on coercive control and punishment mechanisms; other normative resources can be used. In any case, collective guidance is required for individual reasons to lead to desirable social patterns which are achieved through the enforcement or proposition of norms in an orderly way. This contradicts in no way the possibility of the emergence of a "spontaneous order"—which exhibits features that can be rationalized as desirable properties (such as we would have them emerge if we had the capacity to ensure it). Spontaneous order is a possibility, maybe it is identifiable in special and interesting examples (Sugden, 1989) but it shouldn't be confused with a magic mantra. No matter how the formula might sound to the ears of those who distrust collective directives, no matter how often it is repeated, it won't solve the problems which mar societies and which happen not to be spontaneously solved. In organized social life, the harmonious aggregation of spontaneous individual behaviors is problematic, and it can hardly be considered a paradigm of the dominant individual-society connection. Regulation, however, usually presupposes a degree of spontaneous mutual adjustment of actions, as in the kind of coordination which is facilitated by prices, by a shared cultural experience, by accepted constitutive rules of practices (which make it possible to enjoy much personal freedom in social interaction), etc.

The usefulness of collective guidance (in a variety of formats, from friendly "nudges" or advice to "hard law") is probably a challenge rather than a limitation for methodological individualism. Provided there are enforced or successfully proposed norms, which are supposed to be conducive to the achievement of social or collective objectives, it matters that they can be understood and—hopefully—followed in an appropriate manner, in the light of adequate and correctly interpreted principles and values, which in turn enable individuals to act in a meaningful way. Thus, the analysis and explanatory use of a sufficiently rich array of individual reasons (as part of the explanans of rule-based interactions) turn out to be essential.

Once implemented, the sought-for static or dynamic order of things should also be recognized as such. This identification problem also calls requires normative guidance in collective-level explanation and individual understanding. Obviously, cultural and linguistic realities play a role here but, as Vilfredo Pareto (1905) once noted, the "social" is internal to the individual after all. Insofar as methodological individualism is not about a replacement of social realities by non-social ones (and surely it shouldn't), there is nothing paradoxical in stating that collective normative guidance can be addressed through the lenses of individual reasons which are partly the outcome of hermeneutic tasks.

Taking a "regulatory" view of the usefulness of putting norms into operation, it is expected that regulatory action can be instrumental in letting society reflect the kind of order or social model which is associated with the values or principles which are being taken for granted in the first place. The kind of order which is brought about by norm-based regulatory decision-making is usually best understood in the light of the problems which would occur, should the norms fail to play their part. Let us think, for example, of air pollution as it results from transportation choices: norm-based regulatory action is concerned with restoring the normal quality of air, and the seriousness and adequacy of our collective reliance on norms depend on the ability of the latter to protect or restore the quality of air. Should the chosen strategies fail, society as a whole would experience difficult times.

Regulatory policies and the collective setup of orderly social mechanisms are thus usually presented as solutions to various problems: bad behavior on the part of ordinary individuals or individual corporate agents, "market failures", social problems resulting from behaviors whose aggregate effects turn out to be problematic… A common feature is that the difficulties at hand can be understood by the individuals whose behavior is part of the solution, so that designers of a "solution" should care about the reasons of individuals and adjust to their inner logic somehow. This can be considered a development of the old prediction of a well-organized or "rationalized" future (in Schumpeter's writings for example—see the analysis in Demeulenaere (1996), p. 19) but the quest for technical solutions is, by necessity, closely associated with the investigation of the reasons of individual agents.

However, the "solution" which is being proposed in answer to a regulation or organization problem is not neutral with respect to the problem itself and its framing, in the general case. The solution is tentatively applied in a world where the problematic is usually already framed by social or institutional rules. Individual conduct is structured by explainable reasons that are partly pre-existing in the social world as it is. This makes it necessary to investigate the connections between the reasons which play a guiding role in the existing forms of life, on the one hand, and the reasons attributed to the agents in a way which makes sense from the perspective of the chosen regulation or organization purposes and means. Applied work about the performative effect of theories (Brisset, 2016, 2020; McKenzie, 2003, 2006; McKenzie et al., 2007) has explored these matters, taking a particular interest in the implications for the correspondence (or the lack of it) between the intended effects of the use of models (or theories) and the real consequences, as well as the possible association with other factors (Callon, 2010).

Given the increasingly important role of artificial and complex decision-making procedures (automated algorithms, administrative procedure with a role for self-assessment) and artificial conversation devices (such as those permitted by computer-based collaborative devices or by a scheduled plan for regulatory dialogue between a regulation authority and socio-economic agents), and given the increasingly significant use of abstract or theoretical models in concrete governance procedures (for example economic or financial models, cost–benefit analysis procedures, theories of "objectivity" or "prudence"…), the correlation between actual reasons and postulated reasons (or the lack of it) is an important thematic in its own right.

Collectively Chosen Assumptions

The major assumptions which underlie the use of regulation-oriented models are assumptions about behaviors, and individual behaviors more particularly. Such models focus on the causes or motivations of individual conducts, the problems they generate, the results of their accretion for society. Insofar as these assumptions play a role in policy, they give a picture of the kind of society which is considered as a potentially "regulated" society. Such assumptions play an important role in the paradigm of the regulatory State. They are also implicit in beliefs about the self-regulation of those ways of life which are deemed typical of a "civil society", distinct from the sphere of public authority but correlated with the latter.

The various types of regulation help to shape the interactions themselves and thus ultimately the social, economic and cultural forms of life. In the special case of so-called "self-regulation" hypotheses in political economy, the hands-off approach to social life means that market forces can have a maximal impact on individual motives. This amounts to expressing both a taste for an unplanned kind of organization and a preference for definite motivational patterns (such as the will to acquire private property and money) and the resulting states of society (e.g. very diversified consumption goods, deep inequalities) and the environment (e.g. a rapidly downgrading quality of our natural surroundings and resources). Trust in self-regulation is ultimately based on beliefs about the existence of specific motives which are ascribed to the actors; such beliefs pertain to "interests" which relate to individual, family or corporate matters.

186 E. Picavet

Organizing our knowledge about this is a key to the understanding of the resulting collective or "social" problems, and this involves the relationships between positive and normative matters. The beliefs about private motives relate to empirical facts (attitudes as they are) but also to the value, in the agent's eyes, of the kind of conduct which is oriented by the considered motives. It must be assumed that acting on such motives is sufficiently valuable in the eyes of the agents themselves, which gives relevance to both the positive study of tastes and the identification of influential normative reasonings. In most cases, it is appropriate for social agents to act according to such motives only if their expression in conduct is considered legitimate by themselves. This gives credibility to the prediction that they will act in the way which can be expected, starting from the considered model of behavior. In some cases, it is further assumed that the agents would consider it inappropriate to act otherwise because it can be hypothesized that they try hard to comply with exacting standards of behavior.

Looking at social life from a "regulatory" vantage point may lead social analysts to consider that their grasp of individual motivation enables one to predict social behavior, with a view to drafting appropriate schemes for enhanced social interaction. For the purposes of regulating behavior in this way, beliefs play a role at two distinct levels. First of all, trust in public action or social regulation is conditioned by the belief that the grasp of individual motivation is basically correct. In addition, beliefs are implied in the characterization of the "appropriate" nature of the goals which can be achieved through the designed interaction schemes. In this respect, it is worth noting that the sharing of beliefs about the correctness of objectives and actions can be impacted by the social tools of organization or regulation. Thus, M. de Certeau (1990, p. 263) observed that in social devices, beliefs are sometimes weak, which might be compensated for by administrative means, especially in political organization. In religious institutions, by contrast, it was suggested by De Certeau in the same passage that the weakening of institutions is correlated with the strong and widespread dissemination of the beliefs which are controlled and regulated by the institutions.

For example, let us consider the hypothesis of self-interested optimizing behavior in commonplace economic explanations (not to be confused with the more refined general statements about preference-based optimizing behavior in fundamental economic theory). This set of beliefs about economic agents is operating on two levels (at least). It plays a role both in the behavioral forecasts that serve as a foundation for justifying publicly edited norms. It also plays a role in the elaboration of definite prescriptions, expressed in these norms, insofar as the latter can indeed be considered appropriate with respect to the known features of the social situation.

The duality in the reliance on beliefs for organizational of regulatory purposes is mirrored in individual beliefs as they are elaborated by the individual social agents. On the one hand, individuals entertain beliefs about the appropriateness of their own tendencies or attitudes (for example their greed or altruism). On the other hand, they have a capacity to figure out for themselves whether the social objectives or values, and the social choice of prescriptions in order to implement them, are correct or not. The growth of criticism or revolt is always a possibility, and an important one in real-world human affairs. It can be nurtured by the predominance of a "duty to make profit" when it tends to become a social norm (Demeulenaere, 2003, p. 212). The interpretation of a social situation gives testimony to individual creativity; it also testifies to the limits of prediction based on regulatory models which presuppose the obedience or good will of social agents, who must be credited with capacities for critical analysis and reasoning about legitimacy issues. These limits are reached when the underlying principles seem inadequate.

A possible seduction of the "self-interest" social conceptions is their easy coupling with the pretense at organizing social life in an orderly way (Bridel, 2009; Hirschman, 1977). If people are moved by self-interest by and large rather than erratic and potentially dangerous passions (political or religious passions for example), the study of social design boils down to the analysis of incentive mechanisms: seeing to it that social agents are confronted with appropriate incentives is the safest way to implement social goals. This, however, has credibility only insofar as the individuals themselves are comfortable with acting out of their self-interest as their rulers think of it. Should it not be the case, confidence in the hypothetical component of regulation would be undermined. As a matter of fact, some minimal agreement on the logic of self-interest and competition seems required in order safely to rely on interest-maximization hypotheses for the sake of regulatory purposes. Such a minimal agreement, however, cannot be taken for granted and its desirability is doubtful in the first place.

Thus, some neoliberal views of political action, which rely on hypotheses about the predominance of self-interest in human motivation (unlike other neoliberal views which allow for a richer picture of motives), are often perceived as oppressive because individuals do not identify themselves with the anthropological representation of man which is encoded in the underlying views. Indeed, self-interest assumptions are often perceived as misguided

188   E. Picavet

and potentially dangerous if we collectively act on them in regulation tasks, with induced effects on the motives of individual behavior. The critical literature about social life and political economy examines a number of basic reasons why narrowly economic perspectives in life are wrongheaded (Arnsperger, 2005) and the sources of the possible wrongfulness of an all-too-narrow economic or business-minded education (Lacroix, 2009). The impact of model-based regulation on ways of life and on the associated forms of self-consciousness gives rise to criticism.

In an approach of social normativity which is centered on the spontaneous emergence of agreements in the community, the meaning of action is a problematic which associates individual (or decentralized) and collective dimensions. Indeed, the implementation of general principles or objectives is achieved through the framing of agreements or contracts (which result from the social roles of the concerned partners) into a more general system of interactions (investments, market relations and competition regime, markets for consumer choices, etc.). The detailed transactions make sense only in the light of the general social arrangements, which means that the explanation of individual action depends upon the individual representations of social proceedings. For explanatory purposes at least, the gap between the individual and the collective is partly bridged in this way. However, it might be the case that the individuals do not consider that the general social framework of their actions is an acceptable one. This has a negative impact on predictability in implementation processes.

3   The Individualistic Approach to responsibility and the Use of Models or Representations

From Prediction to Prescription

Government in the age of the so-called "regulatory State" (see Majone, 1995; Tirole, 2016) usually intends to act in a way that corrects or remedies problems that arise from market relationships and the hypothesized self-organization of "civil" society, as distinguished from political society in order to stress the role of purportedly non-political actors (such as firms and associations). These elements, which are assumed to be already present in society must be both explained (to predict the effects of the policies which are chosen) and understood in a way that makes the tasks of normative justification possible. A system of beliefs about individuals comes into play, as a consequence.

For instance, the assumption which represents self-interest or the satisfaction of personal preferences as the dominant motivation of economic agents is usually instrumental in the elaboration of the behavioral predictions that serve as the basis for justifying norms (as answers to understandable social problems such as a lack of efficiency on a "market"), and in the prescriptions that these norms contain (which must be consistent with the ambition of reaching good social results, given the assumed motivations of individuals). Here, adapting to relationships and reasons that purportedly already exist in society and in the reasonings of social actors proves essential and it creates responsibility toward the public (in political debate) and the population generally speaking.

Insofar as political initiative can alter or restore forms of life which are partially shaped by beliefs in self-organization and in the virtues of the "market", the process has a "utopian" dimension: reality must adapt to beliefs rather than beliefs to reality, and this makes it necessary to take a "performative" dimension of the use of models and representations (or theories) into account. This utopian challenge is rather rarely discussed in the case of economic organization, probably because self-organization assumptions are often associated with beliefs in the explanatory predominance of self-interest, and the latter seems anti-utopian on the face of it.

Beliefs are put to use in public action and therefore the relationships between method and beliefs must be investigated, from a rationalist perspective. The tools that shape the beliefs of "regulators" have theoretical underpinnings in many cases. This is apparent in economic modeling strategies which rely on individual rationality assumptions and concepts of collective equilibrium. In addition, the operational benchmark models in social or economic regulation provide a basis for the ongoing treatment of information and it is quite clear that the answers to organizational problems have a learning dimension with respect to the flows of information across time. Adhering to doctrines (and possibly to full-fledged theories) involves a risk of rigidity, of course, and the seriousness of this risk must be weighed against the benefits of consistent, non-arbitrary ways of tackling new situations, possibly based on traditions (Popper, 1963, Chapter 4) or routines (Heiner, 1983). Indeed, the backbone of a doctrine gives consistency and predictability to action and attitudes, partly because it enables actors to connect new cases to older ones in an orderly way, through the use of rules which are backed by a more or less theoretical approach.

Behavioral hypotheses which are built into regulatory models deserve to be discussed for themselves with a critical mind. Such hypotheses express principles of rationality, but also—and quite often nowadays—systematic deviations from rationality models, that is to say a number of well-documented "biases". The chosen set of behavioral hypotheses may also comprise standards of behavior which originate in social habits or expectations—such is the case of norms of "correct" investor behavior (for the purposes of competition control), norms which pertain to the consistent promotion of industrial interest in a competitive way (in contrast with agreement with competitors), norms which indicate the best interests of represented agents, etc.

An important case is that of prudential norms which are oriented toward the prevention, mitigation or control of risks, or the deployment of precautionary strategies. Interest, here, isn't just a prediction tool which is relevant for human conduct and society at large can be described as a "society of risk" (to borrow a phrase from Ulrich Beck, 1986) only because the combination of individual interests (or preferences), collective structures and governance lead to a widespread consent to risk-bearing in the face of uncertainty or novelty. Interest is an object of prescription: in collective regulation, social actors are expected to behave in a way which makes sense—for the purpose of justification and public debate—in the light of the hypotheses about their own beliefs, desires or motivations. This kind of prescription makes it possible to assume that behaviors will be aligned with the norm, provided that compliance can be made effective through an effective network of institutional incentives, controls and steering devices (which can be difficult to implement in practice).

In the ideal case of a full implementation, behavior can then be predicted on the basis of behavioral hypotheses that have the same content as prescriptions. This can be used for seeking normative guidance or achieving effective regulation through the characterization of social problems and purported solutions. It can be noted that prescriptions are, at a certain level of analysis, combined with the observation of behavioral trends: people care about their interests in various ways, and these dispositions or tendencies must be taken seriously and used as benchmarks whenever a particular conduct is being prescribed, even though a full-fledged social determinism isn't postulated (and it is typically not postulated in the tradition of methodological individualism as opposed to naturalism and other epistemological attitudes—see Boudon, 2013, p. 45 sq.).

Responsibility Issues: The Cognitive and the Extracognitive

Professional accountability is in principle amenable to a treatment which makes use of the emerging theories of epistemic responsibility (such as Lamy, 2022), which refer to the qualities of practices that aim at producing true beliefs (Pritchard, 2021, p. 5515). These theories are part of the ongoing development of analyses about "cognitive" virtues and the responsibility to foster them and abstain from misguiding delegation of their exercise (De Bruin, 2013, 2015—a thematic also underlying Walter, 2010). They typically make use of epistemological (sometimes historical-epistemology) reasonings and epistemological contributions are relevant even though their individualistic, agent-centered character is sometimes criticized as a limitation. From the perspective of methodological individualism, on the contrary, it may seem desirable that our models of reasonable social action be somehow founded on accounts of *appropriate* individual cognition. This will not apply to all varieties or uses of methodological individualism: a prudent neutrality with respect to the nature of mental operations might appear preferable to some theorists, or in a given research field. This being said, referring to appropriate individual cognition can be a useful benchmark in order to make full use of our objective knowledge of good reasons which are accessible to the agents.

Epistemic responsibility appears to be an integral component of the defensible use of benchmark models which play a structural role in the guidance of behavior. In the most important cases (in economic policy for instance), it assumes an institutional form, and it must be stressed that institutions have their own way to deal with models and put them into operation. They do not "think" in the usual sense but they shape collective attitudes in a decisive way and this applies to the attitudes which have to do with using models so that actions and interactions can follow a definite, valued path.

Of course, the impact of regulatory (or organization-oriented) models on human ways of life poses challenges well beyond the epistemic operations to do with knowledge acquisition and the selection of beliefs. Insofar as ways of life are impacted, so are individual interests and expectations, and the opportunities in life. Moreover, dialogue in regulation or organization can be a source of shifts in power, and conflicts. The cognitive and extra-cognitive aspects turn out to be closely associated. Beliefs and the commitment to let models or theories guide us in the elaboration of these beliefs are at the root of the acceptability of those consequences which result from the development of models or theories which are a guiding force in the regulation of society.

The awareness of specific social problems isn't always spontaneous. It must be secured through the creation of appropriate procedures within organizations. For example, the National Contact Points for Responsible Business Conduct within OECD play such a role when it comes to compliance with the guidelines of Corporate Social Responsibility (CSR) as exemplified in the OECD "principles" addressing the needs and duties of multinational corporations. Procedural treatment, here, cannot be severed from interpretation issues, pertaining to the principles of CSR. The hermeneutic dimension is to some degree already captured at the stage of the formulation of corporate duties for the firms. Thus, the "due diligence" imperative and the precautionary guidelines appear to be closely connected with the procedural dimensions of treatment, prevention and remediation routines, which are sketched at the level of the general recommendations themselves.

In this example, it is quite obvious that beliefs about correct interpretations are being translated into decision-making procedures. First of all, there is a role for positive beliefs (about the functionings of capitalist societies where foreign investment plays a capital role) and normative beliefs (about what precaution, diligence and objectivity—among other benchmarks—rightly mean and which duties they create). These various beliefs help give a shape to institutional procedures; for example, routines for the amalgamation and treatment of information in the presence of emerging risks, or procedures which ensure the independence and impartiality of judgment, proceed from important positive and normative beliefs. In a second step, judgment and interpretation are involved in institutional proceedings, once the institutional devices are created. Then individual agents usually come to the forefront: their own integrity in the interpretation of norms and their ability to translate normative beliefs into practice are at the root of their ability to strike sensible compromises, to issue judgments, etc.

Should the involved beliefs prove rigid or inadequate at either of the two stages, should it become obvious that they are truly alien to the real motivations of the actors, the public image and the efficiency of common institutions could be endangered. This easily happens in the case of those codes of behavior which are meant to help regulate both a system of interaction and individual conduct, if possible on the basis of "internalized" concerns. For example, scientific integrity in science-oriented communities, groups and networks is promoted after this pattern. In order for the selected benchmark principles to be implemented in the right way, they should have ethical significance, beyond the detailed obligations associated with the deontology of given professions and statuses. It is intended that, by turning proclaimed duties into personal affairs, institutional agents in the world of science will be able to foster personal and collective virtues which have a capacity to induce trust in science on the part of the general public, investors, etc. Personal virtue is stressed in a particular way, being crucial to the seriousness and respectability of scientific practice. As a consequence, those procedures which are intended to express a serious institutional commitment to scientific integrity usually focus on individual violations of ethical duties. Insofar as personal responsibility is relied upon (rather than merely the compliance with institutional obligations), it is furthermore often expected that the procedures are not mixed up with legal procedures, such as those which are put into operation by jurisdictions.

Thus, the format of general ethical duties plays a role in the kind of institutional process which is being favored and in the selection of the goal—namely, the development of personal (and collective) virtues in the scientific world. The very generality of the underlined duties, however, is a problem. On the one hand, it is necessary for the internalization process to take place in an ethical perspective. On the other hand, general formulations open the door to arbitrary hermeneutic choices in procedures which have real consequences for individuals. They also favor arbitrarily chosen "examples" of "illustrations" which are sometimes frightening for freedom in research, in scientific writing and in publishing choices. Examples or illustrations cannot be dispensed with, it is often assumed, because it makes sense to keep general guidelines sufficiently general in the first place.

Arbitrariness is a real problem, then, and so is the possible rigidity associated with examples, illustrations and favored interpretations which are gradually being consolidated as central, dominant or "generally accepted". This is potentially harmful to institutions because a degree of flexibility proves essential in order to accommodate new situations or kinds of situations. Paradoxically, then, it is possible that institutional damage results from serious and painstaking efforts to encourage virtues and mutual trust through a responsibility-based kind of collective regulation.

Commitment to Using Norms and Some Challenges of Interpretation

According to a common and rather simplifying perspective, the preference for a contractual vision of things—or (as in the case of R. Edward Freeman, 1984) a quasi-contractual one—would be the privileged means of expressing values or commitments in a responsible way, in relation to the things in which we collectively take an interest as individuals or as social or corporate agents.

The homology between the predominantly contractual organization of so-called "liberal" societies and a decentralized view of collective commitments would then be both inspiring and re-assuring. But is this vision of things tenable? Isn't it the case that more centralized operations of organization, interpretation and implementation in fact play a crucial role in determining the meaning of principles as they are expressed and trusted in collective life?

Should the decentralized view prevail, the collective expression of principles would be considerably underdetermined. Owing to the thickness inherent to the complexity of organizations, there is a significant distance between the commitment to norms and the contribution, through effective action (down to the level of individuals), to their effective expression in institutional life. Social and political theory has long warned of the reality of this distance, because of the complexities of interactions and organizational forms.

For example, in Hobbes, the denunciation of the potentially harmful role of moralists, in *De Cive*, illustrates the possibility of negative effects of commitment to seductive principles. Even though personal commitment might appear virtuous and a cause of trust, the aggregate effects in political society at large can be dangerous, especially because it can impact negatively the trust the people place in rulers. In Spinoza's critique of laws which ban the ostensible use of luxury goods (in *Tractatus politicus*), the problem is the deterioration of the status of the law, when one superficially hopes to implement ethical norms through law.

It can be added that norms that express important underlying principles or objectives and are legitimized by agreements serve as a medium for dialogue between organizations or institutions. Commitment to these norms is not the end of the story; rather, it is a beginning, as further normative developments will follow from the institutional exchanges they help structure. For these and other reasons, the expression of principles in agreements—effective or not, adequate or not—fully depends on the complexity of social and institutional interactions.

In the regulation of systems considered as "markets", the procedural and contractual dimension of the emergence of agreements is often privileged today when we aim at promoting objectives with substance in terms of social or intergenerational justice. The dominant theories of the regulatory state are precisely those conceptions of the role of public power which happen to be compatible with this approach, or even promote it. One can think here of the implementation of relatively consensual principles (such as those of the Sustainable Development Goals) which is important, in particular, for inter-organizational aspects, such as extrafinancial communication about corporate performance.

It is an approach that has been integral to the historical Corporate Social Responsibility movement, with its search for appropriate and effective channels through which we can encourage the emergence of decentralized agreements on standards and practices in all spheres of society. It is seen as a way to help social life reflect a number of principles that are given a positive value (those of CSR). There is thus an explicit link between the ambition of justice, organizational reflection (which is partly strategic in character) and the explicit (but problematically effective) handling of complexity as such.

The implementation of general principles or objectives often involves inserting "meaningful" agreements or contracts (or those typical of a "mission" of the partners concerned) into a more general system of interactions (investments, market relations, consumer/user choices, etc.) or into "systems of action" as in Coleman, 1990). However, it is quite difficult to reconcile a contractual approach to the actions of agents (such as firms) with a global approach to the correct standards of economic organization. In other words, committing oneself to norms through the medium of agreements between stakeholders (in a contractual or quasi-contractual format) is not immediately the same thing as encouraging the implementation of these norms.

Inserting agreements or contracts into an overall system of normative commitments is not without its problems, all things considered. In the first place, the sincere formation of choices about alternatives does not coincide, in the generality of cases, with the simple and clear expression of commitment to values about precise states of the world. This is a lesson that can be drawn from some of the well-known paradoxes in social choice theory (Sen, 1970; Gibbard, 1974). Moreover, behavioral incentives, invitations to reach agreements (which are sources of legitimacy), and changes in the information regime may give rise to perverse effects by causing undesired states of affairs, due to the unexpected intervention of certain social mechanisms. Another problem has to do with the fact that the background principles that structure the interaction systems to be regulated (e.g. competition principles) may impose limits on the credibility and effectiveness—as well as the institutional and legal robustness—of the agreements that are reached (as in "greenwashing" and strategic interpretations). Institutional initiatives themselves can be oriented toward given ends and objectively serve *other* ends because of the interplay between institutional procedures (in cases of "oblique politics", Picavet & Hadhazy, 2014).

Normative guidance is inserted into interaction systems, and with many problems, but this does not make it invisible or entirely dissolved into complex interactions. The reference to values or principles gives interactions

196   E. Picavet

a particular color, by specifying types of action, relevant controls that one commits to (bringing into play "conformity" and "quality" approaches), and also long-term commitments that have real consequences. These commitments are a contribution to the identity and image of the actors, and also to their ways of getting along with standards, labels and a grammar of external communication in a world of institutions. Institutional complexity connects up with interpretative complexity and this has an impact on institutional communication and the perception of the acts of individual actors in collectives.

Individual Reasons, Collective Capacities and Responsibility in the Face of Risks

The sudden materialization of risks for the community in times of crisis requires an examination of the setup of collective responses. The capacities for collective action are therefore at stake. We may think, for example, of events as different from one another as major nuclear accidents, the 2008 crash, the revelation by the Rana Plaza fire of the endangerment of the lives of others at the heart of the so-called "globalization" of economic exchanges and investments, or the health crisis of covid-19. In each case, the capacity to provide effective and consistent answers is a major concern even in societies where dominant ideologies stress that contractual relations (without a global plan), privatizations, a distant and "regulatory" State, etc., should be valued and even cherished. In many areas, it is the opposite of concertation and collective organization which has been promoted in recent decades, in sharp contrast with the discourse of progress and modernity in the postwar era, individualism as a norm, competition, the confinement of civic ambitions and State action, the repression of solidarity have been important tendencies in the capitalist Western States and in other parts of the world since the 1980s.

The confrontation with reality that major crises induce does not always turn to the advantage of the prevailing, decentralized forms of organization. Therefore, the loss of confidence in the authorities is also a risk that can materialize and it can bring about a loss of organizational capacity, hence a loss of capacities in collective action. Complexity is added by public debt as a constraint on public action, in connection with the conditions faced by States in the financial world as it is. On the one hand, it can prevent the State from being ambitious; on the other hand, the very nature of debt in the political sphere, and the pros and cons of canceling debts, are frequent topics in ordinary political discussion. The adoption of models of role allocation, responsibility taking, institutional watchfulness and collective adaptation to risk is a major avenue for facing this risk. The coupling of efficiency and legitimacy is very strong.

The crisis reveals a dimension of our dependence on models: they contribute to shaping collective attitudes toward risks. They are therefore at the heart of due diligence and responsibility as they are collectively implemented in procedural schemes. The models which play a role in governance (e.g. economic models) interact with the models we elaborate (and use in practice) about our own behaviors, beliefs and expectations, about our own attitudes and mutually recognized roles, and about our own interactions with others. Thus, the general framework of methodological individualism is *a priori* capable of yielding insights about the interplay of collective models of responsibility and individual reasonings about responsibility. The study of this interplay is at the heart of collective progress in social regulation. Crises lead us to question the models of collective and institutional action, and the models of individual conduct, that are supposed to allow us to deal with risks and to face the unexpected.

There is a problem of choice and method, then, concerning the social mechanisms to be chosen and the role to be given to individuals in them. Possible choices include giving a priority to legal responsibility (calling for control, sanctions and institutionalized arbitration), or to economic responsibility—calling for incentives, possibly supplemented by "nudges"— or else, to ethical responsibility (calling for a strengthening of education and professional/political awareness) and mechanisms which favor norm-based reputation. In each case, focusing on the collective or on the individual dimensions of social life proves crucial. For example, the individual capacity to give explanations and provide reasons for the committing of epistemic faults is at the root of the proposed mechanism in Lamy (2022). This appears to be closely associated with a conception of epistemic responsibility which concentrates on personal virtues and blameworthiness (or praiseworthiness) at the individual level, rather than structural traits in institutions. Then the individual-based development of an "integrative framework" for understanding epistemic virtues, which focuses on the reasons for complying with an epistemic variety of ethics, becomes an institutional affair, provided the appropriate initiatives are taken at a collective level.

4 Conclusion

In the contemporary socio-economic world, efforts to regulate social life in the presence of complexity involve the search for agreement around general principles which find a minimal expression in systems of interactions that they allow to emerge spontaneously to a certain extent. This is probably the reason why the politics of big principles, nowadays, is conducive to inter-organizational developments, which save significant margins of freedom of choice for organizations.

This general context helps to explain the centrality, in the eyes of authors who are attentive to the evolution of the State and the frameworks of organized collective action, of the figure of the regulatory State. This form—or should we say *stage* ?—of the modern State inevitably leads to disappointments in connection with collective action and the realization of guarantees that constitute rights. Thus, we are not doing anything very serious about global warming, rich countries are not eradicating poverty and are often experiencing a decline in life expectancy in at least part of the population, poor housing (or the absence of housing) is not being remedied despite the very high level of taxes, the right to a healthy environment is ineffective, etc. In spite of all this, the aforementioned context does not at all eliminate the aspiration to master complexity in order to make organized society truly expressive of principles that we hold dear.

Methodological individualism provides insights for a useful synthesis of the positive, normative and hermeneutic dimensions of the problems associated with dealing with individual reasons in the associated processes. However, not just any picture of the individual will do; useful insights presuppose that we start from a sufficiently rich or complex notion of individual reasons, to avoid misrepresentations which follow from arbitrary over-simplifications.

References

Arnsperger, C. (2005). *Critique de l'existence capitaliste : pour une éthique existentielle de l'économie*. Cerf.

Beck, U. (1986). *Risikogesellschaft. Auf dem Weg in eine andere Moderne*. Frankfurt am Main : Suhrkamp. Engl. Transl. M. Ritter, *Risk Society: Towards a New Modernity*. Sage Publications, 1992.

Boudon, R. (1979). *La Logique du social: introduction à l'analyse sociologique*. Hachette. Engl. Transl. D. Silverman, *The Logic of Social Action*. Routledge and Kegan Paul, 1981.

Boudon, R. (2013). *Le Rouet de Montaigne: une théorie du croire*. Hermann.

Bridel, P. (2009). Passions et intérêts revisités. La suppression des "sentiments" est-elle à l'origine de l'économie politique? *Revue européenne des sciences sociales/ European Journal of Social Sciences, 47*(144), 135–150.

Brisset, N. (2016). Performativité et autoréalisation: le cas de la finance. *Revue Européenne des Sciences Sociales/European Journal of Social Sciences, 54*(1), 37–73.

Brisset, N. (2020). *Economics and performativity: Exploring limits, theories and cases.* Routledge.

Callon, M. (2010). Performativity, misfires and politics. *Journal of Cultural Economy, 3*(2), 163–169.

Coleman, J. S. (1990). *Foundations of social theory,* The Belknap Press of Harvard University Press.

De Bruin, B. (2013). Epistemic virtues in business. *Journal of Business Ethics, 113*(4), 583–595.

De Bruin, B. (2015). *Ethics and the global financial crisis.* Cambridge University Press.

De Certeau, M. (1990). *L'invention du quotidien. 1. Arts de faire.* Gallimard. Engl. Transl. S. Rendall, *The practice of everyday life.* University of California Press, 1984.

Demeulenaere, P. (1996). *Homo oeconomicus.* Presses Universitaires de France.

Demeulenaere, P. (2003). *Les Normes sociales. Entre accords et désaccords.* Presses Universitaires de France.

Freeman, R. E. (1984). *Strategic managemen: A stakeholder approach.* Pitman.

Gibbard, A. (1974). A Pareto-consistent libertarian claim. *Journal of Economic Theory, 7*(4), 388–410.

Heiner, R. (1983). The origin of predictable behavior. *The American Economic Review, 73*(4), 560–595.

Hirschman, A. (1977). *The passions and the interests: Political arguments for capitalism before its Triumph.* Princeton University Press.

Klein, D. (2022). *Regulation in einer solidarischen Gesellschaft. Wie eine sozial-ökologische Transformation funktionieren könnte.* VSA Verlag.

Lacroix, A. (2009). *Critique de la raison économiste : l'économie n'est pas une science morale.* Liber.

Lamy, E. (2022). Epistemic responsibility in business: An integrative framework for an epistemic ethics. *Journal of Business Ethics,* March.

Livet, P., & Conein, B. (2020). *Processus sociaux et types d'interactions.* Hermann.

MacKenzie, D. (2003). An equation and its worlds : Bricolage, exemplars, disunity and performativity in financial economics. *Social Studies of Science, 33,* 831–868.

MacKenzie, D. (2006). Is economics performative ? Option theory and the construction of derivatives markets. *Journal of the History of Economic Thought, 28*(1), 29–55.

MacKenzie, D., Muniesa, F., & Siu, L., (Eds.). (2007). *Do economists make markets?.* Princeton University Press.

Majone, G. (1995). *The European community as a regulatory state.* Martinus Nijhoff.

Padioleau, J.-G. (2000). Praxis d'une science sociale de l'action publique. In Baechler, J., Chazel, F. & Kamrane, R (Eds.), *L'acteur et ses raisons. Mélanges en l'honneur de Raymond Boudon*, (pp. 340–350). Presses Universitaires de France.

Pareto, V. (1905). L'individuel et le social. Congrès International de Philosophie. Incl. In V. Pareto, & G. Busino (Eds.), *Mythes et idéologies*, (pp. 259–265). Genève: Droz, 1984.

Picavet, E., & Hadhazy, A. (2014). Sur l'internationalisation des enjeux éthico-politiques et les aspirations européennes. *Incidence*, 10, 155–176.

Popper, K. (1963). *Conjectures and refutations*. Routledge and Kegan Paul.

Pritchard, D. (2021). Intellectual virtues and the epistemic value of truth. *Synthese*, *198*(6), 5515–5528.

Reynaud, B. (2002). *Operating rules in organizations: Macroeconomic and microeconomic analyses*. Palgrave Macmillan.

Sen, A. K. (1970). *Collective choice and social welfare*. Oliver & Boyd, and Amsterdam: North Holland.

Sugden, R. (1989). Spontaneous order. *Journal of Economic Perspectives*, *3*(4), 85–97.

Tirole, J. (2016). *Economie du bien commun*. Presses Universitaires de France. Engl. Transl. S. Rendall, *Economics for the common good*. Princeton University Press, 2017.

Walter, C. (Ed.). (2010). *Nouvelles normes financières*. Springer.

Analytical Tools and Exemplar Case Studies

Examples of Sociological Explanation in Terms of Methodological Individualism

Raymond Boudon

1 Introduction: The Sociological Object

Raymond Boudon (1979, p. 53)

We can now acknowledge that the unity of sociology clearly resides in the specific character of its language. This language can be defined by three fundamental postulates, as follows: the logical atom of analysis is constituted by the individual social actor; the rationality of actors is generally of a complex type (one cannot generally account for logical actions in Pareto's sense with the aid of a single scheme); the actors are included in systems of interaction whose structure fixes certain constraints upon their action (other constraints being, for example, their cognitive or economic resources). In this chapter, we will have ample opportunity to clarify these assertions and check their validity.

If you watch sociologists at work, you will note that their research is motivated by varying questions. In certain cases, the initial concern takes the form of a general investigation: for instance, how best to explain variations in the rates of suicide in time and space (Durkheim)? In other cases, the sociologist's attention is drawn toward an object rather than a question.

R. Boudon (✉)

Department of Sociology, Academy of Moral and Political Sciences, Sorbonne University, Paris, France

e-mail: nathalie.bulle@cnrs.fr

© The Author(s), under exclusive license to Springer Nature Switzerland AG 2023

N. Bulle and F. Di Iorio (eds.), *The Palgrave Handbook of Methodological Individualism*, https://doi.org/10.1007/978-3-031-41508-1_9

204   R. Boudon

Therefore, he might be intrigued by a particular concrete system of interaction without being in a position, at least at the start of his research, to formulate precise questions or hypotheses: the adolescent gang observed by Whyte (1943) in *Street Corner Society*, or the industrial monopoly in Crozier (1963)'s *Bureaucratic Phenomenon*. At other times, he concerns himself with what might be called "systems of interdependence", like the post-Second World War educational market studied by Boudon (1973/1974) in *Inequality of Opportunity*. In yet more cases, he is concerned with *processes*: *The division of labour* (Durkheim, 1893/1960), the development of the family toward the nuclear model in industrial societies (Parsons, 1951), or changes in the types of personality accompanying certain institutional and social changes (Inkeles, 1959; Merton, 1949; Riesman, Glazer & Denney, 1950).

But quite often, research begins with a *particular fact* (an event or datum) in the logical sense and, possibly, in the double sense of a particular word. For example[1]:

1- Why did Athens' allies defect in the Peloponnesian War?
2- When does social organization aim at eliminating unintended effects?
3- Why does the rule of unanimity often prevail in traditional village societies?
4- Why do members of an unorganized group tend to defect?
5- Why are collective powers often governed by the iron law of oligarchy?
6- Why did capitalist agriculture develop much more slowly in France than in England in the eighteenth century?
7- Why has the immoral character of interest lending disappeared in modern societies?
8- Why is there no socialism in the United States?
9- Why do economic booms seem to be associated with higher suicide rates?
10- Why does the diffusion of an innovation follow a chain reaction process in situations where interpersonal influence is greater?
11- Why were Mithra cult and Freemasonry, respectively, successful in Ancient Rome and modern Prussia?
12- Why were the peasants in ancient Rome hostile to monotheism?
13- Why did the French intellectuals of the late eighteenth-century worship Reason?
14- Why did Indian peasants not adopt the birth control measures advocated by the Indian administration?

---

[1] Translator's note: The questions here refer to those discussed in the following and not to those of Boudon (1979). The titles have been adapted.

15- Why do conflicts between employees in a Taylorized firm tend to be more violent?

These few examples suffice to invalidate the well-supported argument that sociology should be principally concerned with the search for laws and should represent, in the social world, a science equivalent to physics in the natural world—at least in terms of its ambitions.[2] The examples above, which could easily be multiplied, reveal that the sociologist's questions are quite often comparable to those of the historian.[3] If pushed, one could even say that the questions just listed concern the historian as much as the sociologist. In every case, the wording of the question is appropriate for an examination of the causes(s) of a *particular* state of affairs, whether it is an *event* [...], a particular *trait* [...], or a *datum* located in history [...].

We will examine several responses that sociologists have given to the preceding questions. This examination will give us an opportunity to grasp the specific character of sociological analysis when it is concerned with topics drawn from the areas of interest of both the historian and the sociologist.

2 The Structure of Explanation in Methodological Individualism

Raymond Boudon (1984a, **p. 40**)

Let us consider any social or economic phenomenon, M, that we are trying to explain. M must be interpreted as a function $M(m_i)$ of a set of individual actions m_i. As for the individual actions m_i, they are themselves, under conditions and in a manner to be specified, functions m_i (Si) of the structureSi of the situation in which the agents or social actors are found. The function (in the mathematical sense) $m_i(Si)$ must be interpreted as having for the actor i a function *of adaptation* to the situation Si. Weber would have said that the action m_i must be *understandable*. The structure Si is, on its side, a function Si(M') of a set M' of data defined at a macrosocial level or at least at the level of the *system* inside which the phenomenon M develops.

Explaining M is, in short, according to this general paradigm [Methodological Individualism], specifying the terms of $M = M \{m[S(M')]\}$.

---

[2] See Stark (1962), Part Two.

[3] The question of the relation between history and sociology is a classic. See for example, Braudel (1962); see also Elias (1977).

3 Why Athens' Allies Defected in the Peloponnesian War?: Thucydides

Raymond Boudon (1992, pp. 47–48)

Thucydides demonstrates that the setbacks experienced by Athens during the Peloponnesian War were caused to a large extent by the poor organization of its alliance system, which facilitated the emergence of undesirable compositional effects by encouraging the allied cities of Asia Minor to "go it alone". In the vocabulary of game theory, the structure identified by Thucydides is called the "prisoner's dilemma", due to the fable with which it is illustrated.

Consider a two-person interaction situation and denote by CC, CD, DC, DD the four situations created by the choice of two actors between cooperation and defection: (Fig. 1).

The situation CD is, for example, the one where the first actor cooperates (C) and the second one defaults (D).

In the "prisoner's dilemma" game:

Actor 1: DC > CC > DD > CD.

Actor 2: CD > CC > DD > DC.

[...] Defection is the "best" choice in an unconditional way.[5] This is the example of Thucydides: The allied cities thought that Athens had, in any

		Strategies of Actor II	
		C	D
Strategies of Actor I	C	C, C	C, D
	D	D, C	D, D

Fig. 1 The structure of the Prisoner's Dilemma[4]

---

[4] Translator's note: The figures indicate the preference-order. Those before the comma relate to the first actor, those after the comma relate to the second actor.

[5] Translator's note: As Boudon (1979, pp. 66–67) explains, the perverseness of this structure stems from the fact that CC (the situation where the two actors would choose the cooperation strategy) is not ranked first by either of the two actors. Thus, each could be tempted to choose D (defection), while hoping that the other would choose C (cooperation). The structure thus encourages the two actors to use strategy (defection). But, if they both do, they would both obtain an unsatisfactory result since the result (DD) is only ranked third in their system of preferences. It should be noted as well that the worst situation for each of the actors is where one of them is cooperation while the other is defection. As a result, it will be to each one's advantage to choose the strategy of defection in a case where one has reason to fear that the other might be defective himself.

case an interest in maintaining its alliance system and that they could consequently abstain from any contribution. By doing so, they contributed to weaken the system at stake, played the game of the adversary, and harmed themselves.

[…] The example of Thucydides suffices to show that the interaction structures studied by game theory do occur in reality. To take the "prisoner" dilemma effect as an example, one can find many illustrations of it in social life. It explains many banal phenomena of everyday social life, such as the queues that form in front of cinemas. But it also accounts for more significant phenomena, such as the arms race, or advertising waste: In many cases, a businessman must advertise as soon as his competitor does; but, even supposing that advertising has an effect, it may happen that neither of them benefits from it, so the advertising expenditure corresponds to a dead loss. Such an effect is not only undesirable, but absurd; yet, it results from the aggregation of rational behavior in the narrowest sense.

4   When Does Social Organization Aim at Eliminating Unintended Effects?: James Buchanan and Gordon Tullock

Raymond Boudon (1977, pp. 52–54)

Buchanan and Tullock (1965) book, *The Calculus of Consent,* considers the problem posed by Jean-Jacques Rousseau in the *Social Contract*: What type of representative organization should be adopted so that the unintended effects of the prisoner's dilemma type are eliminated? Buchanan and Tullock's answer consists first of all in noting that when it comes to eliminating an unintended effect, one must consider the costs of this elimination. As we know from Rousseau, these costs can be high: In order to gain access to civil liberty, the savages must abandon their natural liberty, that is, agree to submit to the constraints and sanctions provided for by the contract. The interest of Buchanan and Tullock's book is that it puts Rousseau's proposals into perspective: In some cases, the elimination of an unintended effect may be more unpleasant and costly than the unintended effect itself. For example, below a certain traffic threshold, it is preferable not to install red lights: Apart from the costs of installation, the unnecessary waits at red lights may be more difficult to bear in total than the inconvenience of not installing them. Of course, beyond this threshold, the situation is reversed.

[…] Similar considerations apply to more complex organizational problems than traffic lights. When a small number n of individuals cooperate in the realization of a task, a democratic type of organization where each has a fraction *1/n* of the decision power can be efficient, in the sense that it minimizes the costs of interdependence: The task is distributed adequately; the elimination of differences of opinion is not too costly (in time, for example). On the other hand, it is clear that as *n* increases, agreement becomes much more difficult to obtain. An asymmetric distribution of decision-making power, which amounts to the establishment of an authority mechanism, will then correspond to a lower level of interdependence costs than the "democratic" solution. For example, if one wanted to adopt the democratic solution of seeking a consensus of motorists rather than the generally adopted authoritarian solution (red lights) for traffic regulation, the interdependence costs would be so high that they would soon become unsustainable.

Naturally, whenever an organizational problem arises, we encounter the difficulty mentioned above, namely that the optimum type of organization can only be determined in extreme situations.

[…] The basic assumption of the study by Buchanan and Tullock is, in short, that any organization aims to eliminate the costs that each person imposes on others in any situation of interdependence. This amounts to saying that the main function of social organization is the elimination of unintended effects. But this elimination is never free. It entails costs that vary with the nature of the perverse effect to be eliminated and with the type of organization chosen. Unfortunately, it is necessary to add to the proposals of Buchanan and Tullock the proposition that the determination of costs cannot, with some exceptions, be established "objectively". The result is that this solution can only be obtained on a practical level through the opposition and confrontation of points of view. The Leibnizian ideal, non disputemus, sed calculemus, is inapplicable, despite the misleading title of Buchanan and Tullock's book: *The Calculus of Consent*.

5 Why Does the Unanimity Rule Often Prevail in Traditional Village Societies?: Samuel Popkin

Raymond Boudon (2006)

An American author, Popkin (1979), wonders why, in traditional village societies, such as those of Vietnam in the 1920s, which he studied in particular, but also in traditional African societies, the rule for determining the collective will is commonly that of unanimity. His answer is of a finalist nature.[6] According to him, the rule of unanimity is retained in this type of society because any other rule would entail unbearable potential risks for certain categories of citizens and would therefore be rejected. Any rule for determining the collective will involves (Buchanan & Tullock, 1965) two types of costs, characterized by the property that one cannot be mitigated without increasing the other, since the greater the number of individuals whose agreement is required before their opinion can be considered as having the force of law, the more difficult it is to obtain a collective decision, but on the other hand, the fewer the number of those who risk having a measure of which they disapprove imposed on them. If the rule requires the agreement of all, it ensures that no one will have a measure imposed on him/her that is contrary to his/her wishes, but the decision is likely to be interminable. This is the situation described by the film *Twelve Angry Men*: It features an American jury whose decisions are subject to the unanimity rule. If a simple majority is left to decide, the decision-making process is likely to be faster, but in the worst case it risks imposing decisions on half of the individuals that they disapprove of. The question then is why traditional village societies commonly choose unanimity rule and modern societies majority rule? In a case like that of Vietnamese village societies in the 1920s, we are dealing with autarkic societies, whose economy functions at a low level and where, for this reason, free time is not counted: Collective decision-making therefore easily takes up a lot of time. On the other hand, in such societies, any institutional change can seriously threaten the weaker citizens. For example, if a majority were tempted to abolish the right to glean in the rice fields and could impose its will, this would amount to a death sentence for those who survive by gleaning. But a constitution that might allow decisions to be made that would be fatal to a part of the population would be considered illegitimate. Conversely, it is because the rule of unanimity is the only one that presents a real guarantee against the serious risks that the weakest would incur that it is commonly adopted in this type of society.

6 Why Do Members of an Unorganized Group Tend to Defect?: Mancur Olson

Raymond Boudon (1984b, pp. 271–272)

Olson's (1965) basic idea is simple and consists in principle of a generalization of notions that I believe are well known in economics, which is not

---

[6] Translator's note: in opposition to an explanation of a causal nature.

210   R. Boudon

surprising since Olson is an economist by training. Olson starts from the notion of a latent group, which he defines as a set of unorganized individuals with a common interest in the production of a collective good. But it must be specified, firstly, that Olson takes the notion of collective good in a very general sense: It can be an institutional measure as well as a material good; secondly, this good is collective from the moment it benefits all the members of a group, regardless of the group. Of course, a latent group is subject to the free-rider effect insofar as everyone has, under general conditions, an interest in refraining from participating in the collective action likely to produce the good in question. From this simple idea, Olson drew consequences that ultimately contributed to renewing sociological analysis on many topics. Thus, his theory explains that intellectuals are often brought to play a key role in the development of social movements. Indeed, a large, atomized, latent group is likely to be subject to the Olson effect and, thus, to be characterized by the fact that its members are both powerless and frustrated. Participation in collective action is, as long as this group is not organized, desirable and irrational. More precisely, its members are in such a situation that each of them desires the results that collective action could bring about, but is at the same time dissuaded from participating in that action. A latent group of this type thus constitutes an exploitable market for the intellectual and more generally for all those who have the means to express themselves and to be heard. This is why, according to Olson, intellectuals played a considerable role in the development of socialist movements in the nineteenth century, for example. This is also why a man like Ralf Nader played a crucial role in the development of consumer movements.

7   Why Are Collective Powers often Governed by the Iron Law of Oligarchy?: Robert Michels

Raymond Boudon (1977, **p. 45**)

Perhaps the most interesting consequence of Olson's theory from the point of view of social change is the proposition to which Robert Michels gave the name of the iron law of oligarchy. Let us assume that a large number of unorganized individuals have an interest in the production of a collective good. According to Olson's theory, under very general conditions, these individuals will not be able to produce this good. If this is indeed the case, the collective need thus created may tempt an entrepreneur in the Schumpeterian sense. The latter will set up an organization intended to exploit, and to satisfy, this collective need. Coercion or the provision of parallel individual goods will allow the entrepreneur to capture his potential public. But once the organization is created, there is no guarantee that the interests of the individuals it is supposed to represent will be satisfied. Indeed, assuming that members of the organization take actions or adopt a policy that their constituents would disapprove of if they were consulted, we can deduce from Olson's theorem that, unless they are explicitly consulted, they will not express their opposition to the policy being pursued: Such an expression would correspond to the production of a collective good. But the theorem shows that an unorganized group is, under general conditions, incapable of producing a collective good. And the group of constituents is indeed inorganized in its relations with its organization (if we disregard the mechanisms of electoral control of the organization, by the "base", which are by essence intermittent and often symbolic).[7] In other words, in the case where the organization that represents them follows a policy that deviates significantly from the interests of its constituents, the latter are, under general conditions, unable to manifest their opposition. The "solution" to this latent state of crisis is often embodied in a new entrepreneur, who exploits the "market" created by the discordance for his or her own benefit and creates either a new organization oriented toward the same potential audience or an internal opposition to the old organization. In many cases, Olson's theorem shows that the result is oligarchic management of the interests and, in the best case, rivalry between competing oligarchies.

8 Why Did Capitalist Agriculture Develop Much More Slowly in France Than in England in the Eighteenth Century?: Alexis De Tocqueville

Raymond Boudon (1979, pp. 75–77)

In *The Old Regime and the Revolution*, Tocqueville tries to explain why, at the end of the eighteenth century, capitalist agriculture and commerce did not develop in France at the same speed as in England. The principal reason for this, he argues, is that, in the France of the *ancien régime*, the high degree of administrative centralization made the State more prestigious than

---

[7] Moreover, the form of the consultation is usually beyond the control of the constituents. In general, particular conditions must be met for a general consultation to be effectively used by the constituents to correct the line followed by the oligarchy.

212   R. Boudon

in England. It also meant that state offices were more numerous and more sought after in France. As a consequence, when a landowner was faced with the choice of staying on his property and trying to increase its production or acquiring a royal office in town, he would prefer, in general, the second alternative.

> Offices under the *Ancien Régime* were not always the same as our own, but there were even more, I think; the number of small ones was almost endless. In the period from 1693 to 1709 alone, it can be calculated that forty thousand of them were created, almost all within the reach of the lesser bourgeoisie. The eagerness of the bourgeoisie to fill these offices was quite unequalled. As soon as one of them felt in possession of a little capital, instead of using it in trade, he would immediately make use of it to buy an office. This wretched ambition did even more harm to the progress of agriculture and commerce in France than taxes. (Tocqueville, 1856/1952, p. 171)

As this last phrase reminds us, we should not overlook the attraction of the tax exemptions that resulted from settling in the town:

> The bourgeoisie, collected in the towns, had a thousand means of lessening the burden of their taxes and often of escaping them altogether. None of them would have had these means available as individuals if they had stayed on their property... This is, it can be said in passing, one of the reasons that made France more full of towns, and above all of small towns, than the majority of other European countries. (Tocqueville, 1856/1952, p. 171)

This analysis is a good example of the way in which Tocqueville reasons in general. In these analyses, the behavior of individuals is always interpreted as intentional. Social agents are, in other words, described as trying to best serve their own interests, while taking into account the constraints resulting from the context or the system of interaction to which they belong. The multiplicity of offices that were offered, the tax advantages, and possible ennoblement that were attached to them, had the effect of ensuring that the strategy of mobility had much greater *visibility*, *accessibility*, and *attraction* in France than in England. On average, the landowner would therefore choose to leave his land more often in France than in England. The combination of these individual choices thus produced an emergent *macrosociological* effect, namely the under-development of commerce and agriculture.

9 Why Has the Immoral Character of Interest Lending Disappeared in Modern Societies?: Karl Mannheim

Raymond Boudon (1992, p. 506)

Mannheim (1929/1936) explains the disappearance of the belief in the immoral character of interest-bearing loans by the fact that the increasing complexity of economic exchanges makes the charging of interest indispensable. As long as exchanges take place, as in traditional societies, between people who meet and rub shoulders with each other, the loan granted today by A to B can be balanced tomorrow by a loan or a service granted by B to A. But this solution to the problem of balancing the exchanges presupposes that the exchanges are personal. As soon as they are impersonal, it becomes impracticable: The commitment by the borrower to remunerate the service rendered by the lender is then the only way to balance the exchange. In a modern society, the social actor thus has good reasons to find interest-based lending good, whereas the social actor in traditional societies has good reasons to find it bad and immoral. Of course, these reasons may not be immediately apparent, and it must be remembered that perception of change presupposes adequate categories for thinking about it. This is why the Church, which was for a long time an essential source of the conceptualization of the world, was able to resist this inversion of values for a long time and continued to condemn interest lending. This resistance must, of course, be analyzed according to Mannheim as having its own reasons.

10 Why There is no Socialism in the United States?: Werner Sombart

Raymond Boudon (1984b, p. 270)

In a famous book, also from the turn of the century, Sombart (1906/1976) asks «Why is there no socialism in the United States?" Why, alone among the great industrialized nations, did the United States not have significant socialist movements in the nineteenth century? The answer, which I will simply summarize very briefly, can be translated into the language of A. Hirschman (1970). Social mobility and geographic mobility are, due to the border phenomenon, much more developed in the United States than in the old European countries, explains Sombart. As a result, an individual who is unhappy with his or her lot is more likely in the United States to use an individual strategy of *exit*, of *defection,* than a *collective* strategy of *protest,* of *voice* in Hirschman's sense. Moreover, the attenuation of status symbols in the United States means that the possibility of changing status is perceived as easier for an American than for a European, who is confronted daily with the symbols of distinction between classes. In short, the individual strategy of *exit* is likely to appear more attractive to the American than to the European, who, on the contrary, will be more likely to conceive that the improvement of his or her condition depends on the improvement of the condition of the group to which they belong. In short, the benefits of *exit* appear to the individual to be greater in the United States, while the costs of this strategy are lower. On the contrary, the benefits of protest, of collective action, are perceived as higher in Europe and the costs as lower. Yet, Sombart continues, socialist ideologies share a common point beyond the diversity of their forms: they are ideologies aiming at legitimizing the defense of disadvantaged groups in the name of the general interest. Therefore, they can only attract a clientele in situations where individuals tend to prefer the strategy of protest to that of *exit*. Of course, this is a model, i.e. a deliberately simplified interpretation. And Lipset (1983) has shown that this model, while perfectly acceptable, can be enriched by a more systematic comparative analysis than that of Sombart (1906/1976).

11 Why Do Economic Booms Seem to Be Associated with Higher Suicide Rates?: Emile Durkheim

Raymond Boudon (1989**, pp. 24–25;** 1998**, pp. 119–122**)

Durkheim rejects "psychology" in his doctrinal texts, but it is omnipresent in his analyses. Not only does he not eliminate in practice the "abstract psychology" of Simmel-Weber, but one can find behind many of Durkheim's analyses, not only in *Forms*, but also in *Suicide*, an application of the methodology of understanding sociology as defined by Weber.

[…] For Durkheim, "egoism" is a global, social variable. It is the societies that are, in his vocabulary, more or less egoistic. But his choice of indicators of egoism is based on psychological hypotheses. Why are variables such as marital status or sex promoted to the rank of indicators of egoism? Because, Durkheim suggests, the single man is, by virtue of his social situation, more tormented than the married man, the widower more anxious, the woman more oriented toward domestic worries. The same is true for indicators of "anomie": the industrialist is doomed to take risky bets on the future; the intellectual is professionally condemned to doubt.

[…] This is clearly seen in one of the most brilliant analyses contained in this work, the one in which Durkheim attempts to account for the paradoxes raised by the correlations between suicide cycles and economic cycles:

> When Durkheim (1897/1962) tries to explain in his *Suicide* why economic booms seem to be associated with higher rates of suicide, he introduces an explanation of the subjective rationality type, not very far away from ideas developed by A. Hirschman (1980) in his theory of the tunnel effect. Durkheim's assumption is namely that the anticipations and expectations of social actors are grounded on good reasons: During a period of stable economic development, they tend to start from the principle that they can expect, say, for the year to come, the same gains as the year before, while during an economic boom, when the situation of many people appears to be getting better, they will change their conjectures as to which objectives can be reached and aimed at.
>
> Durkheim introduces implicitly at this point a very brilliant hypothesis, namely that people would extrapolate from the tangent to the curve at each point of time. So that in the first part of the ascending phase, before the inflection point, their expectations would tend to be under-optimistic, while in the second part they would be over-optimistic. This is at least my interpretation as to why Durkheim predicts an increase in disillusion and consequently in suicide rates in the second part of the ascending phase of the business cycle, but not in the first. (Boudon, 1989, pp. 24–25)

I could make the same remarks about the difference that Durkheim introduces between political and economic crises. One can only understand that one indicator is placed on the side of anomie and the other on the side of egoism if one sees the psychology behind this distinction. What Durkheim tells us about political crises and the fact that they provoke a renewed interest in public affairs anticipates very accurately what Hirschman says in his very Weberian analyses of *Private Happiness, Public Action* (Hirschman, 1982).

12 Why Does the Diffusion of an Innovation Follow a Chain Reaction Process in Situations Where Interpersonal Influence is Higher?: James Coleman & al.

Raymond Boudon (1984a, pp. 46–47)

Coleman (1957) and colleagues' study of the diffusion of pharmaceutical innovations reported a mysterious result. When we consider the population of physicians working in hospitals, the diffusion process follows a characteristic pattern. It is very slow at the beginning: The number of physicians adopting the novelty grows very slowly. Over time, the process accelerates: The number of converts increases more and more rapidly. The speed of the process is maximum when about one in two physicians is converted. From that point on, the pace of the conversion process slows down steadily and becomes extremely slow again when *almost* everyone has adopted the new product. When the process is plotted in a Cartesian diagram, with time on the x-axis and the cumulative number of converts at each unit of time on the y-axis, the process takes on a *sigmoidal* shape: It has the form of an "S" curve. Why is this so? Why, at the aggregate or collective level, that is, at the level of the population considered as a whole, do we observe that the process has a characteristic "S-curve" structure? In the previous symbolism, this characteristic structure represents the aggregate phenomenon M that needs to be explained. The explanation of M as provided by Coleman consists, as we shall see, in specifying the terms of:

M = M {m [S (M')]}, briefly M = MmSM'.

Of course, everyone can verify that the explanation is not immediate. Why this sigmoid structure? The mystery thickens when we note that this structure, characteristic of doctors working in hospitals, does not apply to doctors in private practice. In the latter case, the number of physicians adopting the novelty grows very rapidly at first. Then the rate at which new physicians are converted steadily decreases. It becomes slower and slower as more physicians are won over and tends to zero when *almost* all are won over. When the process is plotted in a Cartesian diagram, again with time on the x-axis and the number of converts at each successive time on the y-axis, the result is not an "S" curve, but an arc curve.

Excerpt from Coleman & al (1957, pp. 261–262):

> In Figure 2 the upper curve (which is roughly similar in shape to the curve for the integrated doctors) represents a snowball process in which those who have introduced pass on the innovation to their colleagues. This curve is described

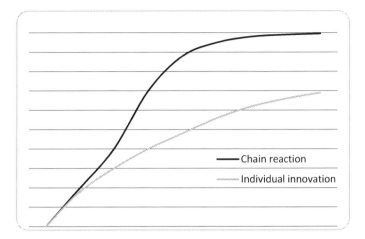

Fig. 2 Comparison of model of "chain-reaction" innovation with model of individual innovation: Evolution of cumulative proportion of individuals who have introduced the innovation Months after start of process (*Source* Coleman, Katz & Menzel [1957])

by an equation which has been used to characterize rates of population growth, certain chemical reactions, and other phenomena which obey a chain reaction process. The lower curve in Fig. 2 is still the individual innovation process. Technically, the individual and snowball processes are described by equations on the graphs, which can be paraphrased as follows: [1] Individual process- the number of doctors introducing the new drug each month would remain a constant percentage of those who have not already adopted the drug. [2] Snowball process--the number of doctors introducing the new drug each month would increase in proportion to those who have already been converted.

$$\frac{dy}{dt} = k(1-y) \qquad [1]$$

$$\frac{dy}{dt} = ky(1-y) \qquad [2]$$

In the fundamental formula that describes the Weberian paradigm, the structure of the situation S is presented as dependent on macrosociological data, or at least on data defined at the level of the social *system* considered in the analysis: S = S(M'). This is indeed the case here; the hospital structure has the effect of providing the doctor with access to quasi-free information of greater credibility than that provided by impersonal sources: the opinions of his colleagues. Moreover, and this is also the result of the hospital structure and the intercommunication networks that it allows, the doctor can consider

218   R. Boudon

himself in a position to assign a greater or lesser degree of credibility to the opinions of his colleagues.[8]

The situation of the office physician is *structurally* different. [...] a contrast appears as soon as it is a question of the office doctor having access to additional information. Indeed, while the hospital doctor can mobilize easily accessible sources of information, the isolated practitioner appears to be deprived. [...] He is therefore likely to rely much more heavily than his hospital colleague on impersonal sources ofinformation. The structure of the situation (S) in which he finds himself is different; his behavior $m(S)$ is therefore likely to be different.

What does this mean at the aggregate level? [...] At each point in time, the increase in the number of office-based practitioners who choose to accept the innovation will be greater the total of those who *have not yet converted*. In contrast, in the case of hospital doctors, because of the interpersonal influence made possible by the hospital structure, this growth is, at each moment, greater the number of nonconverters *and* the greater the number of converts. In the case of individual practitioners, this phenomenon of interpersonal influence is of much more modest importance.

13   Why Were Mithra Cult and Freemasonry, Respectively, Successful in Ancient Rome and Modern Prussia?: Max Weber

Raymond Boudon (1999a, pp. 140–141)

Why, in ancient Rome, in modern Prussia and elsewhere, do civil servants, soldiers, and politicians tend to feel attracted to cults which, like Mithraism or Freemasonry, propose a disembodied vision of transcendence, which see transcendence as subject to regimes that go beyond it and conceive of the community of the faithful as hierarchical under the effect of initiatory rituals? Why did the Roman emperors, from Commodus to Julian, protect Mithraism, just as the kings of Prussia protected Freemasonry? Because the articles of faith of these cults are congruent with what can be called the social and political philosophy of the civil servants, the military, or the politicians. They believe that a social system can only function under the control of a central authority perceived as legitimate; that said authority must be driven by impersonal rules; they adhere to a functional and hierarchical vision of

---

[8] On the decisive role of interpersonal influence in the credibility attached by the actor to impersonal messages, see the classic study by Katz and Lazarsfeld (1955).

society; at the same time, they believe that this hierarchization must, as is indeed the case in the Roman or Prussian state, be based on competences controlled by formalized procedures. The principles of political organization of the State seem to them to translate a just political philosophy; as for the initiatory rituals of Mithraism or Freemasonry, they perceive them as expressing the same principles in a metaphysical-religious mode. Of course, other factors must be taken into account to explain the expansion of these cults, and it should be noted, for example, that Mithraism was not opposed, unlike Christianity, to the practice of other cults by its initiates. Particular factors explain why French Freemasonry did not find the same support from the monarchy as Prussian or English Freemasonry. But what primarily explains the success of these cults is that they contain visions of the social order that appear acceptable to certain categories of social actors.[9]

14 Why, in Ancient Rome, Were the Peasants Hostile to Monotheism?: Max Weber

Raymond Boudon (2013, pp. 89–90)

Why, asks Max Weber (1920/1993), were Roman officers and civil servants attracted to monotheistic cults imported from the Middle East, such as the cult of Mithra, while the peasants were permanently hostile to them and remained faithful to the traditional polytheistic religion? The hostility of the latter to Christianity was so deep that the word paganus (peasant) came to designate the opponents of Christianity, the pagans.

As regards the peasants, Weber explains that the unpredictability of natural phenomena which dominates their activity seems to them incompatible with the idea that the order of things can be subjected to a single will, necessarily endowed with a minimum of coherence. They accepted Christianity more easily as soon as a crowd of saints gave it a polytheistic character more compatible with their experience.

In a word, the Roman peasants based their beliefs on a system of reasons that seemed to them valid. It is the same for officers and the civil servants: With its hierarchical side and its impersonal rites of initiation, the cult of Mithra seemed to them to faithfully reflect the organization of the Roman Empire in a symbolic register. This was to facilitate the diffusion of Christianity among them. The contextuality of these ordinary beliefs does not imply that they are irrational.

---

[9] Translator's note: Boudon relies here possibly on the analysis of Max Weber (1922/1968).

15 Why Did French Intellectuals of the Late Eighteenth-Century Worship Reason?: Alexis De Tocqueville

Raymond Boudon (2010, pp. 63–64)

Tocqueville systematically explains the collective beliefs that intrigue him on the assumption that they derive from an argument that none of the real individuals may have ever literally developed, but which can in all likelihood be attributed to an ideal individual: a shining example of abstract psychology.

Why do French intellectuals at the end of the eighteenth century swear by Reason and why do their vocabulary and ideas spread so easily? The question is enigmatic, because the same phenomenon was not observed at the same time in England or the United States, Tocqueville's favorite comparative poles. Therefore, one cannot be satisfied with seeing in it the expression of general tendencies characterizing all modern societies.

In fact, explains Tocqueville (1856), the intellectuals of the second half of the eighteenth century, the philosophers, have reasons to believe in Reason. In France of this time, the tradition seems to many as the source of all the evils. The nobility and the high clergy are discredited. They consume their authority in Versailles. They do not participate in local political affairs nor in the economic life. The small nobles who camped on their lands clung all the more stubbornly to their privileges as they were penniless. This is why they are called hobereaux, named after a small bird of prey. The lower clergy is perceived as having ties with the nobles. In contrast, the English aristocrat has every reason, especially if he has the ambition to be elected to Westminster, to appear to his constituents as contributing to the smooth running of local affairs. From all of this, the feeling emerges in the minds of many French people that the Ancien Régime states have no raison d'être and that, like many other existing institutions, they owe their survival solely to the authority of tradition. Finally, this argument leads to the conviction that a new society must be erected based on the opposite of Tradition: Reason. This is why philosophers of the Enlightenment develop an artificialist vision of societies, proposing to replace the society of their time with a society conceived on plan in the philosopher's cabinet.

16 Why Did Indian Peasants not Adopt the Birth Control Measures Advocated by the Indian Administration?: Peter Berger

Raymond Boudon (1999b, **p. 84**)

At one point, the Indian administration wanted to convince Indian peasants that having few children was necessary if the country's peasantry was to be freed from poverty. The diagnosis was undoubtedly correct. Therefore, the Indian administration engaged the services of a group of Western researchers who designed a campaign to convince Indian women to use contraceptive methods.[10] The campaign repeatedly failed for an interesting reason: These researchers simply thought that the large number of children Indian families had on average was due to the traditional idea that a large family was a gift from heaven. And they thought that it would be enough to open their eyes, so to speak, to show them that these large families were the cause of their poverty and to inculcate contraceptive methods. In other words, they started with the seemingly simple idea that the problem was to counterbalance the weight of tradition by bringing to the attention of the peasants a set of data that seemed convincing to the researchers themselves.

However, as I said, it was a repeated failure. It failed because the peasants saw the message through their own existential interests. In fact, they had told the researchers from the beginning that they did not want to use the contraceptive pills generously offered by the researchers to Indian women for one simple reason: Under present conditions, a large family represents an insurance against all kinds of risk, including those related to health, age, and income. Moreover, a large family provides services to its members that are not otherwise provided. In these circumstances, it could not be expected that they would easily accept a change in behavior that would be detrimental to each of them, even if it turned out to be to the long-term benefit of all of them.

In this example, the sender of the message did not realize that his message would be processed by the receiver through his existential interests. The message did not work because it had been conceived by the sender without taking into account the point of view of the receiver himself, and in particular the structuring of this point of view by the existential interests.[11]

---

[10] Translator's note: This birth control experiment carried out during 1956–1960 by Harvard School of Public Health, with funds of the Rockefeller Foundation and the Indian government, in a group of villages in the Penjab. The follow-up study was conducted in 1969 (Berger, 1974, p. 205).

[11] Translator's note: Here is Berger's conclusion about this experiment: "If the staff of the project had listened to the villagers, instead of having an anthropologist study their alleged superstitions,

17 Why Do Conflicts Between Employees in a Taylor-Based Firm Tend to Be More Violent?: Charles Wright Mills

Raymond Boudon (1995, pp. 283–284)

In a famous passage from *White Collar*, the American sociologist C. W. Mills (1956) describes office workers in a Taylorized company. They are all working on the same task. They are all located in a large room and have the same individual space, furnished and equipped in the same way. Violent conflicts often arise over issues that are easily perceived by the observer as minor. For example, the right distance from light sources or walls is treated as a coveted commodity. Arguments seemingly out of proportion to these issues appear on a regular basis. Generally speaking, these employees are treated the same, but any sign of preference for one of them on the part of the management is perceived as serious, unfair, and unbearable.

Why do we have these conflicts? A superficial observation could lead to explanations by "susceptibility" or by resentment. In other words, it would be easy to imagine, in order to elucidate the mystery, one of those tautological explanations that spontaneous sociology easily produces, and that I denounced a short time ago. In fact, these conflicts reflect only to a small extent individual idiosyncrasies. In this sense, they are different in their very nature from the disputes that may arise between passengers on a train over a window seat. For one thing, travelers are only stuck in the same space for a short time. On the whole, they are simply juxtaposed by the effect of chance, which wanted all of them to take the same train that day for the same destination. Apart from that, there is nothing in common between them. They do not belong, as Mills' employees, to an interacting system. This is why their reactions are so much stronger and, above all, why they have a completely different meaning: In a production-oriented system, where the individual contributions are identical, the rewards must be equal. Consequently, any difference is perceived as profoundly unfair, as betraying the basic rules of the interaction system. Whereas the traveler's reaction is utilitarian (except in

---

they would have had no difficulty understanding this. […] Humanism, from the Renaissance on, has meant a respect for the place of values and meanings in the affairs of men. The humanities have been the disciplines that have studied human events from within, as it were- from within the subjective perceptions of reality that animate actors on the historical scene and that make their actions intelligible to an outside observer. Humanism in this sense has been widely dismissed as unscientific in the ambience of the social sciences, particularly in Anglo-Saxon countries. The discussion of this chapter indicates that this dismissal may have unfortunate consequences. A humanistic approach to development policy (and just as much to the other areas of politically controlled social change) will be based on the insight that no social process can succeed unless it is illuminated with meaning from within".

the case where he has reserved a place by the window and an intruder has moved in), the Mills employee's reaction is inspired by axiological reasons: her protest expresses a reaction of moral indignation against an injustice; the traveler's protest reflects a simple feeling of annoyance.

References

Berger, P. (1974). *Pyramids of sacrifice: Political ethics and social change.* Basic Books.

Boudon, R. (1973). *L'inégalité des chances.* Armand Colin [Transl. 1974. *Education, Opportunity and Social Inequality.* Wiley]

Boudon, R. (1977). *Effets pervers et ordre social.* PUF. [Transl. 1982. *The Unintended Consequences of Social Action.* Macmillan.ç]

Boudon, R. (1979). *La logique du social.* Hachette. [Transl. 1981. The logic of social action. An introduction to sociological analysis. Routledge & Kegan].

Boudon, R. (1984a). *La place du désordre : Critique des théories du changement social.* PUF [Transl. 1991. Theories of social change: A critical appraisal. Polity Press].

Boudon, R. (1984b). L'individualisme méthodologique en sociologie. *Commentaire, 26*(2), 268–277.

Boudon, R. (1989). *Subjective rationality and the explanation of social behavior* [Report]. Max Planck Institute for the Study of Societies. https://ideas.repec.org/p/zbw/mpifgd/896.html

Boudon, R. (Ed.) (1992). *Traité de sociologie.* [*Treatise of Sociology*]. PUF.

Boudon R. (1995). *Le juste et le vrai : études sur l'objectivité des valeurs et de la connaissance.* Fayard. [Transl. 2001. The origin of values. Transaction].

Boudon, R. (1998). *Etudes sur les sociologies classiques.* [Studies on classical sociologists]. PUF.

Boudon, R. (1999a). *Le sens des valeurs.* [The Meaning of Values]. PUF.

Boudon, R. (1999b). Les principaux enseignements des sciences sociales au sujet de la persuasion de masse. [Key social science lessons about mass persuasion] *European Journal of Social Sciences, 37*(114), 83–96.

Boudon, R. (2006). *Quelle théorie du comportement pour les sciences sociales ?* [What theory of behavior for the social sciences?] Open Edition.

Boudon, R. (2010). La sociologie comme science. [Sociology as a Science]. La Découverte.

Boudon, R. (2013). *Le Rouet de Montaigne.* [Montaigne's Spinning Wheel]. Hermann

Braudel, F. (1962). Histoire et sociologie [History and sociology]. In Georges Gurvitch (Ed.), *Traité de sociologie* [Sociology treatise] (pp. 83–98). Presses Universitaires de France.

Buchanan, J., & Tullock, G. (1965). *The calculus of consent.* The University of Michigan Press.

Coleman, J., Katz, E., & Menzel, H. (1957). The diffusion of an innovation among physicians. *Sociometry, 20*(4), 253–270.

Crozier, M. (1963). *Le phénomène bureaucratique.* Le Seuil. [Transl. 2010. *The Bureaucratic Phenomenon* (with a new introduction by Erhard Friedberg) Transactions Publishers].

Durkheim, E. (1893/1997). *The division of labour in society.* The Free Press.

Durkheim, E. (1897/1962). *Suicide, a study in sociology.* The Free Press.

Elias, N. (1977). Zur Grundlegung einer theorie sozialer prozesse. [On the foundation of a theory of social processes]. *Zeitschrift für Sociologie, 6,* 127–49.

Hirschman, A. O. (1970). *Exit, voice and loyalty. Responses to decline in firms, organizations and states.* Harvard University Press.

Hirschman, A. O. (1980). The changing tolerance for income inequality in the course of economic development. In A. Hirschman (Ed.), *Essays in trespassing* (pp. 39–58). Cambridge University Press.

Hirschman, A. O. (1982). *Shifting involvements. Private interest and public action.* Princeton University Press.

Inkeles, A. (1959). Personality and social structure. In Robert Merton et al. (Eds.), *Sociology today* (pp. 249–276). Basic Books.

Katz, E., & Lazarsfeld, P. (1955). *Personal influence. The part played by people in the flow of mass communications.* The Free Press.

Lipset, S. M. (1983). Radicalism or reformism: The sources of working-class politics. *The American Political Science Review, 77*(1), 1–18.

Merton, M. (1949/1968). *Social theory and social structure.* The Free Press.

Mills, C. W. (1956). *White collar. The American middle class.* Oxford University Press.

Mannheim, K. (1929/1936). *Ideology and utopia.* Routledge.

Michels, R. (1959). *Political parties.* Dover.

Olson, M. (1965). *The logic of collective action.* Harvard University Press.

Parsons, T. (1951). *The Social System.* The Free Press.

Popkin, S. (1979). *The rational peasant. The political economy of rural society in Vietnam.* University of California Press.

Riesman, D., Glazer, N., & Denney, R. (1950). *The lonely crowd—A study of the changing American character.* Yale University Press.

Sombart, W. (1906/1976). *Why is there no socialism in the United States?* Macmillan.

Stark, W. (1962). *The fundamental forms of social thought.* Routledge & Kegan Paul.

Tocqueville, A. de. (1856/1952). *L'Ancien Régime et la Révolution.* Gallimard. [Transl. 1998. The Old Regime and the Revolution. Chigago: University of Chicago Press].

Weber, M. (1920/1993). *Gesammelte Aufsätze zur Religionssoziologie.* Mohr. [Transl. 1993. *Sociology of Religion.* Beacon Press].

Weber, M. (1922/1968). *Economy and society: An outline of interpretive sociology.* University of California Press.

Whyte, W. (1943). *Street corner society.* University of Chicago Press.

Collective Action

Anthony Oberschall

Much of the time in everyday life, each person pursues private goals that benefit them and their closest kin and associates. If uniformity of behaviors occurs, it is because everyone responds to similar incentives and constraints, wants, beliefs, and sentiments, shared with thousands of others such as getting up in the morning at roughly the same time, driving to work at roughly the same time, taking an umbrella when the weather forecast predicts rain, and many other behaviors like buying chicken instead of beef when beef prices increase. Uniformity of behavior also results from a central authority that all conform to, like soldiers marching, citizens paying their income taxes at a due date, a religious congregation praying and singing in unison, students taking a test, and many other organized top-down activities.

In some circumstances, people decide voluntarily on joint action for goals and goods that organizations and institutions fail to provide, or prevent them for getting, and that they are unable to obtain individually. They pool their resources and efforts and coordinate their actions. They form lobbies for influencing legislators; they organize a strike or boycott against businesses who refuse to raise wages and who violate labor laws; they participate in mass marches and demonstrations against war, for civil rights, and other public causes.

A. Oberschall (✉)
Department of Sociology, University of North Carolina, Chapel Hill, NC, USA
e-mail: tonob@live.unc.edu

© The Author(s), under exclusive license to Springer Nature
Switzerland AG 2023
N. Bulle and F. Di Iorio (eds.), *The Palgrave Handbook of Methodological Individualism*,
https://doi.org/10.1007/978-3-031-41508-1_10

These are instances of collective action. The unity of behavior is voluntary and purposive and not a byproduct of individual routines or conformity to authority. Success or failure will depend on how many join and how many freeride, how committed and capable they are to assume the risks and costs of contention, on public support for their cause, and on the strength and resolve of opponents.

This essay deals with collective actions that face active opposition. Other situations analyzed by Elinor Ostrom (1990) concern why and how ordinary people join in collective action and create institutions when governments and markets fail to provide collective goods. The basic principles and MI methodology are the same for all collective action, but active opposition by powerful groups necessitates additions to the baseline theory.

At the heart of a theory of collective action are several questions: under what conditions does an issue, such as women's rights, become a public controversy and attract supporters; what sorts of people and groups participate in collective action, both for and against an issue, and what are their beliefs and motivations; how do participants mobilize resources for action; what is the role of leaders; what explains the character and modes of collective action, which can range from local and episodic events to sustained nationwide drives, from spontaneous to organized actions, from law abiding to violent confrontations; what is the reaction of targets and bystanders, be they groups, organizations or government; what measure of success is achieved? (Oberschall, 1993, Chapter 1).

These questions about collective action are salient for the past as well as the present. One should not assume that the past was different because of television, mass literacy, global capitalism, and the internet. Historians of the Protestant Reformation describe the upsurge of religious mass action that swept Europe and diffused in the 1520s and 1530s across some two hundred cities of the Holy Roman Empire (Ozment, 1975). There was public contention over beliefs and theology, ritual and forms of worship, ecclesiastical organization, public and private morality, public debates before thousands of citizens, invasion and sacking of churches and monasteries, the destruction of religious icons and statuary, public weddings of celibate clergy, public trials against heretics, expulsion of opponents from a city, massacres, executions, and other collective violence, all with high rates of participation by ordinary people. Anyone studying civil strife and ethnic conflicts in the contemporary world will recognize much similarity with collective actions in the Reformation. The theory of collective action applies widely to the human record.

1 Methodological Individualism and the Theory of Collective Action

Methodological individualism (MI) is a mode of theorizing about human behavior and institutions. Known by various labels in different sub-disciplines of the social sciences, such as new institutional economics, behavioral economics, public choice, transaction cost analysis, and rational choice, its central assumptions are that thinking, decision-making, beliefs, attitudes, emotions, and actions are anchored in the ultimate reality of individuals. From this micro-reality, MI explains individual behaviors and interactions, but also norms, values, ideologies, markets, organizations, and other meso and macro-level systems of social action, including the institutions of collective action [James Coleman, *Foundations of Social Theory*, 1990].

In sociology, MI has been central to the Weberian tradition of social action. In an important statement, Weber (1904/1964, p. 110) wrote "Ueber einige Kategorien der verstehenden Soziologie" {my translation]: "Verstehen" (comprehension of meanings) is ultimately the reason why...sociology (Verstehende Soziologie) treats the individual and his actions as its basic building bloc, its atom. Concepts like the "state," "association, 'feudalism' and the like designate for sociology...categories for particular types of human interactions, and it is sociology's task to reduce these concepts to 'meaningful' action, which means without exception the actions ...of partaking individual human beings."

For some purposes, macro-level theory yields interesting results, as in international relations where a state is usefully assumed to be a unitary actor. For collective action, many insights and results cannot be gained at the group level alone: the free rider dilemma and how it is overcome; the diffusion of protest in the absence of a common leadership and organization; how powerful autocratic regimes are toppled by mass demonstrations; and other topics in this essay.

A common confusion about MI is that it endorses individualism as the highest value over and above social welfare, public goods, and selfless devotion to a higher ideal. Weber himself warned against this fallacy and argued that **commitment to social values** and goods such as patriotism and humanitarian assistance for the needy are within the scope of "Verstehende Soziologie." Selfishness and self-interest are not the exclusive motivating force of human behavior, nor is it assumed in MI.

An important assumption of the political economy tradition and MI is **rationality** in decision-making, or rational choice (RC). RC assumes that preferences are consistent, (if a is preferred to b, and b to c, then a is also preferred to c), and that choices are "adaptive," i.e., satisfy some criterion of net benefit, subject to constraints. Because many constraints stem from social norms and interpersonal influences, **behavior** for MI is **adaptive and normative**. Rationality does not describe the intentions of the actors but the consequences of their actions. Non-adaptive behavior in the long run self-destructs when others engage in adaptive behavior, and deviants pay a price for anti-social behavior in a normative pro-social environment.

It is a fallacy to think of rationality as ordinary humans making cost–benefit analyses like a computer. MI assumes only a simple **heuristic** for decision-making under uncertainty because information contains errors and is costly, the future holds unexpected contingencies, time for decisions is scarce and costly, nature and human beings are only partly predictable, and actors influence one another strategically. MI rationality is limited and influenced by one's peers and social milieu. Man is a social animal, as Adam Smith expressed it in the *Theory of Moral Sentiments* with his social "mirror" metaphor: unlike a solitary Robinson Crusoe, in society man is provided with the "countenance and behavior of those he lives with," a "mirror" from which he learns the "propriety or demerits of his sentiments and conduct," "the propriety and impropriety of his passions," and "the beauty and deformity of his own mind."

MI views the **aggregation** of individual behaviors to a collective outcome by summation as an oversimplification of causation. Summation assumes that the macro is a large-scale replica of the micro like the identical Russian dolls lined up from small to large. War between states is not a large-scale replica of interpersonal hostility and aggression. To the contrary, the paradox is that young men who are strangers and don't hate one another are made to fight one another to death: "Individual aggression plays little part in total war…war may cause aggression, but aggression does not cause war" (Hinde, 1997). The institutions for war cause war: a military-industrial complex, arms race with rivals, warmongering by patriotic media, political leaders promoting nationalism and xenophobia, security concerns due to rival states' actions. Although some men volunteer for war, many others are compelled by the state to enlist in the military and fight the enemy regardless of their personal disposition. As Mancur Olson (1965) has shown for a large population, collective goods, taxation and war service among them, will not be provided by voluntary participation because of free riding. The state has to coerce participation.

2 Other Theories of Collective Action

A theory of civil strife by Ted Gurr (1970) aggregates from micro to macro. Social breakdown causes relative deprivation and is the basic precondition of civil strife. Deprivation causes frustration, anger, and aggression. In the aggregate, as frustration and anger increases and spreads in a population, it will lead to more aggression and civil strife.

There are intervening conditions, such as the "coercive potential" of the regime. Just as punishment inhibits individual aggression, social control and security agents will have a deterrent effect. The deterrent hypothesis is also based on simple aggregation: more social control, less rebellion. In this theory, there is no start up mechanism: at what threshold does frustration and anger become collective aggression rather than individual anti-social behavior? There is also no termination and satiation mechanism: will reforms aimed at the causes of social breakdown and frustration restrain civil strife or will they unintentionally increase expectations for more reforms and stimulate more civil strife?

In contrast, MI applied to collective action embeds individual dispositions and behaviors in a social context. People become angered by the police use of excessive and lethal force and other denials of social justice not only because they themselves experience it but because of fellow feeling and **solidarity** with victims. An incident of police brutality and injustice can have a **"multiplier"** effect that angers many at the same time and triggers a collective response. The intervening variable between personal disposition and collective outcome is group solidarity, which is central in the theory of collective action.

Another mode of theorizing about collective action is to start with an extreme outcome, like a race riot or insurrection, look backward for antecedent causes, and then apply these explanations to the entire category of collective events, as Gustave Le Bon (1895/1990) did. His three most important principles were the "law of mental unity" of crowds, the loss of rational faculties, and hero worship and blind submission to a leader. Le Bon's view of collective action still resonates with the public, the media, and politicians and is repeated in films, novels, anecdotes, and commentaries.

According to the **law of mental unity** "whoever be the individuals that compose the crowd… the crowd puts them in possession of a collective mind which makes them feel, think and act in a manner quite different from what …they feel, thing and act in a state of isolation." Contrary to Le Bon, researchers find much **heterogeneity** of both participants and behaviors in political crowds. At the core are **leaders and activists** who organize demonstrations on repeated occasions; **transitory teams** are part-timer attracted episodically by the issue; the **conscience constituency** who may not be physically present contributes public opinion support and makes financial donations; **curious bystanders** do not want to miss an event they can tell their friends about and taking photo ops; **radicals** seek to provoke violence by assaulting the police under cover of the crowd; **"commodity rioters"** on the fringes see a criminal opportunity to break into businesses and loot when the police is otherwise occupied. Most crowds are heterogeneous and the diverse behaviors of crowds result from that heterogeneity.

For instance, observers and the police know that when it starts raining a hostile crowd will disperse although teargas and other forceful means failed to do so. It is not because the rain cools anger but because it alters the benefits and costs of participation differently for members of the crowd. The expected cost of participation is a function of the number of participants, with large numbers expected to lower the risk of arrest and other costs. When it starts to rain, the least committed to the cause, the bystanders, leave because getting wet is not worth the benefit of staying. When the crowd diminishes, transitory teams expect the risk of arrest or getting hurt to become too great, and they leave. As the crowd diminishes even more, radicals and looters no longer experience safety in numbers, and also leave. With only small numbers of hard core activists remaining, some risks arrest but others abandon the confrontation. The mental unity of crowds assumed by Le Bon does not explain such melting away by rain.

Teargas may have the opposite effect from rain. Transitory part timers and bystanders who are law abiding get gassed and become angered and more committed to the confrontation because police repression has become added to the original issue. Others get radicalized due to the multiplier effect by observing aggressive police actions against peaceful demonstrators and become more willing to pay higher costs for participation than initially.

It is true that some crowds or groups within a crowd manifest a great deal of unity in belief and action, but in some situations such unity precedes participation, as in the large pro-democracy demonstrations in East Europe that ended communist rule. Thousands of individuals who already harbored strong democracy and civil liberties sentiments converged to historic locations in their cities to hear speakers demanding the end of communist regimes (Oberschall, 1994). Unity results from the convergence of the already united. Le Bon's idea of mental unity has to be specified to allow for many different outcomes and dynamics.

Similarly for the second Le Bon law on the **loss of rational faculties and moral sense** that make individuals, like barbarians, "descend several rungs in the ladder of civilization…and possess the spontaneity, the violence, the ferocity, and also the enthusiasm and heroism of primitive beings." Research has documented that atrocities, crimes against humanity, genocides, and ethnic cleansing have been perpetrated by organized militias, paramilitaries, and cadres trained and indoctrinated to hate their victims and obey their superiors, and not by crowds, although sometimes crowds have committed atrocities (Staub, 1989).

It is obvious that crowds manifest moral sentiments and emotions, sometimes extreme emotions like anger and hatred, but so do individuals in everyday life. Emotions are an integral part of the human condition. Contemporaries recognize and make sense of the moral sentiments and emotions in classical Greek tragedy and of peoples in other cultures. Some like **James Jasper** [*The Art of Moral Protest*, 1997], and before him Le Bon, conclude that **emotions like anger have a causal force that overrides instrumental actions**. The MI baseline model of adaptive and normative behavior is compatible with and incorporates social psychological variables and emotions like commitment, solidarity, fear of arrest and injury, anger at injustice, hope and despair about success and failure, the we-feeling and exhilaration of thousands who topple a dictator. Whether and how far emotions exercise an independent causal force for explaining social action and crowd behavior is a matter for research.

In my research on ethnic conflict in the former Yugoslavia, I found that some Serbs in Banja Luka were in favor of ethnic cleansing Croats and other minorities from their city even though they had good personal relations with Croat neighbors. They shared the fears and security concerns spread by their leaders and conformed to the majority opinion in their social circles. One emotion, personal liking, was overridden by another emotion, fear, which prevailed in their ethnic group. It is often the case that instrumental choices are embedded in a different mix of emotions. What has to be explained is why one set of emotions prevails over others.

At the Metrovica bridge in Kosovo, the city was divided by a river into Serb and Kosovar sections after the war and fighting ended. Serbs and Kosovars were nightly throwing stones at one another over the heads of a thin line of peacekeepers. Both groups were angry. The Kosovars because they were prevented from returning to their homes; the Serbs because they feared losing their last refuge in Kosovo. They did not throw stones because they were angry; they were both angry and threw stones because their homes and security were threatened. Was anger and stone throwing a substitute for instrumental action? Anger was not distorting their perception of the situation. To the contrary, they had an accurate understanding of homelessness and of their predicament. There had been repeated and futile negotiations by international and national leaders and mediators and nothing was happening about the divided city. Both groups were putting pressure on these bodies for a resolution of the impasse by nightly civil strife in the full view of the international television and news media, which remained the only course open to them.

The third law **hero worship** and submitting to leaders blindly has been confirmed in the historic record. The cause may be charisma, faith, loyalty, and devotion, but in autocratic regimes there is also the huge party apparatus and propaganda campaigns for creating and perpetuating the cult of personality by Mao, Hitler, Stalin, and their fellow dictators, and for coercing unbelievers and skeptics into silence or ritualized surface conformity. More common is how Martin Luther King Jr. rose to prominence in the civil rights movement and became the national spokesperson for African Americans. Despite his profound sense of mission, he had doubts about his leadership role and was frequently challenged within the black movement. It was the Washington administration's search for a moderate black leader to negotiate with during crises, the media news practice of highlighting a "star" identified with an issue, and black activists backing a non-threatening spokesperson for explaining their concerns to a white public that pushed King into the role of leader he assumed and filled so well (Raines, 1977).

Le Bon's hero worship principle is an extreme form of loyalty and bonding between leaders and followers that needs context and specifications. It is also found for sports superstars and the loyalty of sports fans to teams, for film stars and popular music performers, and sometimes even for non-human artifacts like a specific make of motorcycle.

Le Bon's ideas about crowd and collective behavior **resonate with the public, news media, and politicians** because unusual, offensive, provocative, and violent events attract selective media and public attention. Evidence-based MI theory of collective action has only been recently formulated and takes more effort and work to understand than anecdotes and exaggerated narratives. Opportunist political leaders find it useful to shift blame for their own failings to "irrational and immoral" crowds. It is estimated that Donald Trump's 2016 campaign for the presidency got 1.9 billion dollars in free television coverage because of his provocative and outrageous discourse viewed against the backdrop of MAGA hats wearing, flag waving followers chanting "lock her up," referring to rival candidate Hillary Clinton, who got

$750 million free media.[1] Media framing of Trump rallies as political theater attracted a huge audience. The "oxygen" of publicity confirmed the Trump hero worship image. It also assured huge profits for media organizations by way of advertising revenues from enormous audiences and user numbers.

The symbiosis between political theater and media is not recent. **Media** became a **substitute for grassroots organizing** in the 1960s anti-establishment movements of the New Left, the Black Panthers, the anti-Vietnam war campus building occupiers, and other occasions. In one of the first campus protests, at the University of California, students seized the administration building in a sit-down occupation and chanted "the whole world is watching" at the TV cameras. The world did watch, thanks to the television coverage. Student organizers discovered that in lieu of a nation-wide network of local organizations, free access to media had a low cost for spreading protests to national and international constituencies. There was a price to pay for media dependence. As a leader of the Students for a Democratic Society noted about the tactics of the student movement, "the course was set by the New York Times and the TV station, so that whatever was most dramatic was emulated" (Sale, 1973, p. 528).

In the Arab Spring anti-regime democracy movements in Egypt, hundreds of thousand demonstrators converged and occupied Tahrir square due to dissidents messaging on electronic media, in particular the video of a computer programmer who was beaten to death in the street by the police which went viral and outraged hundreds of thousand Egyptians who had remained silent for years. Nevertheless, in the longer run, the media fueled spontaneous anti-regime coalition succumbed to internal factions and feuds also fueled by social media. Anti-regime activist Weal Ghonmin lamented that "we built 100,000 followers...but then the euphoria faded, we failed to build consensus... the political struggle led to intense polarization. The social media quickly became a battleground ...the same tools that united us to topple dictators eventually drove us apart" [*The Guardian* 1/25/16]. New information technology was a two-edged sword in Egypt, as was true in the New Left student movement in the U.S.

---

[1] *Harvard Kennedy School Working Paper* 16–050, 2016.

3   Mobilization for Collective Action

In the 1960s, there was an upsurge of collective action worldwide. Social scientists had opportunities to study protests, marches, demonstrations, and violent confrontations as these events were unfolding and directly with the participants themselves, in addition to media accounts, government, and court records. They concluded that the existing intellectual framework and public narratives about collective action did not account for, and at times even misinterpreted, what the evidence showed. They developed new ideas and theories on neglected dimensions of collective action, in particular the capacity and political opportunity to act collectively, which became known as mobilization theory.

All theories, be they Gurr's, Le Bon's, and the new mobilization theory, recognized the importance of deteriorating life conditions and betrayed ideals that produce discontent, insecurity, grievances, and moral outrage and create a potential for collective action. That potential increases when institutional leaders fail to take ameliorative steps and reforms, and as often happens, are held responsible for worsening life conditions and denied ideals.

Many of the largest and longest collective actions have at their core not only material conditions and class interests but **political and cultural values** like democracy, equality, civil liberties, the rights of minorities and disadvantaged groups, anti-secular religious pursuits like sharia law, the defense of traditional culture against immigrants and foreigners, moral revival, and nationalism. A time series analysis of the standard of living, economic hardship, urban migration, industrialization, and class conflict indices in France over an entire century found only small correlations with strikes, collective violence, and personal distress like suicide. Religious freedom of Catholics against the French secular state, the form of government, civil-military relations, and other non-material interests caused more demonstrations, protest participants, and mass actions than the labor movement for workers' rights and class conflict (Tilly et al., 1975).

In these instances, some supporters of collective action come from groups and nationalities that may not personally experience hardship and oppression. They form a moral **conscience constituency** for ideal beliefs and interests, as fellow travelers did in international communist movement in the 1930s, white liberals did in U.S. civil rights movement in the 1960s, jihadi fighters in al Qaeda and ISIS did for the restoration of radical Islam, and over a century ago the abolitionists did for ending slavery.

The study of collective action links up with theories of social and cultural change, and its findings in turn contribute to understanding interests, belief

Collective Action 235

systems, ideologies, and moral ideals in social action. Mobilization theory further explains how potential demand for change and reform is transformed into actual collective action, and what modes of action result from the strategic interaction between the participants, termed challengers and contenders, and the opposition, which is often the authorities or targets protected by them.

Structural theory without the agency of MI is an incomplete explanation for collective action (Rucht et al., 1998). Its methodology consists of finding statistically significant covariates in large data sets, e.g., economic deprivation is correlated to civil strife and collective protests. The agency that translates deprivation into collective action has to be specified and can lead to very different outcomes. In Russia, in February 1917, the Tsar was forced to resign after a week of unorganized and spontaneous protests by women over bread prices and food shortage that gradually grew into a general strike. Hardship led to large protests by ordinary people, insurrection by soldiers, and eventually the Kerensky government. Two years later, after the Bolshevik seizure of power, there was again great hunger in St. Petersburg and other cities. Food shortages greater than in 1917 bordered on famine, but hunger did not lead to food riots and insurrection. The government requisitioned all food and introduced food rationing. Supporters of the Bolshevik regime received just enough food coupons to survive, but others did not get enough to live on. In desperation, city people sought out villages and illegally bartered their furniture, clothes, and belongings for food. According to Pitrim Sorokin (1975), "in no time the Tsarist civil service and intelligentsia was loyally working for the Bolsheviks, and in return got food coupons sufficient for survival." Mobilization theory explains how agency turns the potential for revolt against the regime due to deprivation into the different outcomes of support for the regime (Tilly, 1978).

There are **obstacles for collective action**. There is no incentive for individuals to contribute voluntarily to a collective good in a large population because they can free ride, i.e., get the benefits without contributing. Free riding can be overcome by personal benefits from participation, termed **selective incentives**, like the social esteem and approval of peers and the affirmation of one's beliefs and values by joint action with social intimates. Social benefits are unlikely in large, anonymous, disconnected masses of people; they are provided in groups with social networks and shared identities. Free rider tendencies in collective action are overcome by **solidarity**.

Starting collective action is also problematic because of innovator-loss dynamics (Coleman, 1990, Chapter 9). The first individuals who challenge the status quo will be negatively sanctioned and are pressured to give up before others join. If that keeps happening, a majority disposed to challenge over an issue may never find out it is the majority. The obstacle is overcome by **"bloc mobilization"** when an entire group decides to challenge, as was the case of church congregations in the Montgomery bus boycott. Preexisting social units with shared beliefs, leaders, and mutual support are far more capable of starting and sustaining collective action than isolated individuals.

A powerful **start-up mechanism** is for a number of preexisting groups to form a coalition, as dozens of student organizations did at Berkeley, all opposed to California state laws and administrative prohibitions against on-campus political activity. The coalition became shortly the Free Speech Movement that thousands of students joined in mass demonstrations. Each coalition entity drew participants from its already existing activists and adherents, a process termed **"bloc recruitment,"** to form a core of activists whose public advocacy attracted large transitory teams at huge rallies.

Mayer Zald and John McCarthy (1987) formulated a **resource mobilization** theory for the capacity for collective action highlighting **social movement organizations** (SMO). Their research showed among others that when Pro-Life mass participation in the 1960s declined following the 1973 Supreme Court decision on abortion rights, the movement stayed alive through state and local chapters run by paid staff, annual dues paying membership, and links with its constituency by newsletters, much like other interest groups. Pro-life SMOs and an identifiable membership base gave the activists legitimacy for lobbying legislatures and the capacity to reactivate mass demonstrations on short notice. William Gamson (1990) researched 53 protest groups and found that persistent and successful mobilization for collective action is more likely when a challenger creates at least a rudimentary organization structure, i.e., a social movement organization or SMO, with a name, leadership, fund raising, recruitment, and publicity capabilities, and a plan and program that appeals to a public.

Public issues are subject to an **issue-attention cycle** (Downs, 1972). An issue may engage intense media discourse and advocacy and draw the enthusiastic participation of a large public. Reforms and policies are then implemented and do improve the situation. Media attention and public advocacy wane and turn to other issues, as happened with Pro-Choice. What SMOs do is keep the issue alive after the mass participation phase, and maintain a capacity for mass mobilization when the issue becomes problematic again.

In autocratic and totalitarian regimes and in police states, a further obstacle to anti-regime mobilization is the surveillance and **repression of opposition**. Freedom of association and speech curtail the formation of SMOs, although small numbers of dissidents persist despite the loss of jobs, detention, police surveillance, and informers. The capacity for mass mobilization is practically nil. Nevertheless, as happened in Eastern Europe in the late 1980s and before that in the 1848 European revolutions and on other historic occasions, powerful regimes are toppled by freedom and democracy movements. This paradox is explained by mobilization dynamics with political opportunity.

When faced with economic demands for liberalization of autocratic rule, regimes engage in erratic reformism which opens a window of political opportunity for regime opponents. A faction of the ruling group, which in Czechoslovakia in the 1960s and the 1980s was the communist party, seeks reforms opposed by the hardliners who control the organs of government and of the party. The reformists believe they cannot prevail against hardliners unless they enlist popular support for change. The reformists soften party control in associations for youth and students, writers and artists, professional associations, media, and religious organizations and encourage people to speak out for the reforms. The reform movement attracts many people who had been fearful of political engagement and had conformed silently to the regime. Loosening of social control lowers the cost of regime criticism and opposition and can trigger a mass movement against the regime. The reformists can lose control of the popular movement which demands regime change and multi-party democracy and not just incremental reforms.

It happened in Czechoslovakia in the Prague Spring of 1968 when Dubcek and his reformers lost their grip on the reform movement they themselves had mobilized. It happened again twenty years later in the velvet revolution of 1989 when the hardline Husak regime, reacting to the successes of Solidarity in Poland and the shaking foundation of the East German regime, loosened social control and enabled the subculture of dissidents in Prague to form Civic Forum, which became the core SMO of the democracy movement (Ash, 1990).

The different outcomes of 1968 and 1989 were due to **political opportunity**: erratic reformism allowed mobilization against the regime on both occasions, but outsider repression happened only in 1968. In 1968, the Soviet leaders were totally opposed to regime change in Eastern Europe and exit from the Warsaw Pact; consequently, the Soviet army crushed the democracy movement. In 1989, the "Gorbachev factor" enabled success. Gorbachev had instituted reforms in the Soviet Union and hoped for the same in other dependent communist countries. He had made it known to European communist regimes that they would have to manage reforms on their own without Soviet intervention. As the anti-regime protesters increased into the hundreds of thousands, communist regimes lost all legitimacy and without

238   A. Oberschall

Soviet assistance they fell like dominos. The two mobilizations were similar, but not political opportunity. Both dimensions of collective action theory have to be joined for explaining outcomes.

4   Participation in Collective Action

The baseline MI model of participation in collective action has four variables: V stands for the values and interest of the collective action for a participant; P is the production function for V and depends on the type of action and on other variables like N, the number of participants or contributors; V times P is the expectation of goal attainment, i.e., the expectation of benefit from participation; S stands for the selective incentives or personal benefits obtained by participation and cannot be gotten by non-participants or freeriders; and C is the cost function for participation, which also depends on N and on the type of regime or target. In line with the MI hypothesis that behavior is adaptive and normative, V times P plus S is the participant's estimate of benefits, and C is the estimate of cost: VP + S - C is the net benefit; if it is greater than zero, the choice is to participate; if zero or less, don't participate. The core of participation theory is to explain the processes that **reduce** the **uncertainties** in **V, P, C, and N configurations** for decisions (Klandermans, 2013).

A social movement, like democracy movements, the civil rights movement, or the environmental movement, consists of hundreds of local mobilizations and episodic confrontations, in many locations over a period of time, by teams of activists and part-time supporters, and occasional strategic aggregation into massive demonstrations and marches, like the annual March for Life in Washington, D.C. or at some other central place with high visibility. The participation model is applicable to particular episodes of local collective action, as well as to the ensemble of large-scale mobilizations.

For a local anti-abortion initiative, like picketing a facility that provides abortion services, V is blocking access and some secondary goals like publicity and fund raising. Together with many other local initiatives, it is a stepping stone to the long-term goal of national legislation and Supreme Court decisions on abortion, V for the entire movement. Local and supra-local N, V, P, S, and C configurations differ and have to be adjusted to what participants and the public perceive and expect. In a locality, thirty activists taking turns can conduct successful picketing for many days, and turning away a dozen abortion clients may be perceived by activists as success. For the annual Washington march, a turnout of a million is the goal, and a Supreme Court

Collective Action   239

decision affirming the constitutionality of restrictive state laws that impede thousands of abortion seekers is a success.

In addition to unconventional collective actions, the supporters of public issues engage in **institutionalized political activities** for goal attainment, as in campaigning for political officials who support their values and goals. Important decisions for movement leaders and organizers are the mix of conventional politics and unconventional collective action to engage in. In autocratic regimes conventional politics, such as free and fair elections, is impossible, and opposition to the regime and policies necessitates prohibited collective action with much higher C and lower P than in democracies.

V, P, C, and N functions derive from research, but it is helpful to start with hypotheses covering some typical and recurring collective actions. For the potential participant, V, P, C, and N are shrouded in uncertainty. The organizers of collective action and their opponents, who are often the government or are protected by the authorities, influence public perceptions of V, P, C, and N so as to tip net benefit in their favor. When confrontations are taking place, the experiences of participants, similar participants elsewhere and media reporting enable the revision of initial expectations. As estimates of V, P, C, and N change, participants decide to persist, others to join, or on the contrary protesters exit and others stay away, and the challenge wanes or ends.

Political opportunity, as in erratic reformism and the loosening of social control by autocratic regimes, are modeled by simultaneous shifts in the V, P, and C functions: people increase V because hope for change overcomes apathy; P shifts higher because reformers inside the regime lift barriers to voicing political views and to participation; and C shifts down because speech and assembly are safer.

Selective incentives are often provided by the organizers of collective action. A peace activist in a distant city can watch a mass march in New York on television. Organizers counter free riding with incentives for live participation with a bus ride to New York with fellow local activists, lodging overnight in the basement of a participating church, a march displaying colorful banners and symbols by thousands along Fifth Avenue to Central Park, speeches at the park by prominent public figures and music by folk artists and rock band with thousands clapping and cheering, and an evening of revelry with friends in New York City, all for under $100. Because these incentives for personal attendance were added by the organizers to the expected benefit PV, for the June 12, 1982 national peace coalition rally in New York, about half the estimated million participants came from outside the New York metropolitan area [research by Anthony Oberschall and students].

Collective action theorists research the causes of these perceptions and expectations and how they change during a particular event or a campaign of confrontations. For instance, organized campaigns for and against abortion, like pro-life and pro-choice, have annual mass marches in Washington, D. C. To reduce the risk of destructive and violent events, police arrests, and teargas, the organizers deploy trained monitors along the edges of the crowd as a buffer for separating hostile counterdemonstrators from the marchers. Monitors also prevent provocateurs from using the cover of the crowd for assaulting the police and destroying property. Peace monitoring lowers expected costs and thousands participate who otherwise would not. The authorities signal high expected costs for law violation by issuing warnings, massive presence of police, and the weaponry displayed.

P, C, and N configurations differ by **collective action tactics**. For a hunger strike, The P/N graph is shaped like an inverted J. A few hunger strikers are sufficient to attract visibility and have a high probability for getting media attention and a response (V) by the authorities. More hunger strikers have a decreasing impact. The hunger strike has a high cost because of suffering and even death and is therefore not a suitable mass tactic. A boycott of businesses on the other hand has a J production function: with only a few boycotters, one expects no effect on V, but when numbers shoot up businesses targeted suffer losses and may decide to negotiate about the issues that caused the selective buying. Consumer boycott has a low cost to participants because there are usually alternative sources for buying and anonymity is protection against target retaliation, i.e., it is a low-cost tactic with positive net benefit for large numbers of participants.

For typical marches and demonstrations, P has a tilted S shape: a few participants will have no impact on goal attainment; in the middle N range additional participants increase V, but beyond a critical mass N, more participants add only marginally to success. On the cost side, C is expected to decrease with N, i.e., there is an expectation of safety in numbers.

Collective action theory is mindful of context for hypotheses about the configuration of V, P, C, and N. Democracies do not repress large hostile crowds by shooting from military weapons, although they resort to teargas. In an autocratic regime, however, as the size of non-violent demonstrations against the regime comes to number hundreds of thousands, as in the democratic movements in China and in Egypt, the likelihood of lethal repression by the military increases rather than decreases. Hundreds were killed in Tiananmen Square in Beijing and in Tahrir Square in Cairo.

Collective Action   241

When boycott participants have no alternative access to essential goods and services, as in the yearlong 1955–56 Montgomery, Alabama, bus boycott for desegregation, the cost is high: auto pooling was outlawed and 17,000 black riders had to walk to work (Luther King, 1964). Black churches and ministers not vulnerable to financial pressures by segregationists mobilized extraordinary commitment by African Americans in nightly church meetings. When King told an old lady she should ride a bus to work because she is too old for walking, she answered, "Oh no, I'm gonna walk…my feets is tired, but my soul is rested." Such high V and solidarity benefits offset the high personal sacrifice for African Americans in Montgomery (High V in VP plus S compensating for high C). 100% participation made a substantial financial loss to municipal transportation keeping half empty buses rolling for white riders. Nevertheless, the white Montgomery leaders sustained large losses of revenue rather than yield on segregation of public facilities, which had a very high value for Southern whites. In the end, the strategy of persisting until the Federal courts got involved on a constitutional issue paid off and a U. S. District Court declared Alabama's state and municipal laws on segregation on buses unconstitutional.

Collective action is adaptive and strategic. Both sides learn from experience by trial and error and by observing similar challenges elsewhere. They devise **new modes of challenge and social control** for tipping the configuration of V, P, C, and N in their favor, and the target group reacts likewise (McAdam, 1983). The civil rights movement learned a great deal about effective collective action for desegregation when the Montgomery bus boycott achieved but limited success at a very high cost because only buses were desegregated and not other public facilities. A new mode of collective action was sit-ins against segregated public facilities and had a better chance of success because of a more favorable P, C, N, and V configuration. Sit-ins required fewer participants at a time (low N); the arrested were replaced by a small number of others and those arrested were bailed from jail for relatively small sums of money (low C); the media covered the sit-in as a continuous news story with high visibility and the sit-ins spread without prior organization to many southern localities. Segregationist civic leaders were vulnerable to sit-ins because a poor image of their city due to racial confrontations inhibited much-needed Northern businesses investments. The aggressive actions of segregationist counterdemonstrators and the police against peaceful black protesters outraged the non-Southern liberal conscience constituency who reacted with financial contributions to civil rights organizations and legal expenses. P increased due to outsider resources compensating for C by local social control agents.

Considerable success was achieved by direct action. A year and a half after its start, 75,000 participated in direct action and sit-ins (Heacock, 1965). A Justice Department report indicated that even before the Civil Rights Act of 1964, in 560 southern state and border communities, 391 had at least one desegrated public facility such as a movie theater, restaurant, motel, and lunch counter, and that in 180 cities civic and human relations committees were dealing with race relations.

Uncertainty reduction explains the **clustering and diffusion of collective action in time and space**, like the outbreak of revolutions and nationalist movements in European cities in 1848–1849, the toppling of East European communist regimes in the late 1980s, the sit-in movement against segregation in 1960, the New Left and anti-Vietnam war university building seizures by students in the late 1960s, and the race riot waves in the U.S. in World War I and again in the 1960s. A key idea is **signaling** information about a favorable V, P, C, N configuration from collective action by a pacesetter group to similar potential challengers who are inhibited by uncertainty but who then revise their estimates and act. The **diffusion** is not the result of organized coordination nor "contagion" but of information from the media that reduces uncertainties in collective action. The success of demonstrations by one group signals to other groups that participants will be more numerous and more committed than expected, and that the authorities will respond to demands rather than simply repress. As more groups experience a measure of success, expected V, P, and N keep increasing and C decreasing, tipping net benefits for more groups that on their own would not have challenged, and creating a bandwagon for diffusing collective action.

Signaling is especially effective when the **pacesetting** challenger occupies a **focal point** and is successful, as was true in Paris in 1848 when it was the center of republicanism and progressivism ever since the French revolution, and Republicans elsewhere in Europe looked to it for leadership. In early 1848 there were insurrections in Sicily and Tuscany demanding a constitution. They might have remained isolated, but then the February revolution in Paris ousted King Louis Philippe and proclaimed the Second French Republic. That success triggered the diffusion of insurrections and revolutions against autocratic governments throughout Germany and the Habsburg-ruled territories in Austria, Italy, Bohemia, and Hungary. When the Prussian monarchy made concessions and promised reforms to Berlin demonstrators, other kings and princes in lesser German states followed the Prussian lead even when they were not threatened by demonstrators. As strong targets make concessions to challengers, weaker targets are also more likely to concede to demands rather than repress, and thus fuel diffusion. The speed of diffusion in 1848 took weeks rather than days or hours because railways and telegraphs did not yet link European cities.

Similarly, the New Left student movement at top prestigious state universities like Berkeley and the Ivy League set the pace of anti-Vietnam war protests against university administrations' links to the military, like the draft, ROTC, and Defense Department funding of faculty research. Events at these prominent universities signaled favorable V, P, C, and N configurations at peripheral colleges. The Columbia University building seizures in April 1968 by anti-war students, followed by failed negotiations, police bust, and student strike, held the attention of the news media and public for the better part of two weeks. Then in May, similar building occupations, demands, and campus confrontations occurred at 35 colleges and universities. They ended at the end of May when students dispersed for Summer recess. The April 1969 Harvard building seizures, campus strike, and police bust, also drawing national attention, was followed by 69 similar campus disturbances at institutions of higher learning (DeNardo, 1972). Although in both instances there were many arrests and some students were injured in police confrontations, the activists were not disciplined and expelled from their universities because their amnesty demands for rule and law breaking were met. The anti-war movement perceived the building seizures, administration and faculty concessions, the radicalization of thousands of students, and national anti-war publicity as a great success: N was huge, commitment to anti-war issues (V) increased, C was low, and P became favorable even at peripheral colleges. Pacesetters created a favorable configuration of factors for success and diffusion.

Mass **anti-regime demonstrations** have occurred in **autocratic regimes** that ban opposition collective action and harass and detain dissidents. The capacity for mobilization is limited but has been overcome by **shared political culture** that includes time and place for protesting oppressive governments learned in school, in literature and films, and embodied in street names, national holidays and parades, monuments, and statuary. In Czechoslovakia it was Wenceslas Square in Prague, August 21, the anniversary of the 1968 Soviet invasion, and October 28, independence-day; in Budapest it was March 15, independence-day and Parliament Square; in Egypt it is always Tahrir Square in Cairo. Everyone knows this, and so do the authorities. The silent majority who comply with the regime knows that if the protest is to happen, it will be at those dates and sites. It also knows that every year a small band of dissidents protests there and get arrested. In ordinary times, no one pays much attention, but when there is a crisis, the people sense an opportunity for reforms and change through mass demonstrations, as happened at other historic moments. What is uncertain is how many will join to make it an effective and successful collective action and how much force the authorities will use.

People spontaneously engage in actions that **reduce uncertainty**. Preceding anti-regime demonstrations, one observes at work places, neighborhoods, taverns, cafes, and in friendship circles intense political discussion in which people probe each other on their political mood and commitment to opposition and encourage each other. They exchange information about which groups and numbers to expect, e.g., whether the transportation workers and steel workers, large groups that normally comply with the regime, are likely to join. They also fraternize with the police and the military, offer them cigarettes, hot coffee, and sports newspapers, and probe them on their disposition for cracking down on peaceful protesters. Ordinary people who have been apathetic and silent thus form favorable expectations on the V, P, C, and N configuration for successful collective action and discount the regime propaganda of intimidation.

On the historic anniversary, at the historic site, thousands decide to join the usual band of dissidents for mass protest. Many bring flags, banners that proudly identify their group or neighborhood, placards with key demands for freedom, and the resignation of hardline regime leaders. They wear hats, armbands, sashes, and lapel pins whose meanings are well-known in the political culture. They create a collective action repertoire that signals unity and commitment, like the Prague demonstrators in Wenceslas Square jingling thousands of key chains for an awesome sound heard all around. The Belgrade public supported the street demonstrators for media freedom in March 1991 against the Milosevic regime by banging pots together at open windows during the state TV evening news hour and making a deafening noise for half an hour, days on end. Everyone knew it meant "we don't want to hear your propaganda lies, we want media freedom." Shared political culture can overcoming formidable obstacles to mobilization when the window of political opportunity opens.

5   Conclusion

MI is an intellectual framework for explaining human behavior including the institutions for producing collective goods by individuals joining together voluntarily for shared values and goals. Compared to the state and other highly organized social units, the social formations for such pursuit are typically loosely organized crowds, protest groups, demonstrations, mass rallies,

Collective Action   245

and social movements. The collective goods sought can range from subsistence issues like food security, wages, and the rights of labor to cultural values like political freedoms, religious rights, democracy, and ethnic autonomy. The arenas of contention range from small social milieus like neighborhoods and college campuses to national political and cultural arenas at the seat of government, in democratic societies and in autocratic regimes.

Because the pursuit of collective goods by some threatens the vested interests and status quo for others, including often the government and powerful organizations it protects, collective action necessitates costly and uncertain contentious challenges because of opposition. As well, voluntary participation for collective goods is problematic because of free riding. Collective action has to overcome these and other obstacles.

Behavior is adaptive and normative. The decision to participate in collective actions depends on the expected net benefit, which in turn depends on the choices of others and the actions of opponents, both of which are uncertain. The pivotal variables in participation decisions are the public support for the value and issue V, the expected number of joiners N, the probability of success of the collective action P, the incentives and benefits of participation S, and the expected costs C. The strategic interaction of challengers and opponents, like tactics of confrontation and social control responses, determines the particular configuration of values of the pivotal variables.

Mobilization theory explains how these uncertainties are reduced to tip net benefit to participation: group recruitment taking advantage of bonds and solidarity in preexisting social circles for commitment to a cause; a heterogeneous coalition of activists, part timers, and conscience constituency who are incentivized by different configurations of the pivotal variables; social movements organization for leadership, resources, and communications with participants and the public; bloc recruitment of contending groups to form a federated structure extending influence to different social milieus; selective incentives from participation that free riders can't get; multiplier effects for raising V when opponents commit acts of injustice and brutality; choice of tactics for positive benefit/cost balance; use of mass media for mobilization and public support; shared political culture and signaling that coordinates the actions of thousands without leadership and prior planning; pacesetter-follower dynamics for diffusion of collective action. Just like David fought Goliath against large odds, by these means a weaker group increases its capacity and willingness for challenging a powerful opponent.

Collective action theory is an ongoing social science enterprise. As societies and cultures change, new modes of mobilization and social control are invented and adapted by challengers, opponents, and governments. The revolution in communications due to information technology and hundreds of millions users of social media like Facebook and Twitter have made for costless access and dissemination of information about signaling and coordinating collective action, amplifying the multiplier effects from outrageous incidents that go viral, and recruiting and networking supporters in both live groups and virtual communities in cyberspace. At the same time governments increased the efficacy of surveillance, and the marketplace of ideas fuels polarization of the public and of counter-movements against movements with misinformation, falsehood, hate-mongering, and false accusations. Research findings about these changes will have to be systematically and continuously integrated with collective action theory.

References

Ash, T. G. (1990, January 18). The revolution of the magic lantern. *New York Review of Books*.

Coleman, J. (1990). *Foundations of social theory*. Harvard University Press.

DeNardo, J. (1972). The diffusion of civil disturbance. Unpublished manuscript. *Yale Political Science Department paper*.

Downs, A. (1972). Up and down with ecology: The issue-attention cycle. *Public Interest, 28*(Summer).

Gamson, W. (1990). *The strategy of social protest*. Belmont.

Gurr, T. (1970). Why men rebel. Princeton University Press. *Journal of Social History, 4*(4).

Heacock, R. T. (1965). *Understanding the Negro Protest*. Pageant

Hinde, R. (1997). The psychological bases of war. Paper presented at Study of War Conference, Wheaton Il. *American Diplomacy, 3*(2).

Klandermans, B. (2013). The social psychology of protest. *Current Sociology, 61*, 886–905.

Le Bon, G. (1895/1990). *The crowd (Psychologie des Foules)*. Viking Press.

King, M. L. Jr. (1964). Stride towards freedom. In Gilbert Osofsky (Ed.), *The burden of race* (pp. 15–24). Harper and Row.

McAdam, D. (1983). Tactical innovation and the pace of insurgency, *American Sociological Review, 48*(6), 735–754.

Oberschall, A. (1993). *Social movements. Ideologies, interests, and identities*. Transaction Publisher.

Oberschall, A. (1994). Protest demonstrations and the end of communist regimes in 1989. *Research in Social Movements, Conflicts and Change, 17*, 1–24.

Olson, M. (1965). *The logic of collective action*. Schocken Books.

Ostrom, E. (1990). *Governing the commons. The evolution of institutions for collective action*. Cambridge University Press?

Ozment, S. (1975). *The reformation in the cities.* Yale University Press.

Raines, H. (1977). *My soul is rested.* Bantam.

Rucht, D. et al. (1998). *Acts of dissent.* Rowman & Littlefield Publishers.

Sale, K. (1973). *SDS.* Vintage Books.

Sorokin, P. (1975). *Hunger as a factor in human affairs.* University Press of Florida.

Staub, E. (1989). *The roots of evil, the origins of genocide and other group violence.* Cambridge University Press.

Tilly, C. et al. (1975). *The rebellious century, 1830–1930.* Harvard University Press.

Tilly, C. (1978). *From mobilization to revolution.* Addison Wesley.

Weber, M. (1904/1964). *Soziologie, analysen, politik.* Kröner Verlag.

Zald, M., & McCarthy, J. (1987). *Social movements in an organizational society.* Transaction Books.

Methodological Individualism in Weber's Sociology of Religion

David d'Avray

A good recent survey of Methodological Individualism, by Joseph Heath, presents Max Weber as its fountainhead and Rational Choice Theory as its most influential recent form (Heath, 2020). Weber certainly lays a clear foundation for the approach that underlies the present volume. In the conceptual introduction to his *Economy and Society* Weber wrote that social constructions such as the 'State' or a limited company are 'purely the outcomes of and the interrelations between the specific actions of *individual* humans, since only in these can we find meaningful action which is comprehensible by us' (Weber, *SG*, 1984, p. 30)[1]. Weber goes on to say that for Sociology (and this would include his kind of sociological History) a collective personality does not perform actions. 'When [Sociology] speaks of "State" or "nation" or "Limited Company" or "Family" or "Army Corps" or similar "constructs",[2]

---

[1] 'Für die verstehende Deutung des Handelns durch die Soziologie sind dagegen diese Gebilde lediglich Abläufe und Zusammenhänge spezifischen Handelns *einzelner* Menschen, da diese allein für uns verständliche Träger von sinnhaft orientiertem Handeln sind.' In my translations of extracts from Weber I follow the principle: 'as close as possible to, and and near as necessary from the original'. One can get quite close to his convoluted syntax and telegraphic moments even if it is not the style one would choose for oneself.

[2] 'Gebilde'.

D. d'Avray (✉)
Department of History and Sociology, University College London, London, UK
e-mail: d.d'avray@ucl.ac.uk

© The Author(s), under exclusive license to Springer Nature
Switzerland AG 2023
N. Bulle and F. Di Iorio (eds.), *The Palgrave Handbook of Methodological Individualism*,
https://doi.org/10.1007/978-3-031-41508-1_11

249

it means by this, rather, nothing but a particular kind of course followed by the social actions of individuals, either in fact or as deemed to be possible …'(Weber, *SG*, p. 31).[3]

Weber goes on to argue that methodological individualism does not rule out the sociological reality of such constructs, but he puts them back into the heads of individuals: 'The interpretation of action must not ignore the fundamentally important fact that those collective constructs which belong to everyday thinking, or to legal or other specialised types of thought, are *conceptions*[4] in the heads of real people (not only of judges and civil servants, but also of "the public"), ideas partly about how things are and partly about how they should be, ideas which determine the orientation of action, and that as such they have an extremely powerful, often decisive causal significance for the course of action of real people. Above all, conceptions about how things should or should not be.[5]' (Weber, *SG*, p. 31).[6]

If Weber is the fountainhead of modern methodological individualism, its strongest running stream in the last half century has been rational choice theory. Joseph Heath observes that 'methodological individualism became synonymous in many quarters with the commitment to rational choice theory. Such an equation generally fails to distinguish what were for Weber two distinct methodological issues: the commitment to providing explanations at an action-theoretic level, and the specific model of rational action that one proposes to use at that level (i.e., the ideal type)'. To unpack that a little: by the 'action-theoretic level' Heath means that 'social phenomena must be explained by showing how they result from individual actions, which in turn must be explained through reference to the intentional states that motivate the individual actors'. The phrase 'the specific model of rational action' is meant to indicate that there are possibilities other than purely instrumental motivation in a rational choice theory framework—Heath mentions Habermas's theory of Communicative Action as one alternative possibility.

---

[3] '… gibt es für sie [Sociology] keine "handelnde" Kollektivpersönlichkeit. Wenn sie [Sociology] von "Staat" oder von "Nation" oder von "Aktiengesellschaft" oder von "Familie" oder von "Armeekorps" oder von ähnlichen "Gebilden"spricht, so meint sie damit vielmehr *lediglich* einen bestimmt gearteten Ablauf tatsächlichen, oder als möglich konstruierten, sozialen Handelns Einzelner … '.

[4] 'Vorstellungen'.

[5] 'Gelten-(oder auch: *Nicht*-Gelten-)*Sollendem*'.

[6] 'Die Deutung des Handelns muß von der grundlegend wichtigen Tatsache Notiz nehmen: daß jene dem Alltagsdenken oder dem juristischen (oder anderem Fach-)Denken angehörigen Kollektivgebilde *Vorstellungen* von etwas teils Seiendem, teils Geltens sollendem in den Köpfen realer Menschen (der Richter und Beamten nicht nur, sondern auch des "Publikums") sind, an denen sich deren Handeln *orientiert*, und daß sie als solche eine ganz gewaltige, oft geradezu beherrschende, kausale Bedeutung für die Art des Ablaufs des Handelns der realen Menschen haben. Vor allem als Vorstellungen von etwas Gelten- (oder auch: *Nicht*-Gelten)*Sollendem*'.

That makes sense but there will be more to be said below about the difference between Weber's and the Rational Choice versions of methodological individualism, particularly when it comes to religion. Weber's version of methodological individualism is integrated with analyses and explanations of the similarities and differences between religions, and his awareness of how much these mattered for the whole character of a society (Weber, *GAR*, i, p. 112, note 4 continued from p. 111).[7] He wrote book-length analyses of the religions of China, of India, and of Ancient Israel, in addition to his more famous study of the Protestant Ethic. His opus magnum *Wirtschaft und Gesellchaft* (henceforth Weber, *W&G*) includes brilliant and compressed passages on the foregoing, as well as on Roman Catholicism, the history of which he interprets with great insight, though he was not a religious believer so far as we know (we don't, really), and though his own background was Protestant. Probably, had he not died so young, he would have devoted a full-length study to Islam. In all these cases he is studying beliefs that were held collectively, and, furthermore, beliefs specific to each of the great systems.

How to bridge the gap between Weber's studies of collectively held religious beliefs and his methodological individualism? The problem is to make the jump from personal motivation to collective religious phenomena and power structures. Then: structure or agency? It is one of the antitheses sociologists like to discuss. The structure is conceived as constraining individual options, while advocates of agency think that man makes history in conditions of his own making. The antithesis is overcome by Weber's argument that widely shared motivations create structures of behaviour. When many minds share much the same motives, structural patterns result.

Here it will be argued that Weber's version of methodological individualism is a powerful tool for elucidating the history and sociology of religions especially if we pay attention to the interplay between value and instrumental rationality that is so central in his thinking. *Nisi fallor*, Weber never spelled out his thought on these issues explicitly. To answer our questions from his work, one has to draw out the implications from his texts: Weber's thought is highly concentrated. One of his paragraphs has the content of an article, and one of his chapters is equivalent to a book. One needs to read between the lines to get his full meaning. The exercise is worth the effort because he is very successful in relating the phenomenology of religion to methodological individualism. Weber has deep insight into the dynamics of collective motivation and the types of religious systems that result from different kinds of

---

[7] '... es gerade praktisch von der allerhöchsten Wichtigkeit ist, von welcher Art das Gedankensystem ist, welches das unmittelbar religiös "Erlebte" ... in seine Bahnen lenkt ... so wichtige Unterschiede in den ethischen Konsequenzen, wie sie zwischen den verschiedenen Religionen der Erde bestehen.'

252 D. d'Avray

collective motivation. Characteristically, he is the opposite of monocausal in his explanations, and offers a range of ideal types of collective religious motivation. Read carefully, Weber's insight into what moves individuals in groups is astounding in its penetration.

1 Irrationality and Predestination

Weber is a virtuoso analyst not only of rational but also of irrational motivation. Self-deception and irrationality are not generally included in the range of themes associated with Max Weber's social theory, but in fact, it is crucial to one strand of his most famous thesis of an 'elective affinity' between Protestantism and Capitalism (cf. Weber, *GAR*, i, p. 83.)[8] He distinguishes between Lutheranism and Calvinism, and also between Calvin's theology and Calvinism as preached and practised by his followers (Weber, *GAR*, i, pp. 91–113, 125). The ordinary Calvinist had an attitude different from that of Calvin himself. For Calvin, one cannot deduce from behaviour whether someone is saved (Weber, *GAR*, i, p. 103)[9] while for later Calvinists, dedication to one's work in the world was a way to achieve certainty of salvation (Weber, *GAR*, i, p. 105)[10] and escape anxiety about one's eternal fate (Weber, *GAR*, i, p. 106).[11]

Weber argues that there were motives that Calvinists themselves did not understand—motivation one can call 'irrational', if one calls actions irrational when their real motives or other causes are different from the reasons the actors give to themselves. According to Weber this strand of 'practical Calvinism' saw prosperity, and the honest pursuit of one's profession, as a *symptom* of election, a manifestation of the fact that the person in question has been chosen from all eternity as one of the saved. An election could not be merited, but with those whom God has chosen, observable consequences will follow. God enables those whom he has predestined for salvation to be virtuous and successful in business.

Weber suggests, and it seems highly plausible, that some at least of those who believe this must have striven to prove to themselves that they were among the elect. Their own account of the cause of their success, however, is not the real explanation of it, because in reality the extra efforts they choose

---

[8] '… untersucht wird, ob und in welchen Punkten bestimmte "Wahlverwandtschaften" zwischen gewissen Formen des religiösen Glaubens und der Berufsethik erkennbar sind.'

[9] 'Er verwirft … einzudringen.'

[10] 'Und andererseits wurde … eingeschärft' (and whole context).

[11] 'das geeignete Mittel zum Abreagieren der religiösen Angstaffekte.'

to make result from free will, not divine election. Weber's explanation is essentially psychological: an example of his understanding of *irrationality* as a cause. But it is individual irrationality. Those who are not moved to prove themselves that they are saved either give up the hard-line doctrine or think maybe they are damned—or perhaps hope that success and, with it, assurance of salvation, will eventually come to them.

Weber analyses the psychology thus:

> Only a person who is one of the elect truly has "efficient grace", only he is capable of increasing the glory of God through works that are good in reality, not just in appearance, thanks to being born again (*regeneratio*) and to his whole life being made holy as a result (*sanctificatio*). And in that he is conscious that the transformation of his life[12]—at least in its fundamental character and persevering intention (*propositum oebedientiae*)—is based on a power, living within him, to increase the glory of God, one which is therefore not only willed by God, but also caused[13] by God, he attains to the highest good for which this kind of religious sentiment strives: the certainty that one is in a state of grace.... Thus, however utterly incapable good works are of serving as means to obtain beatitude—for even the elect are creatures, and everything that they do falls infinitely far below what God demands—good works are no less indispensable as signs that one has been predestined. (Weber, *GAR*, i, pp. 110, 112–113, 124, 162–163)[14]

Weber explicitly pointed to the gap between real psychological motivation and the logic of predestinarianism, and spelled it out (in a footnote, not untypically) with clinical precision:

> Logically, a fatalistic attitude could naturally be deduced as a conclusion from predestination. The psychological effect however was precisely the opposite, because of the intervention of the concept of "testifying by one's life" ('"Bewährungs"-Gedankens'; cf. Kalberg, 2001, p. 55) ... Involvement with practical interests[15] cut short the fatalistic conclusions to which logical inference would lead (and which could incidentally in fact occasionally make an appearance).... (Weber, *GAR*, i, p. 111 note 4)[16]

---

[12] 'sein Wandel'.

[13] 'gewirkt'.

[14] p. 110: 'Nur ein erwählter ... Zeichen der Erwählung'; pp. 112–113: 'Denn es hat vielleicht nie ... Anhängern erzeugte'; p124: 'Wir haben bisher auf dem Boden ... Lebensführung vorausgesetzt'; p. 162–163: 'Entscheidend aber ... Durchdringung.'

[15] 'Die praktische Interessenverschlingung'.

[16] ' Logisch wäre natürlich ... fatalistische Konsequenzen.'

254   D. d'Avray

Weber's understanding of irrational motivation, as instanced by his account of Calvinism—there are plenty of other examples all over his work—has been insufficiently explored. It deserves more study, for it goes beyond the idea of a logical gap such as he observed within Calvinism. Weber proposes for instance that 'Religious experience (*Erlebnis*) as such, like any experience, is naturally irrational': the context is a discussion of William James's *Varieties of Religious Experience* in which Weber develops the idea of an inverse proportion between the content of religious experience and the conceptual formulation of it (Weber, *GAR*, i, p. 112, note 4 continued from p. 111).[17] So the irrationality Weber has in mind here is a subtle concept, applicable to any kind of direct, unmediated experience before it is put into a conceptual shape. It is not a negative concept. He would probably have said that all love is irrational in that sense. Such ideas about 'irrationality' of various sorts need to be teased out of the *oeuvre*. They do not seem to be in the current Baedeker of Weber studies.

Interesting though his ideas about irrational motivation are (is there anything of comparable psychological finesse in Rational Choice Theory?), the fact remains that Weber has much more to say explicitly about rationality and its types. It is the investigation of the interplay of instrumental and value rationality, even more than his sensitivity to irrationality, that marks his brand of methodological individualism off from the Rational Choice variety.

2   Rational Choice and Religion

Weber's *oeuvre* in this respect offers much more, for Rational Choice theory hardly attempts to relate its basic postulates to the variety of religious systems in history. True, there is a Rational Choice account of religion (Stark & Bainbridge, 1996), but Rational Choice Theory is not good at explaining differences between major religions, and indeed between non-religious systems of conviction such as liberalism and communism. This is where Weber's ideas help. On the highest level of abstraction, his concept of value or conviction rationality, *Wertrationalität*, provides on a theoretical level the missing link between methodological individualism and the phenomenology of religious differences; some key paragraphs from his *oeuvre* show a profound understanding of the ways in which religious systems

---

[17] 'Das religiöse Erlebnis als solches … die begriffliche Formulierung vorschreitet.'

Methodological Individualism in Weber's Sociology ...   255

start and spread; finally, his specific analyses of particular religions implicitly involve a granular methodological individualism which can be picked out if one looks for it.

3   Value Rationality

Rational choice theory lacks an equivalent to Weber's concept of value rationality, which postulates individuals whose action is guided to a greater or lesser degree by deeply held convictions.

> A person acts purely in accordance with value rationality (*wertrational*) when he or she acts without thought for the foreseeable consequences, in the service of his or her conviction of what seems to be demanded by duty, self-respect (*Würde*), beauty, religious teaching, piety, or the importance attached to a "cause" of whatever kind. In the sense attached to it by our terminology, value-rational action is always action in response to "commandments" or "demands" to which the person doing the action believes that they must respond. Only insofar as human action is oriented towards such demands—and this is the case only with a fraction that varies greatly and is mostly very modest in size—do we want to speak of Value Rationality. As will become apparent, it is important enough to be distinguished as a special type, although otherwise no attempt is being made here to give classification of the types of action which is in any way exhaustive. (Weber, *W&G*, i, pp. 12–13)[18]

Rational Choice theory tries to classify values under 'preferences', but there are two problems with this. Firstly, values often take the form of (deeply held) principles that will be applied to concrete circumstances that the actor has not foreseen. What the principle then imposes may be the opposite of a preference. For instance, someone could be a convinced ethical consequentialist, and find that their calculation of consequences leads them to sacrifice someone they love. To call the terrible thing the conviction tells them to do to someone they love a 'preference' is playing with words in an unhelpful way. Secondly, it leaves the putative preferences within a black box and makes no attempt to explain how they come about. Rational choice is not very good at explaining religion and downright bad at explaining the differences between religions.

---

[18] Passage beginning 'Rein wertrational handelt, wer' and ending 'zu geben versucht wird'. I use this edition because it is the highly intelligent synthesis, by his wife and best interpreter, Marianne Weber, and Johannes Winckelmann, of the various parts of Weber's unfinished *opus magnum*, and because it is much more widely accessible than the *Gesamtausgabe* edition of the writings on religion: Kippenberg, 2005).

4   Instrumental Rationality

This is not to say that value rationality alone can explain religious phenomena—far from it. Generally speaking, it is some combination of value and instrumental rationality that accounts for the *explicanda*. At one end of the scale, religion merges into what one might appropriately call magic (d'Avray, 2009, especially pp. 52–53),[19] when the ritual or prayer is purely to get something:

> Naturally, ... the specific elements of "divine service", prayer and sacrifice, are also originally magical in origin. With prayer, the border between a magic formula and supplication is fluid, and, especially, technically rationalised prayer activity[20] in the form of prayer wheels and similar types of techical apparatus, of prayer strips hanging in the wind or fastened to images of gods or attached to pictures of saints, or feats of rosary praying measured in purely quantitative terms ... are everywhere closer to the former [the magical formula] than to the latter [supplication]. (Weber, *W&G*, i, p. 258)[21]

Or again:

> Religious or magical motivated action is ... precisely in its original form, an at least relatively rational kind of action: even if it is not necessarily action matching means and ends, then it is at least in accordance with rules based on experience. Just as twirling [a pointed stick] gets the spark to come out of a piece of wood, so too do the 'magical' gestures of the the expert get the rain to fall from heaven. And the spark that is produced by twirling a stick is just as much a product of magic as the rain which results from the expert actions of the rain-maker. Religious or "magical" action or thinking should therefore by no means be classified as different from everyday instrumental action, especially since its aims themselves, too, are predominantly economic in character. (Weber, *W&G*, i, p. 245)[22]

It would however be inept indiscriminately to characterise all religious phenomena as instrumentally rational. As suggested above, a key to understanding Weber's application to religion of his methodical individualism is to analyse the variety of ways in which instrumental and value rationality interact with each other.

---

[19] Weber uses magic in a broader sense to mean any actions which get results beyond what the laws of nature make possible.

[20] 'Gebetsbetrieb'.

[21] 'Natürlich sind auch ... der letzteren nahe.'

[22] 'Religiös oder magisch ... ökonomische sind.'

5 Collective Motivations

Still to be explained are the nature of Value Rationality's relation to instrumental rationality, and the phenomena of collectively held beliefs: how do all these individuals come to share the same ideas and attitudes—to religion or anything else? We may start with the latter question. Key passages from Weber's writings about religion show how he managed to steer a middle course between a materialist explanation of religious conviction and an explanation purely in terms of the influence of ideas.

The most obvious answer to anyone familiar with Weber's *oeuvre* is likely to be: people come to share the same beliefs under the influence of a charismatic leader. Weber's concept of 'Charisma' is an excellent example of his skill in the formation of analytical concepts. He did not use the word in its current everyday sense of 'personal magnetism'. Weber defined charismatic leaders as people who are believed by their adherents to be 'endowed with specific gifts of body and mind thought to be more than natural (in the sense of being unavailable to the ordinary person)' (Weber, *W&G*, ii, p. 654).[23] The beauty of Weber's definition is that it is absolutely value-free. Sociology does not ask if the gifts are real or fake: the belief in them by the followers is what makes the leader charismatic. (Weber, *W&G*, ii, 654).[24] Charismatic leadership can be found in any sphere of life, although it is often found in its purest form in religions (Weber, *W&G*, ii, p. 655).[25]

'In its purest form': but probably it is never found in a completely pure form—it may be added—if one thinks like Weber. It is an ideal type, distilled conceptually for analytical purposes but in historical reality mixed to a greater or lesser degree with other kinds of power. The purpose of 'purifying' concepts like 'charismatic', 'bureaucratic' and 'patrimonial' rulership is not to assign historical structures to one box or another, but to see how the different types are combined and in what admixture.

Charismatic leadership certainly does map well onto a good deal of religious history: the movements started by Christ and Muhammed in particular. The belief by early Christians in Christ's divinity, and by early Islam in the direct inspiration of Muhammed's by God, account for their respective motivations. It is hard to deny that these two leaders and a few others have changed the course of History. A host of Hindu and Protestant sects are,

---

[23] '... Träger spezifischer, als übernatürlich (im Sinne von: nicht jedermann zugänglich) gedachter Gaben des Körpers und Geistes.'

[24] From 'Dabei wird der Begriff "Charisma" hier gänzlich "wertfrei" gebraucht.... bewährten sich in dem Glauben ihrer Anhänger als charismatisch Begabte.'

[25] 'auf religiösem Gebiet oft am reinsten ausgeprägt.'

258   D. d'Avray

on a more modest scale, the outcome of charismatic leadership by gurus or gifted pastors who felt that their reading of the sacred texts had been guided by God and that they were called to found a Church or movement that would get it right. Weber's powerful ideal type points to a straightforward causal link between the perceived charisma and the religious movement, all within the framework of methodological individualism.

Taken on its own, Weber's concept of charisma could leave us with a kind of 'great man' view of religious history which allows little place for social-structural explanation. But Weber explicitly links the two—the leader and the socio-economic structure. On Islam and the warrior orders that arose within it (and analogous phenomena) he writes that it was not that:

> ... the conditions of life of the Beduins and semi-nomads had "produced" the founding of an order, say as the "ideological expression [*Exponent*]" of the conditions of their economic existence. This kind of materialistic construction of history misses the mark just as much here as elsewhere. Rather: when such a foundation occurred, then it had, in the living conditions of these social classes [*Schichten*], by far the strongest chances of outlasting the other more unstable political formations in the struggle for the survival of the fittest [*Auslesekampf*]. But whether they arose depended on entirely concrete circumstances and destinies of religious history, which were often highly personal. Then once the religious brotherhoods had by their achievements proved themselves and been recognised as a political and economic instrument of power, that naturally contributed enormously to their expansion. The messages preached by Muhammed or Jenonadab ben Rechab are not to be "explained" as products of demographic or economic conditions, however much their content may have been conditioned by these factors, among others. But they were expressions of personal experiences and intentions. But the intellectual and social means that they employed, and, furthermore, the fact that it is precisely creations of this kind that enjoy great success, can indeed be understood in terms of those living conditions. The same for Ancient Israel. (Weber, *GAR*, iii, pp. 88–89)[26]

Thus Weber offers a widely applicable model for religious change, one underpinned by methodological individualism. The messages of a charismatic leader, highly individual in origin, catch on and spread when they happen to strike a chord with the social and economic circumstances around them, i.e., when large numbers of people within their reach are predisposed by their circumstance—motivated—to find the message attractive.

Weber's long analysis of the social and economic structures of Ancient Israel aims to show that it was fertile soil for the message of the prophets. The

---

[26] '...die Lebensbedingungen …. Ebenso für Altisrael.'

social structure of ancient Israel depended on many contracts between groups: land-owning warrior tribes with legally protected outsiders such as itinerant shepherds, artisans, merchants and priests. There was an affinity between such contracts, on the one hand, and, on the other, the bond between the people of Israel with God, held together by a covenant (Weber, *GAR*, iii, pp. 87, 89–90).

6 Peripheral Creativity

Another factor predisposing the emergence of a distinctive Jewish religion according to Weber was its peripheral cultural position. Weber offers an explanation for collective religious creativity (of the Hebrew prophets, Jesus Christ, Augustine of Hippo, Francis of Assisi, and Protestant reformers):

> Hardly ever have entirely fresh religious ideas arisen in those places which were at the centre of the rational cultures of their time. It was not in Babylon, Athens, Alexandria, Rome, Paris, London, Cologne, Hamburg, or Vienna, but in pre-exilic Jerusalem, in Galilee of the late Jewish period, in the late Roman province of Africa, in Assisi, in Wittenberg, Zürich, Geneva and in the peripheral regions of the Dutch and English cultural zones, such as Friesland and New England, that innovative rational prophetic or reforming ideas are first conceived. Never, indeed, without the influence of and the impression made by a neighbouring rational culture. The reason is everywhere one and the same: to enable the creation of new religious conceptions, people cannot have lost the habit of bringing their own questions to bear on what happens in the world. It is precisely the person living away from the great cultural centres who has cause to do so when the influence of those happenings begins to affect or to threaten him in his central concerns. Once somebody is living in the middle of a culturally saturated area and employs its technology as a matter of course, he does not ask such questions of the environment any more than, say, a child who is accustomed to travel by tram every day would spontaneously come up with the question of how it is actually possible for it to be set in motion in the first place. The capacity for wonder[27] at the course of the world is a necessary condition for the possibility of questioning its meaning. (Weber, *GAR*, iii, pp. 220–221)[28]

Weber goes on to list some of the events that started the Jewish people thinking, from the wars of independence that built up Yahveh's prestige as

---

[27] 'Erstaunens'.

[28] 'Kaum je sind ... des Fragens nach ihrem Sinn.'

260   D. d'Avray

a god of war to the imminent threat of the collapse of the kingdom, which raised the questions of how to justify the ways of God to man, i.e., the theodicy problem (Weber, *GAR*, iii, p. 221).[29]

7   Particularity of Jewish Religion

One of the answers to the 'theodicy' question shaped the specificity of the Jewish religion. It was that the Jewish people had a special obligation, because of the sworn Covenant, to obey God's moral law. By the end of time, even non-Jews would be able to do that. That was in the distant future. More immediately, the Jewish people must keep the law contained in the Torah and be collectively punished as a people if they did not. That the Jews would suffer disaster as a nation if they did not live by the moral law was the message of the prophets (Weber, *GAR*, iii, p. 311).[30]

The prophets worked within the framework of the orthodox Judaism of their day:

> ... not one of the prophets had any thought of founding a "community". The fact that everything required for that, in particular the creation of a new community of cult, such as the cult of the Lord Christ offered, was absent, and had to be absent from the mental horizon of the prophets, is a sociologically decisive difference from early Christian prophecy. The prophets had their place in the middle of a political and national community, whose destinies were the focus of their interest. And their interest was purely ethical, not to do with cult, by contrast with the Christian missionaries, who, above all, brought the Lord's supper as the channel of grace. In this respect in fact we see a feature of early Christianity, deriving from the communities of late Antique mystery religion, which was completely alien to the prophets. (Weber, *GAR*, iii, pp. 313–314)

8   The Prophets

Weber explains the appeal of the message in various ways—that Israel was on a cultural periphery, that it was a good 'fit' with social class interests, that it was politically relevant to a small people surrounded by the great Egyptian and Mesopotamian empires—but he does not think that its content can be reduced to or ultimately explained by any of those things. In another key

---

[29] 'Jene Erlebnisse nun … Zusammenbruch des Reiches auf.'
[30] 'Auf das sittlich richtige Handeln … Heil alles an.'

Methodological Individualism in Weber's Sociology ... 261

passage, he unpacks the 'personal experiences and intentions' of the prophets of old Israel, showing that the behaviour of the prophets is so idiosyncratically individual as to be irreducible to a general sociological pattern:

> The widest variety of pathological conditions and pathological actions accompany their ecstasies or precede them. There can be no doubt, that precisely these conditions were recognized as the most important guarantee of the authenticity of prophetic charisma, and that they were also to be found, even if in a milder form, when nothing about them is transmitted to us. Even so, some of the prophets give us explicit accounts of them. The hand of Jahwe "weighs heavily" on them. The Spirit "took hold" of them. Ezekiel ... struck his hands together, hit himself on his sides, and stamped the ground. Jeremiah ... became like a drunken man and trembled in all his limbs. The face of the prophets become distorted when the Spirit comes over them, their breath fails them, sometimes they fall down to the ground as if drugged, temporarily robbed of the powers off sight and speech, they writhe convulsively ... After one of his visions Ezekiel ... suffered from paralysis. The prophets do strange things, thought to be full of ominous significance. Ezekiel, like a child, made a siege game using bricks and an iron pan. Jeremiah smashes a jug in public, buries a belt and then digs it up when it was rotten, walks around with a yoke on his neck, other prophets walk around with iron horns or, like Isaiah, go around naked for a long period. ... Above all they hear sounds ... voices ... around them, individual voices and dialogue, but, especially often: words and commands directed to themselves.... Like Ezekiel, they undergo auto-hypnotic states. There are cases of acting and above all speaking under compulsion. Jeremiah felt that he had a split personality. He pleads with his God to allow him not to speak. He does not want to say, but he has to say, what he feels has been given to him to say, rather than what comes from himself, and he indeed feels it is a fearful fate to be compelled to say it... If he does not speak then he suffers fearful agonies, a burning heat comes over him, and the is unable to bear the heavy pressure without unburdening himself.[31] For him, a person is no prophet at all if he does not know this condition, and speaks not under such compulsion but "from his own heart". (Weber, *GAR*, iii, pp. 300–302)[32]

Though the phenomenology of ecstatic prophecy is, for its era and region, a distinctive characteristic of the religion of ancient Israel, Weber thinks it could have developed elsewhere if it had not been prevented: it is not ecstatic prophecy as such which is entirely unique, but a context in which it could develop.

---

[31] 'ohne Entlastung'.
[32] 'Pathologische Zuständlichkeiten ... überhaupt kein Prophet.'

262   D. d'Avray

To date no evidence of this kind of ecstatic oracle prophecy has been found for Egypt and Mesopotamia, nor for pre-Islamic Arabia, but, in the lands near Israel, only in Phoenica (in the form of prophecy by kings as in Israel) and, strictly controlled and interpreted by priests, in the oracles of the Greeks. Nowhere however do we know anything of a free demagogy of prophesying ecstatics like the prophets of Israel. Without doubt, this was not because the relevant states[33] would not have existed. But because, in the bureaucratic kingdoms (Weber, *GAR*, iii, p. 282[34]) as with the Romans, the religious police would have intervened, while by the Greeks, in historical times, these states were no longer regarded as holy, but as sicknesses, and worthless, and only the traditional oracles, regulated by priests, were recognized. In Egypt, ecstatic prophecy appears only in the Ptolomaic period, and in Arabia only in the time of Muhammed. (Weber, *GAR*, iii, pp. 302[35])

9   Methodological Individualism and Functionalism

The Hebrew prophets were often prophets of doom so that the disasters that befell their people did not undermine their charismatic authority, but when they prophesied success, and success followed, that certainly helped. Weber wrote that 'The historical accident that, in defiance of all probability, the unshakeable prophetic belief of Isaiah that his God would not let Jerusalem fall into the hands of the Assyrian army if only the king stood fast, was actually borne out by events, was subsequently the unshakeable basis for the standing of both this God and his prophets' (Weber, *W&G*, i, 261).[36] Again, nothing succeeded like the astounding military successes of early Islam. On the aggregate, individuals naturally thought that God must be on the side of Muhammed's message.

This points to a way of reconciling methodological individualism with what might appear at first view to be a rival framework for social theory: classic functionalism. In its classic Durkheimian form, this appears to sideline individual choice in favour of 'social facts' (Durkheim, 1895).[37] It does

---

[33] 'Zuständlichkeiten': i.e. psychological states.

[34] '… es entspricht den Verhältnissen der bürokratischen Staaten, daß die öffentliche Prophetie dort ausdrücklich verboten war.'

[35] 'Eine solche ekstatische Orakelprophetie … in Muhammeds Zeit auf.'

[36] 'Der historische Zufall … sowohl wie seiner Propheten.'

[37] For my assessment of Functionalism see d'Avray (2010), pp. 94–96. Nathalie Bulle points out to me however that the 'institutional interpretation of social facts in Durkheim may … be compared to the constraints represented by Weber's structural patterns developed by the shared motivations'

not have to do so, as the philosopher Gerry Cohen perceptively explained. 'Imagine ten godless communities, each, because it lacks a religion, teetering on the brink of disintegration. A prophet visits all ten, but only one of them accepts his teaching. The other nine subsequently perish, and the single believing society survives. But they took up religion because they liked the prophet's looks, and not because they needed a religion (though they did need a religion)' (Cohen, 2000, pp. 281–282; d'Avray, RiH 2010, p. 95). Accepting the prophet was a collective choice by the individuals in the tribe, one which as a side effect favoured the cohesion and the flourishing of the tribe.

A similar analysis works well, *mutatis mutandis*, for Weber's analysis of Protestant sects in America and of Jainism: a side effect of sect membership could be economic trust, not just within the sect but also of sect members by outsiders. He illustrates his argument from his experiences in the USA, during a long journey in what was at that time 'Indian territory'. A travelling salesman in the 'Undertakers' hardware' line said to him: '"Sir, for all I care anyone can believe or not believe whatever they like, but when I see a farmer or a merchant who doesn't belong to any Church, then he isn't worth 50 cents to me:—what can make him pay me, if he doesn't believe in anything?"' (Weber, *GAR*, i, p. 209).[38] Weber goes on to report a story by a German-born ear and throat specialist about his first patient in a practice he had established in a large city on the Ohio river.

> When the doctor told the patient to lie down on the sofa for the investigation ... the latter first sat up again and observed with dignity and emphasis: "Sir, I am a member of the ... Baptist Church in ... Street." At a loss to guess what this fact could possibly have to do with the nose complaint or its treatment, he (the doctor) asked an American colleague of his acquaintance about it, in confidence, and was informed with a smile: "All that means is that you need not worry about your bill being paid". (Weber, *GAR* , i, p. 209)[39]

---

evoked above. Durkheims 'social facts' also have an affinity with Niklas Luhmann's social systems, in that both are distinguished from the individuals without which they could not exist, though Luhmann posits an almost unlimited number of interlocking and interpenetrating social systems at any moment in any society. There is more to be done in the way of building bridges between these great social theorists, especially if one does not find Talcott Parson's way of synthesizing Durkheim and Weber—in a nutshell, through 'norms'—sufficient.

[38] '"Herr, meinethalben ... in anything? [*sic*]"'.

[39] 'Sich auf Aufforderung ... ohne Sorgen.')

264 D. d'Avray

To spell out his point, Weber treats the reader to another vivid description:

On a beautiful sunny afternoon in early October I attended a Baptist Baptism ceremony together with some relatives—farmers in the countryside[40] a few miles from M. [capital of a county] in North Carolina—at a pool through which flowed one of the streams from the Blue-Ridge mountains, visible in the distance. It was cold and had dropped below freezing in the night. Around the slopes of the hill stood crowds of farmer families, who had come in their light two-wheeled carts from the neighbourhood but also from afar. In the pool up to the waist stood the preacher, dressed in black. Into the pool, after various preliminaries, waded around ten people of both sexes in their best clothes, who made their profession of faith,[41] and then immersed themselves entirely under water, the women on the arm of the preacher, came up again gasping for air, got out of the pond shivering with their clothes clinging to their bodies, received congratulations from all and sundry while they were quickly wrapped in thick plaids, and driven away back home. One of my relatives, who was standing beside me, and, in the German tradition, not the church-going type, spat as he watched contemptuously,[42] had his attention caught by the immersion of one of the young men. "Look at him,—I told you so!"—When I asked him (after the ceremony): How come you thought so, as you said?—the reply was "Because he wants to open a bank in M..."—Are there so many Baptists in the neighbourhood then, enough to make a living from them?—"Certainly not, but now that he is baptised he will get the custom of the whole surrounding district and will knock out all the competition." The further questions: why? and, how come? got the answer: that admission into the local Baptist community—still strongly observant of religious tradition —which took place only after the most careful "testing"[43] and painstaking investigations, going back to his early childhood, of his "way of life"[44] ("disorderly conduct"? Bars? Dancing? Theatre? Cardplaying? unpunctuality in paying bills? other kinds of frivolity?), was accepted as such an absolute guarantee of the ethical and above all the business qualities of a gentleman that the deposits of the entire neighbourhood and unlimited credit would be at the disposal of the man in question, without

---

[40] 'Buschwald.'

[41] 'wurden auf den Glauben verpflichtet'.

[42] Here Weber has the following footnote in which he does not translate the English into German: 'He said to one of the baptised: "Hallo Bill, was'n't the water pretty cool?' and got the very serious answer: 'Jeff, I thought of some pretty hot place (Hell!) and so I did'n't care for the cool water.'"

[43] 'Erprobung'.

[44] 'Wandel'.

anyone else having a hope of competing. He was a 'made man'. Further observation showed that this or very similar phenomena were repeated in the widest range of places. The men who made it in business, and in general those men only, were the ones who belonged to the Methodist or Baptist or other sects (or sect-like conventicles). If the member of a sect moved to another place or if he was a travelling businessman, he would take with him the certificate or membership of his community, and through it he not only had an *entrée* with other members of the sect, but, above all, credit with everyone. (Weber, *GAR* , i, pp. 209–211)[45]

We note that the entire analysis is in the spirit of methodological individualism: Weber is thinking in terms of individual motivation, or rather, of the common aggregate motivation of many individuals in similar situations. Note also that Weber is not claiming that commercial ambition was the only motive for joining a sect, or even the principal one. The reason why his account can be called 'functionalist' is that the 'virtuous circle' of commercial success does not depend on the desire for commercial success. (Functionalism is hardly needed as a social theory if the function is the explicitly intended aim of the institution: to say that the function of T.V. police dramas is to entertain is true but trivial; to say that it is to cement allegiance to the state less so, and true irrespective of the intentions of the producers.) In practice, there was doubtless a spectrum of motives, from the purely religious to the cynically instrumental. In any case, it is a generalisation about the motivation of people outside the sect as much as about people within it. In a nutshell, the ideal type is that groups held to the highest moral standards get rich because people trust them in business dealings: 'a functional, economic effect is thus developed as an unintended consequence of individuals adhesion to Protestant sects'.[46]

Always in Weber's mind was the question: how unique is a given phenomenon in world history? In that spirit, he develops in a different study (on Hinduism and Buddhism) a comparison between Christian Protestant sects and Jains. Here we find a very similar implicitly functionalist explanation of economic success: religious motivation which has nothing to do with it nonetheless has the lasting unintended consequence of furthering that success. For in Jainism the pursuit of wealth is regarded as a threat to salvation—more so probably than in the Protestant sects studied by Weber—yet Jains tended to accumulate great wealth.

---

[45] 'An einem schönen klaren Sonntag … Kredit bei aller Welt.'

[46] Nathalie Bulle, personal communication.

266   **D. d'Avray**

Again Weber is worth quoting:

> ... the possession of riches, over and above what one needs for survival, is in itself a danger to salvation. One should give the surplus away to temples or animal hospitals, in order to gain merit. And this was practiced in the highest degree in Jain communities, which were renowned for their charitable activity. Note well: the acquisition of riches in itself was by no means forbidden, only, in much the same way as with ascetic Protestantism in the West, the striving to acquire riches and attachment to them. As with ascetic Protestantism, the reprehensible element was the delight in possessions (parigraha), not at all possession or acquisition *per se*. And the similarity goes further. The prohibition, which the Jains took extremely seriously, of saying anything false or exaggerated, and their absolute probity in business dealings, the prohibition of any deception (maya)[47] and of any dishonorable acquisition, meaning above all any acquisition by means of smuggling, bribery and any variety of dubious financial dealings (adattu dama), on the one hand excluded the sect from the involvement which was so common in the Orient in 'political capitalism' (the accumulation of property by officials, tax farmers, and government contractors), and on the other hand had the effect with them—with both Jains and Parsees—that it did with the Quakers in the West, in accordance with the (early capitalistic) motto: "honesty is the best policy". The probity of Jain traders was famous. And so too was their wealth: it was said in former times that more than half the trade of India passed through their hands. (Weber, *GAR*, ii, p. 212)[48]

The parallel with Protestant sects, goes further. Puritans and Jains had this extra element in common: the austerity of their life meant that they kept their wealth, rather than frittering it away on pleasure.

> The "ascetic compulsion to save", familiar to us from the economic history of Puritanism, also had the same effect with them in the sense that they used the possessions they accumulated as working capital rather than using it up or living off the income.[49] ... Their successful accumulation of wealth, to which the commandment that they must not keep more than what was 'necessary' ... set only very elastic limits, was favoured, as it was with the Puritans, by the strictly methodical character of the way of life prescribed for them. Avoidance of drugs/intoxicating substances, of indulgence in meat or honey, absolute avoidance of any sexual incontinence and strict marital fidelity, avoidance of

---

[47] 'He who practices deception will be reborn as a woman.'

[48] '... ist der Besitz von Reichtum ... durch ihre Hände gehe.'

[49] 'Rentenvermögen.'

pride in their status,[50] of anger and of all passions are commandments taken for granted by them, as by all high-class[51] Hindus. (Weber, *GAR*, ii, p. 213)[52]

10 Motives for Religious Adherence

The foregoing has suggested a series of motives shared by groups for adhering to this or that religion: a good 'fit' between the message of a charismatic religious leader and social conditions, giving a motive to follow the leader; the ability of the beliefs to suggest solutions to fundamental questions asked about life and how to live well, questions asked especially in regions peripheral to the great cultural centres, whose assumptions are an influence but not taken for granted; and success of one form or another as authentication of beliefs or as a 'functionalist' mechanism which ensures that the group continues to flourish. To these, we should now add the conscious use of instrumental rationality to ensure that the religion continues.

11 Instrumental Rationality in the Service of Religious Convictions

The famous 'routinisation of charisma' schema can be brought under that heading. A prophet's mission cannot normally survive the prophet's demise without the help of a more normal and 'everyday' sort of system (*Veralltäglichung*, usually translated as 'routinisation', means 'making into an everyday thing'). A religious community based around prophecy:

> ... only comes into being at all ... as the product of routinisation, in that either the prophet himself or his disciples ensure on a long term basis the continuation of the proclamation of the message and dispensation of grace, and hence also ensure the lasting continuation of the economic existence of the dispensation of grace and those who administer it .... (Weber, *W&G*, i, 275–276.[53])

---

[50] 'ständischem Stolz.'

[51] 'vornehmen'.

[52] 'Der aus der Wirtschaftsgeschichte ... begünstigt.'

[53] 'Sie entsteht ... dauernd sicherstellen.'

268   D. d'Avray

Obviously, Weber is not suggesting that only movements started by prophets need this kind of community. (Weber, *W&G*, i, 275).[54]

This is a simple example of Weber's approach to causation. His ideal types are quasi-laws, disjunctive laws if you like in that they narrow down sharply the number of outcomes without reducing them to one. In this case, the disjunctive law is: either a movement started by a charismatic leader gets organised, or it eventually dies away. Which outcome transpires depends on the agency of the followers of the prophet.

12   Spirituality and Devotion as Instrumental Rationality in the Service of Values

Organisation is one kind of institutional rationalisation. Another kind is the use of instrumental technique to ensure the continued flourishing of a religion, by preaching, printing, images, songs, and rituals, etc. I have discussed this 'interface' between value and instrumental rationality elsewhere, for the most part not so much summarising Weber as applying his insights to fresh material (d'Avray, MiE 2010, pp. 112–134; d'Avray, MRR 2010, pp. 94–112). Some of these techniques are perhaps reminiscent of Foucault's 'technologies of the self'. Two examples which are taken directly from Weber are worth mentioning because they show instrumental rationalisation in a spiritual sphere with which it is not much associated: Hindu asceticism and contemplation.

In a key passage, Weber comments as follows on Hindu asceticism:

> Indian asceticism was technically probably the rationally most highly developed in the world. There is hardly any ascetic method which was not exercised in virtuoso fashion in India and very often rationalised into a theoretical technical science, and many forms were only here carried all the way to their ultimate logical conclusions, which are often simply grotesque for us. (Weber, *GAR*, ii, p. 149; [55] d'Avray, 2020, p. 126)

This shows how rational choice is coloured and transformed by the values it serves. Western countries may have developed a more rational form of capitalism (alongside other sorts, high-risk capitalism and state-contract capitalism, which are commoner in world history) but when it comes to the rationalisation of asceticism, Hinduism is more rational. The techniques

---

[54] 'Die "Gemeinde" in diesem religiösem Sinnn ... taucht ... nicht nur bei Prophetie ... auf ...').
[55] 'Die indische Askese ... hineingesteigertworden'.

result from rational choices based on beliefs that need to be understood before the practices become comprehensible.

The search for mystical enlightenment creates a different kind of 'systematic activity':

> ... the specific good which constitutes salvation is not an active quality characterising behaviour, so: not the consciousness of the fulfilment of a divine will, but a a specific sort of state of being. In its pre-eminent form it is: "mystical enlightenment". This too can be achieved only by a minority of specially qualified persons and only through a systematic activity of a particular sort: Contemplation. To achieve its goal, contemplation always needs to eliminate day-to-day preoccupations. Only when everything that is creaturely in man falls altogether silent can God speak inside the soul, as the experience of the Quakers has it, and all contemplative mysticism from Lao-Tsu and Buddha to Tauler is probably in agreement with this experience in substance though not in their verbal formulations. (Weber, *W&G*, i, 330)[56]

The self-emptying is not the end, but a technology for achieving enlightenment.

13   Conclusions

Weber was indeed a methodological individualist in that his starting point was individual motivation. Not for him reified classes, or 'society' as anterior to individual agency. But his understanding of individual choices is infinitely richer than a simple 'ranked preferences x risk assessment' approach. When many individuals have similar motives, we have regularities that can be called structures. Psychological motivation may be at odds with the reasons social actors give to themselves: i.e., actors who tell themselves a story to explain their behaviour may not understand the causes that really lie behind it. So: puritan businessmen subconsciously worked extra hard to prove by their success that they were among the elect, even though their own theology made God, not extra willpower and effort, the crucial determinant of election and the success that manifested it. Charismatic leaders inject into history personal values which may be irreducible to any prior social cause (e.g. Hebrew prophets), but which may change the course of history under certain circumstances. Those circumstances include a good fit between the new message and social conditions around the leader (e.g. Islam), side effects which are

---

[56] 'das spezifische Heilsgut ... zu Tauler, übereinstimmt.'

270   D. d'Avray

conducive to military or business prosperity, and successful instrumental techniques for propagating and transmitting devotion to the religion. What lies behind the contrast between Weber's methodological individualism and that of rational choice theory is his deep understanding of particularities of world history and a broader, richer insight into the different kinds of human motivation.

Abbreviations and Bibliography

For a thorough list of English translations of Weber's works, including all those cited below in the convenient German editions, see G. Roth & C. Wittig, eds., Max Weber, *Economy and Society,* 2 vols., (Berkeley, 1978), i, pp. xxv–xxviii

Cohen, G. (2000). *Karl marx's theory of history: A defence.* Oxford University Press.

d'Avray, D. L. (2019). The concept of magic. In S. Page & C. Rider (Eds.), *The Routledge history of medieval magic,* (pp. 48–56). Routledge.

d'Avray, D. L. (2010). *Medieval religious rationalities. A weberian analysis.* Cambridge University Press. [MRR]

d'Avray, D. L. (2010). *Rationalities in history. A weberian essay in comparison.* Cambridge University Press. [RiR]

Durkheim, E. (1895). *Les règles de la méthode sociologique.* Félix Alcan.

Heath, J. (2020). Methodological individualism. The Stanford encyclopedia of philosophy. https://plato.stanford.edu/archives/sum2020/entries/methodological-individualism/

Kalberg, S. (2001). The "spirit of capitalism" revisited: On the new translation of weber's protestant ethic (1920). *Max Weber Studies 2,* 41–58.

Kippenberg, H. G., Schilm, P., & Niemeier, J. (2005). *Max Weber, Wirtschaft und Gesellschaft. Die Wirtschaft und die gesellschaftlichen Ordnungen und Mächte. Nachlaß*, Teilband 2: *Religiöse Gemeinschaften. Studienausgabe der Max Weber-Gesamtausgabe*, Band I/22–2. J. C. B. Mohr.

Stark, R., & Bainbridge, W. S. (1996). *A theory of religion.* Rutgers University Press.

Weber, M. (1920–1921/1988). *Gesammelte Aufsätze zur Religionssoziologie* i, *Die protestantische Ethik und der Geist des Kapitalismus.* J. C. B. Mohr. [GAR, i]

Weber, M. (1920–1921/1988). *Gesammelte Aufsätze zur Religionssoziologie,* ii. J. C. B. Mohr. [GAR, ii]

Weber, M. (1920–1921/1988). *Die Wirtschaftsethik der Weltreligionen. Das antike Judentum* in *Gesammelte Aufsätze zur Religionssoziologie*, iii. J. C. B. Mohr. [GAR, iii]

Weber, M. (1921/1984). *Soziologische Grundbegriffe* (6th, revised, edition). J. C. B. Mohr. [SG]

Weber, M. (1922/1976). *Wirtschaft und Gesellschaft. Grundriss der verstehenden Soziologie* (5th edition, two vol.). J. C. B. Mohr. [W&G]

Unintended Consequences

Karras J. Lambert and Christopher J. Coyne

1 Introduction

Lionel Robbins (1932, p. 15) famously defined economics as "the science which studies human behaviour as a relationship between ends and scarce means which have alternative uses." Robbins admitted that his definition was heavily influenced by the principle theoretical work of Carl Menger (1881) as well as various writings of Ludwig von Mises, Menger's follower and Robbins's' contemporary. In his later writings, Mises (1949) uses the term "praxeology" to denote the general science of human action of which economics, conceived narrowly as the science of action which can make use of economic calculation, is a major part. This broader theoretical science of praxeology concerns, as Robbins's definition of economics suggests, all choices between competing means perceived as suitable for the attainment of desired ends.

With this means-ends structure of human action in mind, it is clear that all individual actions involve intentionality. Scarce means perceived as suitable to attain desired ends are adopted purposively by every human actor through an

---

K. J. Lambert · C. J. Coyne (✉)
Department of Economics, George Mason University, Fairfax, VA, USA
e-mail: ccoyne3@gmu.edu

K. J. Lambert
e-mail: klamber7@gmu.edu

© The Author(s), under exclusive license to Springer Nature Switzerland AG 2023
N. Bulle and F. Di Iorio (eds.), *The Palgrave Handbook of Methodological Individualism*,
https://doi.org/10.1007/978-3-031-41508-1_12

exercise of choice. Unintended consequences, then, entail the attainment of ends not originally envisioned when deciding to pursue a particular course of action; the employment of particular means, expected to produce a certain desired consequence, actually produced different or additional consequences not intended *ex-ante* by the human actor.

The term "unintended consequences" typically carries a negative connotation, but there is no reason unintended consequences cannot be harmless or even recognized as desirable by the actor. Adam Smith famously noted the *positive* unintended consequences of commercial action in a market setting in his famous passages invoking the "invisible hand." Smith (1776, p. 456) argues that in a market, an individual often makes decisions about what to buy or sell only with self-interest in mind, yet nevertheless these exchanges "promote an end which was no part of his intention." Far from considering the "unintended" aspect of the mutual betterment following voluntary market exchange as troubling, Smith recognizes that an individual acting out of self-interest, "frequently promotes that of the society more effectually than when he really intends to promote it" (p. 456). After all, as Smith (p. 27) writes, "It is not from the benevolence of the butcher, the brewer, or the baker, that we expect our dinner, but from their regard to their own interest."[1]

Virtually every human action will likely involve the generation of *some* consequences which were not intended. As Mises (1960, p. 35) writes, "man too is as far outside the effective range of his action as a reed in the wind." Although the magnitude and scope of consequences from the private action of an individual will tend to be limited, governmental actions can more easily generate more systemic effects, by intervening in the price system or legal institutions, which more directly affect all individuals in the society in some way. Perhaps for this reason, the academic literature on unintended consequences tends to deal with the *negative* unintended consequences of government interventions.

This chapter will cover both the underlying theory of unintended consequences and a number of empirical studies detailing the negative unintended consequences of interventions generally presumed to be conceived and instituted out of the benevolent motives of government actors. While we recognize that there are arguments that governmental interventions tend to have private interests behind them and that any "unintended" consequences are not necessarily "unforeseen" or undesirable from particular points of view, we follow a conventional economic approach toward unintended consequences.

---

[1] Smith's insight has led to a scholarly literature on invisible-hand explanations of a variety social phenomena including norms, money, rules, and discrimination. For an overview of invisible-hand explanations by economists, see Aydinonat 2008.

According to this conventional view, the declared statements of the planners, or at least the commonly understood intentions behind the interventions, are taken at face value. That is, we do not scrutinize the possible *true* intentions of policymakers, but instead take them at their word regarding the stated ends of policies they advocate.

Since it is impossible to conceive causality and therefore match consequences with causes without reference to theory, Sect. 2 describes some notable theoretical approaches toward the consequences of government interventions in economics. Sections 3 and 4 then present an overview of some notable empirical literature on the apparent negative unintended consequences of particular government interventions in both domestic and international contexts respectively. Section 5 concludes.

2 Unintended Consequences in Theory

Merton on Unintended Consequences

Robert Merton's "The Unanticipated Consequences of Purposive Social Action," published in 1936, is considered a seminal paper in the study of unintended consequences. Merton (1936, p. 894) justifies attention to the subject by noting that "In some one of its numerous forms, the problem of the unanticipated consequences of purposive action has been treated by virtually every substantial contributor to the long history of social thought." This is hardly an overstatement, as all sciences seek to uncover causal laws that, once grasped, will enable individuals to more appropriately match the means they choose to adopt with the ends desired. Only correct scientific understanding can inform us whether means are suitable for the attainment of some end.

Merton (1936) identifies three major factors underlying the realization of unanticipated consequences, namely lack of knowledge, error, and "imperious immediacy of interest." Insofar as knowledge of past conditions does not hold in the future, actions, which are by nature forward-looking, may have consequences that are not foreseen due to ignorance of some aspects of the situations in which the actions take place. An error may occur "in any phase of purposive action: we may err in our appraisal of the present situation, in our inference from this to the future objective situation, in our selection of a course of action, or finally in the execution of the action chosen" (Merton

274   K. J. Lambert and C. J. Coyne

1936, p. 901). Finally, an actor may simply be fixated on attaining a particular end and therefore ignore additional consequences that may be reasonably expected to result from the pursuit of some action.

Merton primarily uses the phrase "unanticipated consequences" in the piece, which has a similar, though distinct meaning from consequences which are strictly *unintended*. Although Merton would later treat the two as synonyms, De Zwart (2015, p. 284) finds the conflation unfortunate, as he argues that consequences may be unintended yet nevertheless anticipated as "policy makers foresee more than we give them credit for" and can accept the possibility of unwelcome side effects when pursuing a given intervention.

Bastiat on the Unseen

Nearly a hundred years before Merton's article, the French economist Frederic Bastiat (1850) published an important work highlighting the counterfactual nature of economic laws. Bastiat emphasized that every action has not only an obvious visible effect but also effects that, in a sense, cannot be directly seen due to being counterfactual. Instead, the qualitative nature of the counterfactuals must be grasped through logical reasoning. It is necessary to study economic theory in order to *foresee* the more remote and counterfactual effects of certain actions.

Bastiat gives an example of a boy carelessly breaking a store window. Some onlookers console the owner of the store by noting that now the glazier will now enjoy extra business and the "economy" is better off than it would have otherwise been due to the concomitant increase in the circulation of money. However, Bastiat makes clear that the money spent to pay the glazier for window repair services is now money that cannot be spent by the store owner in alternative ways. For example, now a shoemaker or some other merchant who otherwise would have received the store owner's patronage will have to do without business.[2] Bastiat's story illustrates the simple fact that, when accounting for counterfactual possibilities, destruction is not an economic blessing but simply entails a loss of wealth.[3]

---

[2] It is beyond our scope to thoroughly counter the objection that Bastiat's lesson would not hold if the store owner would have engaged in saving instead of spending out the same sum paid for the window repair. However, the key is to realize that such savings do benefit everyone in society in the sense that the purchasing power of all other monetary units will be counterfactually higher than in the case without the saving and, furthermore, that an act of saving enables the investment in capital goods which results in higher standards of living over time.

[3] Higgs (1992) applies similar reasoning to explain the "myth of wartime prosperity" in the United Sates during World War II. In doing so he makes a broader point about the counterfactual use of resources dedicated to preparing for, and executing, violent conflicts.

Bastiat applies his insight into counterfactual reasoning to taxation, noting that the benefits often touted are plain to see, such as the construction of public works projects, while the disadvantages are unseen because they never had the chance to be pursued. The latter counterfactual case requires one to imagine how taxpayers might have spent the taxed money and how the resources purchased by the government may have been utilized in alternative lines of production. Advocates of taxation and public works projects may proclaim the benefits to be derived from particular projects that only governments are deemed able to pursue, but the unintended consequences may be a counterfactually lower standard of living, as considered from the subjective valuations of the citizens, than might have otherwise been possible. Given the time in which he was writing, Bastiat also raises the problem of opposition to labor-saving machines, which may prevent temporary unemployment but at the expense of preventing increases in prosperity that would result through the utilization of new technology that frees up workers to devote their efforts to satisfying an even greater range of desired human ends.

Mises on Interventionism

In his various methodological writings, Ludwig von Mises (1949, 1957, 1960, 1962) further clarified the counterfactual nature of economic theory and demonstrated how economic laws, such as the law of marginal utility and the law of returns, can be derived from the nature of human action. Mises also dealt directly with the economic logic of government interventionism in a 1929 German work eventually translated into English in 1977 as *A Critique of Interventionism*. In that work, Mises emphasized the importance of developing correct economic theory in understanding the consequences of particular policies. For example, concerning the theory of trade Mises (1977, p. 5) writes, "The purpose of the theorists of free trade was not to demonstrate that tariffs are impractical or harmful, but that they have unforeseen consequences and do not, nor can they, achieve what their advocates expect of them." The conclusion that tariffs produce "unforeseen consequences" from the standpoint of their advocates follows from the knowledge that restricting exchange also results in counterfactually lower wealth than would have otherwise been possible. Mises acknowledges that these consequences may very well be understood, and in that case, it would not be considered "unforeseen" or unintended. However, "The fact is that all production restrictions are supported wholly or partially by arguments that are to prove that they raise productivity, not lower it" (Mises 1977, p. 6).

Mises illustrates the way that economic theory applies to particular government interventions in describing the effects of price controls. He assumes that governments impose a maximum price, also known as a "price ceiling," with the intention of enabling more consumers to afford the price-controlled good. However, he argues that price control as a means is unsuitable to attain the desired end. The reason is that an effective maximum price below the price that would prevail absent the control produces shortage since there will be more prospective buyers than sellers. In the face of this shortage, some method of rationing must be pursued, whether first-come first-serve, or on some other arbitrary basis. If the government wants everyone who wishes to buy at the counterfactually lower controlled price to do so, it must enact further interventions in an attempt to make future production economical, namely controlling prices of relevant factors of production to prevent marginal sellers from leaving the market due to unprofitability. The same logic will apply again and again until either the interveners stop short of their ultimate goal or markets are entirely abolished.

An example Mises often provides, such as in a lecture concerning interventionism given in Argentina in 1959, involves price controls on milk. Mises (1979) assumes that a government declares an effective maximum price on milk so that poor parents can buy as much milk as their children want. It is true that some people who would not have been able to afford milk at a higher price now can do so. However, some marginal producers who previously could operate profitably now suffer losses at the artificially low price and must shift production away from milk. Instead, they may now use the milk to create different products that are not price controlled or employ the factors of production relevant to milk production, such as cows, for entirely different, albeit still profitable, uses.

The result of the price control then, is that the quantity of milk demanded is higher and the quantity of milk supplied is lower than they otherwise would have been. Children as a whole are now getting less milk than before the controls and some form of arbitrary rationing is necessary to solve the shortage. If the government is resolute about getting more people milk at lower prices, they then must try to control the prices of the relevant factors of production so that milk production is profitable at low prices. Controls on those factors of production will beget the same effects and the choice facing interveners will be whether to try to control the prices of every good in the economy for the purpose of achieving the goal to get parents and children more milk or to give up entirely.

Extending Mises: Some Contributions of Hayek, Rothbard, Kirzner, Ikeda, and Bylund

F. A. Hayek picked up on his mentor's analysis of interventionism and described the, often negative, unintended consequences that follow from attempts at large-scale social control in a number of his works. In *The Road to Serfdom* (1944), Hayek assumes the best of intentions on the part of advocates of government planning but nevertheless concludes that such planning constitutes a threat to basic freedoms, as economic planning leads step by step to totalitarianism.[4] Echoing Mises, Hayek (1944, p. 137) recognizes that "the close interdependence of all economic phenomena makes it difficult to stop planning just where we wish and... once the free working of the market is impeded beyond a certain degree, the planner will be forced to extend his controls until they become all-comprehensive." In another famous publication, Hayek (1945) frames the problem facing interveners primarily as one of being unable to replicate the contextual and often inarticulate knowledge on which private individual actions are undertaken.

Murray Rothbard and Israel Kirzner, two students of Mises in the United States, also published theoretical works dedicated to extending Mises's analysis of interventionism. Rothbard (1970) develops a tripartite classification of interventions as being autistic, binary, or tertiary. Autistic interventions involve a restriction on the use of an individual's property such as in prohibitions or mandates. Binary interventions establish a hegemonic relationship between an individual and the intervener such as in taxation or conscription. Tertiary interventions create a hegemonic relationship between an intervener and a pair of subjects. Most interventions analyzed in the academic economic literature are examples of tertiary interventions, including price and product controls, tariffs, minimum wages, and antitrust laws.[5]

Kirzner (1985) develops a typology of the consequences of intervention in terms of how it affects the entrepreneurial discovery process of the market. Echoing Bastiat's emphasis on the "unseen," Kirzner draws attention to four distinct reasons why interventions can produce results less desirable than those that would have been attained in the absence of the intervention. They include the "undiscovered discovery process," the "unsimulated discovery process," the "stifled discovery process," and the "superfluous discovery

---

[4] On why Hayek's argument regarding interventionism under economic planning is not a slippery slope argument, see Boettke (2018), pp. 141–155.

[5] For a discussion of Mises's and Rothbard's contributions to the theory of interventionism, as well as some open areas for further exploration, see Lavoie (1982).

process." In the "undiscovered discovery process," the intervener misidentifies the counterfactual future state. The "unsimulated discovery process" refers to the fact that government planning cannot replicate entrepreneurship in market contexts. The "stifled discovery process" considers the private actions of profit-seeking individuals that are suppressed or crowded out by an intervention. Finally, the "superfluous discovery process" recognizes that intervention creates new and possibly undesirable profit opportunities which would not have otherwise existed.

Integrating the aforementioned ideas of Mises, Hayek, and Kirzner, Sanford Ikeda (1997, 2005, 2015) has written extensively on what he has named the "dynamics of interventionism," or the logic that links together otherwise apparently piece-meal interventionist measures. Like Mises, Ikeda (2015, p. 393) frames the dynamics of intervention in three steps, namely the initial intervention, the realization of negative unintended consequences, and further interventions which attempt to correct the resulting undesired consequences. Ikeda assumes beneficence for analytical purposes without taking a position on whether the stated purposes of interventions are in fact the actual motivations in particular historical situations. Ikeda (2015, p. 399) also notes that there might also be cumulative psychological and behavioral consequences of interventions as new attitudes and practices become entrenched that otherwise would not have been perceived by individuals as the most suitable to attain their ends.

Most recently, Bylund (2016) draws on the contributions of Bastiat, Mises, and Hayek to analyze the unseen consequences of government regulation of the market. His core argument is that government efforts to regulate the market result in unrealized gains in the form of "opportunities that would have been available in the absence of restrictions of the market" (2016, p. 163). The unrealized effect of intervention goes beyond the opportunity cost of the immediate forgone exchanges and affects the broader constellation of economic interactions and exchanges. As Bylund (2016, p. 42) puts it, "a single change causes ripple effects in production just like the waves around a stone thrown into a pond." Regulation hampers the market process along the lines highlighted by Kirzner, resulting in both unseen *and* unrealized consequences. Bylund's synthesis demonstrates the combined power of the ideas of the aforementioned thinkers, which sets the stage for the application of the logic of unintended consequences in practice.

3   Unintended Consequences in Practice: Domestic Interventions

There have been so many empirical studies conducted on the potential unintended consequences of government intervention that it would not be possible to do justice to them all in this chapter. To provide a sample, we cover empirical studies written on four major policy interventions—laws concerning minimum wages, driver safety regulations, drug prohibition, and intellectual property. Although details vary across the cases, a uniting theme is that interventions often produce consequences at variance with their stated or assumed goals. However, due to the counterfactual nature of any theory dealing with causation, there is always room for rebuttals that claim the situation would be *even worse* without the intervention. Therefore, disputes over the consequences of particular interventions often must take place on the level of historical judgment, which involves efforts of persuading others by appealing to established theoretical relations and the correspondence of the conditions under investigation to those assumed in theory.

Minimum Wages

The debate concerning the effects of the minimum wage has become a paradigmatic topic in contemporary economic research. The conventional perspective recognizes minimum wages to be a price floor that will create a "surplus" of workers, or unemployment if instituted above the market-clearing wage. Mises held this position, arguing that effective minimum wages, like all price controls, will result in permanent unemployment of some factors of production and possible capital consumption. If minimum wages are imposed in an effort to raise the standard of living of low-wage workers by permanently raising their wages, the unemployment of some of those workers will be contrary to the stated goals of the intervention. Some workers may enjoy a higher wage than they otherwise would have received without the imposition of a minimum wage, but those who cannot be profitably employed at the effective price floor will now receive no wages instead.

As the method of practicing economics shifted over the course of the twentieth century from the logico-deductive method of Mises and Robbins to an empirical, data-driven approach, the doors opened to attempted refutations of what was long considered economic law. Card and Krueger (1994) have become a touchstone in the economics literature in concluding that an increase in the New Jersey minimum wage did not seem to decrease employment, but even may have *increased* employment. The authors compared changes in employment in fast-food restaurants across state lines between New Jersey and eastern Pennsylvania following a hike in the New Jersey minimum wage. They did not find evidence that employment at fast-food restaurants in New Jersey was affected relative to employment in fast-food restaurants in eastern Pennsylvania, where a similar hike in the minimum wage was not enacted at that time.

Numerous empirical papers have since been published both supporting and rejecting the notion that minimum wages cause unemployment. However, in a recent meta-study, Neumark and Shirley (2022) find that nearly 80% of empirical studies examining the disemployment effects of minimum wages since 1992 support the conventional view at various levels of statistical significance. The bulk of the empirical studies also find that minimum wages tend to have stronger disemployment effects for teens and the less educated, which is what would be expected according to traditional economic understanding.

Perhaps no topic better illustrates the importance of understanding the distinction between theory and history than the debate on the effects of minimum wages. By no means will Neumark and Shirley (2022) or any of the various studies cited therein be the last word on the subject, as there will always be the opportunity for scholars to gather a new data set and attempt anew to refute the conventional economic theory. However, it will be important for such scholars to understand the lesson made clear in Bastiat (1850) concerning the counterfactual nature of economic law. No time-series study can genuinely discover what employment *would have been* absent in a minimum wage law and disprove conceptual counterfactual laws of human action. The fact that most empirical studies *do* tend to find empirical evidence of disemployment effects is unsurprising, but has no bearing on the logical deductions of economic theory. *If* a price floor such as a minimum wage is effective, *then* it will produce a counterfactual disemployment in factors of production, whether in gross labor hours worked or complementary factors of production, relative to a baseline case where such a minimum wage was not instituted.

Driver Safety Regulations

The effect of driver safety regulations is another topic in which the possible unintended consequence of intervention has generated arguments on both sides. Peltzman (1975) opened the debate by arguing that drivers may respond to safety regulations by driving more dangerously. While this may result in some auto occupants surviving who would otherwise have died in

Unintended Consequences   **281**

accidents, the higher rate of accident incidence may result in more pedestrian deaths and nonfatal accidents. Given the volitional nature of human choice, whether drivers *do* in fact tend to drive more dangerously following the institution of safety regulation such as mandatory seat belt laws is a question truly open to empirical investigation. There are also fundamental problems in quantifying "recklessness" and controlling for all relevant factors. Nevertheless, underlying economic logic, which indicates that reducing the costs of actions will tend to result in more of those actions, supports the logic of Peltzman's argument, even if the subjective considerations of individuals prevent the relationship from holding absolutely in all places and times.

As with any empirical question, rebuttals staking an opposing position soon followed Peltman's initial article. Joksch (1976) quickly disputed Peltzman's model and empirical findings, proposing relevant factors that Peltzman did not take into account, such as the size and weight of automobiles, highway improvements, changing seasonal travel patterns, and changing vehicle occupancy, all of which may have biased Petltzman's regression coefficients. Among other criticisms, Joksch also raises issues with state-level data Peltzman used for injury and property damage accident rates since the collected data varies across states and repair costs vary across car models and years.

Peltzman (1976, p. 139) responded that the problems Joksch identified "are present to some degree in every similar statistical investigation." He also raised a point about the counterfactual problem underlying the empirical studies, noting that death rates were falling faster before safety regulations than after they were imposed. Therefore, it would be reasonable to expect them to have continued decreasing more than they did if the regulations were effective and countervailing behavioral adjustments were not present.

Literature supporting and contesting what became known as the "Peltzman Effect" followed, with Robertson (1977) presenting additional criticisms of Peltzman's model and Cohen and Einar (2003) arguing that data shows mandatory seat belt laws reduce driving fatalities without any evidence of compensating behavioral effects on the part of drivers engaging in more reckless driving, and thereby increasing incidents of accidents and non-occupant fatalities. However, more recent empirical studies that investigate the consequence of increased safety regulations in NASCAR races have argued that Peltzman's argument holds, at least among NASCAR drivers.

Sobel and Nesbitt (2007) and Pope and Tollison (2010) both find that NASCAR drivers responded to safety regulations by driving more recklessly, resulting in more on-track accidents and caution laps. Pope and Tollison (2010) focus on the introduction of a head and neck restraint system in

2001 and note that while the new system reduced the chance of serious injury to a driver involved in an accident, it also restricted driver head movements, which is especially necessary when driving on pit roads. The new safety feature may have resulted in an accident in which three crewmen were hit and suffered severe concussions as a result that same year. Safety regulations were then imposed for pit crew members which required the use of helmets and fire suits, an indication that the dynamics of intervention—where one intervention leads to unintended consequences resulting in subsequent interventions—may have been at work.

Drug Prohibition

Empirical investigations inspired by economic theory have also revealed a number of negative unintended consequences of drug prohibition laws. Drug prohibition has been pursued in various ways across countries and generations in an effort to reduce the consumption of drugs and therefore reduce undesirable social consequences such as addiction, crime, and drug-induced deaths. However, according to Thornton (1991), the apparent consequences of these policies have included a rise in imprisonment, creation of black markets and organized crime, corruption of enforcement institutions, spikes in crime related to drugs, increased potency of drug doses, heightened risk of overdose due to dosage variability, substitution to more innovative and dangerous substitutes, monopolization of the medical and drug industries, all without resulting in the intended reduction in drug use and deaths. Other studies of drug prohibition provide further support for a range of unintended consequences associated with drug prohibition (see, Miron 1999, 2001, 2003; Miron and Zwiebel 1995).

Boettke, Coyne, and Hall (2013, p. 1073) trace the United State government's "War on Drugs" and its concomitant effects described above to the 1906 Pure Food and Drug Act, which "required substances containing ingredients like cocaine, morphine, heroin, alcohol, and cannabis to be labeled with information regarding content and dosage." The authors argue that the results of the War on Drugs have been an increase in drug-related violence across multiple countries, a higher likelihood of overdose among drug users, a diversion of funds to incarceration and drug interdiction, the encouragement of cartelization in the drug industry, and the erosion of civil liberties in the United States.

The economic logic underlying their argument is as follows: prohibition raises the cost of production and sale, with marginal sellers forced to leave the market, resulting in fewer suppliers and higher prices. At the same time, quality control is reduced and drug dealers substitute higher potency drugs, both of which lead to more overdoses. The movement toward "harder" drugs also shifts methods of intake of injection, which requires needles. The sharing of needles in turn increases disease transmission, and any follow-on subsidies to clean needle programs simply divert resources from other uses in an effort to keep drug users from contracting and spreading diseases. Drug-related violence has also increased since disputes are forced underground, where means of peaceful resolution are less likely than through legal processes. Higher prices and more violence encourage cartelization, which leads to even higher prices, and so on in a vicious cycle. Furthermore, domestic enforcement of the war on drugs has also resulted in the militarization of police and a wide expansion of government agencies necessary to enforce the prohibition laws.

Redford and Powell (2016) adopt the "dynamics of interventionism" framework in arguing that the Harrison Act of 1914 was meant to address the unintended consequences of previous attempts to prohibit drugs in the United States. The US government enacted policies to discourage the use of opium from the 1880s to 1910s. These policies did not reduce the smoking of opium but did increase smuggling, which led to further laws restricting the sale of medicine without a prescription. A ban on the importation of opium by Chinese residents shifted importation to US residents who then sold it to the Chinese. The result was more Americans involved in the opium importation business. The import ban also led to a rise in the importation of crude opium, which could then be used to produce smoking opium.

After the 1909 Opium Exclusion Act, a ban on the importation of smoking opium, the price of smoking opium more than quadrupled and increased the profitability of growing opium domestically in order to manufacture it into smoking opium. An amendment was later added that specified all smoking opium would be presumed to be imported unless it could be proven otherwise. This further drove up the price of smoking opium and further incentivized smuggling from outside the country.

Finally, Redford (2017) argues that the structure of drug scheduling in the 1970 Controlled Substances Act, part of the broader Comprehensive Drug Abuse Prevention and Control Act, has resulted in the prevalence of synthetic designer drugs created to circumvent legislation with increased variety and potency. Due to the penalties related to how specific named drugs were listed on the drug schedule, Redford (2017, p. 218) notes that drug entrepreneurs find it profitable to "find recreational uses of lower scheduled drugs, seek out and create entirely new or previously unscheduled drugs, and slightly modify the chemical structure of existing, scheduled drugs."

Intellectual Property

Intellectual property (IP) is another area where some scholars have found that government policy has produced considerable negative unintended consequences. Intellectual property laws are commonly considered economically beneficial due to a presumed market failure related to the under-provision of research and development if the benefits of eventual production cannot be secured at least for some time. Otherwise, potential competitors may be able to cheaply replicate the discovery without the need for costly research and development, deterring companies from engaging in such speculative investments in the first place.

At a theoretical level, scholars have noted the potential negative unintended consequences of IP law, including higher prices, broader distortions in resource allocations, and anticommons problems in downstream innovation (see Heller & Eisenberg 1998, Stiglitz 2008, Boldrin & Levine 2008, 2013). Kinsella (2008) argues that IP rights, such as copyrights and patents, are incompatible with ordinary property rights in scarce objects. This is because IP rights necessarily interfere with the property of others, setting arbitrary limitations on how they may use their own resources. He notes that the claims that innovation is spurred by IP law face the usual hurdle of the counterfactual, as we cannot be sure that there would not be even more innovation without the restrictive monopolies granted as a consequence of IP law and the significant sums of money now spent on patents and lawsuits (p. 22).

Cole (2001) surveys publications staking positions both for and against the idea that patents result in increases in productivity through technological innovation and finds little empirical support for the pro-patent position. Not only do the benefits of patents to innovation not appear as large as normally assumed, but the costs are often overlooked. Such costs include not only administrative and legal expenses but a direct hindrance to technological progress. Cole provides a number of historical examples that illustrate the extent to which patents stifle innovation.

For instance, in order to mass produce inexpensive automobile models, Henry Ford was forced to legally spar with the Association of Licensed Automobile Manufacturers, which controlled the 1895 patent on gasoline automobiles awarded to George B. Selden, who were set on producing only high-priced cars. In addition, the patent on airplane wing twisting awarded to Orville and Wilbur Wright in the early 1900s stalled innovation in airplane manufacturing as a result of zealous enforcement and had dubious counterfactual benefits, given how many competitors were also developing airplanes at the time. Finally, James Watt's patent on steam engines arguably stifled innovation for roughly a quarter of a century, with the Industrial Revolution arguably only gaining real momentum after Watt's patent expired in 1785.

Bessen and Meurer (2008) survey the available empirical literature, including evidence from economic history and cross-country studies, to explore whether patents have encouraged or discouraged investment in innovation. While acknowledging that conventional property rights in scarce goods is a major cause of economic growth, they find "a marked difference between the economic importance of general property rights and the economic importance of patents or intellectual property rights more generally" (2008, p. 18). Instead, "the empirical economic evidence strongly rejects simplistic arguments that patents universally spur innovation and economic growth" (p. 18). They conclude that contra the common justification patents, "for public firms in most industries today, patents may actually discourage investment in innovation" (p. 18).

Safner (2016) adopts Kirzner's (1985) framework to investigate the "unsimulated" and "stifled" discovery processes caused by copyright regulation. Safner looks at the 1976 Copyright Act, which substantially expanded the scope and duration of creative works protected as well as the bundle of rights granted, as a base from which further copyright interventions followed to deal with unintended consequences. While we cannot know in advance what future technologies and modes of artistic expression will emerge, no feedback mechanism exists to determine whether a given patent decision yields an economically efficient result or to adapt to the emergence of novel technologies in the future. This means that court decisions will ultimately determine the permissibility of certain actions related to IP litigation claims. Furthermore, the costs of taking certain productive actions are increased, thereby deterring the discovery of new opportunities in the future. Kirzner's "wholly superfluous discovery process," Safner argues, is also evident in the fact that IP rights make it profitable for patent holders to acquire monopoly rights solely to preclude competition and therefore slow innovation.

Finally, the study "Unintended Consequences: Sixteen Years under the DMCA" (2014), published and updated by the Electronic Freedom Foundation, documents a number of cases where "the anti-circumvention provisions of the DMCA [Digital Millennium Copyright Act] have been invoked not against 'pirates,' but against consumers, scientists, and legitimate competitors" (p. 1). Such litigation, the authors of the report argue, creates a chilling effect on scientific research and freedom of expression and sets up impediments to competition and innovation, the opposite result from the stated intentions of advocates of intellectual property.

There is also some experimental evidence that IP law leads to less innovation. Brüggemann et al. (2016) designed an experiment of sequential innovation to explore the implications of intellectual property on the process of innovation. They found that the introduction of IP led to innovations becoming less frequent and more basic, as participants failed to pursue the most welfare-enhancing innovation paths. Moreover, they found that communication among participants did not reduce the negative effect on innovation.

4 Unintended Consequences in Practice: International Humanitarian Action

There is little that separates foreign interventions from domestic interventions in kind. However, the distance over which the interventions occur, whether geographical or cultural, does tend to exacerbate knowledge problems and further reduce the likelihood of attaining the desired results. In this section, we focus on failures of humanitarian actions to produce the ostensible goal of increased standards of living abroad and instead often produce negative consequences in terms of increased violence, corruption, and a stifling of economic development.

Foreign interventions occur in a complex environment of overlapping economic, political, legal, and cultural institutions. Like with domestic interventions, the ultimate effects of foreign interventions may indirectly depend on the strategic behavioral adjustments of individuals living in the societies intervened upon. However, the scale of humanitarian actions means that errors in identifying the solution to problems of economic development will tend to produce more stark negative unintended consequences than relatively smaller-scale domestic interventions.

Furthermore, interveners, both domestic and foreign, tend to conflate technological problems with economic problems (see Coyne, 2013). While technological problems resemble engineering tasks in that they assume constant relations between means and ends and are concerned with finding a solution to what is essentially a constrained maximization problem, the "ends" of economizing individuals are subjectively evaluated by rank orderings over conceivably attainable products within particular contexts. Therefore, the goal of foreign aid to raise standards of living abroad must attempt to target the production of output evaluated more highly from the point of view of the citizens in the nations receiving aid.

P. T. Bauer on the Poverty Trap and Foreign Aid

Economist Peter T. Bauer (1957, 1972, 1981, 2000) spent a considerable part of his career attempting to dispel myths in the field of development economics, particularly regarding the provision of foreign aid. Large-scale institutionalized foreign aid from "developed" to "developing" nations primarily began following the end of World War II and has only increased in magnitude since that time. Such aid was, and is, often justified on grounds of the "poverty trap" hypothesis of economic development. According to this view, individuals in poor countries cannot save enough of their income, because they are living at subsistence levels, to foster the capital accumulation necessary to allow for meaningful economic development. Bauer noted that the logic of this idea would mean that individuals in all countries on earth would have been unable to rise out of poverty; after all, some countries needed to accumulate wealth in the first place to transfer it to others. Instead, he emphasized the need for stable political and market institutions which permit saving, however small initially, and will attract capital investment from abroad.

Although foreign aid is still seen by many as the key to helping poor countries escape the poverty trap, Bauer highlighted many negative unintended consequences of that approach to international economic development. "Foreign aid is demonstrably neither necessary nor sufficient to promote economic progress in the so-called Third World," Bauer (2000, p. 41) argues, "and is indeed much more likely to inhibit economic advance than it is to promote it. That is so because the inflow of foreign aid sets up major adverse effects on the factors behind economic progress." Such "adverse effects" include a heightened risk of conflict as factions within the aid-receiving country make efforts to secure the incoming aid, an increase in the politicization of life in countries receiving aid, and a rise in domestic corruption.

Bauer (2000, p. 43) also points out that aid often goes to governments hostile to their donors, offering examples such as Ghana in the 1950s and Tanzania and Ethiopia in the 1980s. Other examples he identified were the Argentine government receiving aid from the British government through a United Nations program in the midst of the Falklands War in June 1982 and the government of Iraq receiving millions of Western aid throughout the 1980s at the same time that it was building up a military arsenal. Here Bauer was emphasizing the fungibility of aid; in these instances, aid intended to foster development can contribute to conflict, or preparations for conflict.

Bauer also noted that foreign aid subsidizes governments to continue economically damaging interventionist policies such as trade restrictions and price controls. It can also reduce the cost of persecution of productive ethnic minorities. Reversing the common argument for foreign aid, Bauer (2000, p. 46) concludes, "It is official development aid that can create a vicious circle. Poverty is instanced as ground for aid; aid creates dependence and thus keeps people in poverty."

Since Bauer, other economists have also written on the, often perverse, consequences of foreign aid. Easterly (2001, 2006) identifies a range of negative consequences associated with foreign aid due to perverse incentives and knowledge limitations facing both donor governments and recipients. Svensson (2000) finds evidence that foreign aid is associated with increased corruption in countries characterized by competing social groups, with no evidence that aid is allocated to countries with less corruption. Meanwhile, Knack (2001) also finds that foreign aid tends to erode the quality of governance in recipient countries. Knack (2004) finds no evidence that foreign aid promotes democracy. Along similar lines, Djankov, Montalvo, and Reynal-Querol (2006) find that aid both reduces investment and increases government consumption and therefore has a negative impact on economic growth. The same authors (2008) argue in a separate paper, tellingly titled "The curse of aid," that infusions of foreign aid produce rent-seeking behavior in recipient countries similar to and likely worse than that often observed in countries rich in natural resources.

Doing Bad by Doing Good

Coyne (2013) considers various harms done by international intervention despite professed intentions to do good. Governments and NGO bureaucracies face the "planner's problem" in that they operate outside the profit/loss economic calculation system of the market. Therefore, such agencies fly blind, so to speak, with respect to navigating the economic problem of deciding how to allocate scarce resources such that more highly valued ends of recipients are satisfied before less highly valued ends. Since it is impossible to ascertain numerically *ex-ante* or *ex-post* whether a particular course of action is more or less value-added than alternatives, or even value-added at all, outside economic calculation in a market setting, humanitarian actions undertaken by such bureaucracies are largely arbitrary with respect to the ends valued most highly by recipients. Certain desired output measures considered a proxy for economic development may be devised and the aid agency may even be able to attain some or all of these metrics, but there is no necessary relation between those measures and genuine betterment from the point of view of foreign citizens, inclusive of consideration to what could have been attainable with the same resources invested.

The history of military and humanitarian intervention in Afghanistan provides a clear example of the perils and unintended consequences that can result from attempts at nation-building. The US occupation in Afghanistan under the 2001 mission "Operation Enduring Freedom" ended in failure despite an estimated \$2 trillion spent toward reconstruction. Within this intervention, 2010's Operation Moshtarak was intended to not only drive the Taliban out of the Helman Province in southern Afghanistan but also to provide both immediate and long-term humanitarian aid to the region. However, the net result of the operations has been that of creating a short-term dependence on foreign aid that sucked in money from donor countries before the eventual collapse of the country.

Lambert, Coyne, and Goodman (2021) detail findings from the so-called "Afghanistan Papers," internal US government documents released by the Washington Post in 2019, that reveal the extent to which major players in the reconstruction recognized the effort to be a failure. Numerous severe problems stymied the reconstruction efforts in Afghanistan, including a failure to respect local norms and institutions, bureaucratic mission creep as development agencies sought to justify their continued presence in the region, the pursuit of arbitrary and ultimately meaningless output metrics in the absence of market solutions, a rise in local corruption and warlordism, and significant expansion of the regional drug trade.

Previous attempts to raise standards of living in the region, such as the infamous "Helmand Valley Project" which began after World War II, also failed due to engineering mishaps and a failure to take sufficient account of the local context in which the interventions were carried out (see Coyne 2013, pp. 1–7). For example, nomads were forced to resettle lands as stationary farmers and tensions emerged between the local population and those who were resettled by the newly established Helmand Valley Authority. Meanwhile, the land chosen was not suited for agricultural development, which was only made clear later after significant investment in the area. Furthermore, local Afghan farmers were also unable to handle a significant new inflow of water. Planning in the region failed to such an extent that a 1960 article in the New York Times dubbed the project a "comedy of errors" (*New York* Times, 1960, p. 35).

The US government's experience in Afghanistan also shows how the unintended consequences of drug prohibition, discussed in the domestic context in the prior section, apply internationally. A key part of the US government's plan for Afghanistan was to eradicate opium poppy—Afghanistan is the world's leading producer of opium poppy which is used in the production of heroin—to undermine the Taliban and create "alternative livelihoods" for Afghan citizens. The eradication efforts failed terribly with more poppy being produced after the prohibition efforts began. Coyne, Hall Blanco, and Burns (2016) and Coyne (2022, pp. 83–104). Document how the US government's eradication efforts generated a series of negative unintended consequences including regime uncertainty regarding future policies, cartelization of the drug trade which empowered the Taliban, corruption, the criminalization of ordinary life, and violence.

It is also important to note that foreign interventions can also generate negative unintended consequences domestically, within the intervening country. For instance, Coyne and Hall (2018) document how foreign interventions, which require tools of social control to implement plans on foreign populations, can have a "boomerang effect" which reduces freedom within the intervening country. The mechanisms through which this can occur include expansions in the scope and scale of the domestic government as well as innovations in the tools and skills of state-produced social control which can be used against the domestic populace. In the US context, examples include the militarization of police and the evolution of the surveillance state (Coyne & Hall, 2018). This highlights how the range of possible unintended consequences is broad and far-reaching. When intervening in complex systems the intervener can never do just one thing, and the ripple effects can be broad and span across borders.

5 Conclusion

Henry Hazlitt (1946, p. 17) noted that "The art of economics consists in looking not merely at the immediate but at the longer effects of any act or policy; it consists in tracing the consequences of that policy not merely for one group but for all groups." We would broaden this claim beyond economics and extend it to all sciences of human action. A key task of social scientists is exploring the consequences—both intended and unintended—of action. This requires appreciating the seen, the unseen, and the unrealized (Bylund, 2016).

Unintended consequences result from human action in complex and open-ended systems; they can be positive or negative. In the case of negative unintended consequences, the question becomes the ability of people to adapt and adjust. Adaptability, in turn, depends on the institutional environment within which people are embedded. Institutions—the formal and informal rules of the game—produce different epistemic and incentive properties which influence not only the feedback provided to participants in the system, but also their incentive to respond to, or ignore, that feedback. For social scientists, this elevates the importance of comparative institutional analysis (see Boettke, Coyne, & Lesson 2013). This entails the study of different institutional environments—both existing and possible—with an awareness of the adaptability of different systems.

Another implication is an appreciation of what Jacob Viner (1950, p. 2) referred to as "negative knowledge" which refers to an "awareness of the range and depth of our unconquered ignorance." Recognizing how little we know about the world is the reason for humility regarding the design and control of the numerous, overlapping complex systems that constitute the world. At the same time, the recognition of positive unintended consequences, results in an appreciation of the beauty of self-ordering outcomes of voluntary individual action. As James Buchanan (1982), put it, social and economic "order is defined in the process of its emergence." Individuals, pursuing their own interests, must coordinate with others. In doing so, they contribute to a broader order that is beyond the ability of human reason to design or fully grasp.[6] The concept of unintended consequences, therefore, is central to understanding human civilization and the threats to the well-being of that civilization.

References

Aydinonat, E. E. (2008). *The invisible hand in economics: How economists explain unintended social consequences*. Routledge.

Bastiat, C. F. [1850] (2007). That which is seen, and that which is not seen, In *The bastiat collection* (2nd Ed.), Mises Institute, pp. 1–48.

Bauer, P. T. (1957). *Economic analysis and policy in under-developed countries*. Cambridge University Press.

Bauer, P. T. (1972). *Dissent on development*. Harvard University Press.

Bauer, P. T. (1981). *Equality, the third world, and economic delusion*. Harvard University Press.

Bauer, P. T. (2000). *From subsistence to exchange*. Princeton University Press.

Bessen, J., & Meurer, M. J. (2008). Do patents perform like property? *Academy of Management Perspectives, 22*(3), 8–20.

---

[6] For more on the concept of spontaneous order, see D'Amico (2015).

Boettke, P. J. (2018). *F.A. Hayek: Economics, political economy and social philosophy.* Palgrave.

Boettke, P. J., Coyne, C. J., & Hall, A. R. (2013). Keep off the grass: The economics of prohibition and U.S. drug policy. *Oregon Law Review, 91*(4), 1069–1096.

Boettke, Peter J., Coyne, C. J., & Leeson, P. T. (2013). Comparative historical political economy. *Journal of Institutional Economics,* 9(3), 285–301.

Boldrin, M., & Levine, D. K. (2008). *Against intellectual property.* Cambridge University Press.

Boldrin, M., & Levine, D. K. (2013). The case against patents. *Journal of Economic Perspectives, 27*(1), 3–22.

Brüggemann, J., Crosetto, P., Meub, L., & Bizer, K. (2016). Intellectual property rights hinder sequential innovation. Experimental evidence. *Research Policy, 45*(10), 2054–2068.

Buchanan, J. M. (1982). Order defined in the process of its emergence. *Literature on Liberty, 5*(4), 5.

Bylund, P. L. (2016). *The seen, the unseen, and the unrealized: How regulations affect our everyday lives.* Lexington Books.

Card, D. & Krueger, A. B. (1994). Minimum wages and employment: A case study of the fast-food industry in New Jersey and Pennsylvania. *The American Economic Review, 84*(4), 772–793.

Cohen, A. & Einar, L. (2003). The effects of mandatory seat belt laws on driving behavior and traffic fatalities. *The Review of Economics and Statistics, 84*(4), 828–843.

Cole, J. H. (2001). Patents and copyrights: Do the benefits exceed the costs? *Journal of Libertarian Studies, 15*(4), 79–105.

Coyne, C. J. (2010). *Doing bad by doing good: Why humanitarian action fails.* Stanford University Press.

Coyne, C. J. (2022). *In search of monsters to destroy: The folly of American empire and the paths to peace.* Independent Institute.

Coyne, C. J. & Hall, A. R. (2018). *Tyranny comes home: The domestic fate of U.S. militarism.* Stanford University Press.

Coyne, C. J., Hall, A. R., Blanco, A. R., & Burns, S. (2016). The war on drugs in Afghanistan: Another failed experiment with interdiction. *The Independent Review,* 21(1), 95–119.

D'Amico, D. J. (2015). Spontaneous order. In P.J. Boettke & C.J. Coyne (Eds.), *The oxford handbook of Austrian economics* (pp. 115–142). Oxford University Press.

De Zwart, F. (2015). Unintended but not unanticipated consequencecs. *Theory and Society, 44*(3), 283–297.

Djankov, S., Montalvo, J. G. & Reynal-Querol, M. (2006). Does foreign aid help? *Cato Journal, 26*(1), 1–28.

Djankov, S., Montalvo, J. G., & Reynal-Querol, M. (2008). The curse of aid. *Journal of Economic Growth, 13*(3), 169–194.

Easterly, W. (2001). *The elusive quest for growth: Economists' adventures and misadventures in the tropics.* The MIT Press.

Easterly, W. (2006). *The white man's burden: Why the west's efforts to aid the rest have done so much ill and so little good*. Penguin Books.

Electronic Freedom Foundation. (2014). *Unintended consequences: Sixteen years under the DMCA*. https://www.eff.org/files/2014/09/16/unintendedconsequences2014.pdf, Last accessed, July 20, 2022.

Hayek, F. A. [1944] (2007). *The road to serfdom*. University of Chicago Press.

Hayek, F. A. (1945). The use of knowledge in society. *The American Economics Review, 35*(4), 519–530.

Hazlitt, H. [1946] (1979). *Economics in one lesson*. Three Rivers.

Heller, M. A. & Eisenberg, R. S. (1998). Can patents deter innovation? The anticommons in biomedical research. *Science, 280*(5364), 698–701.

Higgs, R. (1992). Wartime prosperity? A reassessment of the US economy in the 1940s. *The Journal of Economic History, 52*(1), 41–60.

Ikeda, S. (1997). *Dynamics of the mixed economy: Toward a theory of Interventionism*. Routledge.

Ikeda, S. (2005). The dynamics of interventionism. In P. Kurrild-Klitgaard (Ed.), *Advances in Austrian economics, Vol. 8* (pp. 21–58). Emerald Publishing Limited.

Ikeda, S. (2015). Dynamics of interventionism. In P.J. Boettke & C.J. Coyne (Eds.), *The oxfordhandbook of Austrian economics* (pp. 393–416). Oxford University Press.

Joksch, H. C. (1976). Critique of Sam Peltzman's study "The effects of automobile safety regulation". *Accident Analysis & Prevention*, 8, 129–137.

Joksch, H. C. (1976). The effects of automobile safety regulation: Comments on Peltzman's reply. *Accident Analysis & Prevention*, 8, 213–214.

Kinsella, S. N. (2008). *Against intellectual property*. Mises Institute.

Kirzner, I. (1985). The perils of regulation. In Israel Kirzner (Ed.), *Discovery and the capitalist process* (pp. 119–149). University of Chicago Press.

Knack, S. (2001). Aid dependence and the quality of governance: Cross-country empirical tests. *Southern Economic Journal, 68*(2), 310–329.

Knack, S. (2004). Does foreign aid promote democracy? *International Studies Quarterly, 48*(1), 251–266.

Lambert, K. J., Coyne, C. J. & Goodman, N. P. (2021). The fatal conceit of foreign intervention: Evidence from the Afghanistan papers. *Peace Economics, Peace Science, and Public Policy, 27*(3), 285–310.

Lavoie, D. (1982). The development of the Misesian theory of interventionism. In Israel M. Kirzner (Ed.), *Method, process, and Austrian economics: Essays in honor of Ludwig von Mises* (pp. 169–184). Heath and Company.

Menger, C. [1881] (2007). *Principles of economics*. Mises Institute.

Merton, R. K. (1936). The Unanticipated Consequences of Purposive Social Action, *American Sociological Review, 1*(6), 894–904.

Miron, J. A. (1999). Violence and the U.S. prohibition of drugs and alcohol. *American Law and Economics Review, 1*(1–2), 78–114.

Miron, J. A. (2001). Violence, guns, and drugs. *Journal of Law and Economics, 2*(4), 615–633.

Miron, J. A. (2003). The effect of drug prohibition on drug prices: Evidence from the markets for cocaine and heroin. *The Review of Economics and Statistics, 85*(3), 522–530.

Miron, J. A. & Zwiebel, J. (1995). The economic case against drug prohibition. *Journal of Economic Perspectives, 9*(4), 175–192.

Mises, L. von. [1960] (2003). *Epistemological problems of economics*, Mises Institute.

Mises, L. von. [1949] (1998). *Human action.* Mises Institute.

Mises, L. von. [1957] (1985). *Theory & history*, Mises Institute.

Mises, L. von. (1962). *The ultimate foundations of economic science.* D. Van Nostrand Co.

Mises, L. von. [1977] (2011). *A critique of interventionism*, Hans F. Sennholz (Trans). Mises Institute.

Mises, L. von. [1979] (2006). *Economic policy: Thoughts for today and tomorrow.* Mises Institute.

Neumark, D. & Shirley, P. (2022). Myth or measurement: What does the new minimum wage research say about minimum wages and job loss in the United States? *Industrial Relations: A Journal of Economy and Society*, forthcoming.

Peltzman, S. (1975). The effects of automobile safety regulation. *Journal of Political Economy, 83*(4), 677–725.

Peltzman, S. (1976). The effects of automobile safety regulation: Reply. *Accident Analysis & Prevention*, 8, 139–142.

Pope, A. T. & Tollison, R. D. (2010). "Rubbin' is racin": Evidence of the Peltzman effect from NASCAR. *Public Choice, 142*(3/4), 507–513.

Redford, A. (2017). Don't eat the brown acid: Induced "malnovation" in drug markets. *Review of Austrian Economics, 30*(2), 215–233.

Redford, A. & Powell, B. (2016). Dynamics of intervention in the war on drugs: The buildup to the Harrison Act of 1914. *The Independent Review, 20*(4), 509–530.

Robbins, L. (1932). *An essay on the nature and significance of economic science.* MacMillan & Co.

Robertson, L. S. (1977). A critical analysis of Peltzman's "the effects of automobile safety regulation". *Journal of Economic Issues, 11*(3), 587–600.

Rothbard, M. N. (2006) [1970]. *Power and market: Government and the economy* (4th Ed.). Mises Institute.

Safner, R. (2016). The perils of copyright regulation. *Review of Austrian Economics, 29*(2), 121–137.

Smith, A. [1776] (1981). *An inquiry into the nature and causes of the wealth of nations.* Liberty Fund.

Sobel, R. S. & Nesbitt, T. M. (2007). Automobile safety regulation and the incentive to drive recklessly: Evidence from NASCAR. *Southern Economic Journal 74*(1), 71–84.

Stiglitz, J. E. (2008). Economic foundations of intellectual property rights. *Duke Law Journal, 57*(6), 1693–1724.

Svensson, J. (2000). Foreign aid and rent-seeking. *Journal of International Economics, 51*(2), 437–461.

The New York Times. (1960, March 13). Mistakes beset Afghan project; Helmand Valley work, which U.S. is aiding lags badly, *The New York Times*, 35.

Thornton, M. (1991). *The economics of prohibition.* University of Utah Press.

Viner, J. (1950). A modest proposal for some stress on scholarship in graduate training. Address before the Graduate Convocation, Brown University, June 3. Brown University Papers XXIV, Providence Rhode Island: Brown University.

Individual Choice and Collective Identities

Günther Schlee

1 Locating the Problem

The dimensions which frame our social space (nation, ethnicity, gender, religion …) and the medium to large-scale collective identifications (meso and macro groups and categories) which instantiate these dimensions (French, Basque, female, Muslim …) are part of our perception of social structure. They are also large-scale phenomena found on the macro level, where supra-individual realities or imaginations are depicted with a broad brush.

Choosing a collective identity, reflecting on it, re-evaluating it, propagating or denouncing it, however, is something strictly individual, at least for all those who believe that a collective mind or a spirit pervading an epoch, or a place are mystifications, or, more politely, ways to speak.

The discussion of choice in the domain of collective identities thus can be located as being part of two classical sociological problems: The micro/macro problem and the problem of structure versus agency.

G. Schlee (✉)
Professor of Social Anthropology, Arba Minch University, Arba Minch, Ethiopia
e-mail: schlee@eth.mpg.de

Director Emeritus at the Max Planck Institute for Social Anthropology, Arba Minch University, Halle/Saale, Germany

© The Author(s), under exclusive license to Springer Nature Switzerland AG 2023
N. Bulle and F. Di Iorio (eds.), *The Palgrave Handbook of Methodological Individualism*,
https://doi.org/10.1007/978-3-031-41508-1_13

Between these poles, we move when we try to model identification, i.e., choices of collective identities, when we explore which interests might have guided the choices of the actors we have observed, when we ask what might have been the pros and cons, their perceptions of the costs and benefits of the options they have considered, both in terms of material wellbeing and according to their value orientations. If one phrases one's research questions in these terms, as an anthropologist one quickly meets the opposition of other anthropologists.

Applying Rational Choice models like cost/benefit calculations to anything, among anthropologists often meets with a kind of scepticism that borders on hostility. If one wonders why, one often gets the answer that the actions of people are not guided by rational choice and that RC theory neglects the irrational. My response that I share the interest in the irrational, but that I would not know how to identify or to measure irrationality other than by applying a Rational Choice model, somehow does not help much in overcoming the reluctance to deal with Rational Choice at all.

Another answer to the question about the location of our problem therefore is: On the fringes of anthropology. The majority of anthropologists cultivate a diction that abstains from Rational Choice categories. "Positionalities" are often discussed in terms of broad categories (like minority status and collective victimhood) rather than being traced to individual calculations and motives.

2 Identification of Violent Conflicts and Situations of Rapid Change

Political anthropologists and students of identification (in a softer variant: belonging) are interested in how people identify with large-scale groups and categories and why. Millions of people, however, live long and happy lives without ever changing their nationality or religion, their gender or political affiliation. This may leave students of identity change frustrated. They therefore turn to where identities are changed, redefined and negotiated. That is in situations of conflict, especially violent conflict, forced migration, oppression, rapid social change…

My own research has for decades had a focus on politicized ethnicities in the Horn of Africa (Schlee, 1989, 2008; Schlee & Watson, 2009a, 2009b, Schlee & Shongolo, 2012a, 2012b, Markakis et al., 2021) and my students and friends have helped me to extend my comparative perspective into West Africa, Central Asia and other parts of the world (Diallo & Schlee, 2000;

Individual Choice and Collective Identities **299**

Knörr & Kohl, 2016; Schlee & Horstmann, 2017). Here are some of the research questions which inspire this kind of research: How do people in a conflict situation define friend and foe? By which categories do they draw the line and how does this line shift, if it does? How do linguistic or religious or geographical or political categories interact in the definition of social identities? We examine religious or linguistic taxonomies and cultural classifications, major groups and their sub-groups, ethnic identifications from wider units to the smaller and smaller locally relevant groups and so on, and we look at the overlapping relationships if these categories cross-cut. Which identifications become more relevant in which context? Do cross-cutting ties enhance social cohesion? Are they a de-escalating factor or can they also play a role in conflict escalation?

3 Structure and Agency

All these taxonomies, classifications, etc., are structures. So, there is no doubt that our[1] starting point is in the domain of structure. One can debate to which extent these structures are cognitive structures. We are of course approaching our field of study primarily by discussing with people and by having a look at their discourses. The discursive and conceptual structures, which emerge in the analysis, are certainly not identical to social structures, but they are in some sort of relationship with them, and they are structures anyhow. So where does agency, the other element of the famous dichotomy "structure and agency", come in? With "agency" we also expect "choice" to enter our field of research interest, because making a choice is a kind of action and all actions imply a choice. They express a preference for alternative courses of action.

Choice, to give just some examples, comes in if you form an alliance. You have to decide with whom to ally, how large your alliance should be, and whom you can exclude. If you make an appeal for solidarity, such an appeal can be phrased with reference to certain characteristics of the group you appeal to and possibly define in this process of characterization. You can appeal to a group identity that you claim to share yourself or one which is associated with a status from which one can derive obligations. You can appeal to a group ethos which is behind the label of group identity. You can choose

---

[1] The collective first-person pronoun refers to the department Integration and Conflict at the Max Planck Institute for Social Anthropology at Halle (Saale). To the extent that the present paper is based on original research, this has been done mostly at that department. In parts of this paper parts of institute's reports are taken up and expanded.

to ignore or deny an identity, of your own or of another, if it does not suit your aims. If you want to affiliate with a group, if you want to re-affiliate, if you quit an alliance—these are situations in which you have problems of choice in the context of social identities. Some social identities are of course ascriptive or they are said to be ascriptive and immutable. Social identities are not a domain of unlimited arbitrariness and choice pervading everything. There are severe limitations to the possibility of choice. But whether identities are relatively stable or subject to opportunistic adjustments, is not a question of dogma but of empirical variation. Some people in fact have virtuosity in manipulating social relationships and others not or less so. One might say they have varying degrees of agency. Real people do not conform to extreme types. We find them somewhere in between, dispersed over the scale which has rational decision makers at one end and role performers guided by habits and internalized norms at the other, and at different times, in different situations, we can find the same actor at different points on this scale.

Some people use choices in manipulating identities whereas others are passively socialized into a given identity and remain within it. These latter possibly do not have the historical knowledge necessary to derive alternative identity constructs from the past, they do not have the social skills necessary for convincing others or lack other prerequisites for changing anything about the identity into which they have been born or socialized. One of these prerequisites is the need to do something about collective identity. Some people never get into a situation in which identity is a problem. In the field of what is given versus what is feasible, of structure versus agency, we need a gradualist perspective. Some structures are more stable than others and some people are better equipped to change structures than others. How exactly individual actions combine to have effects that change structures so that the next round of decisions has to start from a new set of "givens" is another matter. That topic would be big enough for another paper. This combination is more than a simple addition. There are tipping points. If a large number of persons in a given population has been converted to a new or modified identity, the rest might follow. Therefore early converts count more than late converts. Leaders influence followers in a different way from followers influencing leaders etc. In the present chapter, however, we are more concerned with the individual agency as such and have to leave much of the interaction between different actors aside.

One factor deserving special attention in the study of identity choice and the forging of alliances is anticipated group size, the question of how considerations of resulting group sizes influence identity discourses. If a certain perception of an advantage lets it appear desirable to belong to a larger or

Individual Choice and Collective Identities   **301**

smaller group or alliance, how does this translate into inclusion and exclusion strategies or into identity discourses which lead to the desired size? It is obvious that if you are in a strong position you do not need many allies because you get what you want anyhow and you do not want to share the loot[2] with many superfluous helpers. If you are weak you need more allies. Such considerations are very akin to well-known theories like the minimal winning coalition (William Riker) or the theory about "crowding" (overuse of a resource): It is cheaper to use a resource if you are many people sharing the costs of access to the resource or the production of it[3] but then the resource or the site of its use becomes crowded and stress factors set in (Hechter, 1987). If you look at the question of why people might join or quit a group, you, of course, soon discover that members or prospective members of groups are not a homogenous crowd. Therefore, I think that notions like "group interest" or the like are problematic. One of the reasons for this scepticism is that the notion of group interest does not explain the emergence of groups (like ethnogenesis, the formation of ethnic groups). When and why do ethnic groups (religious groups, "cultures", …) emerge and how do they adjust to different configurations? In such contexts, the notion of "group interest" is clearly misplaced. What is after all the "group interest" of a group which undergoes changes in its composition? And what is the group if it is still changing its composition? Must the perceptions of shared interests by its members not also change with variation in group composition as the group emerges? Further, of course, we all know that leaders and followers have quite different calculations in joining or not joining a group and there is a body of theory addressing this problem (e.g., Barth, 1959). The rewards of defection, for example, are much higher for leaders who can bring a group of people along when they are changing sides. Their price will be higher than that of the ordinary follower who might not gain much from his betrayal. In other words: Coming from structural categories like all these classifications we have developed a need for an action theory or to phrase it differently a theory of choice.

Categories and social structures change under the impact of agency in a variety of ways. We can distinguish at least two types of identity change:

---

[2] "Loot" is shorthand for any kind of resource the appropriation of which is the target of the actors under study.

[3] "Resource" in such contexts includes much more than primary resources. A finished product can be a resource in the production of something else. In Hechter's example a country club is used for the production of wellbeing. It has also become usual to speak of immaterial "resources" which may have an instrumental role in the acquisition of material "resources", like power being a resource in a war about oil. This is in perfect agreement with everyday English, where a "resourceful" person is a strong and energetic one.

people disclaim an identity and claim another one, or identity changes over time. The two processes can be independent of one another, like in the case of people moving in and out of a category (like "immigrant" or "lower middle class" or a situationally adopted ethnic category) without the definition and content of that category substantially changing (it just changes its composition in terms of people). Alternatively, the two processes can be interrelated, for example, if the content of identity, under threat of being abandoned, is adjusted to popular demand. Identities evolve under the influence of the discourses about them. They are ideologically redirected and their proponents make use of different kinds of historical material.

So, the discursive strategies in talking about identities are a part of the agency. Slavic-speaking Moslems who belong to ethnic minorities have a religious criterion ("Muslims"), a numerical-demographic criterion and/or a class criterion ("minority") and an ethnic/linguistic criterion at their disposition. In their self-definition in one historical situation they might privilege class in their discourse, in another historical situation they might privilege ethnicity and more recently religion has become favoured, it is the legitimate discourse and the one everyone accepts to explain all sorts of clashes. The group in question (in terms of people who could be listed up if we knew them all, a finite set) might be the same in the three contexts. Speakers just opt for different facets of their identity and by appealing to different labels they also make choices between different value orientations.

This can be illustrated by the Judgement of Paris. Paris, prince of Troy, has to choose which of three goddesses to give the apple for the most beautiful one. Of course, these goddesses might differ in beauty or not, but in Homer's account what finally decides the matter is that they want to bribe him by different means, offering rule over the world, victory in every fight and the love of the most beautiful mortal woman, respectively.

Individual Choice and Collective Identities 303

The judgement of Paris © G. Schlee 2005

These are incommensurable incentives (how much power is better than how much love?), and that models the choice between different dimensions of identification well. Normally you compare religions with religions (you might convert if you think another religion is better than your present one or offers social advantages) and languages with languages (a comparison that might lead you to switch to a language of wider circulation or one which is more appropriate to the situation). In this type of situation, where a large religious community might offer more support than the affiliation to the small community of speakers of a given language (or, in the opposite case, where linguistic nationalism[4] might provide more support than the affiliation to a small sect), however, one changes the register, stressing religion more than language at one time and language more than religion at another time.

History is a big factory for redefining social identities. Historical change of identities can be slow, and it may be hard to determine where in these processes agency comes in. I do not think this question has really been posed often, because people who do systems theory tend to adopt macroscopic perspectives and do not really go down to the interplay of actors which ultimately must be the source of any form of social change. Historians of ideas (*Ideengeschichtler*) treat ideas as evolving from each other and exclude the carriers of those ideas to some extent from their analyses. So, I think there is much to be gained by adopting an individualist perspective and looking

---

[4] Or the appeal to a language family like Panslavism or Turanianism.

at the actual identity of workers and producers of ideology, their incentives, their values, their motivations (hopes and ambitions) and the opportunity structures and constraints within which they live.

4 Structure and Agency in the History of Anthropology

In his contribution to a recent book with the title "*African Political Systems Reconsidered*" (Bošković & Schlee (eds.) 2022), Herbert S. Lewis (2022) describes the original volume (Fortes & Evans-Pritchard (eds.) 1940) as the high point of Oxford structuralism of the Radcliffe-Brown school. Critics of that volume pointed to its structuralist orthodoxy and took up other lines of thinking.

Alternative views to Oxford structuralism go back to Malinowki. Here we can pick out only some highlights of Lewis' account of individual agency, choice, and strategy finding their way back into British anthropology after the dominance of the structuralist paradigm. Prominent figures in this move are Raymond Firth, Malinowki's first PhD student, and Edmund Leach, his last.

Raymond Firth had his first degree in economics and put great emphasis on the role of individual decisions and cost–benefit calculation in the development of societies. Likewise, Fredrik Barth, a student of Leach, studied local politics in a Pathan region of northern Pakistan in terms of opportunistic alliances between leaders and followers joining leaders who offer the best deal.

"A clear pattern from Firth to Barth emerges: structure and agency *avant la lettre*" (Lewis, 2022, p. 31). Lewis derives his comments on Barth mainly from *Political Leadership among Swat Pathans* (1959). To extend this discussion rather than repeating it, I now turn to a book published ten years later, *Ethnic Groups and Boundaries* (Barth ed. 1969a). I examine the role of individual cost–benefit calculations (in the currency of attaining cultural values and gaining recognition) in Bath's analysis of ethnic conversion.

Barth discusses the "Pathan Identity and its Maintenance" (1969b) in a variety of settings where the Pathans (Pashtuns, Pukhtuns, Afghans) come into contact with other ethnic groups. There is ethnic conversion in all these settings, and in all these cases these conversions are explained in terms of positive and negative incentives which make it more attractive to claim one identity than another, (This actor-oriented focus is a far cry from *African Political Systems*. which, as the title suggests, has the system as starting point

Individual Choice and Collective Identities 305

and puts a much greater emphasis on the cohesion of society and the maintenance of a balance of forces and counter forces.) To illustrate this, one example of such a contact situation may be sufficient. Let us look at the southern reaches of the Pathan inhabited areas, where they encounter and live with Baluch.

Baluch tribes are composed of lineages unrelated by origin and individuals following a leader. They share concepts of honour and masculinity with the neighbouring Pathans, belong to the same branch of Islam and practice gender seclusion, wherever possible, with the same rigidity. But in contrast to the Pathans, they have a stronger element of hierarchy in their social organisation. One can move up in that hierarchy by being a more trusted and more important follower of a leader.

To live up to Pathan values, on the other hand, one needs to be independent. This requires one to have one's own piece of land and to be able to practise hospitality on a certain level. The readiness to exact revenge to defend one's honour by one's own means or with the help of immediate kin is another prerequisite for being accepted as a real Pathan. Solidarity among agnates is limited by rivalries over land. If one meets these standards, one can participate as an equal in the council of men (*jirga*), In this council, men sit in a circle, to express their equality, and discuss a matter until they reach unanimity. No one can be overruled. If one is alone with one's view but does not want to block a decision, one can only leave the meeting in protest.

Landless Pathans or people with insufficient holdings, who lack the means of hospitality and whose economic independence is at risk, may be better off if they become clients of a Baluch leader.

"… the client of a Baluch leader [however] cannot speak in any tribal council. To maintain Pathan identity [in such a situation], to declare oneself to be running as a competitor by Pathan value standards, is to condemn oneself in advance to utter failure in performance. By assuming […] Baluch identity, however, a man may, by the same performance, score quite high on the scales which then become relevant." (Barth, 1969a, p. 25).

The scales which become relevant when one becomes a Baluch are not elaborated very much by Barth. His focus is on the difficulty of maintaining Pathan standards and on the Baluch identity as a fall-back option. I assume, however, without having had the opportunity to ask any Baluch, that the Baluch do not perceive their culture as the cheaper option and somehow inferior to that of the Pathans. Titus (1998) provides indications for this.

"…not all the Baloch and Pushtun I met in Quetta [the capital of Pakistani Balochistan] narrow their experiences into stereotypes but those that did characterize the Pushtun as entrepreneurial and religious, and Baloch as having strong tribal values and concern for honor." (Titus, 1998, p. 664).

Titus goes on to explain that, while there may be general agreement on such stereotypes between those who ascribe them and those to whom they are ascribed, the connotations attached to each feature may vary greatly. Entrepreneurial success can be read as venality, "greediness and a lack of honor," tribal values as backwardness. The lack of tribal values attributed to the Pathan (Pushtun) harmonizes with Barth's finding of internecine fighting about land. "For example. one Baloch informant said most Pushtun would kill their own brother for money, an assessment which combines greed, a want of tribal values, and the sense that Pashtun are at war with their relatives" (Titus, 1998, p. 665).

So, rather than concluding that is easier to be Baluch, it is fair to say that the Baluch have a different value orientation and that some people are better at being Baluch than at being Pathan. These would be candidates for ethnic conversion from Pathan to Baluch.[5]

So much about choice and agency in ethnic conversion according to Barth. In connection with the less structuralist, "non-group", more (individual) people-oriented strain in (British) anthropology of the 1950s and 60s, Lewis also mentions Monica Hunter Wilson, Lucy Mair, Fred Bailey and Jeremy Boissevin. Lewis then moves on to America, but here we leave it at that. The deeper we dive into the history of our discipline, the more precursors we find (or are reminded of, if we have come across them earlier). Maybe the views expressed in this paper can claim little originality. We are compensated for this by being able to point to an illustrious ancestry.

5   The Relationship Between Emotions, Rationality and Hard Thinking

Our theorizing about identity choices and exclusion of the "other" shaped by considerations of costs and benefits provokes disagreement. Critics tend to interject that this perspective does not leave enough room for emotions. A mere glance at Homer or Shakespeare or other contributions to world literature will show that people are not concerned mainly with identity and

---

[5] I have not found the discussion fo Baluch becoming Pathan in the literature. Such cases appear to ve rare because of the strong emphasis on purity of descent by the Pathans. It is also hard to imagine which incentives would be attached for a Baluch to such a conversion.

difference but with love and hatred, not re-affiliation but betrayal, not tit-for-tat but revenge—and so on. Acknowledging the prevalence and importance of emotions, however. does not force us to give up Rational Choice explanations, because the outcome of emotion-based action can be as rational or as irrational, as beneficial or as harmful as an action which results from hard thinking and can therefore be studied using the same models. If we say that the outcome of a decision-making process appears to be rational, what we mean by that, and what rational choice theorists mean, is that the outcome is characterized by *Rationalitätsförmigkeit,* i.e., that it looks *as if* it had been shaped by rational considerations. This type of reasoning goes like this: If the actor under study calculates carefully the costs and benefits in view of a target he wants to achieve, his decision will be called rational. But if he were to arrive at the same decision by other means or in some emotionally loaded or intoxicated state, we would still call it rational. This term does not imply anything about the psychological processes actually involved. Useful actions which look quite rational and are rational can result from habituation (doing what one is used to do in such a situation), general experience, which is the same thing as informal statistics (in such situations it has often been good to do this or that), or even purely emotional reactions (such as disgust which keeps one away from a source of infection). Gigerenzer (2007) has shown the efficiency of "gut decisions", in which much information is left aside. Also, cultural learning, including pure imitation of apparently successful models, needs to be considered in this context. Thinking can be greatly simplified or even left to others.

In this way, considering the "rationality" of results rather than of processes, rational choice is even applied in evolutionary biology. The spines of a cactus can be described as representing a positive balance of benefits (protection against being eaten) and costs (the nutrients used for growing them), without attributing a high capacity for deliberation to the cactus. A creeper is a skeleton parasite. By attaching itself to the stem and branches of a tree, it reaches the sunlight without incurring the cost of developing its own skeleton. And a cuckoo that puts its eggs into another bird's nest saves the costs of caring for its progeny, and so on. A cost–benefit-analysis in all these cases can help to explain to us how things have evolved in nature without implying that the organisms in question actually went through a psychological process that we would interpret as rational deliberation. We even speak about the egoism of genes. The ones which spread faster than others are those which produce phenotypes that do a lot to spread them.[6] That makes them

---

[6] They do a lot to spread them without knowing it. Also the phenotypes themselves normally are not aware of the evolutionary processes or the population dynamics in which they are involved. There

look quite rational and makes us look like their instruments, at least from a socio-biological perspective and irrespective of the level of consciousness and the forms of agency involved in each context. Applications of rational choice theory are not based on assumptions about mental states.

From this, we can conclude that choices conforming to the expectations of rational choice theory can be accompanied by the most variable kinds of psychological setups. So, if someone appears highly emotional and offers us an entirely emotional account of his or her motivations, it does not mean that the outcome of his or her actions is not rational. It means only that he or she is motivated by other factors. Of course, emotions can lead us straight into disaster, but also rational calculation can lead to disastrous results (in the case of miscalculation or incomplete information or merely bad luck),[7] while emotional reactions can be quite adaptive. All I want to say here is that emotions cannot explain rationality away or vice versa. Emotions accompany decisions with both rational and irrational outcomes alike. We need to study both the kinds of rationality and the kinds of emotions involved in various situations and one does not exclude the other. Styles of decision-making vary on a similar scale as the one described above (from hard thinkers to role performers), namely a scale that has hard thinkers at one end and sponta-neous, emotional actors at the other. And on this scale, too, the same actors can be found at different points in different situations. For example, a man can be cool and calculating when it comes to money and passionate when it comes to women, or the other way round, or his decision-making style changes with different moods or stages of intoxication.

The relationship between behaviour, including behaviour in violent conflicts, and emotions needs to be clarified. It is an empirical question that needs to be answered with case-specific evidence. To infer from actions to emotions is problematic. To kill does not mean to hate. And, as we have just seen, the types of explanations that involve reasoning about emotional states do not necessarily compete with rational choice types of explanations. The question of whether killing a certain person leads to a rational outcome can be asked and answered independently of the question of whether the

---

are proximate mechanisms. Animals engage in sex because they want sex, not in order to produce progeny. If producing progeny was the aim of these animals, sex would not have to feel good to them.

[7] One may have good information (or the best available, given that the future is unknown), may make correct calculations and conclude that a particular decision is going to lead to success in nine of ten cases and to failure in one. One makes this decision and the outcomne with the 10% probability occurs. This is an illustration of bad luck.

killer hates that person. The two explanations complement each other rather than compete with each other. Looking at emotions does not abolish rational choice but adds complexity.

Here is my favourite anecdote about killings and emotion. It was in northern Kenya. We were visiting a group of Degodia Somali. They used to live peacefully together with Ajuran Somali when I visited them some years earlier. In the meantime, there had been major massacres between the two groups. A young boy explained to us who everyone in the hamlet was: "This is the house of so and so; this is the house of so and so". He also enumerated the people who had died recently in the "tribal clashes" as the press called them. He said: "Well, one brother of my father was killed when he was newly married. So his widow just went back to her family. And that house belonged to the other brother of my father who has been killed. His wife already has two small children and so she stayed with us (…)" And on he went enumerating with the precision of an accountant who in his family had been killed, without showing any emotions. I presume a ten-year-old male Somali is a man and needs to show that he is a man—and a man does not show emotions. He would not answer questions that relate to emotions. We are left alone with these questions. Does he hate his enemies? I am sure that somewhere he must feel hatred and outrage. But at the same time, he was in perfect harmony with himself, because he was growing into the role that was expected of him. He was a future avenger in a supportive environment for future avengers. One cannot have strong emotions all the time. Later, when the time for killing has come, he would have to work himself into an emotional state which would enable him to kill someone in face-to-face combat. The AK 47 (Kalashnikov, a widespread light assault rifle) which allows killing at a distance, however, alleviates this problem somewhat.

The emotions that come with identification can be strong or weak and of many different kinds. There are some assumed identities to which people do not attach a great deal of emotion, for example, those that serve in the process of "passing as". If, when out on the street during the daytime, a person of Moroccan origin living in Belgium likes to be taken for a Belgian, he or she will not be gravely disappointed if unsuccessful in this endeavour, at least not necessarily and in all cases. The relevant others are other Moroccans whom one meets again in the evening, when no pretence is required. (Roosens, 1989) On the other hand, there are identifications for which people die or kill, and there are those which may never be admitted without shame and can never be mentioned without mortal offence. There is not only identification in the sense of categorical inclusion but also bonding. Simons (2000) has explained what makes it possible to mobilize young males for conflict. Social-ization into the military role comprises teaching them to rely on each other, to defend each other, and to die for each other in combat. Certain types of emotional identification are required for close bonding, which is necessary for performing certain tasks successfully or increasing one's chances of surviving in certain situations.[8]

A great deal remains to be done, but we (Donahoe et al., 2009, Eidson et al., 2017) have already attempted at least to develop a set of hypotheses concerning the ways in which different kinds of identification are subjectively experienced.

6   Purely a Matter of Choice?

The question of whether identity is a matter of choice tends to be answered in two ways: yes or no.[9] Some say yes, identification, i.e., the choice of identity, is guided by interests and aims, and is therefore instrumental and located within the domain of human agency. Others tend to say no and stress that there are many collective identities into which the individual is born. Strangely these "given" identities are not called "innate" or "congeni-tal" but "ascriptive". "Ascriptive" here means ascribed by others, beyond the influence of the bearers of such identities.

Personally. I prefer to rephrase the question: To what extent are collective identities a matter of choice? I expect the answer to differ from case to case. In the preceding paragraphs, we have explored differences in the extent to which identifications can be emotionally loaded, and these may not be the only factors that affect their potential to be given up and replaced or reshaped. In other words: identities have higher or lower amenability to choice and other cognitive and discursive processes that affect them.

1. *Limitations to the freedom of identification*

It may be hard to change or manipulate identities or to choose freely from a set of identities. There are, once again, three reasons for this (higher or lower) resistance to change, or for the kind and extent of the difficulty we have in

---

[8] There may be situations in which the defection of one would jeopardize the lives of everyone else or even of everyone including the defector. Military trainers will evoke just this type of situation if they want to create strong bonding.

[9] I am thinking here of the decades-long, rather one-sided debate of constructivists/ instrumental-ists against "primordialism" and "essentialism", which does not deserve to be summarized here. See Donahoe et al. (2009, pp. 8–9).

Individual Choice and Collective Identities 311

moulding identities—the degree of recalcitrance of the material we wish to shape. One reason is the characteristics of identities: there are sticky identities, dense identities, salient identities, emotionally loaded identities and so on, and they all form part of wider taxonomies and are ruled by a kind of grammar that tells us how they can or cannot be combined with each other. Another reason is linked to the characteristics of persons (identity bearers or persons engaged in identity discourses about others): In these persons, we may find political instinct and manipulative drive, knowledge about history and society, and persuasive skills to a higher or lower degree—or these factors may be absent entirely. A third reason is linked to the characteristics of situations. We may also call this the sphere of interaction.[10] Depending on a whole series of situational variables, our claims to an identity may not be plausible to others, or others may have a material interest to deny them. We are not alone in designing our social world.

2. The space of agency left by these limitations

These different limitations imply that different spaces of agency remain. These spaces are also used with different frequency and intensity. Some people are virtuosi who play with social identities with the same ease with which a real virtuoso glides his fingers over the keyboard of a piano. Then again, there will always be those who assume that the collective identities with which they themselves identify are given and fixed and to whom it never occurs that they may be manipulable. The lack of a manipulative attitude towards collective identities may be the result of a lack of capability or a lack of will. An apparent lack of will to engage in identification games may be the absence of initiative or it can itself be an act of will: the will to stick to one's identity as a moral imperative—or scruples against doing otherwise. Further, people may find it quite rewarding to live in harmony with their social environment and never to challenge their place in society or in the system of categories others use to describe them. Identity work and re-identification, whether successful or not, as explained above, often take place in situations of violent conflict or of rapid social and economic change (Schlee, 2003) or "when history accelerates", to borrow from a book title (Hann, 1994). When conditions change and established identifications become unrewarding, also these identities may change. Instead of falling into the trap of dichotomizing "primordial" versus "instrumental" identities (which presumably can be changed ad hoc), we

---

[10] We locate the sphere of interaction where the domain of action (the agency) of actors overlaps with that of other actors and their actions have effects on each other.

should adopt a gradualist perspective and talk about different speeds of identity change as history accelerates or goes through phases of stability. If there are no circumstances that force people to re-negotiate identities, it may be quite rational not to engage in identity games but to enjoy the warmth of conformity and unquestioned belonging.

7 The Information Economics of Identification

Stieglitz (2002),[11] in an article based on the speech he gave when he was awarded the Nobel Prize for Economics, writes about the classical domains of economics: labour relations, the stock market, production firms, insurance and the like. He finds that classical economic models do not describe the world we live in. There is no invisible hand that balances demand and supply and there is much more unemployment than classical theory would predict. In classical economics, an oversupply of labour should cause a decline in the price of labour, which would then meet an increased demand in response to the low price. Therefore wages should vary a great deal more than employment figures. In the real world, the opposite is the case. Stieglitz explains this discrepancy between the classical theory and the empirical data by the neglect of one factor: information. He says we need information economics. Information economics is about what people want other people to believe or to know or to ignore. An employer might prefer to keep his workforce at relatively high wages so that they do not look for higher wages elsewhere. There are strong economic incentives to keep the fluctuation of a workforce low. He would not automatically go for the cheapest labour and allow his own labour force to go for the highest wage offers. Training and the acquisition of job experience are expensive. He may also be concerned with the motivation of his workers, and their expectation of "justice" in order to keep them working properly. On the other hand, top workers might not be promoted (but might be allowed to enjoy some less conspicuous advantages) so as not to attract the attention of competing firms who might offer higher wages. If you have a top worker whose market value is higher than what you pay him, there is no need to let him know. But he should have an incentive to keep his high performance up by having a good salary in comparison to the ones he knows about. All this is about what people know or do not know in not totally transparent markets.

---

[11] I thank Steve Reyna for this reference.

Individual Choice and Collective Identities 313

We cannot dwell much longer here on the economy. What I propose instead is to apply economics of information of the type proposed by Stiglitz to our own research topic, which is identification with collective identities.

In considering the costs and benefits of identities, we have, so far, looked mainly at various groups and the resources they hold. To belong to a small but powerful group that is in secure possession of a rich resource base (so that one does not need to build up strength by admitting more members or seeking alliances with people with whom one would then have to share) is obviously good, because it implies a favourable balance between resource base and group size. Of course, from the point of view of the individual, questions of the uneven distribution of wealth, for example, among leaders and followers, are important. We have also dealt with these differentiations and complications, perhaps not fully in every case, but we have been aware of them. So far, however, we may have had a blind spot when it comes to information. Consider the following questions:

Question 1. What is the value of the information that a given identification is "false" or "fake"?

Let me start with a case history in which this value appears very low. A British colleague, Professor Stewart (the name has been changed), once told me that he is Hungarian by birth. He came to Britain with his parents during the crackdown on the insurgents in the 1956 uprising, and the family then changed its name to Stewart, I suppose for the benefit of the British who might have found the Hungarian family name hard to pronounce. After he had become a senior academic—I think he was a Reader at the time—he was invited by a Stewart Family Association. This association comprised many notable Stewarts who were proud of their ancestry. Of course, he immediately explained that he was not a "real" Stewart and was told that this did not matter. He is a witty person and nice to have at a dinner table.

Why did the information about "real" or "fake" family affiliation matter so little in this context? Obviously, the family name was little more than an excuse to meet interesting people, and he as a person was considered to be an asset. What would have happened if he had been a notorious bore, a zealous diet change activist, a pauper who asks everyone for money, or some other unpleasant kind of person? He probably would not have been asked to attend any meetings, even if he had been a "real" Stewart.

Another story which is at first glance different, but has a very similar "deep structure" is based on my conversation with a Somali businesswoman in Marsabit, Kenya (Schlee, 1989, p. 29). Like others in the Somali business community, she originated from the British colony in the north, the region or country now known as Somaliland. She was of the Eidagalla clan of the Isaaq clan family. To the local Rendille, who have a clan named Elegella, she was known as the "Somali woman of Elegella". Since their arrival in Kenya, the members of her family had claimed that Eidagalla was the same thing as Elegella. I pointed out to her that according, to my research, this was not the case, and that the similarity of the names was just coincidental. This did not really seem to surprise her. Her answer was pragmatic: "It is much better to have people than to be alone".

Her claim to be of the same clan as some Rendille, although she stemmed from a different ethnic group, Somali, is not at first glance implausible. Among the camel nomads of northern Kenya who speak Cushitic languages (Rendille, Gabra, Sakuye and various Somali groups) there are numerous cases in which parts of different ethnic groups share clan identity, Some of these interethnic clan relationships are due to common origins and subsequent ethnic differentiation, others to interethnic re-affiliation after the modern ethnic groups had come into being. In times of need, like the loss of livestock due to drought or raids, one can ask clan brothers in other, even hostile, ethnic groups for help or seek refuge with them. Thus, these interethnic clan relationships are also bridges for the exchange of personnel between different ethnic groups (Schlee, 1989; cf. above, the discussion about Barth and ethnic conversion between Pathan and Baluch).

What this family had done was enter into a contractual relationship that only looked like finding ones long-lost clan brothers in a foreign country. They claimed to be Elegella and gave some preferential treatment to Rendille customers who belonged to that clan. This gave them an entry point into Rendille society. People would call them "abiyo" (mother's brother) or "ingo" (mother's sister) if their mother was from that clan or address them as grandparents if that was the case with their mother's or father's mother. Those people with wives from Elegalla would call them wife's brother or wife's sister, and so on. The arrangement lasted, as it was not costly, and it was convenient to both sides. Critical questions about clan history were out of place.

As a mental experiment we can ask ourselves what would have happened if this family had not come as traders but as impoverished nomads who had suffered drought or had been raided and needed to re-build their herd. What would have happened if they had asked for camels? Then critical questions would not have been out of place. Is Elegella actually the same as Eidagalla? Where is the common origin and where and how have the two groups split from each other?

With reference to these examples, one might conclude that, when identities are based on contractuality, the value of the information that they are "fake" is relatively low. Conversely, the value of this information increases when identities lack a high degree of contractuality. My use of the term "contractuality" might require clarification: Some time ago (Schlee, 2008, pp. 30–33) I argued that, rather than thinking in terms of a dichotomy between given or inherited relations ("status" (Maine (1986 [1861])[12]), on one hand, and contract, on the other, we should view relationships as having more or less of a contractual character, i.e., varying degrees of contractuality: "Speaking of the economics of identity change, an important distinction is whether an identity is thought of as natural or as contractual. Contractual elements can be more strongly or more weakly represented, so we can speak of contractuality as varying along a scale". Let me summarize the argument here briefly. Contractual elements are especially evident when relationships of inclusion and exclusion can be changed, and this applies equally to groups and alliances. In widening the membership of one's group, one includes more people in the category to which the pronouns "we" and "us" refer, while an alliance is always about relations with others. Alliances do not abolish group boundaries; they do not even blur them. On the contrary, boundaries are highlighted in the processes of defining partners and concluding, affirming, and reaffirming an alliance.

There may be implicit contractual arrangements in cases involving a widening of membership in descent groups, although on a scale from "status" to "contract" one would expect descent groups to be way over on the status side. "Such an implicit contract may take the form of a veiled threat of expulsion: "We will treat you as a brother as long as you behave like one". If the group has a formal character, like a voluntary association, membership is of a contractual type. The duties of a member and the obligation of the group towards its members are defined by the by-laws of the association and accepted by the act of joining it. Group membership may thus comprise contractual elements to a higher or lesser degree, depending on the status of a member and the kind of group. In the case of an alliance, however, the relationship is always openly contractual, involving two or more different parties and the form of cooperation they have agreed on" (Schlee, 2008). Parliamentary coalitions are good examples of alliances. The parties which form them remain separate entities. Their main aim is to bundle votes to grant the government they wish to form a safe majority. But even they are not pure contracts in the sense of a purely instrumental, target-oriented relationship between total strangers. There tends to be some degree of "sameness" between coalition partners. "The parties forming a coalition, though remaining separate, have some common denominator. They are part of wider groupings.

---

[12] See also Feaver (1969).

They are all bourgeois or leftist, belonging to one or the other class-based block. In other cases, the moderates of different former blocks, striving for the middle of society where they expect most votes to be, form centrist coalitions against what they perceive as the margins: the parties subscribing to more radical versions of their respective political ideologies. In other words, alliances are always between different groups, but these groups may belong to the same or different wider groupings. Even single-issue ad hoc alliances that are immediately dissolved after the shared aim has been achieved might have identity-related aspects that go beyond the purely contractual" (Schlee, 2008, p. 31f). This shows, I think, that there is variation between two extremes, namely inclusion based on status or sameness (who we are) and inclusion based on contract; but it also illustrated that even the purest, apparently ascribed, status-based, inherited forms of inclusion in a group contain contractual elements (you may be shunned if you break the rules), while, at the other extreme, even the purest contractual forms of inclusion contain elements of group identity.

"To the extent that alliances are not 'purely contractual' but are guided by considerations of social identities and the moral obligations attached to them, one can therefore rephrase the contractual nature of alliances as a question of degree. This is in line with my attempts to rephrase other dichotomies so prevalent in social theory in a gradualist language and to regard them as variables so as to be able to study their variation and co-variation and thereby their interconnections. 'Contractuality' as a variable would have to state to what degree an arrangement is contractual and, by implication, to what degree it is shaped by non-contractual elements. On a contractuality scale from 0 to 1 a given alliance may have a value of, say, 0.6, and this would mean that it is primarily contractual but significantly influenced by other factors. These other factors may be:

1. The partners are not freely chosen. The choice of allies is limited by considerations of closeness or similarity rather than purely strategic considerations like optimal group size for the formation of a minimal winning coalition.
2. The context of agreement shows a mixture of strategic considerations with 'Culture' and 'Custom'". (Schlee, 2008, pp. 32–33)

Many other factors can interfere with the purely contractual character of a relationship.

In the case of one mixed type (the implicit contract in the claim of the Somali woman to belong to the Rendille clan Elegella), we have seen that the

Individual Choice and Collective Identities   **317**

contractual element of the relationship made the question about the historical authenticity of her claim irrelevant. This leads us to our next question.

Question 2. What is the value of information in identifications with different degrees of contractuality?

Professor Stewart played the game. His family history, and the fact that he was not a "real" Stewart, was an amusing item at the dinner table, and that is something that turned it from a liability into an asset. Playing the game is basically a contractual agreement. I am nice to you, as long as you are nice to me. Stewart or not, if he had been boring, arrogant or too critical of his hosts, he would not have been around for long. The Somali woman did a similar thing. She behaved as a clan sister and people reciprocated by treating her as such.

Our initial conclusion, after looking at these case histories, is that, in each case, it was the contractual element that made the relationship negotiable and helped to maintain it, even though both had an element of "fake" descent— in the Stewart case one which was admitted from the start, in the Somali case one which was easy to find out. We may conclude that the more pronounced the contractual character of a relationship, the lower the value of information about whether the identities underlying the relationship are "real" or "fake".

Other kinds of information become valuable in contractual relationships. In such a relationship one needs to know whether the partner fulfils his or her part of the contract. So it is not surprising that there is a huge litera-ture in business administration and experimental social psychology dealing with "monitoring", "auditing", "trust", "free-riding", "control", "defection", etc. All this has to do with information about partners and our inability to look into their brains and to know whether and when they will stop playing the cooperative game and turn against us. What is the price of their loyalty? There is no need to convince anyone of the importance of information in this context. It may have gone unrecognized in economics for a long time but certainly not in business administration.

In the two cases presented here there was, actually, not very much to monitor. One more element shared by both cases is that little was at stake. None of them was about life or death or huge amounts of money. This leads us to question 3.

Question 3. What is at stake in identity claims ?

As Boris Nieswand (2011) has argued in his thesis about Ghanaian migrants in Germany, the single best predictor of the wealth of any given person is nationality. In other words, if you have no information about a person—no name, no picture, no information about age, gender or anything else—and you have to make a blind guess of how much money that person has after asking only one question, the best thing to ask about is his or her nationality. It is an even better indicator than class. After all, the welfare recipients in one country might be better off than the middle class in another. Anything to do with legal entitlements is also a hopelessly bad wealth indicator. History shows that today's organized crime is tomorrow's aristocracy (Comaroffs, 2009).

The enormous income differences from one country to another motivate boundary crossings and exclusionist practices to prevent just that. No doubt hiding information that might lead to deportation or providing information (e.g., in the shape of real or fake documents) that might lead to a secure and legal form of residence is of enormous importance to illegal immigrants, and they are ready to invest a lot in hiding or providing information about their identities.

In addition to what is at stake, the cost of providing knowledge needs to be taken into account in the information economics of identification. Think of an academic whose family name is reminiscent of that of a knight who lived in the year 1300. He faces the following questions: Do I really wish to trace the relationships in church registers and other documents? Would that take me half a year or more or less? Do I take the risk of not finding anything and making a fool of myself? He will possibly conclude that the knowledge of being the offspring of some long-defunct iron-clad predator is not worth the costs of providing it and the risks involved, and that focusing on his academic career is the safer road to fame. The question might be answered differently if the issue involved some real estate or a bank account to be inherited. Then the value of the information might clearly exceed the costs of getting it.

8   Shaping the Past

In the preceding paragraph, we have mentioned an element of the past that might not be worth finding out about. There might be other things that we are pretty sure about, but which we prefer to ignore or deny. The groups, institutions, and communities into which society is organized, articulated, or divided (the social structures) all derive their legitimacy from the past. The present section explores some ways in which present-day agency affects what is known about the past and thereby stabilizes or reshapes or even undermines social structure.

Individual Choice and Collective Identities 319

An often-used metaphor for historical origins is that things stem from their roots. Nothing could be further from the truth. The origin of the metaphor is a plant. But not even plants stem from their roots. A plant grows from its seed on the surface of the soil both upwards and downwards. It drives its roots down into the soil and the future stem sprouts upwards. The roots avoid obstacles like stones and seek nutrients. They are selective. Their direction of growth, their number and density reflect the needs of the plant.

This gets lost if we apply the botanical metaphor loosely and explain the origins of cultures and peoples by pointing to their roots. Collective identities and social institutions, too, do not stem from their "roots". They drive their roots into the past and they are selective about the elements of the past to which they connect.

Picking the right elements from the past, those which provide nutrients to the claimed collective identities in the present is identity work. Much of it is done by historians in the wider sense. Even in societies that are mainly composed of illiterate people and where no written records are kept, we find individuals who are able to handle vast bodies of knowledge which are part of a collective memory, shared in bits and pieces by many but held by only some in more complete, more consistent and authoritative versions.

Through eastern and southern Ethiopia, we find a boundary dividing the Somali Regional State from Oromia, where Oromo are the vast majority and the titular nation. This line finds its continuation in northern Kenya, where the present boundary between Marsabit County and Wajir County goes back to an old colonial division between Oromo and Somali, or to what the British perceived as such a division. The Oromo language and Somali (with marked dialect differentiation in the south) both belong to the Lowland branch of Eastern Cushitic, but they are clearly distinguished and far beyond mutual intelligibility for monolinguals (think about French and Italian or English and German for comparable levels of difference). Other cultural differences, too, are clearly marked. But apart from unquestionable Oromo and unquestionable Somali the area is inhabited by groups that do not fit this binary division. There are Oromo-speaking Somali (with a Somali identity based on genealogical links) and groups who claim never to have been Oromo or Somali but to be something else. Violent competition for pasture, water and political representation is frequent and groups form changing coalitions and re-affiliate in this struggle. In these processes of (re-)identification they selectively refer to shared origins, past conviviality or long-standing enmity.

Some of the history produced in this context is pure fabrication. Genealogical links might be invented and laughed about by everyone who is not part of

320   G. Schlee

the deal and is not interested in maintaining the fiction (Schlee & Shongolo, 2012b, p. 36).

More frequently, however, the invention of "facts" is not needed because there are enough true stories to pick from, maybe relocating some of the emphasis and with slight re-interpretations.

Some subclans of the clan Ajuran claim a long and uninterrupted Islamic tradition and speak Somali. Other subclans speak Oromo among themselves, had in precolonial times affiliated themselves as satellites to a Boran-centred alliance and helped the Somali-speaking newcomers to be accepted there and to get access to pasture and water. At the highest level of segmentation, the Boran are divided into two exogamous moieties, Sabbo and Gon. The Ajuran sent gifts to the ritual head of Gon, the *qallu* of Worr Jidda or Jilitu, and thus ritually became members of the Boran clan Jilitu and thereby part of the Gon moiety.

Enter the British. They perceived, not entirely wrongly, the Ajuran as a branch of the Boran and therefore Oromo. The Ajuran ended up west, on the Oromo side, of the line the British drew to stop the advance of Somali from the east.

If Ajuran drive their roots into the past they have many different things to which they can turn and to which they can connect. There are still widespread use of the Oromo language and the former ritual connection to a Boran Oromo High Priest. On the other side there are genealogies linking them to Somali and there is the Islamic religion, to which some subclans in the past were reputed to adhere with more devotion than others.

Throughout the colonial period and beyond, the identification as Somali became stronger and stronger. Around 1980 my Ajuran interlocutors, mostly middle-aged men, would flatly deny that their clan ever had anything to do with Boran or had even paid ritual tribute to "pagan" dignitaries. I got such a response even from an old man whose name I had found adorned with a Boran title in the colonial archives.[13] It is hard to imagine that children were told anything else about their origins than that they were Somali and have always been Somali.

Obviously, there were religious values and ideological forces at work there. But we cannot fail to see that this "conversion to Somality" took place in a context of Somali expansion and crowding the Boran out from some of their traditional wells and from the pastures within reach for animals watered at these wells, pastures which cannot be used in the dry season without these wells.

---

[13] That is the documents from colonial times preserved in the Kenyan National Archives.

By 2000 competition had turned violent between the Ajuran and "other" Somali, and the Ajuran reconsidered history. The old alliance with the Boran was fondly remembered and revived. Detailed knowledge about the precolonial position of the Ajuran at the side of the Boran re-appeared. Some Somali roots withered and the Ajuran grew roots into the Oromo part of their past instead. There were even stories about having lovers from other ethnic groups and who begot whom. Relationships through the *genitor* rather than the *pater*, the socially recognized father who had solemnized the marriage and paid the bride price, even if known are never used for descent reckoning among these strictly patrilineally organized peoples, a form of organization which needs hard and fast rules about paternity. But to emphasize the closeness to the Boran, even to such relationship's roots were driven to draw identification from them.

Obviously during the period of denial, the Oromo past of the Ajuran has not fallen into complete oblivion. A "backup copy" of historical knowledge had been kept somewhere in somebody's mind. It could be reactivated and spread when needed. In written archives, it would not be unusual for testimonies about the past to be forgotten and rediscovered. But apparently, such rediscoveries are also possible in oral cultures. The Ajuran have such an oral culture in everything but religion. The only book pastoral Ajuran tends to have is the Qur'an. Under these conditions, it is astonishing how much knowledge may remain hidden for a long time without getting lost. (Schlee & Shongolo, 2012b, p. 85).

9   Conclusion

Structures can be a constraint of choice, and it is that aspect of structure which is generally foregrounded in sociological theory. The present chapter, however, puts the emphasis not on what structure does to choice and other forms of agency but on what agency does to structure. It stresses the importance of individual agency, and in particular choice, for the manipulation, re-evaluation and change of structures. The structures examined in this context are categories of social classification and collective identities on a relatively large scale, like nationality, ethnicity, religious affiliation, descent group (clan, tribe), and linguistic affiliation.

References

Barth, F. (1959). *Political leadership among the Swat Pathans*. Athlone Press.

Barth, F. (1969a). Introduction. In F. Barth (Ed.), *Ethnic groups and boundaries: The social organization of culture difference* (pp. 9–38). Universitetsforlaget.

Barth, F. (1969b). Pathan Identity and its Maintenance. In F. Barth (Ed.), *Ethnic groups and boundaries: The social organization of culture difference* (pp. 117–134). Universitetsforlaget.

Bošković, A., & Günther, S. (Eds.) (2022). *African political systems* Reconsidered. Berghahn.

Comaroff, J., & Comaroff, J. L. (2009). Reflections on the anthropology of law, governance and sovereignity. In Franz von Benda-Beckmann, Keebet von Benda-Beckmann & Julia Eckert (Eds.), *Rules of the law and laws of ruling* (pp. 31–59). Ashgate.

Donahoe, B., Eidson, J. R., Feyissa, D., Fuest, V., Hoehne, M., Schlee, G., & Zenker, O. (2009). The formation and mobilization of collective identities in situations of conflict and integration. *Max Planck Institute for Social Anthropology Working Paper No.* 116. Halle/Saale: Max Planck Institute for Social Anthropology.

Diallo, Y. & Schlee, G. (Eds). (2000). *L'ethnicité peule dans des contextes nouveaux: la dynamique des frontières*. Karthala.

Eidson, J. R., Feyissa, D., Fuest, V., Hoehne, M. V., Nieswand, B., Schlee, G. & Zenker, O. (2017). From identification to framing and alignment: A new approach to the comparative analysis of collective identities. *Current Anthropology* 58 (3), 340 – 351.

Esser, H. (1999, February 11). *Inklusion und Exklusion*. Lecture at Bielefeld University.

Feaver, G. (1969). *From Status to Contract: A Biography of Sir Henry Maine 1822–1888*. Longmans.

Fortes, M. & Evans-Pritchard, E. E. (Eds.). (1940). *African Political Systems*. Oxford University Press.

Gigerenzer, G. (2007). *Gut Feelings: the intelligence of the unconscious*. Viking.

Hann, C. (Ed.). (1994). *When History Accelerates: Essays on rapid social change, complexity and creativity*. Athlone.

Hechter, M. (1987). *Principles of group solidarity.*University of California Press.

Knörr, J. & Kohl, C. (Eds.) (2016).*The Upper Guinea coast in global perspective*. Berghahn.

Lewis, H. S. (2022). African Political Systems and Political Anthropology. In: Aleksandar Bošković, & Günther Schlee (Eds.) *African political systems reconsidered* (pp. 15–46). Berghahn.

Maine, H. (1986 [1861]). *Ancient Law*. University of Arizona Press.

Markakis, J., Schlee, G. & Young, J. (2021). *The nation state: A wrong model for the Horn of Africa*. Max Planck Institute for the History of Science. (Open access: https://www.mprl-series.mpg.de/studies/14/index.html)

Nieswand, B. (2011). *Theorising transnational migration. The status paradox of migration*. Routledge.

Riker, W. H. (1962). *The theory of political coalitions*. Yale University Press

Roosens, E. E. (1989). *Creating ethnicity: The process of ethnogenesis*. Sage.

Schlee, G. (1989). *Identities on the move: clanship and pastoralism in northern Kenya*. Manchester University Press.

Schlee, G. (2003). Identification in violent settings and situations of rapid change. *Africa 73*(3), 333–342.

Schlee, G. (2008). *How enemies are made: Towards a theory of ethnic and religious conflicts*. Berghahn Books.

Schlee, G., & Horstmann, A. (Eds.) (2018). *Difference and sameness as modes of integration: Anthropological perspectives on ethnicity and religion*. Berghahn.

Schlee, G., & Shongolo, A. A. (2012a). *Islam and ethnicity in Northern Kenya and Southern Ethiopia*. James Currey.

Schlee, G., & Shongolo, A. A. (2012b). *Pastoralism and politics in Northern Kenya and Southern Ethiopia*. James Currey.

Schlee, G., & Watson, E. E. (Eds.) (2009a). *Changing identifications and alliances in North-East Africa. Ethiopia and Kenya*. Berghahn.

Schlee, G., & Watson, E. E. (Eds.) (2009b). *Changing identifications and alliances in North-East Africa: Sudan, Uganda, and the Ethiopia-Sudan borderlands*. Berghahn.

Simons, A. (2000). Women can never 'Belong' in Combat. *Orbis,* 451–461.

Stiglitz, J. E. (2002). Information and the change in the paradigm in economics. *The American Economic Review 92*(3), 446–501.

Titus, P. (1998). Honor the Baloch, Buy the Pushtun: Stereotypes, social organization and history in Western Pakistan. *Journal of Modern Asian Studies 32*(3), 657–687.

Understanding Religious Radicalization: The Enigma of Beneficial Violence

Hans G. Kippenberg

1 The Problem of Religion and Violence

Since the 1970s, there has been no shortage of news stories feeding the suspicion that religion and violence are closely linked. Upheaval in Iran or the events in Texas around Waco and other American religious communities were the first indications of this. In the first years of the new century, there were others who, following the outrageous attack of young Muslims on the U.S. in 2001, saw in it an example of the violence not only of Islam but of religion in general. There is no lack of publications of all kinds condemning religion in toto. Religion is "God delusion," wrote Richard Dawkins (2006); it poisons the world, we read in Christopher Hitchens (2007); it faces its end in view of the terror it produces, declared Sam Harris (2004). Dawkins (2006, p. 424) agrees with a journalist's view that the "cause of misery, chaos, violence, terror and ignorance is, of course, religion itself"; according to Harris (2004, p. 136), Muslim thought is a web of myths and conspiracy theories; there is "no reason to believe that economic or political improvements … could remedy the malaise." Who would want to doubt it: religions have potential for violence. The traditions of Judaism, Christianity, and Islam contain glorifications of violence. From philosophers of religion such as Thomas Hobbes

---

H. G. Kippenberg (✉)
Comparative Religious Studies, University of Bremen, Bremen, Germany
e-mail: kippen@uni-bremen.de

© The Author(s), under exclusive license to Springer Nature
Switzerland AG 2023
N. Bulle and F. Di Iorio (eds.), *The Palgrave Handbook of Methodological Individualism*,
https://doi.org/10.1007/978-3-031-41508-1_14

326 H. G. Kippenberg

(1588–1679), who saw in the state the only remedy against the devastating religious wars of his time, to scholars of ancient religions such as Walter Burkert (1931–2015) and literary scholars such as René Girard (1923–2015), who saw sacrificing as a salutary act which purified a society of its aggressions, religion, and violence have been closely associated. Powerfully eloquent and sometimes aggressive, this has become a general conception of religion, which continues to be publicly disseminated today.[1]

2   The Case of Islamic Violence: Jihad

Occasionally, violence is inferred from the nature of Islam ("Islam itself is the cause of jihad"). Islamic scholar David Cook (2005) has seen in militant violence the original meaning of jihad, as he elaborates in *Understanding Jihad*. At the end of his account of the military conquests of the early Muslims, who conquered the territories between Egypt and the Arabian Peninsula (called the "Fertile Crescent"), then Central Asia, Afghanistan, and other territories, and seeing this as a gift from God, he concludes: "Because of the miracle of the conquests, jihad emerged as one of the core elements of Islam." (Cook, 2005, p. 13). Global radical jihadism and its martyrdom operations have their roots here (Cook, 2005, pp. 128–131). Cook passes over the possibility that the emergence of Islamic violence may have something to do with the history of violent European and American interventions in Islamic countries (Lüders, 2015, 2017). Contra Cook one cannot isolate individual acts of violence by Muslims and the meaning they give to their actions from these circumstances. Understanding this meaning is the methodological key to making sense of their radicalism and thus explaining their violence. The fact that the originator of the thesis of a global commitment of all Muslims worldwide to jihad, Abdallah Azzam, came from the Palestinian territories occupied by Israel and no longer placed his hopes for the liberation of this country in local Muslims but in Muslim foreign fighters is, according to David Cook, not worthy of further explanation: this fact allows us to understand his acting and thinking (Cook, 2005, p. 128).[2] If one wants to understand the motives of perpetrators of violence, one must turn to the personal subjective experiences of those involved, as Jessica Stern (2009)

---

[1] Walter Burkert (1972); René Girard (1972/1979). On Girard's theory, including its elaboration by others, see the 2007 special issue *Religion* 37(1), pp. 1–84.

[2] The insight into this context is due to Thomas Hegghammer (2006/2008).

does.[3] Murder of family members, expropriation of land, humiliation, experiences of injustice, but also experiences of unexpected victories over the godless form motivations for the violent acts. Anger and revenge are the driving force behind the perpetrators' radicalization. However, these subjective motives are not in themselves a sufficient condition for the execution of the act, especially since not everyone subject to those experiences turns to violence. They must be adopted by "Holy War Organizations" (according to Stern).[4]

Thomas Hegghammer presents six hypotheses for the fact that it was only after 1980 that Islamist propagandists regarded the threat to the Muslim umma from external, and no longer only from internal, enemies and that they advocated an individual duty on the part of all Muslims worldwide for those threatened, without first requiring parents or ulama to give consent to jihad (Hegghammer, 2010/2011). Accordingly, one would have to look at the situation of Islamic countries and their interpretation by Muslims in order to understand the approval of Islamic acts of violence. The violent actions of young, mostly male, Muslims do not follow from the essence of Islam, but from a definition of the situation of the Islamic community that empowers them to act violently. The experience with the military and political interventions of the USA, England, France, and Israel in the Islamic countries of the Middle East has become part of a widespread definition of the situation of Muslims and has contributed to a narrowing and hardening of the *concept of jihad* to the military. On the other hand, this hardening legitimizes the use of violence against infidels. The political history of Islamic countries in the nineteenth and twentieth centuries has influenced central concepts.[5]

The history and systematics of religion belong together, as explained by Jörg Rüpke in his analysis of the way Burkhard Gladigow develops Joachim Wach's approach (Rüpke, 2021): the goal of knowledge can be neither a hermeneutic *comprehension of* the religion's essence nor the scientistic

---

[3] Jessica Stern has found a successful way of dealing with both aspects by dividing her book into two major parts, "Grievances that Give Rise to Holy War" and "Holy war Organization.".

[4] For example, the study of the Hamburg group around Mohammed Atta by Terry McDermott (2006, p. xvi): "The men of September 11 were, regrettably, I think, fairly ordinary men".

[5] In the Treaty of Lausanne of 1922, the victorious powers, France and Great Britain, sealed the dissolution of the Ottoman Empire as a consequence of its military defeat in the First World War. It was replaced by the secular nation-state of Turkey. This had been preceded by major territorial losses of the empire. North Africa had become independent. The Sykes–Picot Agreement of 1916 had separated different colonial interest areas in the Near East. Great Britain was granted dominion over an area corresponding roughly to present-day Jordan, Iraq, and the area around Haifa; France over southeastern Turkey, northern Iraq, Syria, and Lebanon. Egypt had been dependent on foreign loans since the construction of the Suez Canal (1859–1869). To secure the route to India, Great Britain acquired the Egyptian shares in the canal, occupied the country in 1882, and formally made it a protectorate in 1914; in 1922, Egypt became a formally independent kingdom, but, like the other Middle Eastern states, remained the victim of ever more Western interventions—which have continued to the present. See Lüders (2015, 2017).

328   H. G. Kippenberg

reinterpretation of religion in terms of an evolutionary-biological or cognitivistic systematic discipline. It should be rather a more modest or limited reorganization of the options for the action of a religious community.

Narrowing military jihad to an essential feature of Islam has by no means been characteristic of Islam throughout its political history. A strong case for a broader understanding of jihad in the history of Islam is due to Asma Afsaruddin. She expanded the source base of the legal literature, which indeed often understood *jihad* as armed struggle, to include the exegetical literature, which recognized a wider range of meanings of *jihad* as "striving in the way of God" denoting both physical and spiritual efforts (the literal meaning of *jahada*): both the struggle for the healing of humanity through the spread of Islam and the struggle for the betterment of the believers themselves in view of God's commandments. Considering this literature, other actions become equivalent to military fighting. To bear enmity patiently *(ṣabr)* was the Prophet's advice to his followers (Sura 3: 200), especially in situations of confrontation with opponents of the Prophet. An expanded conception of jihad is also held in the hadith collections when they understand the support of widows and the poor as fighting in the way of God (Sura 2: 195). By breaking down the narrow focus on physical struggle in the juristic literature with the help of the more variant semantics of *jihad* and martyrdom *(shahid)* in the Quran and hadith, Afsaruddin shows that patience *(ṣabr)* was valued as an achievement in addition to high regard for the performance of physical *jihad.* Both attitudes reflect the different situations in which the Prophet and the community found themselves in Mecca and Medina. While Muhammad had long been at peace with the infidels during his years in Mecca, this changed after the hijra (Arabic *hiǧra*) in Medina, where the sword verse Sura 9: 5 calls for the fighting and killing of polytheists unless they convert to Islam. The change from toleration of infidels to violence against them is a central theme of Islamic theology. There are some Muslim scholars who assume that the sword verse superseded previous revelations to the contrary, citing Sura 2:106, which admits such a possibility.

At the time of the early Islamic empires of the Umayyads and Abbasids, military importance gradually gained preponderance over the spiritual and moral, without the high esteem of patience and forbearance disappearing altogether.[6] Cook and others with him neglect this aspect. A glance at the table of contents and the index of his study reveals that the alternative concepts of *hudna* (truce) and *sabr* (patience) are not included, although

---

[6] Asma Afsaruddin (2016; more fully 2013); also in Afsaruddin (2015) "War and Peacemaking in the Islamic Tradition" (pp. 115–140), where she discusses intra-Islamic disputes over the legitimacy of military *jihad*.

Understanding Religious Radicalization ...

a silence of arms or exercised patience in enduring injustice are indeed Islamic alternatives to belligerent jihad. Islamic international law acknowledges treaties with non-Muslims, which should by no means therefore all be regarded as tactical and an expression of Islamic intolerance.[7] Even when Muslims do not give up their claim to dominate the contracting party and its territory, they can refrain from taking up arms in the interests of their community, provided the conditions do not contradict Islam.[8]

This is where another reflection on the vital basic goods that must not be put at risk comes in. Abū Ḥāmid al-Ġazālī (1058–1111 A.D.) described this aspect with *maṣlaḥa*.

> Their religion (Din), life, mind, lineage and property shall be preserved. Thus, anything that guarantees the preservation of these five foundations (legal assets) is *maṣlaḥa*, and anything that leads to the impairment of these foundations is perdition.[9]

The linking of existential basic goods of the community (including religion) with the public interest of the community (Opwis, 2007), was updated by the *Ǧamāʿa al-Islāmiyya*— in the 1970s[10]—and we also encounter it again in Yūsuf al-Qaraḍāwī. He incorporated al-Ġazālī's reflections and expanded its list of basic goods. To the five already mentioned have been added honor, peace, rights and freedoms, the institution of justice and shared responsibility in an exemplary community, and everything that makes life easier for Muslims, eliminates oppression, perfects character, and guides Muslims toward what is best in customs, (social) institutions, and interactions.[11] Just as the umma has expanded socially and globally, the respected cleric *(ʿālim)* understands the common good of the community more comprehensively than al-Ghazali (Salvatore, 2009). Tariq Ramadan goes even further when he asserts, "To seek for the good (*maslaha*) of man, in this life and the next, is the very essence of Islamic commandments and prohibitions" (Ramadan, 2004, p. 42). Other scholars have also embraced the concept of a public sphere constituted by communities (rather than the actions of the ruler), as the collection of studies by Salvatore and Eickelman (2004) shows. Finally, Reinhard Schulze (2016) has written the whole *Modern History of the Islamic*

---

[7] As an example of the widespread prejudice: Denis MacEoin (2008).

[8] Rüdiger Lohlker (2006), e.g. 41–42. (see the index to 'Security Treaty').

[9] Source: *islam-pedia.de/index*.php/Masalih Mursala (Arab.: ج المصالح المرسلة‎) means "the consideration of the general interest".

[10] Ivesa Lübben and Issam Fawzi (2004, pp. 4ff., 22). See also Roel Meijer (2009); Mariella Ourghi (2010, pp. 59–72).

[11] Quoted from Muhammad Qasim Zaman (2004, esp. 134–135).

world with an orientation to this concept of "public sphere" (Schulze, 1998/2016, p. 18). If one wants to understand an act of violence by Muslims, then one must abandon any definition of the essence of Islam as violent or even terrorist. One must include Muslims' definition of situations in reconstructing what they do. Those who suspect Islam of terrorism and wage war against it therefore have a stake in Muslims' definition of their situation (Kapitan, 2003). To use a metaphor, they are fighting a "disease" with means that make it an epidemic.

3 Definition of the Situation: The Thomas Theorem

The study of the history of a religious community does not lead to a single or even dominant value that determines its actions, but to a variety of religious evaluations, conditioned by different situations of the community. There are no values, only evaluations, notes Andreas Urs Sommer (2016). This is also true for the case of religions. In the "classical" sociological theory of action of Talcott Parsons, situations are the neutral field of action for extrasituationally conceived values. Today, the Thomas theorem takes the place of Parsons' means-ends scheme in action analysis, according to Hans Joas (1996, p. 235). Even if one ascertains values through questioning, one may not infer a corresponding consistent action. Rather, it is only the situational reference that determines action. This approach has consequences for the method.

Toward the end of the nineteenth century, philosophy in Germany took a turn that also became momentous for religious studies. Wilhelm Windelband (1848–1915), in his rectorate speech—"History and Natural Science" (1894), set philosophy the task of turning to the cognitive processes in the sciences, to fathom their structure and to put them into a general form, and he offered a piece of advice worth taking to heart:

> Never has a fruitful method grown out of abstract construction or purely formal considerations of logicians: to these falls only the task of bringing what has been successfully practiced on the individual to its general form, and then of determining its meaning, its epistemic value, and the limits of its application.(Windelband, 1894/1907, pp. 357–358)

The question of whether historical knowledge could claim the same rank as scientific knowledge seemed particularly urgent to Windelband. Windelband answered in the affirmative. While the natural sciences determined laws in reality, the historical sciences focused on singular events, i.e., actions. The particular structure of "action" is the reason for their difference.

Windelband was not alone in proposing a theory of recognizing history. Max Weber also worked on this problem. In what sense do "objectively valid truths" exist at all on the ground of the sciences of cultural life? he asked in a methodological article in 1904 (Weber, 1904, esp. p. 147). His answer was along the same lines: it is the valuations in the actions that must be examined, distinguishing between recognizing and judging. To judge values was a matter of belief; the scientist, however, had to limit himself to their recognizability in action. The necessity of such a justification of historical and sociological knowledge resulted from the loss of the belief in progress (Dahme, 1988). All those mentioned rejected somehow lawful progress in history. Hartmut Esser, who also advocated this model of explaining action[12] writes at the end of an essay on "The Definition of the Situation" a sentence that considers the subjective expectation inherent in such a definition just as much as the external circumstances:

> The internal selection of the models of the situation ... does not happen in a vacuum. It takes place against the background of immutable *faits sociaux*, albeit in turn socially constructed by *actors*..... People undoubtedly define their situation themselves, but they do so not of their own free will, but under immediately found...circumstances. (Esser, 1996, p. 32)

Neither do circumstances exist independently of the actors' definition of the situation, nor do the intentions to act exist independently of the circumstances. Religious actors choose canonical models as a program of action or script or screenplay (in the terms of Esser) that fitted their definition of the situation. The nature of the conflict is thereby already socially constructed and is further determined by the actors with their choice of the action. This creates a double contingency. "We do not see things as they are, we see them as we are," wrote Anaïs Nin. The representation of things as they are for us is not sufficient to determine their meaning for the actors. Moreover, in addition to their perception of the situation, their practical social action is a mode of its own. Actors can consider the value of responsibility for the consequences, e.g., for the Islamic community, in their actions, or they can make the value of their actions absolute independently.

---

[12] Esser, (1996, 1999), Chapter 2 "The Thomas Theorem," pp. 59–70; 1995: on the Thomas Theorem, its originators, and its diffusion in sociology, see Merton (1995); Esser (1999, pp. 47–50) contrasts the views of Parsons and Mead. On p. 56, he presents his own model of the external and internal conditions for defining situations.

4 Ethics of Responsibility or Ethics of Conviction

In order to analyze the effects of social action, Weber makes a distinction between two types of ethics that are fundamental to a study of the link between religious expectations of meaning and social action (Weber, 1915–1920/1989, pp. 479–522). Agents can renounce the norms that regulate actions in the realms of economics, domination, and law. Weber addressed this possibility in the so-called Intermediate Reflection (*Zwischenbetrachtung*), elaborating so-called ethics of meaning alternatives in the realms of economics, domination, art, and sexuality. The ethics of brotherhood, war of faith, aesthetic experience of art, and eroticism also have their place in the tensions between existing orders and world-rejecting religions of redemption. The belief in universal human rights could be added. Even if actors know themselves as committed to a religious value such as brotherhood, this does not necessarily result in a distinct ethical practice, as Hartmut Esser also assumes (Esser, 2001, pp. 326–327). Rather, a new compulsion to decide arises as to whether the ethical value of action depends on the intrinsic value of the action or its success; whether the value of the *conviction* entitles one to reject responsibility for the consequences of the action, or whether *responsibility for the* consequences also determines the correctness of the choice of action (Weber, 1915–1920/1989, p. 497). An ethics of responsibility, however, remains bound to the same value as an ethics of conviction.

> There can be no ethical action without a value of conviction. (Schluchter, 1988, p. 198)

By refusing to recognize valid norms and thus allowing the inherent laws of existing orders to emerge, ethics creates space for alternative values.

Often, routine relieves the actors of the need for their own situation definition. However, when the plausibility of a particular definition and the belief in the success of action fade, actors may become aware that they have other promising possibilities to define the situation they are in and switch from an "automatic" to a "reflexive" mode. This brings into play the availability of different models of evaluation and action. Hartmut Esser speaks here of "framing: the selection of the frame of reference." (Esser, 2001, pp. 259–334; Esser, 2015). If one examines the actions of religious actors in conflict situations from this point of view, then it is canonical models and not universally valid values that are used to define the situation. Only they are also capable of embedding one's own actions in the salvation history of his/her community. Therefore, one must not assume causality between beliefs and actions but rather take the situation and the available models for responding to circumstances as the object. It is necessary to understand why people have made these role models the prefiguration of their own actions. The teleological model that Talcott Parsons made the principle of his *Structure of Social Action* is based on a pre-ordering of intention—the belief in values—before action. Hans Joas rightly contradicts such a pre-ordering. Perception and cognition could not be pre-ordered to action as values in themselves, "but [they are HGK] to be understood as a phase of action, through which action is guided and redirected in its situational contexts" (Joas, 1996, p. 232).

5 Islamic Canonical Prefigurations of Violent Actions: Ghazwa and Jihad; Ashura

The recent empirical scholarly studies, in accordance with these assumptions (of a plurality of situations; of a plurality of canonical prefigurations; of different ethics), have no longer been oriented to the paradigm of "religious violence," but to "religion and violence." "We have tried to avoid the phrase 'religious violence' since it can be interpreted as saying that religion causes violence. Instead, we discuss 'religion and violence', the relationship between the two that is often more frequent than the peaceful upholder of the traditions would like" write Juergensmeyer, Kitts and Jerryson (217, p. 3).

> Violence in both real and symbolic forms is found in all religious traditions, it can be regarded as a feature of the religious imagination. Almost every major tradition, for example, has some notion of sacrifice and some notion of cosmic war, a grand moral struggle that underlies all reality and can be used to justify acts of real warfare. (Juergensmeyer et al., Violence and the world religions, 2017, p. 6)

Such a repertoire of real and symbolic forms of violence must then be joined by militant communal action that is justified in religious terms. To this end, religious communities must interpret existing conflicts in religious terms, in Jürgen Habermas's words, act as "interpretive communities" (2009, pp. 392–393). Then, in conflict situations, their followers can call their comrades to their aid to fight the unbelieving opponent.

Ghazwa and Jihad

One example is Islam with its concept of *jihad* (Arab. *ğihād,* literally "effort, struggle, endeavor, commitment"), when reduced to a military model in situations of threat to the Islamic community. Contributing to this simplification is a spate of scholarly studies that attribute all violent acts of Muslims (called "terrorism") to the Islamic precept of *jihad*.[13] The term has become through Muslim online propagation a foreign and self-designation of militant subcultures (Lohlker & Abu-Hamdeh 2013; Lohlker, 2016, p. 7).

In Islam, Sura 9 opened the justification of violent acts. Chronologically the last of all suras, it had been written in Medina and denounced the peace treaty with those polytheists who had abandoned the Prophet in his battle with a Byzantine army (631 AD).

> When [...] the holy months have passed, then kill the (trespassing) pagans wherever you find them, capture them, besiege them and lie in wait for them everywhere! But if they repent, perform the ritual prayer properly and pay the poor tax, then let them go their way. God is merciful and ready to forgive. (Sura 9:5)

With the spread of Islam and the defense of the external borders of Islamic states, participation in the fight against the wicked, the *ghazwa* became a religious merit, with five wars of the Prophet (Badr 624 nChr, Uhud 625 nChr, Khandaq 627 nChr, Mecca 630 nChr and Hunayn 630 nChr.)[14] as models for the September 11, 2001, attackers. Accordingly, they understood their act as a "*ghazwa* in the way of God" (Spiritual Manual 1/3; 3/1) (Kippenberg & Seidensticker 2006). The terminology of al-Qaeda tied in with this: the "Manhattan ghazwa" or "The two ghazwa's of New York and Washington" (Fielding & Fouda, 2003, pp. 108, 121, footnote 10).The term was associated with religious preparation for the act: the perpetrators had to prepare for the sacred act with recitations, rituals, and prayers, thus focusing their intention (*niyya*) entirely on the act and only the act as required by Islamic law when performing an act that is supposed to be valid.

> An act of worship without niyya is invalid, and so is niyya without act. (Schacht, 1964, p. 116)

---

[13] Under the keyword 'Jihad' you can find this extensive literature at *amazon.de.*

[14] Fred McGraw Donner (1981) discusses the circumstances and foundations of Muhammad's conquests in the context of his teachings and political successes.

Sayyid Qutb and his brother Mohammad Qutb, who taught at the University of Jidda, where Osama bin Laden was one of their students, applied this maxim to religious militancy.[15] A Muslim's genuine faith must be judged by his inner intention, not his outward action (Steinberg, 2005, p. 41). Only in this way can a Muslim fighter attain the high good of martyrdom (Cook, 2002, pp. 19–21). This is also how the 9/11 manual sees it when it calls the impending attack a *ghazwa* "in the way of God" and prescribes Quran recitations, rituals, and meditations for its success the night before:

> Reciting the surahs *al-Tawba* [= surah 9] and *al-AnfÁl* [surah 8] and considering their meaning and the eternal blessing, God has prepared for the believers, (especially) for the martyrs. One of the Prophet's companions said, "The Messenger of God commanded us to recite them before the ghazwa. We then recited them, gained booty and remained unharmed." (1/3)

Upon boarding the aircraft, the fighters are asked to perform more recitations and prayers:

> When you board the t[airplane][16]:
> The very moment you put in your foot, and even before you really enter it, proceed with the prayers, and consider that it is a raid on the path (of God) as the Prophet has said, "Marching out in the morning and returning in the evening on the path of God is better than this world and what is in it," or how he has said it. (3/1)

In this way, they purified their intention in three stages: the last night—on the way to the airport—when boarding the plane.

Hans Blumenberg called such models "prefiguration." His writing in this regard was published only after his death (in 1996), in 2014.[17] He elaborated on the achievement of such a creative imitation of a canonical model in the context of his work on political myths. Even with a dwindling of metaphysics in modern times, they would have retained a legitimate place as metaphors. A theological narrative such as that of the prophet's wars, with their miraculous victories over the enemies of the faith, could become a prefiguration. Prefiguration gives legitimacy to a decision of an agent. When agents want to be certain in the interpretation of an uncertain situation in

---

[15] On the teachings of Muhammad Qutb compared with those of his brother Sayyid Qutb, see Sabine Damir–Geilsdorf (2003, pp. 294–299).

[16] In the text there is only the letter *t*. This probably means the Arabic word "airplane".

[17] Hans Blumenberg (2014, pp. 9–15), see on this subject: Philipp Stoellger (2000, pp. 3–4, 194–196).

H. G. Kippenberg

which they find themselves and in the rightness of their action, they transform a canonical narrative into a prefiguration of their action. Thus, even in a largely non-metaphysical world, they reduce the arbitrariness and contingency of their actions and assure themselves of their success. The prefigurative narratives become metaphors in a lifeworld horizon.[18]

Ashura

This is what happened in Shiite Islam with the Ashura celebrations. Shiites annually commemorate the violent death of their Imam Husain together with his comrades-in-arms (*mojahedin*) in 680 CE on the 10th of Muharram in the city of Kerbala. At that time, after the death of his father, the caliph Mu^c awiya, Yazid had Husain killed in order to take the caliphate himself, which was Husain's right in the eyes of the Shiites. Husain would have become the savior, the Mahdi, saving the world from evil. Within a few years after these events, Shiites held mourning ceremonies on this day and castigated themselves for abandoning their Imam and thus preventing their salvation by the Mahdi. From Persian sources and European travelogues, we know that Shiites in Iran have staged the Battle of Kerbala in street parades and passion plays every year since the seventeenth century on the 10th of Muharram (the Ashura Day). They bloodily flagellated themselves and imitated battles to express their grief over the victory of the unjust order—an order that will only end with the appearance of the Mahdi at the end of time. "He will fill the world with righteousness and justice, just as it was filled with injustice and unjustness before."[19] The believer still lives in a world filled with injustice (*zulm* = tyranny). Only when Imam Mahdi emerges from his concealment and fills the world with justice *(al-ʿadl)* will this change. This Muharram ritual was performed publicly in post-Safavid times; it created a social space in which the narratives of justice and self-sacrifice, of expressions of grief and misrule were filled with current meanings by the audience, thus re-enacting the ancient ritual (Rahimi, 2012, p. 326).

---

[18] At the same time that Blumenberg was valorizing metaphor and pragmatics at the expense of metaphysics, we see an analogous process in ethnology of religion, albeit in different terminology. Clifford Geertz and Talal Asad each make different kinds of connections between the social lifeworld of believers and their religious relation to the world: Geertz from the religious worldview to action; Asad from social action to the religious worldview. As with Blumenberg, a new means of orientation for action has taken the place of metaphysics.

[19] At the end of the fourteenth century, the North African philosopher Ibn Khaldun compiled in his *Muqaddima* the most important hadiths on the Mahdi expectation, checked their reliability and concluded that they were not genuine. He explained their validity not from theology but otherwise. He interpreted the political uprisings triggered by this expectation also in North Africa as the social power of expanding kinship groups, *asabiyya*. (Ibn Kaldûn, 1967, pp. 257–259).

The rituals were twofold: a commemoration of a historical drama and a public enactment. Therefore, they were always associated with a public discourse about contemporary injustice and the hoped-for just world. The photographs taken by press photographers of the demonstrations against the Shah's regime in 1978/79 show actions borrowed from the Ashura processions and associated passion plays. Not content with venting their anger at the unjust world, the faithful overthrew this order by force to prepare the way for the Mahdi Husain. The Ashura staging of the uprising cannot be rhymed with Max Gluckman's assumption of a cathartic discharge of discontent, which thus preserves the existing order and is incapable of revolution (Gluckman, 1953/1963). Rather, it proves Victor Turner right that a drama is being performed here that humiliates the powerful and empowers the humiliated (Turner, 1969/1989). In the form that was handed down, the faithful recalled the events of old. Now, however, as fellow soldiers of Husain, they are to be ready to die in the struggle for the righteous world.[20] Also in this case one must not stop with the explanation of the meaning model. The Iranian demonstrations took place under the circumstances of the "White Revolution" of the Shah regime. Its implementation destroyed rural social forms, created a class of landless peasants, and brought about a huge growth of destitute migrants in the city of Tehran (Katouzian, 1974; Kazemi, 1980). In the overpopulated cities, it was religious institutions that were most able and willing to assist and absorb the uprooted and disenchanted. Said Amir Arjomand, in a review of the 1960s and 1970s, noted a noticeable increase in religious activities. Religious printed matter and cassettes enjoyed increasing popularity; religious associations, which migrants joined, mushroomed (Arjomand, 1988, pp. 91–94). Shiite institutions still had revenues and assets, even though the Shah's legal reforms had severely weakened them. Thus, the advance of Western modernization in the countryside drove the uprooted in the cities into Shiite networks. Overtaken by Iran's economic development, social Shia became attractive to many of them. Involvement in rural social forms and loyalties was replaced by religious communities.

---

[20] Fischer (1980); Kippenberg (1981); Ourghi (2008).

6 The Zeal of Phinehas: A Jewish and Christian Prefiguration

The Abrahamic religions know similar paradigmatic narratives not only in the case of Islam but also of Judaism and Christianity. Thus, the fourth book of the Bible *Numbers* reports that the people of Israel, after their liberation from Egypt and their covenant with God, befriended pagan Moabite women on their way through the desert and celebrated sacrificial feasts with them. This provoked the wrath of their God. When one of the Israelites, a man named Zimri, then took one of the pagan women into his tent, Phinehas the priest rose up,

> took a lance in his hand, went after the Israelite into the chamber and pierced them both, the Israelite and the woman, through the body. Then the plague among the Israelites was stopped [...]. And the LORD said to Moses, 'Phinehas [...] has turned away my wrath from the Israelites, because he has been jealous among them in my stead. Therefore say, 'Behold, I grant him a covenant of peace: to him and to his descendants shall be granted a covenant of perpetual priesthood, for that he has been jealous for his God and has made atonement for the Israelites'. (Deuteronomy 25)

Thus Phinehas succeeded in averting God's wrath on his people. His exemplary zeal became a model in the revolt of the Jews in the time of the violent Hellenization of Jerusalem in the second century BC against the Seleucid rulers. Jewish fighters, led by the Maccabees, took action against the Greek rededication of the Temple to one dedicated to the Greek god Zeus Olympos (167 BC). In the narrative of the revolt, the example of Phinehas played a key role. The priest Mattathias, the father of Judas Maccabaios, witnessed how a Jewish fellow citizen followed the order of the Greek ruler and wanted to make a sacrifice to Zeus Olympios (Schäfer, 2010, pp. 29–41).

> When Mattathias saw this, he became zealous and his innermost being trembled; rightly did anger rise up in him, he ran and slew him at the altar. At the same time, he killed the king's man who was making the sacrifice and destroyed the altar. Thus he zealously defended the law, as Phinehas had (once) done toward Zimri son of Salu. But in the city [= Modin] Mattathias cried out with a loud voice: 'Everyone who is zealous for the law (and) stands by the covenant, let him follow me.' And he and his sons fled to the mountains; but all their possessions they left in the city. (1 Macc 2: 24–28)

Understanding Religious Radicalization … 339

The fighters were called "zealots" after Phinehas' example.[21] Those who were ready to fight and die against the godless cult and for the Jewish paternal laws were assured of God's help in the battles through His appearance, His *epiphany* (2 Macc 12:28; 14:15), and also of the title of martyr. In the synagogal and ecclesiastical transmission of the Books of Maccabees, these narratives were considered models not only of Jewish but also of Christian resistance to Paganism.

However, even in the situation of the community under the Greek ruler, events did not necessarily always result in a violent uprising against him. Thus, the community of Qumran, the Essenes, separated from the Hellenistic rulers and the Maccabees without the use of violence.

At that time, many who sought justice and righteousness went down to live in the desert. (1 Maccabees 2:29)

The Essenes saw themselves as the holy remnant of Israel with whom God had made his new covenant and hoped to be spared from coming end-time catastrophes thanks to their loyalty to this covenant (Vermes, 1977, Chaps. 6 & 7 ; Kampen, 1988, pp. 66–81). This renunciation of violence was no less rooted in the Jewish religion than the resistance of the activists; only they defined their situation differently under the pagan rulers. Covenant loyalty to their God did not demand rebellion, but the exodus from the kingdom of the ungodly, as Israel once did from Egypt.[22]

When in the first century AD in the Jewish war against Rome of 66–70 AD one of the greater movements bore the designation "Zealots," this was probably not based on an already longer established designation for a religious-political party, which had gone back to the Maccabees, as Martin Hengel assumes,[23] but on an established designation of an exemplary activism similar to Phinehas. This had become the "figura": an amalgamation of old and new meaning, of continuity and innovation (Greenblatt, 2010, pp. 12–15). The exemplary zeal of Phinehas could be adapted to current new conflicts with unbelievers and thus become performative in the sense of Juergensmeyer.

---

[21] Martin Hengel (1961, pp. 61–78 at 77, see also pp. 151–234). Hengel's assumption that this was a religious party already in existence before the conflict is not compelling and has met with rejection. On this point, see Richard A. Horsley (1993, pp. 121–129) concluding (p. 128) that the term 'zeal' "focused primarily on fellow Jews who broke the law in some significant way".

[22] A systematizing treatise on this strategy by the economist and social scientist Albert O. Hirschmann (1970).

[23] See above note 21.

340   H. G. Kippenberg

This also explains the survival of the example. The early Christians, while distancing themselves from the actions of the Zealots of their time, as Richard A. Horsley has shown, paid reverence to the Maccabees. They later ensured the literary transmission of the 1st and 2nd Books of Maccabees[24] as part of the Greek translation of the Hebrew Bible, known as the Septuagint. During the medieval crusades, Christians took this definition of the situation as a model for their acvtivities.[25] In his privilege to the Templars, the Pope motivated the participants with the comparison:

> The milites Templi of Jerusalem, the new Maccabees in the time of grace, have denied the desires of worldliness, have left their possessions behind, have taken up their cross and have followed Christ. ... It is through them that God liberates the Church in the East from the filth of the pagans and casts out the enemies of Christianity. (Auffarth 1994, pp. 368–369)

Like the Maccabees, Christians also fought for the purity of their community, the church.[26] Thus, a Jewish narrative could also become a model in a new situation of the Christian religious community. In the course of centuries, a Christian ideal of a military fighter of faith gradually emerged, who is already carrying out God's judgment on the Gentiles today. In the ancient Church, the Maccabees, leaders of the successful revolt against the Hellenistic rulers, were a living memory. In particular, they were models for the Christian Crusades against the Muslim occupation of the Holy Land. In the Christian tradition, the Maccabees became saints of the Church long before the Crusades (Auffarth, 2002, pp. 131–140). The participants in the Crusades of the tenth and eleventh centuries took their cue from this example (Auffarth, 1994; 2002, pp. 123–148). They felt called on to purify Jerusalem from Islamic desecration. In the twentieth century, Protestant premillenarianism renewed the need for end-time destruction of God's enemies in the Holy Land and supported actions to that end (Hunt, 2001; Boyer, 1992).

The effect of this prefiguration on Jews can even be observed in the assassination of Israeli Prime Minister Rabin in 1994, when, in the course of the Oslo peace process, he was prepared to return to the Palestinians occupied Palestinian land that God had promised to the Jews in the Bible. The assassin,

---

[24] A precise and extensive treatment of the literary-critical and historical problems comes from Christian Habicht (1979).

[25] Christoph Auffarth (1994, pp. 368–369); also Auffarth (2002 Chapter 7: "Auferstanden in den Himmel". Die Makkabäer. jüdische Heilige als Modell für die Kreuzfahrer ["Risen to Heaven". The Maccabees. Jewish Saints as a Model for the Crusaders"] pp. 123–150); Auffarth (2005).

[26] Thomas Sizgorich (2009) explains the militancy in late antiquity not only of Christians but also of Muslims from the violent defense of the identity of their faith communities.

Yigal Amir, saw in Rabin's plan an apostasy from faith in the promise and betrayal of the covenant with God, and shot the prime minister at a peace rally on November 4, 1995 (Karpin & Friedman ,1998). When questioned after his arrest by state investigators, he stated that no rabbi had authorized him to commit this act—although they had presented arguments of justification for it, which he knew—but rather the study of the biblical models of Phinehas and Jael.[27] The young assassin Amir had been inspired by the narrative of Phinehas and his killing of Zimri.[28] It was "culturally mobile."[29]

The Book of Esther records another case of Jewish glorification of violence. In the Persian Empire, persecution and extermination of Jews had once been planned. However, Jews were able to prevent and take revenge with the help of good relations with the court.

> The Jews struck at all their enemies: They slew them with the sword, and killed them, and cut them off; and with them that hated them they dealt as they pleased. (Esther 9:5)

Annually, these events are celebrated in the festival of Purim, in today's Israel sometimes combined with threats and expressions of hatred toward Palestinians (Horrowitz, 2006). Modern conflict trajectories of faith communities confirm basic features of the analysis of previous cases. This is true for the current Middle East conflict between Israelis and Palestinians, which has been defined by both fighting parties with recourse to canonical prefigurations. But Jewish and Arab supporters of peaceful coexistence between Israelis and Palestinians have also invoked prefigurations (Kippenberg, 2016). And Hindus and Buddhists, too, have been guided by prefigurations in conflicts between their religious communities and non-believers; but here, too, not only in their use of, but also in their renunciation of, violence.

In this case, the prefiguration refers to a situation in which a religious community is threatened in its existence by forces of unbelief and believes that it can only avert the danger by acting violently in accordance with a canonical model. The reference to such a prefiguration happens publicly. Mark Juergensmeyer (2017) has spoken of "performative violence" in the handbook *Religion and Violence.* By publicly enacting violent acts in the form of a religious script, the canonical model and the understanding that a religious public has of its own situation enter into an exchange.

---

[27] See on Jael Judges 4:17–24.

[28] On the questioning of Yigal Amir and the unpublished transcripts, see Ehud Sprinzak (1999, p. 281 with notes 77 and 73).

[29] Steven Greenblatt (2010) uses the term *Cultural Mobility* for such discontinuities.

7 From the End of Religious Terrorism

With the linguist Tomis Kapitan, the rhetoric of the fight against terrorism can be summarized thus: those who speak of terrorists cause their listeners to lose the willingness to want to know anything about the reasons for their actions; they divert attention from the fact that their own policies may have contributed to the emergence of the phenomenon; they suggest that it is absurd to negotiate with such people; they separate the act of violence from its rationale; counter-violence and annihilation is the only appropriate response (Kapitan, 2003). I would like to use studies of the end of terrorist groups in general and religious ones, in particular, to examine whether the "war on terror" alone really does make the groups that act in this way disappear. Two detailed studies have critically examined the strategy of a "war on terror" by looking more closely at the end of terrorist groups (Jones & Libicki, 2008; Cronin, 2009). The authors concluded that the military "war on terror" has yielded less than expected. In the vast majority of cases, actions by local forces ended the existence of these groups or the groups changed as they became involved in the political process—assuming the goals the groups sought were specific enough to be subject to negotiation.

For a better understanding of religious violence by jihadist groups, their abandonment of it is particularly interesting. A particularly glaring one leads back to Osama bin Laden's 1998 declaration of war, the *Islamic World Front's call for jihad against Jews and crusaders*, which had also been signed by Egyptian, Pakistani, and Bangladeshi jihadists (Lawrence, 2005, pp. 58–62). The first part raises three accusations against America: the U.S. had occupied the holiest Islamic places in the Arabian Peninsula in order to steal the mineral resources, humiliate the Muslims, and militarily oppress the Muslim peoples; the U.S. had secondly caused serious harm to the Iraqi people through its embargo [after Iraq's war against Kuwait in 1992]; the U.S. thirdly destroyed Iraq and wanted to dissolve the states of the region into defenseless small states in order to guarantee Israel's superiority. This first part justifies the declaration of war with political and military events that are also viewed critically by non-Muslim observers of Western interventions in the Middle East.[30] But to take them as a declaration of war on God and to take the defense of Islam as the supreme duty of all believers—formally cast in the form of a legal judgment (*ḥukm*)—gives the events a religious frame of reference and thus justifies extreme violence against Americans and Jews.

---

[30] See Michael Lüders (2015, esp. 30–31, 42–45; 2017).

It is an individual duty of faith of every Muslim in every country to kill Americans and their allies, civilians and military, to liberate the al-Aqsa Mosque in Jerusalem and the Holy Mosque of Mecca from their grip, so that all their armies withdraw from the Islamic world, defeated and unable to threaten any Muslim yet. (Lawrence 2005, p. 61)

Finally, the declaration exhorts Muslims not to hang on to their lives in the face of the inevitability of war. Do they really prefer life in this world to life in the world to come? Then their salvation would be at stake.

Shortly after Rifᶜai Aḥmad Ṭaha signed the statement as representative of the Egyptian *Ǧamāᶜa al-Islāmiyya*, he was forced to withdraw his signature at the intervention of the organization's leadership. On June 5 of the previous year, 1997, during a military trial of *Ǧamāᶜa activists in* Egypt, one of the defendants had read out a statement proclaiming a unilateral cease-fire. The leadership council (*maǧlis aš-šūrā*) of the *Ǧamāᶜa*, in prison since Sadat's assassination in 1981, had decided on the initiative (*mubādara*) or revision (*murāǧaᶜa*)—with the approval of its emir, the blind Sheikh ᶜUmarᶜAbdar-Raḥmān, who is serving a life sentence in the U.S.[31]

Ivesa Lübben and Issam Fawzi have reconstructed this process in detail. Under the heading "The Violence Dilemma," the authors describe how in the mid-1980s the middle cadres of the *Ǧamāᶜa* were released from prison, built their own mosques, developed charitable initiatives, and at the same time persisted in violent actions and thus in confrontation with the Egyptian state. For their part, Egypt's state organs retaliated with punitive actions, while at the same time members of the grouping decided to participate in the jihad in Afghanistan. Those responsible, many of them in prison, took the initiative for a cease-fire in order to secure the necessary protection for the families of the prisoners, ensure their reintegration into Egyptian society, secure the promotion of the faith (*daᶜwā*) in the interest of Islam, and respond to the challenge facing Egypt and the Islamic world (Lübben & Fawzi, 2004, pp. 18–19). A continuation of violence would have called Egypt's secular and Coptic citizens to action by the state apparatus against Islam; growing social problems (drugs, gang activity, immorality) also spoke to this. This required a new definition of the situation in which one found oneself. The rejection of violence was given Islamic legitimacy. The situation was no longer interpreted baldly as ǧāhiliyya (a time of ignorance as before the revelation of Islam). Instead, the *Ǧamāᶜa* sought to establish a relationship with reality through a concept of law that named prerequisites for making one's own

---

[31] Ivesa Lübben and Issam Fawzi (2004, pp. 4–5.). See also Roel Meijer (2009); Mariella Ourghi (2010).

judgments in practicing sharia. To this end, it fell back on the legal concept of the common good (*maṣlaḥa*), which has been strongly revalued in modern times. The question was whether or not violence served the common good of the Islamic community (*umma*) (Lübben & Fawzi, 2004, p. 22). The term *maṣlaḥa* makes it possible to weigh the implementation of the sharia in terms of its advantages and disadvantages for the common good.[32] Consequences that are detrimental to the community of believers may result from actions that are, in principle, legitimized by sharia. The following legal-theoretical reasons for renouncing violence are cited: the judgment of legal scholars, the weakness of Muslims, a peace treaty with the enemy, and others (Lübben & Fawzi, 2004, p. 21). The two authors of the study see this as only a tactical rejection of violence. A real revision of the discourse on violence should have started from a principled rejection of violence, not from weighing the advantages and disadvantages for the community. In fact, however, the processes in which jihadist groups engage in a political process are grounded in such trade-offs (Cronin, 2009, pp. 35–72). One author who was involved in this revision—Na ǧih Ibrahim—was the focus of Johanna Pink's attention. He expressed his views in columns in various Egyptian newspapers between May 2011 and March 2015 during the Arab Spring. He considered the task of unflinching preaching more important than military combat, interpreting Muhammad in light of the prophet Jesus, also known to the Quran. Even suffering and persecution must not detract from the preaching and do not permit violence against its perpetrators (Pink, 2017).

Summa summarum: There is therefore no shortage of rules and models for religiously justified violence. Therefore, one cannot categorically deny that religions can cause, justify and aggravate violent conflicts. That religions are violent, however, coarsen the facts instead of understanding them. Rather, one will have to call them ambivalent in their manifestations, which results in different and conflicting options for action for individual Muslims and makes the individual the key to the question of whether Islam as a whole has embraced the value of violence (Kippenberg, 2008).

References

Afsaruddin, A. (2013). *Striving in the path of god.* Oxford University Press.
Afsaruddin, A. (2015). *Contemporary issues in Islam.* University Press.

---

[32] For an account of the history of this concept in Islamic legal thought past and present, see: Felicitas Opwis (2007).

Afsaruddin, A. (2016). Jihad and Martyrdom in Islamic thought and history. In *Oxford research encyclopedia religion* (March 2016). Online: https://doi.org/10. 1093/acrefore/9780199340378.013.46. Published online: 03 March 2016.

Arjomand, A. (1988). *The turban for the crown. The Islamic revolution in Iran.* Oxford University Press.

Auffarth, C. (1994). Die Makkabäer als Model der Kreuzfahrer. Usurpationen und Brüche in der Tradition eines jüdischen Heiligenideals [The Maccabees as model of the crusaders. Usurpations and fractures in the tradition of a Jewish ideal of Saints] In Elsas, C. et al. (Eds.), *Tradition und translation. FS Carsten Colpe* (pp. 362–390). Walter de Gruyter.

Auffarth, C. (2002). *Irdische Wege und himmlischer Lohn. Kreuzzug, Jerusalem und Fegefeuer in religionswissenschaftlicher Perspektive* [Earthly ways and heavenly rewards. Crusade, Jerusalem and purgatory in religious studies perspective]. Vandenhoeck & Ruprecht.

Auffarth, C. (2005). Heilsame Gewalt? Darstellung, Begründung und Kritik der Gewalt in den Kreuzzügen [Salvific violence? Representation, justification, and critique of violence in the crusades]. In M. Braun & C. Herberichs (Eds.), *Gewalt im Mittelalter. Realitäten—Imaginationen.* [Violence in the middle ages. Realities—Imaginations] (pp. 251–272). Wilhelm Fink Verlag.

Blumenberg, H. (2014). *Präfiguration. Arbeit am politischen Mythos.* In A. Nicholls & F. Heidenreich (Eds.), [Prefiguration. Working on political myth]. Suhrkamp.

Boyer, P. (1992). *When time shall be no more. Prophecy belief in modern American culture.* Harvard University Press.

Burkert, W. (1972). *Homo Necans. Interpretationen altgriechischer Opferriten und Mythen* [*Homo Necans. Interpretations of ancient greek sacrificial rites and myths*]. Walter de Gruyter.

Cook, D. (2002). Suicide attacks or 'Martyrdom operations' in contemporary jihad literature. *Nova Religio 6*, 7–44.

Cook, D. (2005). *Understanding jihad.* University of California Press.

Cronin, A. K. (2009). *How terrorism ends. Understanding the decline and demise of terrorist campaigns.* Princeton University Press.

Dahme, H.-J. (1988). Der Verlust des Fortschrittglaubens und die Verwissenschaftlichung der Soziologie. Ein Vergleich von Georg Simmel, Ferdinand Tönnies und Max Weber [The loss of faith in progress and the scientificion of sociology. A comparison of Georg Simmel, Ferdinand Tönnies, and Max Weber]. In O. Rammstedt (Ed.), *Simmel und die frühen Soziologen. Nähe und Distanz zu Durkheim, Tönnies und Max Weber* [*Simmel and the early sociologists. Proximity and distance to Durkheim, Tönnies and Max Weber*] (pp. 222–274). Suhrkamp.

Damir-Geilsdorf, S. (2003). *Herrschaft und Gesellschaft. Der islamische Wegbereiter Sayyid Qutb und seine Rezeption.* [*Rule and society. The islamic pioneer Sayyid Qutb and his reception.*]. Ergon.

Dawkins, R. (2006). *The god delusion.* Bantam Press.

Esser, H. (1996). Die Definition der Situation [The definition of the situation]. *Kölner Zeitschrift für Soziologie und Sozialpsychologie [Cologne Journal of Sociology and Social Psychology], 48*, 1–34.

Esser, H. (1999). *Soziologie. Spezielle Grundlagen. Band 1, Situationslogik und Handeln* [*Sociology. Special foundations. Vol. 1, Situational logic and action*]. Campus.

Esser, H. (2001). *Soziologie. Spezielle Grundlagen. Band 6, Sinn und Kultur [Sociology. Special foundations. Vol. 6, meaning and culture].* Campus.

Esser, H., & Kroneberg, C. (2015). *An integrative theory of action: The model of frame selection.* Cambridge University Press.

Fielding, N., & Fouda, Y. (2003). *Masterminds of terror. The truth behind the most devastating terrorist attack the world has ever seen.* Mainstream Publishing.

Fischer, M. (1980). *From religious dispute to revolution.* Harvard University Press.

Girard, R. (1972/1979). *Violence and the sacred.* Johns Hopkins University Press.

Gluckman, M. (1953/1963). Rituals of rebellion in South East Africa. In *Order and rebellion in tribal Africa* (pp. 110–136). Routledge.

Greenblatt, S. (Ed.). (2010). *Cultural mobility. A manifesto.* Cambridge University Press.

Habermas, J. (2009). Die Revitalisierung der Weltreligionen—Herausforderung für ein säkulares Selbstverständnis der Moderne? [The revitalization of world religions—Challenge for a secular self-understanding of modernity?] In J. Habermas (Ed.), *Philosophische Texte, Band 5. Kritik der Vernunft. [Philosophical Texts, Vol. 5. Critique of Reason]* (pp. 387–407). Suhrkamp.

Habicht, C. (1979). Makkabäerbuch [Maccabees book] In: *Jüdische Schriften aus hellenistisch-römischer Zeit* [Jewish writings from hellenistic-roman times], vol. 1 (pp. 167–198). Mohn 1979, 167–198.

Harris, S. (2004). *The end of faith—Religion, terror and the future of reason.* W. W. Norton.

Hegghammer, T. (2006/2008). Abdullah Azzam, the Imam of Jihad. In G. Kepel & J.-P. Milelli (Eds.), *Al Qaeda in its own words* (pp. 81–101 and notes pp. 318–340). Harvard University Press.

Hegghammer, T. (2010/2011). The rise of muslim foreign fighters: Islam and the globalization of Jihad. *International Security 35*, 53–94.

Hengel, M. (1961). *Die Zeloten. Untersuchungen zur jüdischen Freiheitsbewegung nin der Zeit von Herodes* [The zealots. Studies on the jewish freedom movement in the time of herod]. Brill.

Hirschmann, A. O. (1970). *Exit, voice and loyalty. Responses to decline in firms, organizations and states.* Harvard University Press.

Hitchens, C. (2007). *God is not great. How religion poisons everything* . Twelve Books.

Horowitz, E. (2006). *Reckless rites, purim and the legacy of Jewish violence.* Princeton University Press.

Horsley, R. A. (1993). *Jesus and the spiral of violence. Popular jewish resistance in roman palestine.* Fortress.

Hunt, S. (Ed.). (2001). *Christian millenarianism. From the early church to Waco*. Hurst.

Ibn Khaldûn (1967). *An introduction to history. The Muqaddimah*. Trans. Franz Rosenthal and abridged by N. J. Dawood. Routledge and Kegan Paul.

Joas, H. (1996). *Die Kreativität des Handelns* [The creativity of action]. Suhrkamp.

Jones, S. G., & Libicki, M. C. (2008). *How terrorist groups end: Lessons for countering al Qa'ida*. RAND Corporation.

Juergensmeyer, M. (2017). Religious terrorism as performance violence. In M. Juergensmeyer, M. Kitts, & M. Jerryson (Eds.), *The Oxford handbook of religion and violence* (pp. 280–292). Oxford University Press.

Juergensmeyer, M., Kitts, M., & Jerryson, M. (Eds.), (2017). *Violence and the world religions. An introduction*. Oxford University Press.

Kampen, J. (1988). *The hasideans and the origin of pharisaism. A study in 1 and 2 maccabees*. Scholars Press.

Kapitan, T. (2003). The terrorism of 'Terrorism'. In J. Sterba (Ed.), *Terrorism and international justice* (pp. 47–66). Oxford University Press.

Karpin, M., & Friedman, I. (1998). *Murder in the name of god: The plot to kill Yitzhak Rabin*. Granta.

Katouzian, H. (1974). Land reform in Iran. A case study in the political economy of social engineering. *Journal of Peasant Studies, 1*, 220–239.

Kazemi, F. (1980). *Poverty and revolution in Iran. The migrant poor, urban marginality and politics*. New York University Press.

Kippenberg, H. G. (1981). Jeder Tag 'Ashura, jedes Grab Kerbala. Zur Ritualisierung der Straßenkäm¬pfe im Iran. [Every day 'Ashura, Every grave kerbala. On the ritualization of street fighting in Iran]. In K. Greussing (Ed.), *Religion und Politik im Iran. Religion and politics in Iran* (pp. 217–256). Syndikat.

Kippenberg, H. G., & Seidensticker, T. (Eds.). (2006). *The 9/11 handbook. Annotated translation and interpretation of the attackers' spiritual manual*. Equinox.

Kippenberg, H. G. (2008). Die Macht religiöser Vergemeinschaftung als Quelle religiöser Ambivalenz [The power of religious communalization as a source of religious ambivalence]. In B. Oberdorfer & P. Waldmann (Eds.), *Die Ambivalenz des Religiösen. Religionen als Friedenstifter und Gewalterzeuger* [The ambivalence of the religious. Religions as peacemakers and generators of violence] (pp. 53–76). Rombach.

Kippenberg, H. G. (2016). Religious definitions of the Middle East conflict: Frames of reference—violent action scripts—global diffusion. *Journal of Politics, 63*, 65–92.

Lawrence, B. (Ed.). (2005). *Messages to the world. The statements of Osama bin Laden*. Translated by James Howarth. Verso.

Lohlker, R. (2006). *Islamisches Völkerrecht* [Islamic international law]. Kleio Humanities.

Lohlker, R. (2016). *Theologie der Gewalt. Das Beispiel IS*. [Theology of violence. The example of IS]. UTB GmbH.

Lohlker, R., & Abu-Hamdeh, T. (Eds.). (2013). *Jihadi thought and ideology*. Logos.

Lübben, I., & Fawzi, I. (2004). *Die ägyptische Jama'a al-Islamiya und die Revision der Gewaltstrategie* [The Egyptian Jama'a al-Islamiya and the Revision of the strategy of violence]. Deutsches Orient Institut.

Lüders, M. (2015). *Wer den Wind sät. Was westliche Politik im Orient anrichtet* [He who sowed the wind. What Western politics is doing to the Orient]. C.H. Beck.

Lüders, M. (2017). *Reaping the storm. How the West plunged Syria into chaos*. C.H. Beck.

MacEoin, D. (2008). Tactical Hudna and Islamist Intolerance. *Middle East Quarterly, 15*, 39–48.

McDermott, T. (2006). *Perfect soldiers. The 9/11 hijackers: Who they were, why they did it*. Harper Perennial.

McGraw Donner, F. (1981). *The early Islamic conquests*. Princeton University Press.

Meijer, R. (2009). Commanding right and forbidding wrong as a principle of social action: The case of the Egyptian al-Jamaᶜa al-Islamiyya. In R. Meijer (Ed.), *Global Salafism: Islam's new religious movement* (pp. 189–220). Hurst.

Merton, R. K. (1995). The Thomas theorem and the Matthew effect. *Social Forces, 74*, 379–424.

Opwis, F. (2007). Islamic law and legal change: The concept of Maslaha in classical and contemporary Islamic legal theory. In A. Amanat & F. Griffel (Eds.), *Shari'a. Islamic law in the contemporary context* (pp. 62–82). Stanford University Press.

Ourghi, M. (2008). *Schiitischer Messianismus und Mahdi-Glaube in der Neuzeit*. (Shiite Messianism and Mahdi Faith in Modern Times). Ergon.

Ourghi, M. (2010). *Muslimische Positionen zur Berechtigung von Gewalt. Einzelstimmen, Revisionen, Kontroversen* [Muslim positions on the justification of violence. Individual voices, revisions, controversies]. Ergon.

Pink, J. (2017). Helden der Verkündigung, Helden des Kampfes. Naǧih Ibrahim und die ägyptische Ǧama'a islamiyya [Heroes of the Annunciation, Heroes of the Struggle. Naǧih Ibrahim and the Egyptian Ǧama'a islamiyya]. In F. Heinzer, J. Leonhard, & Ralf von den Hoff (Eds.), *Sakralität und Heldentum* [Sacrality and heroism] (pp. 245–262). Ergon.

Rahimi, B. (2012). *Theater state and the formation of early modern public sphere in Iran. Studies on Safawid Muharram Rituals, 1590–1641 CE*. Brill.

Ramadan, T. (2004). *Western muslims and the future of Islam*. Oxford University Press.

Rüpke, J. (2021). 'Systematische Religionswissenschaft und 'Religionsgeschichte'. Von Wach zu Gladigow ['Systematic religious studies' and 'History of religion'. From Wach to Gladigow]. In C. Auffarth, A. Koch, et al. (Eds.), *Religion in Kultur—Kultur in Religion: Burkhard Gladigows Beitrag zum Paradigmen-Wechsel in der Religionswissenschaft* [Religion in culture—Culture in religion: Burkhard Gladigow's contribution to shifting paradigms] (pp. 69–87). University Press.

Salvatore, A. (2009). Qaradawi's maslaha: From ideologue of the islamic awakening to sponsor of transnational public Islam. In J. Skovgaard-Peterson & B.

Gräf (Eds.), *Global Mufti. The phenomenon of Yusuf al-Qaradawi* (pp. 239–255). Hurst.

Salvatore, A., & Eickelman, D. F. (Eds.). (2004). *Public Islam and the common good*. Brill.

Schacht, J. (1964). *An introduction to Islamic law*. Clarendon Press.

Schäfer, P. (2010). *Geschichte der Juden in der Antike. Die Juden Palästinas von Alexander dem Großen bis zur arabischen Eroberung.* . [History of the jews in antiquity. The jews of palestine from Alexander the great to the Arab conquest]. Mohr Siebeck.

Schluchter, W. (1988). Gesinnungsethik und Verantwortungsethik: Probleme einer Unterscheidungn [Ethics of commitment and ethics of responsibility: Problems of a distinction]. In *Religion und Lebensführung* [Religion and Lifestyle]. Vol. 1: *Studien zu Max Webers Kultur- und Werttheorie*. [Studies on Max Weber's theory of culture and values] (pp. 165–200). Suhrkamp Verlag.

Schulze, R. (1998/2016). *A modern history of the Islamic world* [Geschichte der islamischen Welt im 20 Jahrhundert]. I.B. Tauris 2012.

Sizgorich, T. (2009). *Violence and belief in late antiquity. Militant devotion in Christianity and Islam*. University of Pennsylvania Press.

Sommer, A. U. (2016). *Werte: Warum man sie braucht, obwohl es sie nicht gibt* [Values: Why you need them even though they don't exist]. J.B. Metzler.

Sprinzak, E. (1999). *Brother against brother. Violence and extremism in Israeli politics from Altalena to the Rabin assassination*. The Free Press.

Steinberg, G. (2005). *Der nahe und der ferne Feind: Die Netzwerke des islamistischen Terrorismus.* (The near and the far enemy: The networks of islamist terrorism). C.H. Beck.

Stern, J. (2009). *Terror in the name of god. Why religious militants kill*. Ecco 2004.

Stoellger, P. (2000). *Metapher und Lebenswelt. Hans Blumenbergs Metaphorologie als Lebenswelthermeneutik und ihr religionsphänomenologischer Horizont*. [Metaphor and lifeworld. Hans Blumenberg's metaphorology as lifeworld hermeneutics and its religious phenomenological horizon]. Mohr Siebeck.

Turner, V. (1989). *Das Ritual. Struktur und Antistruktur.* (The ritual. Structure and anti-structure). Campus.

Vermes, G. (1977). *The dead sea scrolls. Qumran in Perspective*. Collins 1977.

Weber, M. (1904/1988). Die 'Objektivität' sozialwissenschaftlicher und sozialpolitischer Erkenntnis [The 'objectivity' of social scientific and socio-political knowledge]. In M. Weber (Ed.), *Gesammelte Aufsätze zur Wissenschaftslehre* [Collected essays on the theory of science] (pp. 146–214). J.C.B. Mohr (Paul Siebeck).

Weber, M. (1922/2001). *Wirtschaft und Gesellschaft Vol. 2: Religiöse Gemeinschaften* [Economy & Society. Vol. 2: Religious communities]. In H. G. Kippenberg & P. Schilm (Eds.), with the assistance of Jutta Niemeier. J.C.B. Mohr (Paul Siebeck).

Weber, M. (1922/1968). *Economy and society*. Guenther Roth and Claus Wittich (Eds.), University of California Press.

Weber, M. (1915–1920/1989). *Die Wirtschaftsethik der Weltreligionen. Konfuzianismus und Taoismus*. [The economic ethics of world religions. Confucianism and Taoism]. Edited by Helwig Schmidt-Glintzer in collaboration with Petra Kolonko. Mohr (Siebeck).

Windelband, W. (1894/1907). Geschichte und Naturwissenschaft [History and Natural Science] In Wilhelm Windelband *Präludien. Aufsätze und Reden zur Einleitung in die Philosophie* [Preludes. Essays and speeches on the introduction to philosophy] (pp. 355–379). J.C.B. Mohr.

Zaman, M. Q. (2004). The ulama of contemporary Islam and their conceptions of the common good. In A. Salvatore & F. D. Eickelman (Eds.), *Public Islam and the common good* (pp. 129–155). Brill.

Risk Takers or Rational Conformists: Extending Boudon's Positional Theory to Understand Higher Education Choices in Contemporary China

Ye Liu ⓘ

1 Introduction

Bourdieu's cultural reproduction theory continues to fascinate contemporary sociologists seeking answers to the persistent educational inequality across different social contexts (Liu, 2018; Reay et al., 2009; Van de Werfhorst & Hofstede, 2007). Comparatively speaking, Raymond Boudon's positional theory—which extends Bourdieu's cultural capital thesis to understand social differentials through educational choices—is still under-explored and under-researched in the contemporary sociology of education. In Boudon's (1974) positional theory of 'primary and secondary effects', social reproduction in education occurs through a dual process. Primary social reproduction occurs through the direct influence of a family's cultural capital on the child and his or her ability to achieve in school. However, social reproduction also occurs through secondary effects, whereby the impact of families' cultural capital is

---

This article was previously published: Liu, Y. Choices, risks and rational conformity: extending Boudon's positional theory to understand higher education choices in contemporary China. High Educ 77, 525–540 (2019). https://doi.org/10.1007/s10734-018-0285-7. https://creativecommons.org/licenses/

Y. Liu (✉)
Department of International Development, King's College London, London, UK
e-mail: ye.liu@kcl.ac.uk

© The Author(s), under exclusive license to Springer Nature Switzerland AG 2023
N. Bulle and F. Di Iorio (eds.), *The Palgrave Handbook of Methodological Individualism*, https://doi.org/10.1007/978-3-031-41508-1_15

mediated by choices students make about their educational careers. In turn, these choices influence their future educational outcomes.

Boudon's theoretical approach is of particular relevance to understanding educational choices at a tertiary level for two main reasons. First, the availability of choices is comparatively more abundant at the transition to tertiary education than within earlier stages of schooling. These choices include types of institutions, fields of study, modes of provision, geographical preference and international institutions (Liu, 2017). Second, the expansion of higher education systems since the 2000s has magnified the provision of choices at both the national and international levels (Marginson, 2016). The nature of the transition to higher education allows more room for students' choices to have an impact on their ultimate achievements, thus increasing the space for cultural capital to intensify the process of social reproduction. In an attempt to better understand social inequality in choices and strategies in higher education—a topic overshadowed by the Bourdieuian debate's focus on the rigid cultural reproduction through education—this study extends Boudon's positional theory to the Chinese context and investigates how students from different social origins make choices regarding higher education.

Existing research that extends Boudon's theory of educational inequality in new contexts has focused on comparing the differences between primary effects of cultural capital and secondary effects of educational choices (Boado, 2010; Jackson et al., 2007; Jackson & Jonsson, 2013; Nash, 2006), thereby contributing to the on-going debates regarding whether primary or secondary effects play a larger role in educational inequality. On the one hand, some studies attribute persistent inequality in education to primary effects of cultural reproduction through an individual family's cultural activities and acquisition, since educational choices are conditioned upon academic performance (Boado, 2010; Nash, 2006). On the other hand, some scholars argue that secondary effects play a more important role, since students from lower socioeconomic backgrounds do not translate their academic performance to the same level of ambition as their privileged counterparts through their educational choices (Erikson & Rudolphi, 2010; Jackson et al., 2007; Jackson, 2012).

However, we know very little about how students interpret their socio-cultural positions during the choice-making process and how students from different social origins assess the risks involved in navigating through choice systems. In light of this gap, this research explores how students from different social origins navigate through the choice system and how they interpret risks during the choice-making process in contemporary China. 1 draw upon 71 interviews with university students from the birth cohorts between 1995 and 1997, asking how the students from different backgrounds make choices about university and interpret risks while navigating through the three-choice and quota systems. My data reveal that individuals' family characteristics and geographical origin manifest in the process of choices and strategies. Furthermore, the students from working-class and agricultural families from non-metropolitan areas are doubly punished by a lack of social and cultural resources and by the institutional discrimination hidden in the quota system. When hope and possibility contract, these students reduce to internalize their socioeconomic and geographical disadvantages and come to terms with a lack of equal opportunities in a seemingly meritocratic system.

2　Theoretical Framework

Boudon's positional theory argues that students make different choices 'according to their position in the stratification system' (Boudon, 1974, pp. 36). This position is further elaborated in two dimensions: sociocultural identity and economic rationale. The former means that students make decisions that are shaped and constrained by their family characteristics and identity (Brooks, 2008; Glaesser & Cooper, 2013; Reay et al., 2009). The latter is often argued to be the 'rational' choice, which calculates the economic costs and benefits of a university degree given extant resources and maximum long-term returns (Breen & Goldthorpe, 1997; Boudon, 2006). This dual position is interdependent during the process of decision-making; however, a person's sociocultural circumstances affect his or her rational choices, thus modifying pure economic rationality (Boudon, 2003). Differing from Bourdieu's thesis of rigid cultural reproduction and cultural determinism through cultural capital, habitus and field (Bourdieu & Passeron, 1977), Boudon draws our attention to the choice-making process, which interrogates the competing and complementary roles played by socioeconomic status, cultural capital, identity and rationality in shaping educational choices.

Prior literature on Boudon's theory follows the thesis of the rational choice perspective, which examines the direct impact of individual families' socioeconomic backgrounds and financial and cultural resources on students' choices in higher education (Brooks, 2008; Kleanthous, 2014; Reay et al., 2009). It has been shown that students from professional families tend to choose academic pathways and go to elite universities, while their working-class counterparts, even those of the same ability, are more likely to select vocational courses or less prestigious institutions (Duru-Ballet et al.,

2008; Thomsen et al., 2013). Moreover, students from culturally rich families tend to choose fields of study to strengthen their cultural advantages (Van de Werfhorst et al., 2003). By contrast, students without rich cultural resources, such as those from working-class families and migrant origins in some contexts, tend to choose fields such as engineering or mathematics, which allow them to compensate for a lack of cultural resources (Kleanthous, 2014).

Similarly, some studies have shown that students' choices are motivated by rational cost-return calculations of a degree (Davies et al., 2014; Hartog et al., 2010; Wilkins et al., 2012). This is true particularly for students from working-class families who base their decisions on information regarding the economic returns of a degree (Clark et al., 2015, Thompson & Simmons, 2013). Given the rising cost and commodification of higher education in some countries like the UK (Wilkins et al., 2013), students from working-class backgrounds are concerned with choosing those fields with good 'value for money', according to some recent research (Clark et al., 2015; Thompson & Simmons, 2013). The latent impact of socioeconomic backgrounds and cultural capital can be observed in information-seeking patterns and peer effects on students' choices about higher education (Wilksons et al., 2013).

Another stream of literature has linked cultural identity to the patterns of students' choices in universities and fields of study (Reay et al., 2009; Jetten et al., 2008). In some studies, cultural identity is analysed in relation to individual students' position in the social structure. The French sociologist Duru-Bellat finds that students from working-class backgrounds tend to choose vocational pathways instead of universities mainly because academic pathways are seen to be at odds with their cultural identity (2010). Similar results are confirmed in the research on the educational choices made by children of immigrant origin (Boado, 2010; Kleanthous, 2014). Kleanthous's study shows that students from immigrant origins are much less likely to make ambitious choices than their non-immigrant counterparts, partly because of a lack of identity fit in universities in Cyprus (2014). Boado's research confirms the students from migrant origin were conservative about choices of academic track of elite upper secondary schooling in France (2010). Some research also highlights students' geographical identity in the process of choice-making (Donnelly & Evans, 2016; Holt, 2011). For instance, cultural identity can be interpreted as a strong attachment to geographical identity and locality, which shapes some Welsh students' choices of higher education (Donnelly & Evans, 2016). In the case of China and Australia, rural students have a different cultural identity from their urban

Risk Takers or Rational Conformists: Extending ... 355

counterparts, and this geographical distinction not only affects their choice-making about higher education but also involves constant negotiations of their identity in integrating into urban universities (Holt, 2011; Li, 2012).

We know that students' choices in higher education are affected by the rational calculation of economic and cultural resources, and we also know that their choices are shaped by their cultural or geographical identity. Nevertheless, much less is known about how the two lines of action based on rational calculations and cultural identity complement or contradict each other. In addition, we are not aware of the specific mechanisms through which the risks are assessed by students from different social origins and mediated by cultural identity and rational choice during the decision-making processes. Therefore, I seek to address the aspect of risk assessment in the choice-making process.

Coleman's pioneering research on school cultures in the US demonstrates the role of the 'decision-to-conform' mechanism (Coleman, 1961; Coleman et al., 1966) in shaping the academic differences between minority students in different schools (Coleman, 1990). His empirical analysis shows that minority students achieved better academic outcomes in 'high socioeconomic status schools' where their peers from better-off and well-educated families tended to drive up academic aspirations (Coleman, 1990). By contrast, the adolescent subculture in state schools encouraged minority students to pursue popularity and sports excellence rather than academic achievements (Coleman, 1990). Thus, Coleman argues that the desire to conform to the school culture explains the over-achievement and under-achievement of minority students. The decision-to-conform is further developed statistically by Breen and Goldthorpe as the 'relative risk aversion' (Breen & Goldthorpe, 1997, pp. 283; 1999). Differing from Coleman's qualitative approach, the relative risk aversion formulates the models of educational choices statistically and interprets the educational strategies adopted by both middle-class and working-class families as a way to avoid the risks of status decline and maintain stability in the social structure (Breen & Goldthorpe, 1997, 1999).

The decision-to-conform mechanism and the risk-aversion strategy are further unpacked in Yair's analysis of how the evaluation of risks affects the nature of choices in the course of rational action or expressive action (Yair, 2007). Drawing inspiration from Swidler's culture as a tool kit thesis (Swidler, 1986), Yair highlights the contradictions and compromises between two lines of action based on the rational calculation of instrumental utility and conscious conformity to cultural norms and values (Yair, 2007). These two lines of action are not always complementary; rather, risks arise when the action of rational utility is not informed adequately by cultural norms and values. Therefore, he argues that the rational course of action often gives in to conformity choices in unsettled times (Swidler, 1986), which require constant pragmatic adaptations to changing circumstances (Yair, 2007). Thus, risks associated with different choices are most likely to motivate individuals to conform to their own status and identity instead of taking bold action to maximize their future opportunities.

Risk assessment is particularly relevant in the Chinese context of the choice system, where the penalty for mismatching choices with academic performance has costs not only in terms of the opportunities associated with the prestige of universities or fields but also, and more importantly, the access to higher education. The next section will examine the complex three-choice system in access to higher education in China, how students from different social backgrounds assess the associated risks and how they adopt strategies to negotiate the conflicts between rational choice and cultural conformity. I will begin by highlighting some key features of higher education system and university choices in China, which qualify as a strategic case to examine the theoretical standpoints of Boudon's positional theory.

3 Stratification of Higher Education and University Choices in Contemporary China

China offers some attractive attributes as a case with which to examine the complexity of university choices. First, the massive expansion of higher education since the 1990s has resulted in an increasingly stratified system, with elite universities at the top and a large number of institutions at the provincial level (Liu, 2015). Elite universities have resisted the market pressure to expand their recruitments by free-riding the State's elite programmes, such as 'World-Class Universities' or 'Double First-Class' initiatives (Marginson, 2016; Liu, 2018, 2016). These universities reap substantial financial and opportunity benefits from the State's investment in key fields of study; thus, they further distinguish themselves from the massive production of graduates experienced by the comprehensive and provincial institutions (Carnoy et al., 2014).

Second, in addition to the hierarchical system, the fields of study have become differentiated within the same tier of higher education institutions as well as between different tiers (Li, 2012; Liu, 2015). Technology, natural sciences and engineering are comparatively more selective than Arts, Humanities and Social Sciences, as measured by their enrolment criteria (Liu, 2016) and evidenced by their labour market returns (Guo et al., 2010; Hartog et al.,

2010). The relatively higher value of these fields of study is made further apparent by the fact that nearly 43% of elite universities (985) specialize in technology, science, or engineering (MOE, 2021). Moreover, the same fields of study have varied market returns according to prestige, as the graduates from elite universities have much higher earning potential compared to those from less well-known institutions (Hartog et al., 2010; Marginson 2016).

The increasingly hierarchical differentiation of pathways by different tiers and fields of study allows for more space for choices as well as more risks (Liu, 2018). The complex choice system and the associated risks deserve some contextualization. After senior secondary graduates and other eligible candidates obtain the *Gaokao* (the Entrance Examinations to Higher Education) results, they are required to list their choices in the *University and Field Forms*, which are submitted to the Ministry of Education at the provincial or local level. Table 1 provides a sample of the form, which details the nature of the choices of universities and fields of study. The three-choice system in this article is used to illustrate the choices at both the vertical level of types of universities and the horizontal level of the order of specific fields of study. For the former, the three preferable institutions are identified in each tier of the system—namely, the key universities, the non-key institutions and the non-degree institutions. Under each institution, students are required to list at least three preferred fields of study. However, it should be noted that the number of choices varies across provinces, which is often not limited to three. For instance, in Shanghai, students can submit 4 choices for elite and key universities, 6 for non-key universities and a further 6 choices of fields of study in each category.

In theory, individual institutions choose students based on their academic performance in the *Gaokao*. That is, the higher a student's *Gaokao* scores are, the more likely it is that the student will be accepted. However, the three-choice system complicates the entire choice and selection process. The horizontal sequence of the three institutions listed on each tier is of great importance. The choices regarding the sequence of the three institutions involve risk-taking. The first choice of a university in each tier is crucial. The scenario of risks emerges from this point of decision-making. Students will risk being rejected by their first choice if their *Gaokao* scores fail to meet the selection criteria of their first-choice university (Loyalka et al., 2012). It is also likely that students will not be accepted by universities that they list as their second or third choice. This results from the severe competition between universities in the same tier, as each individual institution prefers to be the first choice (Li et al., 2012). Hence, universities penalize those who list their institutions as the second or third choice by raising the admission threshold

358 Y. Liu

Table 1 A sample of the university and field form

Universities with priority selection rights including military colleges, national security colleges and teachers' training colleges

Institution	Field		
1	1	2	3
2	1	2	3
3	1	2	3
Tier 1 Key Higher Education Institutions (the 985 elite and 211 key universities)			
Institution	Field		
1	1	2	3
2	1	2	3
3	1	2	3
Tier 2 Non-Key Higher Education Institutions			
Institution	Field		
1	1	2	3
2	1	2	3
3	1	2	3
Tier 3 Thee-year Certificate Colleges (Vocational and Technical colleges)			
Institution	Field		
1	1	2	3
2	1	2	3
3	1	2	3

Note The majority of the provinces use similar forms for types of universities and fields of study. However, the Ministry of Education in Shanghai, for instance, includes 4 choices for each institution, and further 6 fields of study for each institution
Source the Ministry of Education in Shanghai 2015

by at least 50 points (Loyalka, 2009). It is of great importance that students are able to make sensible choices of universities; otherwise, they risk not being admitted into either their chosen university or, in some cases, any university in the same tier.

Furthermore, students' choices of university are also constrained by the quota system, which is an estimate of the number of places assigned to different provinces by each institution (Liu, 2016). The quota system is supposedly an instrument for recruitment planning used by individual universities to estimate their capacity for enrolling new students (Liu, 2016). However, the quota system has resulted in a tendency to prioritize candidates from home provinces, since individual institutions rely largely on financial support from the local governments (Liu, 2016). For example, Peking University fixes a quota of 272 new places for home applicants, while only allocating 33 places to applicants from its neighbouring province, Hebei, and

17 for those from the Western province of Ningxia, as of 2009 (Liu, 2016). Therefore, the quota policy affects the choices and mobility of the students.

Given this situation, risk assessment is particularly relevant for students to make choices about university. The risks arise first from matching academic performance with desired institutions and fields of study and then from navigating the prestige of the three-choice system. The risks are further complicated by the presence of the quota system when making choices beyond home provinces.

4   Methodology and Data

The data used for this research are drawn from a total of 71 semi-structured interviews involving undergraduates from a variety of institutions and fields of study. All the respondents were born between 1995 and 1997, with a mean age of 19 years. The city of Shanghai was selected due to its advantage of having a diverse student population from different geographical origins and a variety of universities. Students were selected randomly from different types of universities and fields of study. I recruited students from three main channels. First, I approached students in the canteens, the libraries and the sports centres on campus. Second, my colleagues contacted and recruited students from a variety of social organizations, including the Youth League, film clubs and volunteer associations. The third strategy was the use of social media websites and applications such as WeChat to complement the search for the eligible research population.

The students were targeted to represent four different types of universities in China: one elite university, one key university, one comprehensive university and one university specializing in Finance and Accounting. Students came from a variety of fields. The students' identities and their institutions are anonymized, and pseudonyms are coded instead. Table 2 summarizes in detail the number of in-depth semi-structured interviews by fields of study and types of institutions. The interviews lasted approximately two to two and a half hours. All the interviews were conducted in locations on campus chosen by the respondents. The interviews were conducted in Mandarin Chinese, and audio-recorded with the permission of the respondents, completely transcribed in Chinese and analysed in English. Three rounds of coding were employed in the data analysis. First, I relied on opening coding to discover the themes of choices in the narratives. Second, I focused on identifying the incidents relating to risk assessment during the choice-making. In the final round of coding, I sought to interpret whether risk assessment represented

360   Y. Liu

Table 2 The detailed profile of respondents from different fields of study and types of institutions

Fields of Study	Number and Type of Institutions	Number of Students
Environmental Science	2 (Elite and Key)	9
Medicine	2 (Elite and comprehensive)	11
Engineering	1 (Key)	6
Law	2 (Elite and Comprehensive)	8
Foreign Languages	2 (Key and Comprehensive)	10
Literature and History	1 (Elite)	6
Accounting and Finance	1 (Specialised)	11
Media Studies	2 (Key and comprehensive)	10
		Total 71

rational action, cultural conformity or a re-negotiation of both courses of action.

5   Findings

Table 3 presents the detailed number and profile of respondents' socioeconomic and demographic characteristics. Nearly one-fifth of the respondents had parents in either managerial positions or leading cadres in a managerial position, with another third having parents from professional families. Students from working-class and agricultural backgrounds made up the other half of the respondents. Around 84% of the respondents had parents who had completed secondary schooling or higher education, while only 15% had parents whose educational was below secondary schooling. Female students accounted for 56.3% of the total respondents, while the remaining 43.7% were male students. Moreover, students from Shanghai accounted for nearly two-fifths of the total respondents, while the other three-fifths were from the rest of China. More than 66% of respondents identified themselves as graduates from key schools in contrast to around one-third from regular state schools.

University Choices and Rational Action

I first explore the rational choice discourse in the students' narratives and examine the extent to which students' choices are based on rational calculations of economic or/and opportunity returns. Around 85% of the respondents (60 out of 71) confirm that employability of a degree or a university

Table 3 The detailed number and profile of respondents' socioeconomic and demographic characteristics

	Percentage of the respondents	Number of respondents
Socioeconomic status	100	*Total 71*
Managerial class and cadres in a managerial position	19	14
Professional class	30	21
Working class	37	26
Agricultural working class	14	10
Parental education level	100	*Total 71*
Higher education	25.3	18
Completed secondary schooling	59.1	42
Less than secondary schooling	15.5	11
Gender	100	*Total 71*
Male	43.7	31
Female	56.3	40
Geographical origins	100	*Total 71*
Shanghai	39.4	28
Non-Shanghai	60.5	43
Types of schooling	100	*Total 71*
Key	66.2	47
Normal state	33.8	24

is the key to their decision-making. The remaining 11 students said that personal interests and 'passions' were the most important factors affecting their choices. The majority of the students associated their choices in the fields of study or institutions with good future career plans. They also seem to express this logic consistently and coherently in their narratives. As Liangyan, a 19-year-old girl majoring in Medicine at a key university, stated, *'going to university is about preparing myself for a career with advanced qualifications and skills'*.

As Liangyan was growing up in a medium-sized city in Jiangxi, her father was laid off from his previous 'golden rice-bowl' employment (lifelong job security) at a state-owned enterprise in the 1990s, when the massive wave of redundancies occurred during the market reform (Hanser, 2005; Nee & Opper, 2012). Ever since her father had moved from one job to another in order to make a living. Her particular family circumstances shaped the rationale of Liangyan's university choices. She aimed to choose a field that she described as *'unemployment-proof'*:

> I chose Medicine mainly because of the employability and security associated with this profession. Even during a recession, people still need doctors and

medical care. I simply do not want to experience what happened to my father. I need a job that is safe and secure under any circumstances.

It might be difficult to find a risk-free profession; however, employment opportunities seem to represent a shared rationale for choosing a particular field. The word '*jiuye*' (finding a job) is the most frequently used word in the interview data. Regardless of their social backgrounds, the students all seemed to gather information about the employment prospects associated with a particular field or institution when preparing for their university choices. However, students from different backgrounds have developed their knowledge about employability in very different ways. Those from working-class or agricultural families tend to seek advice from their secondary school teachers and the school alumni network. Chungui, a 19-year-old boy from a village in the agricultural province of Anhui, discussed his choice of Engineering in a key university:

> I asked my school teacher for advice, and I was put into contact with a couple of school alumni who were already in university through QQ. They advised me that the fields with placements and internship would be more employable in the future. So, I chose Engineering here with plenty placement opportunities.

By contrast, the students from professional backgrounds relied significantly either on parental help or parental extended social networks. High-status parents tend to mobilize their networks and resources to inform a rational course of action. Xiaotao, an 18-year-old student from Changsha, went to an elite school in Hunan province. She is from a provincial middle-class family: her mother is a high-ranking civil servant in a district government and her father works as a manager in a Co-Operative company. She recalled how her parents used their contacts to help her to make a sensible decision about university:

> My mum chose for me. Her colleague's daughter studied Finance and Economics several years ago in Shanghai. She got a fantastic job after graduation. She arranged for me to meet this girl, my role model. I was so impressed with her career that I decided to study Finance.

It seems that students both from privileged and disadvantaged backgrounds tend to follow a rational course of calculating employability and labour market outcomes when choosing universities and fields. Both groups' choices are rational in nature; however, the students from working-class and agricultural backgrounds do not have rich resources and networks to inform their choices, so they tend to gather information about job prospects from their school teachers, classmates and school alumni. The source of information and advice does not seem to be consistent, concrete and coherent, which might affect how these students interpret their risks in making choices about higher education.

Choices, Socioeconomic, Cultural and Academic Conformity

Now I examine how the students relate their socioeconomic and cultural backgrounds to their choices of institutions and fields of study. Students from privileged socioeconomic backgrounds and metropolitan areas seem to make choices, consciously or unconsciously, that closely correspond to their parents' socioeconomic status. Lixia, an 18-year-old girl from Shanghai, chose to study German at a key university. She is from a middle-class professional family in Shanghai, with her mother as Mathematics subject teacher in a key school and her father a Party official in the District government. When asked about her choice of German in relation to her family backgrounds, Lixia acknowledged that she did not want to follow her parents' secure jobs but compromised on a choice that would lead to a professional occupation:

> My parents have very secure jobs. My dad is a civil servant and my mum is a teacher. But I don't desire this kind of security. I am more adventurous. I would like to work in international trade. That's why I chose to major in a foreign language. My parents would like me to choose secure jobs. German is a compromise. If I cannot find a job in an international company, I can always teach.

Lixia's self-perceived '*adventurous*' nature is in fact based on solid socioeconomic and cultural resources. Her compromise in her university choice does not represent a departure from her socioeconomic background; rather, it allows her to seek similar options in professional occupations or even maximize future opportunities, either in '*international trade*' or in '*teaching*'. Differing from Lixia's adventurous choices, Xiaoshu, an 18-year-old female student from Shandong, followed in her mother's footsteps and chose a medical school. Xiaoshu's mother is a senior doctor in a military hospital in Dalian. Her mother appreciated her choice but warned her against the '*long and intensive training*' in the medical school. Therefore, Xiaoshu chose the less demanding and competitive field of medical quarantine and prevention, which would allow her to choose either '*a medical career*' or to be '*a civil servant in the National Bureau of Customs*'.

By contrast, students from working-class and agricultural backgrounds are conscious of their socioeconomic, cultural and academic limitations when making their choices. Xinxia recalled her choice of university and field after the 2014 *Gaokao*. Aged 18, Xinxia is from a working-class family in a run-down industrial area in the city of Wuhan. Having been a top student in a state school during her senior secondary stage, Xinxia was reasonably confident about achieving good academic performance in the *Gaokao*, but she was struggling to find a '*suitable*' field that would allow her to '*extend her competitiveness*'. She made a choice to study Law with a focus on civil law, and she explained her decision as follows:

> It is a dream to study in Shanghai. But I knew I would be the 'tail of a phoenix'. I cannot compete with other Shanghai students in English, creativity or presentation skills. My thing is memorization. That is what I am good at. I figured that studying Law would give me a certain advantage.

'*The head of a chicken and the tail of a phoenix*' is a Chinese saying, which describes positional advantages or disadvantages in less or more competitive contexts. Having been the '*head of a chicken*' in secondary schooling, Xinxia was conscious of her advantages and skills—that is, learning from memorization. Meanwhile, she was insecure about a lack of 'modern skills' such as English, presentation and communication skills that would cause her to be the '*the tail of a phoenix*' among her counterparts from privileged backgrounds in university. Yet, the irony here lies in the fact that the legal profession also requires good communication and presentation skills, which was acknowledged in Xinxia's narratives. Her choice of Law seems to suggest that the desire to conform to an extant academic or cultural identity outweighs the long-term rational course of action.

Having scant social and cultural resources, students from working-class and agricultural backgrounds tend to choose fields that would minimize their 'linguistic', 'cultural' or 'academic' disadvantages. Wanggang, a 19-year-old boy from rural Anhui, recalled his choice-making process about university and then his rationale for choosing Engineering. With both parents as agricultural workers, he could not seek help from his parents. He took the responsibility of '*researching the recruitment details on-line*' and '*discussing his choices with his school teachers*':

> It is crucial to estimate your chances. The university gave 40 quotas in 2013 for candidates from Anhui. The quota was reduced slightly in 2014, which

meant my Gaokao scores had to be higher than the year before. It was like gambling. It was a risky choice.

If choosing the university was like '*gambling*', the choice of the field was a painful process of accepting his socioeconomic and cultural disadvantages. He weighed his '*passion*' in the field of international finance against a '*realistic*' choice of engineering:

I really wanted to work in international finance. I never had a chance to travel around. International finance was a dream job. But this field requires a lot of communication skills. I am not confident. My spoken English is also poor. Back in the school, we were not taught (English) properly. My schoolteachers also told me that I need some guanxi (personal contacts) to get internships if I chose finance. I guess engineering suits me better, as the university will arrange placement and internships. As long as I work hard, I can have good job opportunities'.

Wanggang's narrative illustrates the deep-seated inequality that which fails to be captured by the numeric measure of the *Gaokao* scores. Wanggang, a *Gaokao* 'champion' from a rural village in Anhui, made a glorious transition to a key university in Shanghai. His story might paint a rosy picture of meritocratic *Gaokao* selection. However, his choice of field suggests an internalization of the academic and cultural disadvantages of being a rural student. Constrained by a lack of social capital and cultural resources, he instead conforms to his 'academic' identity of working hard in the field where he has a chance.

Risks, Socioeconomic and Geographical Conformity

Finally, I turn to examine how students from different social origins assess the risks involved in making choices, whether risk assessment affects students differently in relation to their backgrounds and how they navigate through the three-choice system. The majority of the students (67 out of 71) confirm that they have considered risks when applying for fields of study and universities. However, there is a distinctive pattern of different interpretations of risks by students from different social backgrounds. Students from Shanghai and from privileged backgrounds seem to be confident about their choices and interpret risks as '*encountering challenges*'. For example, Lijiayan, a 20-year-old female student from a professional family in Shanghai, who majored in Law in an elite university, recalled her choices as '*risk taking*' and '*experiencing challenges*'.

The narrative about risks as encountered challenges is echoed in Zihan's story of making three choices. Zihan, an 18-year-old male student whose father is a chief surgeon from a highly rated state hospital in Shanghai and whose mother is head of finance in the District government, was very bold with his choice of Media Studies in an elite university:

> The risk started from the very first choice in the elite category. Every choice I made would have a knock-on effect on the subsequent choice. I decided to risk my first choice in the elite category. I was not prepared to go to any university ranked lower than my first choice. I did not even submit second or third choices. I was prepared to repeat a whole year if I was not chosen by the elite university. My parents thought I was very stubborn, but they were very supportive. Luckily I dared to take the risk and I was accepted.

Contrary to a rational calculation of chances in all categories of universities and fields of study, Zihan's choice strategy seems to be a deliberate risk-taking, and even a 'stubborn' approach, mainly because he could afford the penalty of his bold decision. If his 'gamble' for his dream university had failed, his family's resources could cushion the effect of failure by supporting him to repeat another year and prepare for the next *Gaokao*. Such economic and emotional resources are unavailable for students from disadvantaged backgrounds. Throughout their narratives, the students from non-Shanghai areas or working-class or agricultural families translate the disadvantages associated with their geographical or social origins into uncertainties and insecurities about their choices. In sharp contrast to Zihan's choice strategy of gaining access to elite opportunities, Shilu, an 18-year-old girl, is more conservative about her choices. Born into a professional family in a small city in Hubei, and having been to a key school throughout her secondary schooling, Shilu described the '*most difficult decision*' of her life:

> My heart went to the first choice of the elite university, but my head told me that my chance was much slimmer than that of those from Shanghai. I did very well in the Gaokao. But it was much more competitive to go to university in Shanghai for someone from outside. I decided to abandon my first choice of the elite institutions, which allowed me to choose a more popular field of study in a less prestigious university. I made a careful second first choice of Accounting and Finance. After all, employment opportunities in this field will pay off.

Shilu's choice strategy illustrates both a rational calculation of her chances and conscious conformity to her geographical identity constrained by the quota system. On the one hand, abandoning the opportunities in the elite universities would win her more chances on the non-elite track and avoid the penalty of losing out on the opportunities, which suggests a rational course of action. On the other hand, the penalty already kicks in for those from non-metropolitan areas before making choices about higher education, as the quota system obscures the opportunity structure for those from different geographical origins. Similar strategies are also adopted by students from working-class and agricultural families. Yaoyu, a 19-year-old male student from a rural area in Henan, explains how the risks prevented him from making a bold decision in the same field in elite institutions:

> The rules of the game are very simple but cruel. The choice system clearly favoured the native Shanghai students. My Gaokao scores would be good enough to go to an elite university. But it might not guarantee a place in Engineering. I didn't want to risk it. I make a safe bet of Engineering in a key institution. My strategy was to prioritize job prospects over the status of the university.

Yaoyun's narratives illustrate both a rational calculation of his chances and a risk-aversion strategy to conform to less ambitious and safer options in the non-elite track. Lina, a 19-year-old girl from a modest working-class family in Jiangsu, discussed the risks of being rejected for English Literature in an elite university and therefore settled on Spanish in a comprehensive university. She admitted the risks of choosing a university in Shanghai but reflected on whether she should have '*high expectations*' about herself. The desire '*not to be a burden to her parents*' made her prioritize a safe choice of university and '*lower*' her expectations, thus giving up her '*dream to study in an elite university*'. Lina discussed how the realization of her '*limitations*' helped her to make a '*sensible*' choice. Like many other students from agricultural or working-class families, Lina seems to have come to terms with a lack of economic, social and cultural resources that might have made her bolder in her choices. When hope and opportunity clash, these students reduce to internalize their disadvantages, adjust their expectations about themselves and the future and accept the less ambitious university choices, which are filtered by the institutionalized choice system and made available to them as meritocratic outcomes.

6   Discussion and Conclusion

This study examines Boudon's positional theory to make sense of different patterns of university choices in contemporary China. It has a number of findings, some of which relate to Boudon's theoretical standpoints and some of which are more broadly relevant to the Chinese context. At the theoretical level, most of the interview data support Boudon's secondary effects thesis in the Chinese context; that is, individuals' family characteristics manifest in the process of choices and strategies. Students from professional families and the metropolitan city of Shanghai benefit from rich economic, social and cultural resources that have enabled them to develop a confident and clear vision about their university and career prospects, which are unavailable to students from working-class and rural families. Even when students from all social origins demonstrate a rational calculation of returns of a degree and employability, the students from working-class and agricultural backgrounds have scant resources and networks to inform their choices, except for their school teachers, classmates and school alumni. The sources of information and advice do not seem to be consistent, concrete and coherent, which might contribute to the less ambitious and confident patterns of choices.

Socioeconomic and cultural inequality is further obscured in the choices of the fields of study. The findings partly confirm Yair's hypothesis that the rational maximization of instrumental utility is compromised by conscious and unconscious conformity (Yair, 2007). Those from privileged backgrounds seem to make conscious or unconscious choices, which are informed by their social and cultural resources and closely correspond to their parents' socioeconomic status. By contrast, those from working-class and agricultural families find themselves constrained by a lack of social capital and cultural resources as well as disadvantaged by a lack of modern skills. Thus, they tend to limit their choices in fields that either require minimal cultural and social capital or allow them to extend their extant academic skills, such as memorization. Students from privileged backgrounds do not seem to be conscious of their choices of the fields that might conform to their socioeconomic and cultural identity. However, those from disadvantaged families are much more conscious of their desire to conform to an extant academic or cultural identity, which in some cases might outweigh their long-term rational course of action.

Furthermore, this study explores students' narratives regarding the risks associated with the three-choice system and the strategies they develop during the choice-making process. On the one hand, the three-choice system seems to allow those from privileged backgrounds like Zihan to take chances that might be considered risky and irrational; this is made possible by their socioeconomic and cultural resources and the geographic advantage of being a Shanghai native. On the other hand, students from working-class and agricultural families, bound by their disadvantages, have adopted risk-aversion strategies; that is, they seek to maximize their opportunities in the desired fields of study instead of choosing top-ranked universities. These students sacrifice their elite opportunities in the most prestigious universities in order to secure a position in a field with higher labour market returns at a less well-known institution. This strategy can be interpreted as rational conformity, which on the one hand gives in the desire for upward mobility in status, but, on the other, maximizes the long-term employment opportunities.

At the contextual level, perhaps the major issue that arises, at least in regard to the distribution of higher education opportunities, is that the underlying rules of the university recruitment system—more specifically, the quota system—reinforce the privileges of those from affluent areas like Shanghai. Previous research suggests an uneven distribution of higher education institutions between eastern metropolitan areas like Beijing and Shanghai and poorer western regions (Liu, 2016). Those from under-developed western and central provinces seek upward social mobility through higher education opportunities, particularly in eastern areas. However, the quota system undermines the meritocratic nature of higher education selection and exacerbates geographical inequality in the opportunity structures.

Constrained by both the three-choice system and the quota system, many students from non-Shanghai areas seem to come to terms with a lack of opportunities equal to those of their Shanghai counterparts. When hope and opportunity clash, these students reduce to internalize their geographical disadvantages, adjust their expectations about themselves and the future and accept the less ambitious university choices, which are filtered by the institutionalized choice system and made available to them as meritocratic outcomes. It can be argued that the quota system is a result of the growing power of the eastern political elites supporting preferential access to higher education for their local populations. Consistent with the findings on geographical inequality as the main stratifier in contemporary China (Liu, 2018; Wu, 2010), this study provides uncomfortable evidence of the creation of legitimized and institutionalized discrimination against those from poorer regions.

References

Boado, H. C. (2011). Primary and secondary effects in the explanation of disadvantage in education: the children of immigrant families in France. *British Journal of Sociology of Education, 32*(3), 407–430.

Bourdieu, P., & Passeron, J.-C. (1977). *Reproduction in education, society and culture*. Sage.

Boudon, R. (1974). *Education, opportunity and social inequality*. Wiley.

Boudon, R. (1998). Limitations of rational choice theory. *American Journal of Sociology, 104*(3), 817–28.

Boudon, R. (2003). Beyond rational choice theory. *Annual Review of Sociology, 29*(1), 1–21.

Boudon, R. (2006). Are we doomed to see the homo sociologicus as a rational or as an irrational idiot? In J. Elster, O. Gjelsvik & A. Hylland (eds.), *Understanding choice, explaining behaviour. essays in honour of Ole-Jorgen Skog* (pp. 25–41). Olso Academic Press.

Breen, R., & Goldthorpe, J. H. (1997). Explaining educational differentials: towards a formal rational action theory. *Rationality and Society, 9*(3), 275–305.

Breen, R., & Goldthorpe, J. H. (1999). Class inequality and meritocracy: A critique of Saunders and an alternative analysis. *British Journal of Sociology, 50*(1), 1–27.

Brooks, R. (2008). Accessing higher education: The influence of cultural and social capital on university choice. *Sociology Compass, 2*(4), 1355–71.

Carnoy, M., Isak F., Prashant L., & Jandhyala, B. G. T. (2014). The concept of public goods, the state, and higher education finance: A view from the BRICs. *Higher Education, 68*(3), 359–378.

Clark, S., Mountford-Zimdars, A., & Francis, B. (2015). Risk, choice and social disadvantage: Young people's decision-making in a marketised higher education system. *Sociological Research Online, 20*(3), 9. https://doi.org/10.5153/sro.3727

Coleman, James S. (1990). *Equality and achievement in education*. Westview Press.

Coleman, J. S. (1961). *The adolescent society*. The Free Press.

Coleman, J. S., & Campbell, E. Q., Hobson, C. F., McPartland, J. M., Mood, A. M., Weinfeld, F. D., & York, R. L. (1966). *Equality of educational opportunity*. U.S. Government Printing Office.

Davies, P., Tian Q., & Neil, M. D. (2014). Cultural and human capital, information and higher education choices. *Journal of Education Policy, 29*(6), 804–825. https://doi.org/10.1080/02680939.2014.891762

Donnelly, M., & Evans. C. (2016). Framing the geographies of higher education participation: Schools, place and national identity. *British Educational Research Journal, 42*(1), 74–92.

Duru-Bellat, M. (2010). Raymond Boudon ou la portée d'un certain « universalisme abstrait » dans l'analyse genrée des inégalités. In Chabaud-Rychter, D., V. Descoutures, E. Varikas & A. M. Devreux (eds.), *Sous les sciences sociales, le genre* (pp. 165–176). Paris, La Decouvert.

Duru-Bellat, M., Annick K., & David, R. (2008). Patterns of social inequalities in access to higher education in France and Germany. *International Journal of Comparative Sociology, 49*(4–5), 347–368.

Erikson, R., & Rudolphi. F. (2010). Change in social selection to upper secondary school-primary and secondary effects in Sweden. *European Sociological Review, 26*(3), 291–305.

Glaesser, J., & Cooper. B. (2013). Using rational action theory and Bourdieu's habitus theory together to account for educational decision-making in England and Germany. *Sociology, 48*(3), 463–481.

Guo, C. B., Mun, C. T., & Xiaohao D. (2010). Gender disparities in science and engineering in Chinese universities. *Economics of Education Review, 29*(2), 225–235.

Hanser, A. (2005). The gendered rice bowl: The sexual politics of service work in Urban China. *Gender and Society, 19*(5), 581–600.

Hartog, J., Sun, Y., & Ding, X. H. (2010). University rank and bachelor's labour market positions in China. *Economics of Education Review, 29*(6), 971–979.

Holt, B. (2012). Identity matters: the centrality of 'conferred identity' as symbolic power and social capital in higher education mobility. *International Journal of Inclusive Education, 16*(9), 929–940.

Jackson, M., & Jonsson, J. O. (2013). Why does inequality of educational opportunity vary across countries? Primary and secondary effects in comparative context. In M. Jackson (ed.), *Determined to succeed? Performance versus choice in educational attainment.* (pp. 306–338). Stanford University Press.

Jackson, M., Erikson, R., Goldthorpe, J. H., & Yaish, M. (2007). Primary and secondary effects in class differentials in educational attainment: The transition to A-level courses in England and Wales. *Acta Sociologica, 50*(3), 211–229.

Jetten, J., Iyer, A., Tsivrikos, D., & Young, B. M. (2008). When is individual mobility costly? The role of economic and social identity factors. *European Journal of Social Psychology, 38*(5), 866–79.

Kleanthous, I. (2014). Indigenous and immigrant students in transition to higher education and perceptions of parental influence: a Bourdieusian perspective. *Policy Futures in Education, 12*(5), 670–680.

Li, J. (2012). World-class higher education and the emerging Chinese model of the university. *Prospects, 42*(3), 319–339. https://doi.org/10.1007/s11125-012-9241-y

Li, H., Meng, L., Shi, X., & Wu, B. (2012). Does attending elite colleges pay in China? *Journal of Comparative Economics, 40*(1), 78–88.

Liu, Y. (2018). When choices become chances: Extending Boudon's positional theory to understand university choices in contemporary China. *Comparative Education Review., 62*(1), 125–146.

Liu, Y. (2017). Women rising as half of the sky? An empirical study on women from the 'one-child' generation and their higher education participation in contemporary China. *Higher Education. 74*(6), 963–978.

Liu, Y. (2016) *Higher education, meritocracy and inequality in China.* Springer.

Liu, Y. (2015). Geographical stratification and the role of the state in access to higher education in contemporary China. *International Journal of Educational Development 44*, 108–117.

Loyalka, P. K., Song, Y., & Wei, J. (2012). The effects of attending selective college tiers in China. *Social Science Research, 41*(2), 287–305.

Loyalka, P. K. (2009). *Three essays on Chinese higher education after expansion and reform: Sorting, financial aid and college selectivity*. PhD thesis. Stanford: Stanford University.

Marginson, S. (2016). High participation systems of higher education. *The Journal of Higher Education, 87*(2), 243–271.

Ministry of Education. (MOE). (2021). *The list of higher education institutions*. Ministry of Education. Available from http://www.moe.gov.cn/jyb_xxgk/s5743/s5744/A03/202206/t20220617_638352.html. Accessed 26 October 2023.

Nash, R. (2006). Controlling for 'Ability': A conceptual and empirical study of primary and secondary effects. *British Journal of Sociology of Education, 27*(2), 157–172.

Nee, V., & Opper, S. (2012). *Capitalism from below markets and institutional change in China*. Harvard University Press.

Reay, D., Crozier, G., & Clayton, J. (2009). Strangers in paradise? Working-class students in elite universities. *Sociology, 43*(6), 1103–21.

Swidler, A. (1986) Culture in action: Symbols and strategies. *American Sociological Review 51*(2), 273–286.

Thompson, R., & Simmons, R. (2013). Social mobility and post-compulsory education: revisiting Boudon's model of social opportunity. *British Journal of Sociology of Education, 34*(5–6), 744–765.

Thomsen, J. P., Munk, M. D., Eiberg-Madsen, M., & Hansen, G. I. (2013). The educational strategies of Danish university students from professional and working-class backgrounds. *Comparative Education Review 57*(3), 457–480.

Van de Werfhorst, G. H., & Hofstede, S. (2007). Cultural capital or relative risk aversion? Two mechanisms for educational inequality compared. *British Journal of Sociology, 58*(3), 391–415.

Van de Werfhorst, H. G., Sullivan, A., & Cheung, S. Y. (2003). Social class, ability and choice of subject in secondary and tertiary education in Britain. *British Educational Research Journal, 29*(1), 41–62.

Wilkins, S., Shams, F., & Huisman, J. (2013). The decision making and changing behavioural dynamics of potential higher education students: the impacts of increasing tuition fees in England. *Educational Studies, 39*(2), 125–141.

Wu, X. G. (2010). Economic transition, school expansion and educational inequality in China, 1990-2000. *Research in Social Stratification and Mobility, 28*(1), 91–108.

Yair, G. (2007). Existential uncertainty and the will to conform: The expressive basis of Coleman's rational choice paradigm. *Sociology, 41*(4), 681–698.

Methodological Individualism and Formal Models

Werner Raub and Arnout van de Rijt

1   Introduction

This chapter addresses the role of formalization and formal model building as a 'tool' for methodological individualism (MI) in the social sciences. With respect to discipline, the chapter is largely, but not exclusively, on sociology. Also, the chapter is on theory construction and does not address statistical modeling or the link between theory construction and statistical modeling.

Briefly, our take-home message is that formalization is not always needed but can be sometimes useful or even necessary, for example, for making assumptions explicit and for the derivation of implications from assumptions. This often includes implications that are in some sense 'deeper' and less obvious. Particularly interesting cases are implications that are unintended and unanticipated from the perspective of the actors involved, sometimes also at first sight counterintuitive from the perspective of the researcher.

W. Raub (✉)
Department of Sociology, Utrecht University, Utrecht, The Netherlands
e-mail: W.Raub@uu.nl

A. van de Rijt
Sociology, European University Institute, Fiesole, Italy
e-mail: Arnout.VanDeRijt@eui.eu

© The Author(s), under exclusive license to Springer Nature Switzerland AG 2023
N. Bulle and F. Di Iorio (eds.), *The Palgrave Handbook of Methodological Individualism*,
https://doi.org/10.1007/978-3-031-41508-1_16

Appreciation for how formalization and formal model building generate scientific headway can perhaps be evoked best by studying examples. Therefore, the chapter is not designed as an abstract discussion of what is meant by 'formalization' and 'formal models' or of pros and cons of using formal models for social science theory formation. Rather we illustrate how formalization can contribute to theory formation and scientific progress using various examples. These examples are from different theoretical approaches and concern different research fields and research problems, including meanwhile 'classic' as well as more recent contributions. For each example, we highlight two features. First, insights that do not require elaborate formalization and second implications that do require some degree of formalization so that they can be derived from the respective theory's assumptions. Notwithstanding the chapter's topic—formal models—the exposition itself is largely informal, leaving out technical detail.

The following section offers a concise summary of key features of theory formation in line with MI, including why and how formalization can be useful for such theory formation. Subsequently, we turn to our examples. We start with three classic examples that are related to rational choice theory and its applications in sociology and the social sciences more generally. Three further examples are from more recent work involving network analysis and agent-based computational models. Some concluding remarks round off the chapter.

2   Methodological Individualism and Formal Models: Some General Remarks[1]

MI attempts, roughly, to explain macro-phenomena and macro-level regularities by linking macro- and micro-levels of analysis and by employing assumptions on micro-level behavioral regularities. Coleman's (1986, 1990, Chap. 1) well-known heuristic diagram (Fig. 1) summarizes key building blocks of such explanations (Coleman's diagram has quite some antecedents, see Raub & Voss, 2017 for details).

In brief, propositions on macro-conditions and macro-outcomes are represented by Nodes A and D in Fig. 1. Propositions on associations

---

[1] A brief sketch will hopefully do. In this *Handbook*, Opp's chapter on 'Methodological individualism and micro-macro modeling' offers more detail on a perspective on MI that is in several though not all respects similar to ours. The references in the present section provide pointers to further information on a range of issues.

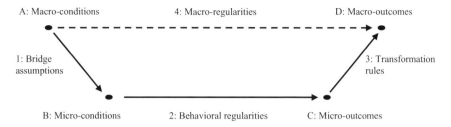

Fig. 1 Coleman's diagram

between macro-conditions and macro-outcomes, such as empirical regularities, are represented by Arrow 4. Macro-outcomes as well as macro-level associations are explananda of MI explanations. Propositions referring to micro-conditions are represented by Node B. Employing common terminology, such propositions refer to 'independent variables' in assumptions about regularities of individual behavior, such as assumptions about actors' incentives and beliefs in the special case of variants of rational choice theory. 'Bridge assumptions' (Wippler & Lindenberg, 1987), represented by Arrow 1, specify how macro-conditions affect micro-conditions. Propositions describing micro-outcomes such as individual behavior and, thus, explananda on the micro-level are represented by Node C. In line with these specifications, Arrow 2 can be seen as representing a micro-theory of behavior or anyway one or more micro-level hypotheses on behavior. Finally, Arrow 3 represents 'transformation rules' (Wippler & Lindenberg, 1987), namely, assumptions on how behavior of actors generates macro-outcomes. In terms of the conventional Hempel-Oppenheim (Hempel, 1965) model of explanation, the diagram indicates that explananda at the micro-level follow from an explanans with assumptions on regularities of individual behavior (Arrow 2), macro- as well as micro-conditions (Nodes A and B), and bridge assumptions (Arrow 1). Macro-level explananda, outcomes (Node D) and regularities (Arrow 4) require an explanans with assumptions on regularities of individual behavior (Arrow 2), macro- as well as micro-conditions (Nodes A and B), bridge assumptions (Arrow 1) and transformation rules (Arrow 3).

Approaches along these lines in sociology include, but are not restricted to, the rational choice approach, Goldthorpe's (2016) sociology as a population science, applications of agent-based models, and various models for the analysis of social networks (for overview see Gërxhani et al., 2022; Raub et al., 2022). Our examples below are closely related to such approaches. We will develop and discuss each example employing the terminology we have introduced for the various 'elements' (nodes and arrows) of Coleman's diagram.

'Formalization' and 'formal theory' refer to explicitly specifying assumptions so that implications can be derived using well-defined and explicit rules of deduction. Implications can be derived analytically or using simulation methods, such as in agent-based modeling. This typically requires employing some mathematics or formal logic. Formalization and formal theory must be distinguished carefully from the use of more or less formal notation. Rather, the key issue is that rules of deduction are available and employed. There are several standard arguments why theory construction in general and more specifically theory construction as in MI often benefits from specifying formal models or from formal specification of some elements of the theory (on formalization and the benefits of formal theory construction, see, for example, Coleman, 1964; Hummell, 1972, 1973; Ziegler, 1972 with a focus on sociology, Rodrik 2015; Tirole, 2017, Chap. 4, with a focus on economics, and Lindenberg, 1992 on adequacy criteria for formal theory). Such arguments include that deriving implications from assumptions is a key feature of theory construction but is also often a far from trivial task, certainly so when theory construction involves macro- and micro-level assumptions as well as assumptions that relate both levels. Formalization can then help in or is even necessary for deriving implications. Likewise, formalization can reveal implicit assumptions that must be made explicit to derive certain implications as well as reveal that certain implications do not at all depend on certain assumptions. Furthermore, formalization facilitates comparing the implications of different sets of assumptions or can be even necessary for such comparisons. This includes contributions of formalization by identifying necessary or sufficient (sets of) assumptions for certain implications. It is clear that such benefits of formalization are also important when it comes to checking for the consistency of sets of assumptions and to analyzing how robust certain implications are to variations in assumptions.

We now move on to discussing examples that make transparent how formalization and formal model building can be useful for MI. Among other things, we will address, per example, those elements of Coleman's diagram for which formalization becomes important. This will highlight, too, differences between examples with respect to the specific elements that benefit from or are in need of formalization. For lack of space, we provide only rough sketches that aim at providing sufficient intuition, rather than in-depth treatments. We brush over technical details that can better be studied by consulting the primary literature to which we refer. Throughout, we highlight one specific and fundamental[2] benefit of formalization for theory construction, namely,

---

[2] 'Fundamental' in the sense of our characterization of the notion of 'formalization'.

enabling the derivation of implications that would be hard or impossible to derive employing exclusively informal reasoning. For each example, we also indicate how implications that can only be derived using formalization contribute to empirical research. In the concluding section of the chapter, for some of our examples, we briefly address how they also shed light on various other advantages of formal theory construction mentioned above.

3 Group Size Effects on Collective Good Production in the Volunteer's Dilemma

At least since Olson's (1971) meanwhile classic contribution, group size effects have been a key topic of research on collective good production. Importantly, actors not contributing to collective good production cannot be excluded from consumption of the good. Therefore, actors typically face incentives not to contribute ('free riding'). According to Olson, group size will often be negatively related to collective good production. Olson discussed various conditions that affect this macro-level association. One of his key points is 'selective incentives': additional individual benefits for an actor that are obtained only if the actor contributes to collective good production. Roughly, Olson claimed that without selective incentives collective good production becomes more problematic in large(r) groups. Also, the production function for the collective good affects the macro-level association (see, for example, Sandler, 1992 for further discussion of these and other conditions). Diekmann's (1985) Volunteer's Dilemma (VOD) is an instructive formal model that allows for deriving a group size effect under a well-specified set of conditions.[3]

VOD is a noncooperative game (see an accessible, yet careful textbook like Rasmusen, 2007 for details on game-theoretic concepts and assumptions). Hence, binding and enforceable agreements or unilateral commitments are not feasible for the actors involved. The game is played by N actors. Figure 2 specifies the normal form of VOD. Rows represent the pure strategies of a focal actor: to contribute (CONTR) or not to contribute (DON'T) to the production of the collective good. Actors choose simultaneously and independently. Columns show the number of other actors who contribute. Cells specify the focal actor's payoffs as a function of the actor's own strategy

---

[3] Diekmann (1985) can be consulted on technical details for VOD. In the following, we also build on Raub and Voss (2017, pp. 14–18) for a 'reconstruction' of the VOD model in terms of Coleman's diagram.

	Number of other actors choosing CONTR				
	0	1	2	...	$N-1$
CONTR	$U-K$	$U-K$	$U-K$	...	$U-K$
DON'T	0	U	U	...	U

Fig. 2 Diekmann's (1985) Volunteer's Dilemma ($U > K > 0$; $N \geq 2$)

and the number of contributing others. $K > 0$ indicates the costs of individual contributions. The production function is so that the good will be produced if and only if at least one actor (a 'volunteer') contributes. Contributions by more than one actor imply that each of these actors pays the full costs of providing the good, while contributions of more than one actor do not further improve the utility level of any actor. Note that the costs K of individual contributions are smaller than the gains U from the good. Note furthermore that there are no selective incentives associated with individual contributions.

In terms of Coleman's diagram, the macro-outcome of interest, represented by Node D, is the probability P that the collective good will be provided. Group size N is the key macro-condition (Node A). Arrow 4 then represents the relation between N and P, our macro-level explanandum. In line with MI, the macro-level probability of collective good production and group size effects are derived by employing game-theoretic analysis that links macro-level conditions and outcomes with micro-level individual behavior of actors. More specifically, given the normal form of VOD, standard game-theoretic assumptions on behavior in noncooperative games allow, first, for deriving micro-level outcomes, namely, individual probabilities p of contributing to collective good production and for deriving implications of group size effects on these micro-level probabilities. Second, given the micro-level probabilities, the normal form of the game allows for also deriving the macro-level probability of collective good production and for deriving group size effects on the macro-level probability. Note that the normal form of VOD specifies bridge assumptions (Arrow 1 in Coleman's diagram) on macro-micro transitions. Namely, the normal form shows how each actor's payoffs depend on own choices as well as those of all other actors—that is, the normal form specifies the structure of actors' interdependence. Also, the normal form specifies transformation rules (Arrow 3) since it allows to derive the macro-level probability of collective good production from the micro-level probabilities of individual contributions.

Standard game-theoretic assumptions such as the assumption of equilibrium behavior in noncooperative games are micro-level assumptions represented by Arrow 2 in Coleman's diagram. In an equilibrium, each actor's

Methodological Individualism and Formal Models 379

strategy maximizes own payoffs, given the strategies of the other actors. One can easily verify that VOD has N equilibria in pure strategies, namely, the strategy combinations with exactly one volunteer choosing CONTR, while all other actors choose DON'T. The collective good is provided with certainty in each of these equilibria. From the perspective of game-theoretic rationality assumptions, the problem is that each actor prefers the equilibria with another actor as the volunteer to the equilibrium with own volunteering. Also, it is a standard assumption that rational actors, in a symmetric game like VOD, play a symmetric equilibrium in the sense of choosing the same strategies. VOD has a unique symmetric equilibrium in mixed strategies. In this equilibrium, each actor chooses CONTR with probability $p* = 1 - (\frac{K}{U})^{\frac{1}{N-1}}$. Employing a game-theoretic approach, this equilibrium is a plausible 'solution' of VOD. Thus, p^* is represented by Node C in the diagram. Note that $0 < p^* < 1$.

By now, it becomes transparent how formalization supports the analysis of group size effects on collective good production in VOD. It is clear that formalization, in the sense of explicit specification of VOD and its parameters, is needed for deriving equilibria of VOD and, more specifically, for deriving the symmetric equilibrium in mixed strategies as the solution of the game. Formalization is therefore also needed for deriving micro-level contribution probabilities. Furthermore, an implication follows concerning group size effects on micro-level contribution probabilities: these probabilities decline with increasing group size N, with $\lim_{N \to \infty} \left[1 - (\frac{K}{U})^{\frac{1}{N-1}} \right] = 0$. It is straightforward to test this implication empirically (see already Diekmann, 1986 for experimental work on VOD and Tutić, 2014 for a summary of results of later experimental work). This shows, too, that formal theoretical models, in contrast with an intuition one sometimes encounters among sociologists, need not be untestable. On the contrary, formalization can be necessary to derive testable implications of theory.

What about the role of formalization when it comes to the macro-level probability of collective good production, including group size effects on that probability? In VOD, under game-theoretic assumptions, group size is related to collective good provision through two different mechanisms. It is sufficient for the good to be provided that one single actor bears the costs. All actors are bearing the costs with positive probability in the symmetric mixed equilibrium. Therefore, increasing group size exerts a positive influence, since the number of actors increases who may decide to contribute. Conversely, there is also a negative influence of increasing group size, since each actor's individual probability p^* to contribute decreases with increasing N. Formalization is clearly not needed to recognize both mechanisms and to see that they imply opposite effects on the macro-level probability. However, formal analysis is needed to decide which of these two 'forces' prevails and to derive the total effect. Given the symmetric mixed equilibrium, the probability that the collective good will be provided because there is at least one volunteer, is $P^* = 1 - (\frac{K}{U})^{\frac{N}{N-1}}$. This is the macro-outcome in VOD (Node D). Since $P^* < 1$, it follows that production of the collective good may fail. Moreover, as can be seen from the right-hand side of the equation, the negative effect of increasing group size outweighs the positive effect. Hence, a further testable implication follows, represented by Arrow 4 in Coleman's diagram and concerning a macro-level association: with increasing N, the probability decreases that the collective good is provided. Since $\lim_{N \to \infty} \left[1 - (\frac{K}{U})^{\frac{N}{N-1}} \right] = 1 - \frac{K}{U}$, the probability of collective good provision approaches $1 - \frac{K}{U}$ for increasing N. It should be clear that the assumptions employed for analyzing VOD are consistent with failure of collective good production being an unintended as well as unanticipated outcome for the actors.

In addition, it is useful to observe that formalization in the sense of explicit specification of the normal form of VOD has also provided specifications of bridge assumptions and transformation rules. In general, specification of a game in terms of the normal form[4] makes explicit how macro-conditions affect micro-level conditions such as actors' incentives and beliefs and how actors' behavior, in turn, affects macro-outcomes. This is important in light of Coleman's (1987, 1993) arguments that linking macro- and micro-level analyses through careful specification of bridge assumptions and transformation rules is a key task of sociology as well as a typically non-trivial task, and that sociological theory is also often deficient precisely when it comes to specifying bridge assumptions and transformation rules. The VOD example shows how formal tools can be employed to improve theory construction in these respects.

4　Deriving Micro-Level Theorems: Utility Theory

Our next example is on how formalization contributes to improving a key component of explanations in line with MI, namely, a micro-level theory of behavior, represented by Arrow 2 in Coleman's diagram. MI as such is

---

[4] Using the extensive form representation of a game, thus including the game tree, rather than only the normal form, would also make the game's information structure explicit and would further contribute to specifying macro-to-micro as well as micro-to-macro links.

'neutral' with respect to micro-level theories in the sense that very different such theories could be employed in principle. Nevertheless, variants of rational choice theories, notwithstanding that they are controversial in some quarters of sociology, are often employed in MI. This is so even though it is broadly acknowledged that there are 'anomalies': empirical regularities of behavior in a variety of contexts that are hard to reconcile with rational choice assumptions (Camerer, 2003; Kahneman & Tversky, 2000). It stands to reason (see, for example, Raub, 2021) that the prominence of rational choice theories for MI in spite of anomalies is due on the one hand to a range of successful applications of rational choice theories, including both micro- and macro-level explananda in key domains of sociology, in terms of testable predictions and corroborating evidence, certainly in comparison with competing approaches (Wittek et al., 2013 provides a useful overview). On the other hand, the prominence of rational choice theories seems due to their suitability for generating implications, including macro-level implications, from MI models, certainly when one aims at deriving implications analytically, rather than with simulation methods. This is a result of parsimonious assumptions on behavioral regularities, thus preserving tractability of models. Because of this feature, rational choice theories can be employed in theoretical models for the explanation of micro- and macro-phenomena in rather different domains of sociology. One thus obtains a common core of such models with respect to assumptions on micro-level behavior (see Diekmann & Voss, 2004, p. 20). This enhances coherence and family resemblance over a series of models and facilitates cumulative growth of knowledge.

For our example, we focus on a specific version of rational choice theory, namely, utility theory (see Harsanyi, 1977, Chap. 3 for a clear and systematic exposition and Diekmann, 2022 for a concise discussion from a sociological perspective).[5] It is important to appreciate that the theory, just like other variants of rational choice theory, has axiomatic foundations. The axioms of the theory comprise a 'primary definition' (Harsanyi, 1977, pp. 10–11) of rational behavior in terms of a small and parsimonious set of rationality postulates. Basically, these postulates define 'consistency' of an actor's behavior in terms of properties of the actor's preferences (and, in decisions under risk in the sense of Harsanyi, 1977, consistency of the actor's subjective probabilities that represent beliefs). Transitivity of preferences is an example of consistency requirements: if the actor prefers A over B and B over C, the

---

[5] Note that there are different axiom systems, including different systems for different kinds of decision situations such as decisions under certainty, risk, and uncertainty (Harsanyi, 1977, Chap. 3). In this sense, there are different versions of rational choice theory. There are also formalized versions, sometimes including axiomatic foundations, for alternatives to rational choice theory, prospect theory being a prominent example (see Wakker, 2010 for a comprehensive textbook).

actor likewise prefers A over C. The important and at first sight surprising result is that the axioms imply the existence of a utility function and that behavior in accordance with the rationality postulates can be characterized as maximizing the utility function. One can furthermore derive properties of the utility function, such as uniqueness up to order-preserving transformations ('ordinal utility') in the case of decisions under certainty, or order-preserving linear transformations ('cardinal utility') in the case of decisions under risk and uncertainty. These implications constitute a 'secondary definition' of rational behavior. Hence, the theory *does not start from* an assumption of utility maximization. Rather, it is the other way around: the theory starts from assumptions on observable behavior in line with the rationality postulates. Utility maximizing behavior is then an *implication* of the postulates. Note that this likewise makes it clear that we need not assume that actors consciously calculate utilities for alternative courses of their behavior. Rather, and in a strict sense, the rationality postulates imply only that actors behaving according to the theory also behave *as if* they are maximizing a utility function.

What are lessons from our example on benefits of formalizing theory? Denote a strict preference of A over B by $A \succ B$ and assume that an actor has strict preferences $A_1 \succ A_2 \succ \ldots \succ A_i \succ \ldots \succ A_n$ over alternatives A_i, with $i = 1, \ldots, n$. Clearly, simple notation already suffices to see that the utilities $U(A_i)$ for these alternatives should satisfy $U(A_1) > U(A_2) > \ldots > U(A_i) > \ldots > U(A_n)$ if we want to guarantee that the utilities 'somehow' correspond to the preferences. Formalization in the sense of elaborate formal reasoning is not needed to derive this insight. However, this is very different from the much stronger result briefly sketched above that a utility function with certain properties exists and the actor behaves as if maximizing that function if the actor's behavior obeys the rationality postulates. Namely, this result does require an elaborate formal proof such as provided by Harsanyi (1977, pp. 22–46).

The axiomatic foundation of rational choice theory, together with the implications on the existence and properties of utility functions, yields a precise characterization of what is meant by 'rational choice', including a well-defined measurement theory for utilities and subjective probabilities, based on observing an actor's choices. This has important advantages. First, we obtain testable implications. In fact, 'anomalies' such as the Allais Paradox or the Ellsberg Paradox, and many other 'biases' of human behavior show that behavior in certain decision situations typically violates the axioms. Second, rational choice theory can be employed as a micro-theory in paradigmatic examples of micro-macro models such as the perfect market model of neoclassical economics. That model shows that under macro-conditions characterizing a perfect market and given rational behavior of producers and consumers, macro-outcomes of micro-level exchange can be derived, including the existence of a Pareto-optimal equilibrium with equilibrium prices and an equilibrium distribution of goods (for a sociological perspective, see Coleman, 1990, pp. 40–41 and passim).

Rational choice theory does assume, for example, that an actor preferring A over B and preferring B over C also prefers A over C (transitivity of preferences). However, rational choice theory *as such* is silent about whether the actor does in fact prefer A over B and B over C. Also, the theory as such does not make assumptions about specific properties of A, B, and C that cause A to be preferred over B and B to be preferred over C. In particular, the theory as such does not assume that actors are purely self-regarding in the sense of 'utility = own money' or in the sense of maximizing (or minimizing) certain material consequences for themselves. Of course, empirical applications require additional 'substantive' assumptions, alongside the rationality postulates, on actors' preferences and beliefs. However, these are indeed additional assumptions, not to be confused with the primary and secondary definition of rationality.

This also shows that rational choice theory is indeed 'wide' in Opp's (2023) sense by allowing in principle for very different substantive assumptions on preferences and beliefs. However, the rationality postulates do require consistency of those preferences and beliefs as sketched above and also imply maximizing behavior so that the theory is indeed inconsistent with micro-level theories such as assumptions on 'bounded' rationality or prospect theory. This is not in line with Opp's 'wide version' (see also Diekmann, 2022, p. 104).

5   Deriving Theorems on Micro-to-Macro Relations: Social Choice Theory

Social choice theory comprises a meanwhile vast amount of work, often employing formal modeling, on collective decision-making and procedures for such decision-making (see Sen, 2017 for comprehensive treatment and List, 2022 for a concise overview). One can see social choice theory as a branch of welfare economics, with ramifications for ethics, political philosophy, and political theory, including normative as well as empirical research problems. Here we briefly focus on how Arrow's (1963) impossibility theorem, a key contribution, presumably *the* classic contribution, to social choice theory, sheds light on how formalization can contribute to theory formation in line with MI.

Arrow considered methods for aggregating the individual preferences of actors over a set of alternatives into a collective preference of the group composed of these actors. This is not only a topic for economists, political scientists, and philosophers, but also for sociologists. For example, Arrow's result has implications for decision-making procedures in formal and informal organizations as well as for effects of such procedures on individual members and their behavior. That Coleman has repeatedly published on problems in the field of social choice theory (including various chapters in 1990, Part III and Part IV) is therefore less surprising than the scarcity of contributions by other sociologists.

From the perspective of MI, social choice theory in general can be seen as studying micro-to-macro links (Arrow 3 in Coleman's diagram). More specifically, in our terminology for Coleman's diagram, Arrow considers well-specified examples of transformation rules and his theorem is a general result on properties of a class of transformation rules.

It is rather common in the literature to start expositions and discussions of Arrow's theorem by first referring to an example that provides intuition on what Arrow is studying (as Coleman, 1990, pp. 397–398 does, too). The example is from work pursued already in the eighteenth century, namely Condorcet's paradox. This paradox is also useful for our purposes because it illustrates a useful result that can be obtained without elaborate formal reasoning. Consider the following situation with three actors and strict preferences over three alternatives A, B, and C as follows. For Actor 1, we have $A \succ B \succ C$, the ordering for Actor 2 is $B \succ C \succ A$, while for Actor 3 we have $C \succ A \succ B$. Note that each actor's preferences are rational, namely, consistent in the sense of transitivity (cf. Sect. 4). We now apply a seemingly plausible and perhaps also 'democratic' rule for aggregating the actors' individual preferences into collective preferences for the group: majority rule. According to this rule, the collective preference over alternatives A_i and A_j is $A_i \succ A_j$ if a majority of the actors has individual preferences $A_i \succ A_j$. Note that majority rule can be seen as a transformation rule that specifies how micro-outcomes, in this case individual preferences, are to be aggregated into a macro-outcome, in this case collective preferences of the group of actors. For our example, majority rule yields collective preferences $A \succ B$, $B \succ C$, and $C \succ A$. Hence, the collective preferences involve a cycle and are irrational in the sense of violating transitivity.[6] One sees that the intuitively plausible majority rule implies surprising problems.

Arrow (1963) answered the question if Condorcet's paradox highlights a minor aberration or is an example of a more fundamental problem. Arrow's impossibility theorem demonstrates indeed a fundamental problem. Roughly, Arrow proves that there is no procedure for aggregating individual preferences of two or more actors over three or more alternatives into collective preferences for the group that satisfies a parsimonious set of seemingly plausible axioms. It is not necessary here to specify these axioms precisely (for details, see Arrow, 1963; Sen, 2017, Chap. 3*; List, 2022, Section 3.1). They include requirements such as that all logically possible combinations of individual preference orderings are allowed ('unrestricted domain'), while the collective preferences are complete and transitive.[7] Other axioms require that the collective preference over alternatives A_i and A_j is $A_i \succ A_j$ if each actor has individual preferences $A_i \succ A_j$ ('Pareto principle') and that there is no actor such that the collective preference over alternatives A_i and A_j is $A_i \succ A_j$ if that actor has individual preferences $A_i \succ A_j$, irrespective of the individual preferences of all other actors over A_i and A_j ('non-dictatorship').[8] It goes without saying that proving Arrow's theorem, other than the simple though illustrative result on majority rule, does require thorough formal analysis and derivation.

Arrow's theorem is not only interesting as a theoretical result. It has likewise ramifications for empirical research. For example, the theorem shows that there is room for strategic behavior when it comes to collective decision-making. This includes, but is not restricted to the powers of an agenda-setter to affect the outcome of collective decision-making through strategically designing the alternatives between which actors can choose or the order in which choices are made (for fascinating case studies, see Riker, 1986).

Again, note that Arrow's axioms can be seen as characterizing properties of transformation rules that aggregate preferences of individual actors into collective preferences. We can now also see another and at first sight less obvious reason why Arrow's theorem, like related work in social choice theory, is interesting for sociologists. Sociology, similar to other social sciences such as economics and political science, includes the study of 'corporate actors'

---

[6] Also, given the individual preferences of the three actors, the majority rule does not produce a so-called Condorcet winner, an alternative preferred to every other alternative in pairwise comparison.

[7] 'Ordering' refers to complete and transitive preferences over a set of alternatives. Aggregation rules yielding complete and transitive collective preferences are 'social welfare functions' in the sense of Arrow.

[8] Another (and crucial) axiom is on 'independence of irrelevant alternatives', which is of a more technical nature.

(Coleman, 1990, Part III–IV). Examples of corporate actors are firms and other formal organizations and in other applications also more informal organizations like teams and similar groups. It is sometimes assumed that such corporate actors can indeed be treated as actors in the sense of rational choice theory next to 'natural persons' as actors (see, for example, Coleman's, 1990, pp. 4, 12, comments on neoclassical economics when assuming households and firms as actors and on work in political science and history that considers nations as actors). This would imply, however, that corporate actors have preferences that are consistent in the sense of axioms constituting the primary definition of rationality. These preferences would have to be aggregated from the individual preferences of actors having a 'say' concerning the preferences of the corporate actors. For example, in the case of firms, these could be shareholders or managers. Arrow's theorem—and formal analysis proving the theorem—then shows that specifying transformation rules that aggregate those individual preferences adequately into collective preferences of the corporate actors can be a complex and far from trivial task and that such aggregation may not be feasible in certain contexts. This also shows that care should be applied, also in empirical work, when corporate actors are considered as 'actors' in the sense of rational choice theory.

It is noteworthy that Arrow (2014) himself, in an autobiographical note on how his own ideas and his own work leading to his theorem developed, mentions that thinking about the relations between preferences of shareholders and standard assumptions of neoclassical theory on firms has been important for him in the context of discovery. It could well be, certainly given the background of his own work on social choice, that Coleman's arguments on the need for careful specification of transformation rules have been likewise motivated, at least to some degree, by what can be learned from social choice theory on problems of preference aggregation (see, for example, some remarks in Coleman, 1987, pp. 182–184).

Remark Our examples so far are related to one specific and influential approach within MI, namely, rational choice theory and its applications in the social sciences, broadly conceived. For various reasons (see Sect. 2 of this chapter) it is useful to make it clear that there are other strands in MI that likewise show that formal modeling, while not always needed, sometimes provides answers to questions that could be hardly answered by relying solely on informal reasoning. Again, we turn to some instructive examples.

Methodological Individualism and Formal Models 387

6 The Small-World Problem

In 1967, Stanley Milgram published the results of an experiment (Milgram, 1967). In the experiment, 296 residents of Kansas and Nebraska were given an envelope containing some documents. The intended destination for the envelope was 'Target Person' (TP), someone at an unspecified address in Massachusetts. One of the documents in the folder asked residents to send the envelope to TP if they knew the person or if not, to send the folder to someone they thought was more likely to know TP, who would then in turn be asked to forward the envelope. The envelope also included a roster in which each participant along the chain could write their name, so that the chain was tracked.

At the time there was reason for skepsis that these letters had any reasonable chance of reaching their destinations. There was no Facebook; no one had any data on what the global social network looked like. So no one knew whether a route even existed from most origins to most destinations, let alone that this route be findable through decentralized search without a roadmap. Moreover, even if a route existed in most cases, it would have had to be very short to be successfully traversed: The mail chain at each consecutive step had to overcome a new the risk of prematurely being ended because of an uncooperative subject or failed delivery. Milgram nonetheless found that folders arrived at their destinations in many cases, 64 out of the original 296. The median number of cooperating subjects in a chain, excluding TP, was only six.

How can we understand the surprising success of Milgram's small-world experiment? Without elaborate formalization one can see that if we are all connected to random people then path length must be short, because if you have 100 friends who each have 100 other friends who each have 100 other friends, then in five handshakes you are connected to 10 billion people, more than there exist on this planet. But we also know that connections are not at all to random people. Instead, the friends of friends are often already your friends. Most ties exist in local clusters of people organized around a focus like a geographical location, workplace, or shared interest. The global network must exhibit high local clustering. How can the world then still be small and can envelopes in Milgram's experiment arrive at their faraway destinations in six hops? This is the small-world problem.

A satisfactory solution was found three decades later. A simple and famous model (Watts & Strogatz, 1998) revealed that a network can be 99% local and 1% random and still exhibit short path length. All that is required is that a few connections are to faraway people. Watts and Strogatz (1998) started with a regular ring lattice structure consisting of n nodes, with k ties per node going to the nearest other nodes on the ring (Fig. 3, middle). Each node represents a person and each tie a friendship. This is a very large world: The average path length is $\approx n/(2k)$ and scales linearly with n. Achieving an average path length of six between any two Americans in a 1960s world with a US population of 200 million would require people to have $k = 200 \text{ m}/(2 \times 6) = 17$ million friends. This would then appear to be a terrible model. But all Watts and Strogatz needed to do now to solve the small-world problem, is tweak this model slightly, by allowing a small probability of a random tie. They replaced a small percentage of local ties with random ties (Fig. 3, right). These random ties, it turns out, provide the necessary shortcuts that make an otherwise very large world very small. The perturbed lattice has an average path length that approximates that of a random network (Fig. 3, left) like we sketched in the previous paragraph (Newman et al., 2000), equal to $\ln(n)/\ln(k)$. It scales logarithmically with increasing population size. Now we only need a minimum of $k = (200 \text{ m})^{1/6} = 24$ friends per person to explain Milgram's result.

This model solves the small-world problem. The analysis reveals a micro-macro contrast: Actors have mostly local connections, yet the world itself is a village that connects people far away in remarkably few steps. There is no proportionality in the relationship between the degree of localness of actors and the degree of global connectedness. To appreciate this, consider the situation from the perspective of Coleman's diagram (Fig. 1). The bridge assumption is that transportation costs (macro-condition) negatively impact the cost of random ties (individual incentives), leading actors reacting to such incentives (micro-theory) to entertain fewer of them (individual behavior). The transformation rule makes clear that such changes in individual behavior

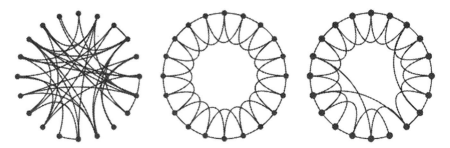

Fig. 3 Watts-Strogatz graphs (Watts & Strogatz, 1998) with $n = 20$ nodes and $k = 4$ ties per node, and random tie probabilities $p = 1$ (left), $p = 0$ (middle), and $p = 0.075$ (right)

will have practically no impact on the smallness of the world (macro-outcome). It only takes a minimal number of random ties to make the world very small. Adding more ties makes it only marginally smaller. Indeed, while the increased affordability of flying and the advent of the Internet have dramatically reduced the costs of maintaining long-distance relationships and have increased such relationships considerably in the past half-century, the world has only become slightly smaller. Facebook has been estimated to have reduced the number of handshakes from six to four (Ugander et al., 2011).

This lack of proportionality in the relationship between the number of long-distance ties in a network and the length of shortest paths is why informal reasoning about the small-world problem could not solve it, why a model was needed. Informal reasoning can surely predict that more random-ness in connectivity will lead to shorter path length, because connections become less redundant so more people can be reached faster. But crucial is what this relationship looks like. Is it gradual or discontinuous, do you need a lot of randomness or only a little? Can we have a world that is mostly clustered and not so random and still have very low path length? Informally it is hard to reason that a subtle, non-gradual relationship between randomness and path length is what it takes to solve the small-world problem.

7 Continuous Versus Threshold Behavior in the Schelling Model

Thomas Schelling (1971) showed that extreme segregation does not require extreme preferences for segregated living. Extreme segregation can also readily emerge when in-group preferences are only moderate. Suppose two types of actors, Xs and Os, live on one side of a street. They are happy with any next-door neighbors as long as they are not both of the opposite type. Then, integrated street patterns like XOXOOXOXXO tend to make at least some actors unhappy (namely actors 2, 3, 6, and 7), while segregated streets like OOOOOXXXXX satisfy everyone. Because the only reason to move can be the presence of too many out-group neighbors, and because moves render both origin and destination neighborhoods more segregated, dynamics cannot run from less integrated streets to more integrated streets. The only hope for integration is a jammed pattern like XOOXXOOXXO, in which everyone has exactly one outgroup neighbor. But here a minimal distur-bance is enough to render someone unhappy whose response would lead the configuration to disintegrate.

The Schelling model predicts a ready tendency for high degrees of segregation in classrooms, cities, sport clubs, and at dining tables. The macro-outcomes will result even when the micro-level incentives to segregate are weak. Applied to black-white segregation in US cities, the Schelling model is consistent with the observation that US citizens grew more tolerant of other-race neighbors in the second half of the twentieth century, while segregation did not diminish (Coleman, 1971). As long as actors (and especially whites) avoid being in a local racial minority, neighborhoods will continue to lack racial diversity. To be sure, the Schelling model is not necessary to explain persistent discrimination in cities. Other mechanisms such as the presence of institutional racism suffice. Their segregation-inducing effects may be self-evident and not require formalization. Rather, the point is that even in the absence of other mechanisms, segregation can continue to be very high. Macro-outcomes are not commensurate to incentives and behavior on the micro-level: mild intolerance and strong intolerance produce similar segregation levels.

Schelling's result has generally been regarded as highly robust to changes in model assumptions, even such changes as permitting a preference for mixed over in-group neighborhoods (e.g., Zhang, 2004). But in 2006 Bruch and Mare (2006) made an important discovery. They found that Schelling's result breaks down when an actor's response function is not a simple threshold—too many outgroup neighbors and an actor will leave—but rather a probabilistic and continuous function of the fraction of outgroup neighbors—the more outgroup neighbors, the more likely an actor will leave.

Formally, Bruch and Mare consider the probability that an actor at time $t + 1$ will move to some neighborhood to be ce^{f_t} where f_t is the fraction of same-type actors living in that neighborhood at time t and c is a normalizing constant that ensures probabilities across all available neighborhoods sum to 1. Simulating the thus altered Schelling model, they find that neighborhoods fail to segregate. Bruch and Mare then consider an analytically more tractable scenario consisting of two disjoint neighborhoods. There are equal numbers of actors of type X and O. Each time t, all actors again choose which of the two neighborhoods they wish to live in and stay or move accordingly. Bruch and Mare assume that the fraction f_{t+1} of actors of type X at time $t + 1$ moving to neighborhood 1 is given by $f_{t+1} = \left(1 + e^{1-2f_t}\right)^{-1}$. This fraction is assumed equal to the fraction of actors of type O moving to neighborhood 2. Bruch and Mare numerically iterate through this process, starting from a perfectly segregated setup with $f_1 = 1$, and find that with increasing t, f_t approaches ½, perfect integration. This is the opposite of what Schelling predicted.

Methodological Individualism and Formal Models 391

Van de Rijt et al. (2009) extend this result by considering the more general case of $f_{t+1} = \left(1 + e^{\beta(1-2f_t)}\right)^{-1}$, adding β as a coefficient for the magnitude of actors' in-group inclinations (McFadden, 1973). Note that $\beta^* = 2$ is a pitchfork bifurcation point: the Bruch and Mare result of long-term integration generalizes from the special case of $\beta = 1$ that Bruch and Mare considered to any $0 \leq \beta \leq 2$. Perfect integration is then the only stable equilibrium. For values $\beta > 2$ they recover the original Schelling result: There are now three equilibria, one stable equilibrium in which X is dominant and its mirror image in which O is dominant, and an integration equilibrium that is an unstable saddle point that can only be reached from initial condition $f_1 = 1$ (somewhat analogous to the XOOXXOOXXO example we started off with). The two segregation equilibria exhibit higher levels of segregation for higher β, and these levels rapidly approach perfect segregation. For example, at $\beta = 3$, stable segregation is already at 93% O (X), 7% X (O). The model thus showed that there exists a critical level of responsiveness to neighborhood composition in actors' decision-making below which Schelling's self-segregating populations become self-integrating.

This formalization raises the possibility that sufficient reductions in in-group preferences or sufficient increases in conflicting motives for mobility (housing prices, proximity to school, work, station or highway) *can* eliminate persistent segregation under appropriate macro-conditions (e.g., absence of institutional racism in the case of black-white residential segregation). Any initial level of segregation is then unable to reproduce itself, instead generating a weakened degree of segregation in the future, which in turn gives rise to an even weaker degree of segregation, with integration as long-run macro-outcome.

This insight did not precede but rather followed the modeling efforts. No one had previously informally reasoned anything like this, as far as we know. Agent-based computational models such as the Schelling model allow the exploration of macro-consequences from the systematic altering of model assumptions. It is through the systematic altering of explicit model assumptions that Bruch and Mare were able to discover the counterintuitive result that when neighborhood preferences are continuous rather than discontinuous in neighborhood composition they can be integration-promoting.

8 Self-Correcting Dynamics in Social Influence Processes

The insight from the formal modeling in Sect. 7 is that self-reinforcing residential segregation dynamics can turn self-correcting once the degree of reinforcement is reduced to below some critical point. For segregation to be self-reinforcing, the probability of members of the group that is already in the majority in neighborhood 1 moving into neighborhood 1 must equal or exceed the fraction of members currently living in that neighborhood. This can only happen at $\beta > 2$. This is for example the case at $\beta = 3$ and $f_t = 0.6$, when $f_{t+1} = \left(1 + e^{\beta(1-2f_t)}\right)^{-1} = \left(1 + e^{-0.6}\right)^{-1} = 0.65 > 0.6 = f_t$, but not at $\beta = 1$ and $f_t = 0.6$, when $f_{t+1} = \left(1 + e^{-0.2}\right)^{-1} = 0.5 < 0.6 = f_t$. In the latter case the cascade moves in opposite direction, toward integration. If the pressure to follow others is weak, then no dominant choice takes hold.

This insight is not restricted to segregation dynamics, where choice alternatives are neighborhoods and actors decide between neighborhoods based on how many others have previously done so. Also, it does not require that there are two types of actors X and O. It readily applies to any binary choice a population of decision-makers might face. Indeed, when applied to the broad interdisciplinary field of social influence research, it produces a conclusion that goes counter to prevailing wisdom. The dominant view in scholarship is that when quality of alternatives is hard to observe, actors look to others for clues about what alternative to choose. Social influence is thought to then lead what is already popular to become ever more popular, even if there is in fact no quality difference. But the formalism we have just seen shows that this does not follow unless one assumes a sufficiently strong degree of social influence. Under weak social influence, a currently more popular alternative will be more likely chosen but to a lesser degree, leading that popularity to be reduced. This now shrunken popularity will reduce later choosers' inclination to opt for that alternative, further diminishing support, and so forth, until the advantage is entirely gone. Equilibrium behavior is a population choosing both alternatives in equal numbers, analogously to neighborhood integration.

Van de Rijt (2019) argues that this conclusion of self-correcting dynamics in social influence processes generalizes to scenarios in which one alternative is strictly qualitatively superior to the other. When actors perfectly observe quality, they will simply choose the best alternative. Self-correction is instant. If quality is imperfectly observed, then actors will trade off their personal sense of which alternative is superior, which is more often right than wrong,

Methodological Individualism and Formal Models 393

against popularity. Whether quality can overcome a popularity disadvantage is determined by the strength of social influence. This can then be modeled simply by adding a quality coefficient $q \geq 1$ to the choice model that represents the ratio of the quality of the superior to the inferior alternative, $f_{t+1} = \left(1 + q^{-1} e^{\beta(1-2f_t)}\right)^{-1}$, f_{t+1} being the fraction of people choosing the qualitatively superior alternative. Under sufficiently weak social influence (β small enough), the only equilibrium is a majority of actors choosing the high-quality alternative. This equilibrium f^* for a given β and q is given by $q = f^*(1 - f^*)^{-1} e^{\beta(1-2f^*)}$ and is stable. For example, for $\beta = 1$ and $q = e$, the stable fraction choosing the superior alternative equals $f^* = 0.66$. Even if the initial fraction of actors choosing the superior alternative f_0 is minimal, a self-correcting dynamic will make the high-quality alternative eventually reach popular dominance.

Van de Rijt (2019) tested this prediction of self-correcting dynamics under weak social influence, considering evidence from seven experimental studies. These studies include conformity research from social psychology, information cascade experiments from economics, political mobilization studies from political science, and research on unpredictability in cultural markets by sociologists. This evidence had previously been interpreted to support the idea of collectively irrational crowds trapped in self-reinforcing processes sustaining popular dominance of inferior choices. Each study was instead shown to exhibit the predicted self-correcting dynamics, pushing the system toward a majority of actors making the correct choice.

This final example also shows how a model developed for one specific social science context—segregation research—revealed an unexpected dynamic (step 1) that in turn gave rise to a novel prediction for a different context—social influence research (step 2). It is hard to imagine that informal reasoning alone would have been able to accomplish either the first or the second step. Concerning the first step, informal reasoning tends to be about what one can see plainly rather than about what has never happened before. Reaching the conclusion that racially biased whites can produce integrated cities requires thinking away many obvious roadblocks to such a hypothetical dynamic, not even considering the negative reception such a conclusion could count on in politicized academic circles. Instead, the main informal critique of the Schelling model has historically been empirical rather than conceptual, namely the observation that both blacks and whites when surveyed indicate they would prefer to avoid living in majority-black neighborhoods (Krysan & Farley, 2002; Massey & Denton, 1993). However, when assumptions on black and white preferences are made more consistent with such preferences, the Schelling model continues to produce high levels of segregation because whites' tolerance of diversity continues to be lower than originally assumed in the Schelling model (Farley et al., 1978; Fossett, 2006; Harris, 2001), leading them to move out when blacks move in. In a world of racism, inequality, and segregation it is a leap to instead consider the counterfactual possibility of whites' in-group preferences being weak enough to kickstart a self-reinforcing spiral of integration. Yet this is precisely what the Bruch and Mare model suggests as a possibility under the right macro-conditions.

About the second step: Even if one would have informally reasoned the possibility for such a self-correcting dynamic in the case of integration, then extending it to social influence dynamics would be another leap. There is no a priori reason for informal theorizing about residential segregation to find applicability to the download dynamics of music markets. There is nothing intuitive about considering the decision about which song to download analogous to the decision on where to live. Not only are the choice alternatives radically different, so are the social processes generating feedback. In the case of segregation, actors respond to whether other actors exhibit a certain unalterable feature they did not choose (race), while in the case of social influence, actors do not care who others were born to be, but rather about what choices others have made. The realization that these seemingly unrelated scenarios would share a previously overlooked feature of social dynamics simply follows from the abstraction provided by modeling. Since the same model could be used for both scenarios, the same dynamics could be observed in both scenarios.

Lastly, the modeling generates an interesting implication for hypothesis testing in the field of social influence studies. Without a model, there is no criterion for the magnitude of effects. The default form of hypothesis testing has thus been resorted to in past work on social influence. Empirical work on social influence has been focused on testing whether the null hypothesis of no social influence can be rejected. Showing that actors care at least a little bit about what others do has always been the standard. Not surprisingly, this has often turned out to be the case. Coming out of the new analysis is a critical level of social influence that must be crossed to produce the self-reinforcing dynamics sought after that allow populations to settle on the 'wrong' choice alternative. This provides a specific, and more ambitious, non-zero value for the null hypothesis in statistical testing in social influence research moving forward (e.g., see Frey & Van de Rijt, 2021 for such a test).

Methodological Individualism and Formal Models   395

9   Conclusion

By sketching various examples from rather different strands of research, this chapter has highlighted the benefits of formalization and of employing formal models for theory construction in line with MI. We have also addressed that formalization, depending on the specific case at hand, can be fruitful for different 'elements' of theory construction—basically, different nodes and arrows of Coleman's diagram—in line with MI.

We have mentioned various benefits of formalization but concentrated on only one such benefit, although a fundamental one: the derivation of implications that would be hard or impossible to derive employing exclusively informal reasoning. For concluding the chapter, it may be helpful to briefly add a highly selective view at how our examples shed light also on other advantages of formalization.

First, consider once again VOD. Employing game theory as a formal tool requires us to make assumptions explicit that might remain implicit otherwise, while these assumptions do affect implications. One such assumption is that VOD is a noncooperative game, meaning that the actors involved cannot engage in contractual agreements or one-sided commitments on whether or not to contribute to the production of the collective good. This assumption is indeed crucial for deriving the solution 'symmetric equilibrium in mixed strategies' and for the implications of that solution. If actors could engage in contractual agreements, well-known rationality postulates for cooperative games like 'joint efficiency' (Harsanyi, 1977, p. 198) would imply that actors would agree to determine one volunteer through a chance mechanism, with expected payoffs $U - K/N > U - K$ for each actor, where $U - K$ is each actor's expected payoff in the symmetric equilibrium in mixed strategies. Further work on VOD, based on careful formalization of assumptions and sometimes rather complex derivation of implications, has moreover shed light on the implications of *different* sets of assumptions. This includes comparison of the standard symmetric version of VOD, also employed in this chapter, with an asymmetric version in the sense that payoffs are not the same for each actor (Diekmann, 1993; Weesie, 1993). It also includes comparing the standard version of VOD with one such that timing of providing help becomes an issue (Weesie, 1993, 1994). This work shows how implications do differ per set of assumptions. Similarly, in models of segregation and social influence we have seen that the assumption that preferences are 'strong' (in the sense sketched above) turns out to be necessary for the implications of, respectively, segregation and inferior lock-in.

Second, at the outset of this chapter we suggested already that formalization could reveal that certain implications do not at all depend on certain assumptions. The small-world problem offers a good example. While the goal of the small-world model was to offer a solution for a social science puzzle, it unexpectedly also achieved something profound and rare: A formalization from social science that could be exported to yield important findings in biological and physical systems. It revealed that the narrow sociological formulation of the problem in terms of acquaintance networks was unnecessarily restrictive. The model used to show the counterintuitive co-existence of high clustering and short path length in theoretical networks does not require the assumption that nodes be people, nor that connections be acquaintanceships, nor that the space in which nodes and ties are located be geographical. This allowed the insight produced by the model to be vastly more general. The model finds applicability in any setting in which there is some notion of connectivity and of distance. The Watts and Strogatz paper empirically demonstrates the breadth of its applicability, from the neural network of a worm to movie-actor networks to power grids. Each is shown to exhibit the small-world property that it is very locally wired yet capable of fast global transmission. Scholarship in the interdisciplinary field of network science, for which this is generally regarded to be one of the foundational papers, has subsequently discovered the combination of high clustering and short path length in a vast universe of networked systems, with applications ranging from the design of efficient computer systems to disease prevention.

This chapter shows that formalization and formal theory construction should be employed with care in MI and social science in general: it is not an aim in itself but is a 'tool' for achieving other aims. It is not always necessary to use the tool. When clear informal reasoning suffices to derive conclusions, time and effort spend on formalization can better be invested otherwise. However, certainly when it comes to theory that is characteristic of MI, namely, involving micro- as well as macro-levels of analysis and assumptions linking those levels, formalization and formal theory construction can be useful by facilitating the derivation of implications from a set of assumptions.

References

Arrow, K. J. (1963). *Social choice and individual values* (2nd ed.). Wiley.

Arrow, K. J. (2014). The origins of the impossibility theorem. In E. Maskin & A. Sen (Eds.), *The Arrow impossibility theorem* (pp. 143–148). Columbia University Press.

Bruch, E. E., & Mare, R. D. (2006). Neighborhood choice and neighborhood change. *American Journal of Sociology, 112*(3), 667–709.

Camerer, C. F. (2003). *Behavioral game theory*. Russell Sage.

Coleman, J. S. (1964). *Introduction to mathematical sociology*. Free Press.

Coleman, J. S. (1971). *Resources for social change: Race in the United States*. Wiley.

Coleman, J. S. (1986). Social theory, social research, and a theory of action. *American Journal of Sociology, 91*(6), 1309–1335.

Coleman, J. S. (1987). Psychological structure and social structure in economic models. In R. M. Hogarth, & M. W. Reder (Eds.), *Rational choice. The contrast between economics and psychology* (pp. 181–185). University of Chicago Press.

Coleman, J. S. (1990). *Foundations of social theory*. Belknap Press of Harvard University Press.

Coleman, J. S. (1993). Reply to Blau, Tuomela, Diekmann and Baurmann. *Analyse & Kritik, 15*, 62–69.

Diekmann, A. (1985). Volunteer's dilemma. *Journal of Conflict Resolution, 29*(4), 605–610.

Diekmann, A. (1986). Volunteer's dilemma. A social trap without a dominant strategy and some empirical results. In A. Diekmann, & P. Mitter (Eds.), *Paradoxical effects of social behavior* (pp. 187–197). Physica.

Diekmann, A. (1993). Cooperation in an asymmetric Volunteer's Dilemma game. *International Journal of Game Theory, 22*, 75–85.

Diekmann, A. (2022). Rational choice sociology: Heuristic potential, applications, and limitations. In K. Gërxhani, N. D. De Graaf, & W. Raub (Eds.), *Handbook of sociological science: Contributions to rigorous sociology* (pp. 100–119). Edward Elgar.

Diekmann, A., & Voss, T. (2004). Die Theorie rationalen Handelns. Stand und Perspektiven [The theory of rational action. State of the art and perspectives]. In A. Diekmann, & T. Voss (Eds.), *Rational-Choice-Theorie in den Sozialwissenschaften. Anwendungen und Perspektiven* (pp. 13–29). Oldenbourg.

Farley, R., Schuman, H., Bianchi, S., Colasanto, D., & Hatchett, S. (1978). "Chocolate city, vanilla suburbs:" Will the trend toward racially separate communities continue? *Social Science Research, 7*(4), 319–344.

Fossett, M. (2006). Ethnic preferences, social distance dynamics, and residential segregation: Theoretical explorations using simulation analysis. *Journal of Mathematical Sociology, 30*(3–4), 185–273.

Frey, V., & Van de Rijt, A. (2021). Social influence undermines the wisdom of the crowd in sequential decision making. *Management Science, 67*(7), 4273–4286.

Gërxhani, K., De Graaf, N. D., & Raub, W. (Eds.). (2022). *Handbook of sociological science: Contributions to rigorous sociology*. Edward Elgar.

Goldthorpe, J. (2016). *Sociology as a population science*. Cambridge University Press.

Harris, D. R. (2001). Why are whites and blacks averse to black neighbors? *Social Science Research, 30*(1), 100–116.

Harsanyi, J. C. (1977). *Rational behavior and bargaining equilibrium in games and social situations*. Cambridge University Press.

Hempel, C. G. (1965). *Aspects of scientific explanation*. Free Press.

Hummell, H. J. (1972). Zur Problematik der Ableitung in sozialwissenschaftlichen Aussagensystemen [On the problem of deduction in social science systems of propositions]. *Zeitschrift für Soziologie, 1*, 31–46, 118–138.

Hummell, H. J. (1973). Methodologischer Individualismus, Struktureffekte und Systemkonsequenzen [Methodological individualism, structural effects, and system level consequences]. In K.-D. Opp, & H. J. Hummell (Eds.), *Probleme der Erklärung sozialer Prozesse II: Soziales Verhalten und soziale Systeme* (pp. 61–134). Athenäum.

Kahneman, D., & Tversky, A. (Eds.). (2000). *Choices, values and frames*. Cambridge University Press.

Krysan, M., & Farley, R. (2002). The residential preferences of blacks: Do they explain persistent segregation? *Social Forces, 80*(3), 937–980.

Lindenberg, S. (1992). The method of decreasing abstraction. In J. S. Coleman, & T. J. Fararo (Eds.), *Rational choice theory: Advocacy and critique* (pp. 3–20). Sage.

List, C. (2022). Social choice theory. In E. N. Zalta (Ed.), *The Stanford encyclopedia of philosophy* (Spring 2022 edition). https://plato.stanford.edu/archives/spr2022/entries/social-choice/.

Massey, D., & Denton, N. (1993). *American Apartheid: Segregation and the making of the underclass*. Harvard University Press.

McFadden, D. (1973). Conditional logit analysis of qualitative choice behavior. In P. Zarembka (Ed.), *Frontiers in econometrics* (pp. 105–142). Academic Press.

Milgram, S. (1967). The small-world problem. *Psychology Today, 1*(1), 60–67.

Newman, M. E., Moore, C., & Watts, D. J. (2000). Mean-field solution of the small-world network model. *Physical Review Letters, 84*(14), 3201.

Olson, M. (1971). *The logic of collective action* (2nd ed.). Harvard University Press.

Opp, K.-D. (2023). Methodological individualism and micro-macro modeling. In N. Bulle, & F. Di Iorio (Eds.), *Palgrave handbook of methodological individualism*. Palgrave Macmillan.

Rasmusen, E. (2007). *Games and information: An introduction to game theory* (4th ed.). Blackwell.

Raub, W. (2021). Rational choice theory in the social sciences. In M. Knauff, & W. Spohn (Eds.), *The handbook of rationality* (pp. 611–623). MIT Press.

Raub, W., De Graaf, N. D., & Gërxhani, K. (2022). Rigorous sociology. In K. Gërxhani, N. D. de Graaf, & W. Raub (Eds.), *Handbook of sociological science: Contributions to rigorous sociology* (pp. 2–19). Edward Elgar.

Raub, W., & Voss, T. (2017). Micro-macro models in sociology: Antecedents of Coleman's diagram. In B. Jann, & W. Przepiorka (Eds.), *Social dilemmas, institutions, and the evolution of cooperation* (pp. 11–36). De Gruyter.

Van de Rijt, A. (2019). Self-correcting dynamics of social influence processes. *American Journal of Sociology, 124*(5), 1468–1495.

Van de Rijt, A., Siegel, D., & Macy, M. (2009). Neighborhood chance and neighborhood change: A comment on Bruch and Mare. *American Journal of Sociology, 114*(4), 1166–1180.

Riker, W. H. (1986). *The art of political manipulation*. Yale University Press.

Rodrik, D. (2015). *Economics rules. Why economics works, when it fails, and how to tell the difference*. Oxford University Press.

Sandler, T. (1992). *Collective action*. University of Michigan Press.

Schelling, T. C. (1971). Dynamic models of segregation. *Journal of Mathematical Sociology, 1*(2), 143–186.

Sen, A. (2017). *Collective choice and social welfare* (Expanded ed.). Harvard University Press.

Tirole, J. (2017). *Economics for the common good*. Princeton University Press.

Tutić, A. (2014). Procedurally rational volunteers. *Journal of Mathematical Sociology, 38*(3), 219–232.

Ugander, J., Karrer, B., Backstrom, L., & Marlow, C. (2011). *The anatomy of the Facebook social graph*. arXiv:1111.4503.

Wakker, P. P. (2010). *Prospect theory: For risk and ambiguity*. Cambridge University Press.

Watts, D. J., & Strogatz, S. H. (1998). Collective dynamics of 'small-world' networks. *Nature, 393*(6684), 440–442.

Weesie, J. (1993). Asymmetry and timing in the Volunteer's Dilemma. *Journal of Conflict Resolution, 37*, 569–590.

Weesie, J. (1994). Incomplete information and timing in the Volunteer's Dilemma. A comparison of four models. *Journal of Conflict Resolution, 38*, 557–585.

Wippler, R., & Lindenberg, S. (1987). Collective phenomena and rational choice. In J. C. Alexander, B. Giesen, R. Münch, & N. J. Smelser (Eds.), *The micro-macro link* (pp. 135–152). University of California Press.

Wittek, R., Snijders, T. A. B., & Nee, V. (Eds.). (2013). *Handbook of rational choice social research*. Stanford University Press.

Zhang, J. (2004). Residential segregation in an all-integrationist world. *Journal of Economic Behavior & Organization, 54*(4), 533–550.

Ziegler, R. (1972). *Theorie und Modell* [Theory and model]. Oldenbourg.

Controversial Issues Surrounding Methodological Individualism

Holistic Bias in Sociology: Contemporary Trends

Ieva Zake

1 Introduction

In his seminal critique of social sciences Karl Popper argued that the sociological perspectives, which root themselves in the ideas of Plato, Hegel, and Marx suffer from the negative influence of historicism and its affiliated idea of holism. His criticism was perceived as both controversial and devastating at the time (see, for example, Watkins, 1957, p. 107) as it outlined the intellectual threats of historicism and holism to the future of not only the social sciences, but also the entire political and moral project of an open and nontotalitarian society. Popper's critique of historicism and holism in the social sciences is as relevant today as it was when he wrote "The Open Society and its Enemies" and "The Poverty of Historicism." In this chapter, I will apply Popper's critical conceptual tools to demonstrate and discuss the presence of holism in contemporary sociological perspectives. By analyzing examples of recent sociological research, I will argue that the consequences of holism persist in today's social sciences as they did when Popper first highlighted them. The problems that plagued and weakened historicist social science in Popper's time are no less worrisome today.

I. Zake (✉)
Department of Humanities and Social Sciences, Millersville University, Millersville, PA, USA
e-mail: Ieva.Zake@millersville.edu

© The Author(s), under exclusive license to Springer Nature Switzerland AG 2023
N. Bulle and F. Di Iorio (eds.), *The Palgrave Handbook of Methodological Individualism*, https://doi.org/10.1007/978-3-031-41508-1_17

404   I. Zake

2   Popper's Critique of Historicism and Holism

Karl Popper was deeply concerned that theories and methods of social sciences, which were rooted in Platonian, Hegelian, and Marxist ideas were detrimental to the pursuit of rational research. He also saw their potential to undermine and threaten the foundational Western doctrine that prioritized the value of individual and was necessary to guarantee the possibility of an open and non-totalitarian society. What made these ideas dangerous was the way in which they promoted the intertwined and ultimately flawed mindsets of historicism and holism. Popper's warning was that thanks to Plato, Hegel, and Marx, historicism and holism were spreading wildly throughout politics, philosophy, and social sciences with detrimental results.

Historicism, according to Popper, was a broad, overarching conceptual and also ideological framework that interpreted history as a macro-level movement that followed its own logic or so-called historical-dialectical laws. Moreover, it posited that history evolved independently of the social actors on the ground thus rendering individual agency inconsequential. When applied in social sciences, historicist research aimed to generate abstract and deliberately vague forecasts about predetermined historical developments. They insisted that application of science-based and rational research methods such as generalization, experiment, falsifiability, objectivity, neutrality, and even pursuit of true knowledge was not appropriate for social sciences. Instead, historicist social scientists dedicated themselves to uncovering the inevitable and predetermined "rules" or "forces" of History. Specifically, Marx, whom Popper identified as one of the most influential historicists, trained his followers among social scientists to look "upon the human actors on the stage of history, including the 'big' ones, as mere puppets, irresistibly pulled by economic wires – by historical forces over which they have no control. The stage of history, he taught, is set in a social system which binds us all; it is set in the 'kingdom of necessity'" (Popper, 1971b, p. 101). Following these Marxist guidelines to see individual actions as predetermined by the "rules of History," historicist research produced analyses of social movements and class conflicts that explained them within the context of all-powerful macro-level processes. No matter what individuals and groups would set out to act upon or do, their contributions to the historical change were seen by historicists as merely feeding into the grand and self-perpetuating engine of historical process. Individuals' actions and choices were meaningful and valuable only when seen through the ever-expanding lens of the macro-level historical narrative. In other words, historicism taught social scientists that only those social and political activities that fulfilled the impending historical plan were expected to generate any noteworthy results (Popper, 1944a, p. 101). As a result, historicism led social scientists away from conceptual and methodological interest in the choices and social impact of individuals and their activities.

At the instruction of Marx, historicist social scientists were deeply committed to studying, to the point of glorifying, large-scale social and political changes such as revolutions and other radical political movements. In their view, such processes reflected History's inevitable and true logic (without any individual "interruptions") in the most clear and articulate way. In fact, historicist social scientists showed a distinct desire to seek out, confirm, and re-affirm indications of impending revolutions, radical upheavals, and socio-political conflict as the ultimate mission of their research. They not only focused on identifying and analyzing dramatic transformations, but also in many cases advocated for them and even joined in the movements as "scholar-activists" involved in the so-called "praxis." This was due to their self-perception as interpreters of social conflicts and change into the language of overarching, predetermined, and thus ultimately unchangeable course of History (Popper, 1944a, p. 102). That is, on the one hand, historicists passionately wanted to see themselves as revolutionaries and contrarians to the existing social arrangements; on the other hand, according to historicists themselves, their role was just to hasten the arrival of the already pre-determined historical outcomes (ibid.: 88; also 1944b, p. 125). To explain this paradox or contradiction, Popper concluded that historicists sought comfort in the world that was continuously changing by "clinging to the notion that change can be foreseen because it is ruled by an unchanging law" (Popper, 1945, p. 89). Historicist social scientists worshipped History as an irrational force, thus proving there was not much rationality and science in their conceptual framework.

Methodologically, historicism postulated that societies could be understood only through change. In fact, any society could reveal its essence and potential solely through radical transformation. This led historicist social scientists directly to a conclusion that radical, disruptive transformations were morally superior, while tradition and incremental social and political changes were corrupt, ineffective, and counter-historical. Popper was concerned that this mindset was making social scientists prone to falling for ideologies promoted by idealistic extremists and dictators if they were convincing enough in presenting themselves as agents of the "laws" of pre-ordained History's grand scheme.

The next conceptual step that resulted from the focus on overarching History was the historicist social science's insistence that the only valid history was that of the social "wholes"—social groups, classes, and societies at large (Popper, 1944a, p. 91). Historicism in social sciences was intrinsically tied to what Popper called "holism." It meant that social collectives or entire societies were not seen as constituted by individuals and social relations among them. Instead, all societies were treated as integrated organisms that *as wholes* were moving along a certain historically predetermined trajectory (Popper, 1945, p. 72). Historicist/holist social scientists approached the social wholes as entities or beings that followed their own traditions, exhibited unique characteristics, and even had their own spirits or personalities (Popper, 1944a, p. 92). Popper was highly critical of this holist bias or "confusion" (Popper, 1945, p. 92) in social sciences for multiple reasons.

First, Popper argued that holism in social sciences was due to the prevalent influence of Plato's moral and political views, which Popper was exceedingly concerned about. According to him, Plato authored the fundamental tenant of holism, namely, that **individual must be "created for the sake of the whole, and not the whole for the sake" of the individual** (Popper, 1971a, pp. 80–81). Popper found this to be a deeply troubling position that grew out of Plato's hatred for the individual, his freedom, and his "varying particular experiences" (ibid., pp. 103–104). Instead, Plato missed "the lost unity of tribal life. A life of change, in the midst of a social revolution, appeared to him unreal. Only a stable whole, the permanent collective, had reality, not the passing individuals. It is 'natural' for the individual to subserve the whole, which is no mere assembly of individuals, but a 'natural' unit of higher order" (ibid., p. 80). In Plato's view every individual was assigned a role in a social collective that determined their place in the society. The collectives were organized in a way that perpetuated and protected the interests of the society as a whole, not those of individuals. True justice was "nothing but the health, unity, and stability of the collective body" (ibid., p. 106). Plato's ideas about individuals, the state, and the relations between them were collectivist and ultimately totalitarian, and Popper called Plato's philosophy "the morality of a closed society – of the group, or of the tribe; it is not individual selfishness, but it is collective selfishness" (ibid., p. 108). By highlighting this, Popper criticized Plato's holism in opposition to those who perceived Platonian social, moral, and political philosophy "as humane, as unselfish, as altruistic, and as Christian" (ibid., p. 104). Although Popper acknowledged Plato's philosophy as a tremendous achievement and expressed respect for him as "the greatest of all philosophers" (Popper, 1963, p. 335), he was clear about the dangers of Plato's moral and political ideas. As discussed in this

Holistic Bias in Sociology: Contemporary Trends   **407**

chapter, Popper's warnings about the "magic spell" of Plato were right on target as these moral and political positions have increasingly found home in both epistemology and methodology of contemporary social sciences.

Second, as shown by Popper, holism was deepened by the Hegelian idea that the unique spirit of the unified social whole could find its ultimate true realization in the powerful state that acted "on the stage of history" to compete "for world domination" (Popper, 1971b, p. 57). Through Hegel, holist conception **replaced individuals as agents of history with the supreme national or collective spirit and made the will of the state into the driving force of history** (ibid., p. 58). This elevation of the political (or other social) collectivity to the role of a historical agent established a direct link between historicism and holism. It added further philosophical ammunition to Plato's totalitarianism in its "perennial revolt against freedom" (ibid., p. 62).

Third, Popper warned that when Platonian and Hegelian holism was taken up by historicist social scientists, they conceived of and proceeded to study social systems as integrated living organisms that existed independently from their constituting parts, concretely, individual social actors. They approached the social wholes as **self-generating and as able to determine the characteristics and actions of its members**. In the influential holist writings of Marx, "men – i.e. human minds, the needs, the hopes, fears, and expectations, the motives and aspirations of human individuals – are, if anything, the products of life in society rather than its creators" (Popper, 1971b, p. 93). Following this principle, holist social scientists developed methodologies that focused on social processes and collective identities as opposed to studying nuances of individuals' experiences. Holists used collective behavior as the sole source of explanation for individuals' actions because these social scientists assumed from the start that individual choices and decisions were derived from the social whole. Holists saw people as operating in a world where even their most idiosyncratic proclivities could be traced back to the social groups and collectives that they belonged to. Meanwhile, the organically integrated social wholes themselves were "governed by macro-laws which are essentially *sociological* in the sense that they are *sui generis* and not to be explained as mere regularities or tendencies resulting from the behavior of interacting individuals. On the contrary, the behavior of individuals should (according to sociological holism) be explained at least partly of such laws" (Watkins, 1957, p. 106).

In fact, holist sociologists and other social scientists looked at collectives as if they were "real entities, which must be comprehended, intuitively, as wholes, in order that their place in the historical process be explained"

(Cohen, 1963, p. 250). For example, when sociologists proclaimed that a social class "behaved" one or another way, this statement was clearly based on a presumption of a unified or generalized behavior of all individuals constituting this class, rather than an empirical observation of actions taken by the members of this collective. In fact, the empirical characteristics of social "class" itself were vague and lacked concreteness in holist social sciences. Therefore, Popper's fourth critical point was that holist social sciences were nothing more than a form of mysticism because **holists assigned irrational, quasi-spiritual, and super-human capabilities to the social wholes.** These mysterious capabilities could not be studied or measured empirically as they were quite impossible to pin down. Instead, they were to be understood by the social scientists intuitively (Popper, 1944a, p. 91). Holist social sciences failed to effectively describe any concrete, actually existing social wholes because "in every such case it would always be easy to point out aspects which have been neglected, in spite of the fact that they may be important to some context or other" (Popper, 1944b, p. 127). In other words, every social scientist could define and analyze a social whole in their own way based on whatever description fits their research goal. Moreover, in holist social sciences, social collectives were essentially deified. Consequently, holist research about social or political collectivities made prophecies, rather than scientific predictions (Popper 1945, p. 77).

Holist social sciences, argued Popper, were closer to utopianism than rational methodology for studying social processes (Popper, 1944b, p. 126). Holism in social sciences operated as a quasi-religious conviction or bias, putting in question both its scholarly deliverables and potential social policy recommendations. Holism produced policy ideas that privileged large-scale and whole-sale transformations where one type of organic social system would be destroyed in order to be replaced with a different one. In conjunction with historicism, holism in social sciences led to a rejection of incremental, gradual, and careful change in favor of idealistic radical revolutions. In Popper's mind, however, such ideas of social change were dangerous because their deeper conceptual roots, as pointed out above, were fully planted in Platonian totalitarianism. In the following, I will apply these Popperian criticisms to analyze the impact of holism in contemporary social sciences.

3 Holism in Contemporary Sociology

As it will become evident, the bias of holism in social sciences persists and arguably has even deepened since Popper's time. Not heeding Popper's warnings, social scientists have failed to recognize holism's pitfalls; instead, they have applied it to their analyses of modern society with zeal and dedication. Today, holism is not limited to one sociological perspective, but rather presents itself in the full range of theories and approaches. Using Popper's critical tools discussed above, I will identify the specific ways in which the holist bias is currently present in the study of types of social wholes and collectivities that were not considered at the time of Popper's writing. This evidence will lead to a conclusion that social scientists do not appear to be sufficiently concerned about how the prevalence of holism is making social sciences vulnerable to irrationalism.

Sociology of Gender

A new area of social science research that has grown extensively in the recent decades is sociological study of gender. It has become pervasive, even dominant, and it almost exclusively utilizes feminist theoretical principles. Using feminist theory, social scientists who study gender are laser-focused on disconnecting gender identity and roles from any possible biological or natural roots, that is, they demand that gender is seen as unnatural and entirely social. This separation between the socially constructed gender and the facts of biological sex was introduced by feminist social scientists in the mid-twentieth century. Recently, they have advanced an additional notion that one's biological make-up is not actually an unchangeable fact either. Instead, it is a "sex category" that can be adjusted according to the norms of social belonging (see, for example, Gilbert, 2009, p. 109). In other words, one's body is as socially constructed as gender identity.

Importantly, all feminist social scientists reject the analysis of gender through the perspective of individual variation and experience. Rather, it is primarily theorized and studied as a self-generating social whole (this is where holist bias in feminist social science becomes very important). Specifically, classical feminist sociologists and theorists such as Dorothy Smith (1990) and Judith Butler (1990) argued that all aspects of an individual's identity, experience, consciousness, and life-course are determined by socially regulated belonging to the collectivity of gender. Butler stated that "persons only become intelligible through becoming gendered in conformity with recognizable standards of gender intelligibility" (Butler, 1990, p. 122), that is, one becomes recognized as a person in the society only by following socially constructed rules of gender "performance." Meanwhile, Smith focused on differentiated roles for men and women that were created by gendered division of labor. She argued that gender roles were the key element in shaping individual lives as they allowed men to pursue conceptual and theoretical work, while women took care of the physical reality and material needs (cooking, cleaning, caretaking, etc.).

Moreover, these influential gender theorists concluded that scientific concepts, theories, and methodologies that were developed by men were hardly universal or neutral because no one, including scientists, could ever escape being products of socially constructed gender roles and being "performers" of pre-determined gender rules. Sociologists (just like any other scholars) did their scientific work in the context of the material reality of men's lives and their position in the unequal system of gender power. In general, argued feminist social scientists, no knowledge, including the scientific one, could ever be gender-less—it was inevitably always gendered and ideological because it arose from and reflected imbalances of power between the two genders. And because women's experiences were shaped by a different kind of gendered labor or gender "performance," any study of women or by women required alternative methodology, objectivity, and epistemology, that is, one that would account for everyday realities of women's lives. Smith in particular called on the social scientists not to try to reduce, but rather embrace and highlight the politicized nature of knowledge and objectivity.

Building upon these ideas, feminist sociologists such as Donna Haraway, Helen Longino, Sandra Harding, and others utilized the concept of "standpoint epistemology" and demanded that social scientists should require the scientific community to recognize all allegedly "universal" knowledge as androcentric (see Harding, 1992). They also asserted that the androcentric science had to be systematically and deliberately replaced with research generated by women and about women. Feminist social scientists advocated for a forced transformation of social sciences to ensure that research with "women-friendly" bias was seen as "objective" and "scientifically sound" scholarship. In the subsequent decades, this gendered epistemology movement in social sciences continued to grow with feminist sociologists passionately studying how "social identities operate as epistemic resources, yielding reliable knowledge about how power works in the social world" (Olson & Gillman, 2013, p. 66).

Applying Popper's analytical tools, allows us to see that these ideas of feminist social sciences are built on fundamentally holist and historicist principles. Feminist sociological research is founded on a premise, which is accepted,

Holistic Bias in Sociology: Contemporary Trends   411

not proven, that gender is an all-powerful social whole that perpetuates and reproduces itself through struggle for social power. Feminist social scientists posit that individuals cannot escape the collectivity of gender. But it also cannot be empirically characterized or fully identified because it is actually disconnected from the facts of reality, especially that of biology. Gender is a socially and historically contextual, changing, and performative construct. Yet it possesses tremendous superpowers that control and even subsume individuals and their personal differences. Feminist sociologists see belonging to the socially constructed gender as something that determines every individual's world view, attitudes, and also experiences. An individual always knows the world and speaks from the perspective of their social location as determined by their gender. As a result, for feminist social scientists, there is no reason to look at social processes through the lens of an individual because no one is ever able to escape the all-encompassing influence of the socially pre-determined gender identity anyway. In fact, individual social scientists themselves are so controlled by their gender identity that they have no other option, but to study the world through their gendered location within the social system. Feminist social scientists, consistently with their holist vision, conclude that universally shared, objective, neutral, or cross-individual knowledge is impossible.

Feminist-based social sciences agree that gender cannot and should not be analyzed on the individual level because it is experienced collectively, as a social whole. Moreover, individual-based findings could possibly reveal significant variations in how, for example, women relate to or make sense of their femininity or that they may not perennially feel victimized by womanhood or that femininity may appear more natural, rather than performative, to them. In feminist sociology such inquiries and findings are typically inadmissible because, first, they could question the assumption of gender as a controlling social whole; and, second, they could undermine the goal of whole-sale replacement of the androcentric social reality with the feminine or feminist one. These limitations in the practice of feminist social science betray its holist bias, that is, feminist sociology sees the social collective as the primary force in shaping individual experience. The unfortunate outcome of the holist bias is that although feminist social scientists lionize themselves as liberators of those who had been erased by the androcentric science, in reality they are as likely to omit or ignore actual individual experiences and agency. Because of their holistic emphasis on the collective over individual, feminist social scientists end up following ideological, rather than scientific goals. The holist-based epistemology and methodology make feminist social science vulnerable to becoming an instrument of shifting political winds.

Sociology of Race and Ethnicity

Holism has caused similar issues in the contemporary social science of race and ethnicity that has grown significantly in the recent decades, especially in the US (see, for example, Beshara, 2019; Bonilla-Silva & Dietrich, 2011; Bryant, 2011; Grier-Reed et al., 2018; Hurst, 2015; Hughes et al., 2015; Orey et al., 2013; Pyke, 2010; Smith, 2018). Both feminist and race/ethnicity scholars start out with an assumption that individuals' self-understanding and actions are determined by a socially constructed whole. For feminists, it is gender, while for race and ethnicity scholars, it is the racial/ethnic group or collective. Ultimately, there is no escape from the power of one's socially constructed ethnic/racial identity. If any individuals in fact do not see their lives as controlled by their ethnicity, these social scientists then actively advocate reintroducing one to their ethnic collective. For example, a fairly typical study of this genre by Trieu and Lee (2018) analyzes the experience of what they call "internalized racial oppression" using qualitative interview data from predominately college-educated Asian Americans who are 1.5 and 2 generation immigrants in the Midwest. The study is launched from a premise that its subjects suffer from internalized oppression that manifests itself when individuals do not wish to be always associated with their racial or ethnic group or when they express willingness to integrate into the larger supra-ethnic social context. In particular, the urge to integrate is interpreted as a form of "self-hatred" that inevitably reproduces racial inequality (p. 68). Using the framework of internalized oppression, the study proceeds to focus on the stories of individuals who were exposed to ethnic/racial history of their heritage group on college campuses either in classes or ethnic identity-based organizations. These experiences, the study argues, led the subjects to "recognize their oppression and intervene in their reality through actions" (p. 79), that is, they developed a type of critical consciousness of Asian-ness, which allowed them to "resist the white racial frame of whiteness as normative" (ibid.). Through this experience, the previously non-committal identity was replaced with a self-perception that was critical of ethnic and racial relations in American society and allowed the subjects to break out of their internalized oppression.

As indicated by this study, the goal of rediscovering one's "authentic" ethnic identity is to ensure that individuals learn to belong to a social whole. The underlying principle is that only through group belonging one can find a place in the larger society—these identities are omnipotent and essential for the individuals' lives to "make sense." Those who fail to be fully invested in their ethnic identity and do not see their place in the larger ethnic/racial context of power distribution, are treated as "unaware" of themselves. If individuals fail to recognize their place within the social whole, they are described as afflicted by a kind of "false consciousness" (or "internalized oppression") that is mistaken and must be overcome.

Yet, similarly to the feminist holists, race and ethnicity scholars tend to be vague in providing empirically measurable parameters of racial and ethnic groupings that they analyze. And because the racial/ethnic social wholes are not clearly defined, but rather described as social constructions (see, for example, Lightfoot, 2010), holist scholarship does not tend to study the different paths that individuals take within these racial and ethnic collectivities. Instead, they have made a theoretical and methodological commitment to analyze how these vague social wholes control and regulate subjects' selfhood. They are not looking for how individuals might be making sense of, rejecting or actively reproducing their social identities. For example, research that addresses race and ethnicity in education makes an argument that students are locked into historical and persistent inequalities of housing, healthcare, and other socioeconomic structures that lead to disparate learning outcomes for urban, particularly African American, students (see Ladson-Billings, 2017). This research firmly rejects individualistic explanations of educational outcomes among African American students because "we cannot reduce their academic problems to individual failings" (p. 86). All explanations must concentrate on the racialized injustice where "the entire society is arranged against poor urban students" (ibid.). As holists, these scholars prioritize the structure over individual and yet, when talking about her personal story, Ladson-Billings herself proceeds to describe her parents as "strivers" (p. 85) who were driven by a certain set of personal goals that helped them ultimately raise two academically accomplished children.

This principled focus on ethnic/racial social wholes as opposed to studying individual experiences ensures that holist social scientists center their research around power imbalances and inequality among social groups. In fact, these sociologists presuppose from the outset that one's ethnic/racial identity is worth studying because it creates a pathway through which individuals enter the struggle for social power that characterizes the larger social "organism" of the American society. This larger social whole, according to the race and ethnicity scholars, has its own inherent nature or essence, specifically, it is irreparably unjust, competitive, and flawed. "The spirit" or "the character" of the American social whole is corrupt and it facilitates series of interlocking social ladders and barriers to create a system of oppression that privileges and uplifts some groups, while pushing down others. Holist social scientists see the pervasive oppression within the social whole of the American society and explain it as due to "Americanness" as such.

To demonstrate this approach, a study on emergence and closure of charter schools[1] (Paino et al., 2017) begins from a holist premise that these schools should be analyzed through the perspective of the racial identity of their students. Looking at the data, the authors do indeed find some evidence that charter schools with more black students appear to get closed at a higher rate. But other factors such as the school's age and size seem to have a correlation with closures as well. Nevertheless, the study explicitly discounts these other factors and proceeds to conclude that because charter schools exist within the oppressive American social system, one has to believe that school closures are indeed race-based. Thus, the researchers address disparities in the context of the social whole's pre-determined characteristics and specific factors explaining school closures are reduced to the ones that can be attributed to the "personality" of the whole of the American society.

The implication of this type of holist analysis is that individuals are not able to escape outcomes predetermined by the nature or essence of the social whole and how it structures the collectivities within it. Consequently, the ability to evolve, to change and, in fact, to have history belongs to the system, not individuals as agents of history. In another example (Allen et al., 2018), sociologists analyze enrollment and completion data for African American students in public 4-year colleges between mid-1970s and mid-2010s. They note some, but not sufficient, gains for African American students and conclude that racism "runs deep and wide in the DNA of higher education, forming symbiotic relationship with other institutions in U.S. society" (p. 67). These social scientists presuppose that individuals' lives are ultimately shaped by the features of the social whole they live in. Differences in people's life outcomes are explained as due to a "social DNA" or racism's "invisible touch" (Tate, 2016) that is rather vaguely defined and short on empirical metrics. The holist bias guides social scientists to give a lot of influence and power to the generalized social whole of the entire American society and its nature, instead of identifying concrete, measurable failures of opportunity in relevant social arrangements. As a result, holism produces research that puts less emphasis on evidence, but more on ideological critique of "Americanness."

---

[1] Charter schools in the US are distinct public education institutions that have entered into a written charter with the state to achieve specific educational goals (improved instruction, innovative teaching methods, increased learning opportunities, etc.) to be permitted to operate and utilize public funds. They are "schools of choice," which means that families choose to enroll students in these schools regardless of their school district. Charter schools have significantly smaller burden of state or local regulations regarding management and operation.

Holistic Bias in Sociology: Contemporary Trends   415

To summarize, Popper's critical tools help us see the influence of holist tendencies on the current research of race and ethnicity. Its impact can be detected when social science research is not interested in discussing individuals as agents or including multiplicity of social factors in their analyses; when individuals lives are seen as entirely pre-determined by the social wholes; when abstract "personality" of the social whole is analyzed as having unquestionable control over individual actors' experience as social beings. Holist bias produces social science that treats the social whole as existing within itself, outside of the individuals' realities and as self-confirming. Holist social scientists prioritize the bigger story of the social whole rather than individual experiences or group variations. This leaves such research vulnerable to criticism that it is based on preconceived conclusions and functions as an ideology and policy statement, rather than a scientific pursuit.

Environmental Sociology

Another area of contemporary social sciences that has been impacted by holism is the so-called environmental sociology that combines research and activism on climate change. In a broad sense, it presents itself as an attempt to explain why the climate crisis persists "amid the growth of environmental attention and concern" (Stoner and Melathopoulous, 2015 p. 22 quoted in Gundeson, 2017, p. 282). To answer this question, environmental sociologists turn to macro-level analyses of ideologies and social forces that, according to them, fuel the conflict between capitalism and the need to preserve a generally defined "environment." Thus, Ryan Gundeson (2017) argues that sociologists "should investigate ideologies that deny or muffle the changes necessary for addressing the environmental crisis through the method of ideology critique" (p. 282). This method, rooted in Frankfurt School's critical theory, focuses on socio-political ideologies not as made or held by individuals, but as expressed in public policy, political decisions, and public discourse. Environmental sociologists argue that ideologies possess extensive powers to either end or continue environmental issues. To explain the reason for why current policies and practices do not enforce requirements of "sustainable environment," these sociologists blame the wrong ideologies that "conceal contradictions through legitimation and/or reification of the social order" (p. 263). These ideologies exist to preserve the socio-political *status quo* of the larger social whole, and individuals not only accept these ideologies, but also reaffirm them in their everyday lives (p. 270).

Environmental sociologists tend to approach individuals as predetermined by larger frameworks of thinking and acting set by the power of the social whole (namely, capitalism). In their view, even if individuals are consciously aware of the frames through which they are forced to operate, they are unable to escape them. These sociologists analyze environmental policies and political decisions as driven by forces that are larger and more powerful than individual agents or groups on their own. Individuals are rather like pawns in the hands of the ruling ideas and interests of the social whole. Environmental sociologists see the social whole as self-interested and self-perpetuating, and it is not constituted by the individuals and their actions. It generates and preserves itself (see, for example, Foster, 1999, 2012; Munshi, 2000; Pellow & Brehm, 2013). So the general logic of this scholarly approach is that environment-related policies are ultimately made by the larger social whole because it controls social agents who do not speak for themselves but for the forces that constitute the whole. Such framing of research has clear indications of holist bias, especially in terms of seeing individuals as mouthpieces of the social whole, not as social actors. It deifies ideologies as tremendously powerful, even mysterious forces that permeate individuals' lives and thoughts. In sum, environmental sociologists who work from within the critical theory perspective fall under holist social sciences as defined by Popper.

Self-Decolonization of Sociology

As a complement to the holist ideas discussed before, over the last few years sociology has developed a self-critique that calls for decolonization of the social sciences as "part of the global economy of knowledge that grew out of the imperial traffic in knowledge" (Connell, 2018, p. 400). This decolonization perspective sees all knowledge in social sciences as produced through "structural division of labor between periphery and metropole" whereby "the colonized work was, first and foremost, a source of data" (ibid.), while the imperial center processed this data to generate formalized knowledge, theories and subsequently applied sciences (p. 401). As a result, sociological knowledge represents very specific perspectives "that arise from the social formations of the global North, *because of* their historical position in imperialism and their current core position in the neoliberal world economy" (p. 402, emphasis in the original). In other words, sociological ideas that had been treated as universal were actually products of the geopolitical "North" and "West." Moreover, these ideas had exploited or suppressed potential social science research coming out of the colonized "East" and "South." Decolonization of sociology argues that moving forward, social science has to correct "the distortions and exclusion produced by empire and global inequality" and reshape "the discipline in a democratic direction *on a world scale*" (ibid., emphasis in the original).

To achieve these goals, sociologists should deliberately seek out and elevate ideas and research from the parts of the world that have suffered from intellectual colonialism; sociological theoretical frameworks and methods have to be reconceptualized to make them "usable for the social groups marginalized by empire" (p. 403); and sociology has to recognize that knowledge from outside of "metropole" is actually rooted in a completely different epistemology (p. 404). Taken together, these steps should advance the idea that sociological knowledge "is a socially embedded and practiced episteme" (p. 405) that is produced by a certain type of "workforce" residing in particular locations. These producers of sociological knowledge cannot escape their historically determined position in the context of past colonization. In other words, sociology is an intellectual product of individuals who are either colonizers or the colonized, and their position in either side of the historical struggle determines what methods they use, and how and what knowledge they generate. Decolonization of social sciences requires de-emphasizing the Western and Northern ideas and elevating the Eastern and Southern ones.

The impetus behind this demand for change in sociology has elements of holism. As with other holist approaches of contemporary sociology, the decolonization perspective starts out with a premise that individuals cannot escape identities created by the social wholes that they belong to (ethnic or racial, national, gender, class, etc.). The call to decolonize sociology is rooted in the assumption that identities regulate individuals' intellectual pursuits, perspectives, and research—ideas are either those of the oppressor or of the oppressed. Both the oppressing and the oppressed social scientists are not seen as capable of raising above their pre-programmed social reality. Individuals are products of the social, political, and historical contexts that make their knowledge either possible or not. Importantly, to move to decolonize sociology does not advocate for transcending social and political disparities and allowing scholars to generate knowledge that is not predetermined by the social wholes that they are a part of. Instead, decolonizing means reversing the current situation and grating the South and the East a privileged position in the knowledge hierarchy. The fact of the social whole controlling scholarship will remain in place. From Popperian perspective, such an approach would not actually advance scientific knowledge, but rather institute irrational, politically driven controls and regulations over scholarly work.

4   Conclusion

In the recent years, some theorists have tried to find a middle ground in the debate between holism and methodological individualism. They argue that there is an opportunity for social sciences to make a choice between the two approaches on the basis of objective rational criteria such as their usefulness in providing the best possible explanations of the social world (Zahle & Collin, 2014, p. 194). Others have proposed the idea of "explanatory pluralism," where social scientists would alternate between utilizing either holistic or individualistic methodology and epistemology. It is argued that the discussion about holism vs. individualism should move "away from a winner-takes-all debate to a debate about how different approaches should be combined, related or interact" (ibid., p. 172). In addition, it is argued that the different explanatory strategies might work on different levels of the social analysis therefore social scientists need to figure out how these approaches can co-exist and even integrate with each other. Typically, such ideas originate with theorists who focus on the linguistic and reductionist nature of the difference between methodological holism and methodological individualism. Deeper theoretical problems and misunderstandings about methodological individualism exhibited in this approach are discussed elsewhere in this volume. In the context of the present chapter, however, two points are important.

First, these theorists assume that a relatively equal conversation between methodological individualism and holism is possible. But such an assumption is not supported by the reality of contemporary social sciences, as demonstrated here. A careful look at the actual contemporary sociology reveals that social sciences are in fact tilted toward holism with less and less room left for explanations generated by methodological individualism. Holism's influence extends from ideas about how research should be conducted to who is doing it because researchers themselves are seen as inescapably controlled by their social identities. As Popper had warned decades ago, such a situation can make the project of sociology as a scientific pursuit untenable.

Second, the theorists who argue about increasing irrelevance of the debate of holism vs. methodological individualism insist that "today, it is widely held that there is no necessary linkage between being either a methodological individualist or holist, and having a certain political orientation. Thus, discussions of methodological individualism and holism typically take place without any reference to political values of any form" (ibid., p. 7). This statement references an idea that methodological individualism is linked to a vision of a society associated with political liberalism, while holism connects with collectivistic and even totalitarian political beliefs. These theorists argue that social scientists who pursued one or another methodological framework do so simply because they think that one is methodologically superior to the other. This is not a convincing argument considering that choosing holism can produce skewed research outcomes.

In fact, holist bias in social sciences has a strong tendency to politicize or make ideological the research that it produces. Popper was concerned that social scientists' fascination with holism was leading them toward making conclusions and taking positions that were in fact antagonistic to individual liberty in the name of a collectivistic view of the world. Holistic methodology allowed social scientists to confirm, reinforce, and justify as "scientific" their own particular, often political vision. In other words, holist social science advanced research, which could not be justified or substantiated through actual scientific means. It had to be taken at face value as one was expected to approach an ideology. The review of the contemporary uses of holism here should raise flags of caution on this issue today as well.

Consequently, as demonstrated by the presented examples from contemporary social sciences, the importance of seeing a difference between holism and methodological individualism is not and cannot be over. They are disparate as one does in fact promote a collectivistic view and the other focuses on individual-based understanding of the social world. While one sees social reality as a self-generating and powerful whole, the other analyzes social reality through the multitude of complex and unique factors. It cannot be denied that privileging the social whole above the individual has impact on the results on one's research. In reality, holist bias tends to silence attempts to focus on how individuals as social actors that actively construct and change the social world around them, and therefore conversations about broader implications of social sciences utilizing holism are still warranted. Popper's warnings about non-scientific irrationalism and political goals overtaking scientific ones in social sciences are far from having been resolved.

References

Allen, W., McLewis, C., Jones, C., & Harris, D. (2018). From Bakke to Fisher: African American students in U.S. higher education over forty years. *RSF: The Russell Sage Foundation Journal of the Social Sciences, 4*(6), 41–72.

Beshara, R. K. (2019). From virtual Internment to actual liberation: The epistemic and ontic resistance of US Muslims to the ideology of (counter)terrorism Islamophobia/Islamophilia. *Islamophobia Studies Journal, 5*(1), 76–84.

Bonilla-Silva, E., & Dietrich, D. (2011). The sweet enchantment of color-blind racism in Obamerica. *The Annals of the American Academy of Political and Social Science, 634*, 190–206.

Bryant, W. W. (2011). Internalized racism's association with African American male youth's propensity for violence. *Journal of Black Studies, 42*(4), 690–707.

Butler, J. (1990). *Gender trouble: Feminism and the subversion of identity*. Routledge.

Cohen, P. S. (1963). Review of Karl Popper's "The poverty of historicism". *The British Journal for the Philosophy of Science, 14*(55), (November), 246–261.

Connell, R. (2018). Decolonizing sociology. *Contemporary Sociology, 47*(4), 399–407.

Foster, J. B. (1999). Marx's theory of metabolic rift: Classical foundations for environmental sociology. *American Journal of Sociology, 105*(2), 366–405.

Foster, J. B. (2012). The planetary rift and the new human exemptionalism: A political-economic critique of ecological modernization theory. *Organization & Environment, 25*(3), 211–237.

Gilbert, M. A. (2009). Defeating bigenderism: Changing gender assumptions in the twenty-first century. *Hypatia, 24*(3), 93–112.

Grier-Reed, T., Gagner, N., & Ajayi, A. (2018). (En)countering a white racial frame at a predominantly white institution: The case of the African American student network. *Journal Committed to Social Change on Race and Ethnicity, 4*(2), 65–89.

Gundeson, R. (2017). Ideology critique for the environmental social sciences. *Nature and Culture, 12*(3), 263–289.

Harding, S. (1992). Rethinking standpoint epistemology: What is "strong objectivity"? *Centennial Review, 36*(3), 437–470.

Hughes, M., Kiecolt, K. J., Keith, V. M., & Demo, D. H. (2015). Racial identity and well-being among African Americans. *Social Psychology Quarterly, 78*(1), 25–48.

Hurst, T. E. (2015). Internalized racism and Commercial Sexual Exploitation of Children (CSEC). *Race, Gender & Class, 22*(1–2), 90–101.

Ladson-Billings, G. (2017). Makes me wanna holler. *The Annals of the American Academy of Political and Social Sciences, 673*, 80–90.

Lightfoot, J. (2010). Race, class, gender, intelligence and religion perspectives. *Race, Gender & Class, 17*(1/2), 31–38.

Munshi, I. (2000). 'Environment' in sociological theory. *Sociological Bulletin, 49*(2), 253–266.

Olson, P., & Gillman, L. (2013). Combating racialized and gendered ignorance: Theorizing a transactional pedagogy of friendship. *Feminist Formations, 25*(1), 59–83.

Orey, B. D., King, A. M., Lawrence, S. K., & Anderson, B. E. (2013). Black opposition to welfare in the age of Obama. *Race, Gender & Class, 20*(3–4), 114–129.

Paino, M., Boylan, R., & Renzulli, L. A. (2017). The closing door: The effect of race on charter school closures. *Sociological Perspectives, 60*(4), 747–767.

Pellow, D. N., & Brehm, H. N. (2013). An environmental sociology for the twenty-first century. *Annual Review of Sociology, 39*, 229–250.

Popper, K. R. (1944a). The poverty of historicism, I. *Economica, 41*(11), 86–103.

Popper, K. R. (1944b). The poverty of historicism, II. A criticism of historicist methods. *Economica, 43*(11), 119–137.

Popper, K. R. (1945). The poverty of historicism, III. *Economica, 46*(12), 69–89.

Popper, K. R. (1963). *The open society and its enemies. The spell of Plato, Volume I.* Harper & Row Publishers.

Popper, K. R. (1971a). *The open society and its enemies. The spell of Plato, Volume I.* Princeton University Press.

Popper, K. R. (1971b). *The open society and its enemies: The high tide of prophecy: Hegel, Marx, and the aftermath, Volume II.* Princeton University Press.

Pyke, K. D. (2010). What is internalized racial oppression and why don't we study it? Acknowledging racism's hidden injuries. *Sociological Perspectives, 53*(4), 551–572.

Trieu, M., & Lee, H. (2018). Asian Americans and internalized racial oppression: Identified, reproduced and dismantled. *Sociology of Race and Ethnicity, 4*(1), 67–82.

Smith, D. (1990). *The conceptual practices of power: A feminist sociology of knowledge.* Northeastern University Press.

Smith, D. T. (2018). Negotiating Black self-hate within the LDS Church. *Dialogue: A Journal of Mormon Thought, 51*(3), 29–44.

Tate, A. S. (2016). 'I can't quite put my finger on it': Racism's touch. *Ethnicities, 16*(1), 68–85.

Watkins, J. W. N. (1957). Historical explanation in the social sciences. *British Journal for the Philosophy of Science, 8*(30), 104–177.

Zahle, J., & Collin, F. (Eds.). (2014). *Rethinking the individualism-holism debate: Essays in the philosophy of science.* Springer.

Methodological Individualism and Reductionism

Francesco Di Iorio

1 Introduction

This chapter analyzes and critiques an idea that is currently widespread in the philosophy of the social sciences, as well as among some proponents of analytical sociology, namely that methodological individualism (MI) is committed to reductionism.

As understood in this chapter, the concept of reductionism is a rehashed modern version of the ancient concept of atomism and describes the inability of certain sociological and economic approaches to take into account the systemic and socio-cultural constraints that influence individual action. According to the reductionist interpretation of MI, this doctrine is incapable of allowing for the social conditioning of individuals and therefore lacks scientific and epistemological legitimacy. To put it simply, MI fails to

---

An earlier and partly different version of this chapter was published in French in a special issue of *L'Année Sociologique* (2020/2021. Vol. 70, pp. 97–128) guest-edited by Nathalie Bulle. I would like to thank her and the other participants of the conference 'L'individualisme méthodologique aujourd'hui', organized in June 2019 at the French School of Rome, for their valuable comments on this chapter. I would also like to express my gratitude to Jean Petitot and three anonymous referees for their constructive suggestions. Most of this text was translated from the French by Jennifer Clark.

F. Di Iorio (✉)
Department of Philosophy, Nankai University, Tianjin, China
e-mail: francedi.iorio@gmail.com

© The Author(s), under exclusive license to Springer Nature Switzerland AG 2023
N. Bulle and F. Di Iorio (eds.), *The Palgrave Handbook of Methodological Individualism*,
https://doi.org/10.1007/978-3-031-41508-1_18

develop realistic and valid explanations of social phenomena (Di Iorio, 2015, pp. 75–115).

Two important clarifications must be made here. The first is that, while the term 'reductionism' is mainly used in reference to MI with the above-mentioned negative meaning, it is also, though rarely, used in a different way. In particular, Jon Elster (2023) defines his brand of MI as reductionist, but clarifies that this does not mean that he denies the existence of institutional constraints on agents and the possibility to explain social phenomena 'by structural reasons' (see also Elster, 1982; Lukes, 2006, p. 6). It only means that any account of these phenomena must be based on a microfoundationalist research strategy and that explanations in terms of holistic entities that exist over and above the individuals must be rejected (ibid.; 1985). This chapter does not deal with this Elsterian use of the term reductionism, but investigates the legitimacy of the more common use of it with respect to MI. The second clarification that I want to make is that, while MI, as understood by its main theorists, is not committed to reductionism in the negative sense of the word, this is not true for some simplistic variants of individualist explanation that are either based on fictional assumptions (social contract theory, mathematical theory of economic equilibrium) or on psychologism (e.g. Homans' behavioral sociology). In the subsequent pages, MI is used solely to describe the non-atomistic version of this approach——i.e., the version endorsed by its main theorists—while reductionism refers to the widespread and mistaken view that MI is constitutively incapable of accounting for social and systemic constraints on action.

This chapter is structured around three main points. First, it seeks to clarify the precise meaning of the term *reductionism* as it is used in the contemporary literature by critics of MI. Even though those who accuse MI of being reductionist all agree that this approach is incompatible with the systemic analysis of social phenomena and social conditioning, they do not always arrive at this conclusion in the same way. As we will see, there are two types of reductionist interpretations of MI: that of *psychological reductionism*, and that of *semantic reductionism* (Di Iorio & Chen, 2019). In addition, the latter can be divided into a *nominalist variant* and an *anti-nominalist variant*. The variations between these different concepts of reductionism are explained and analyzed in detail, filling a gap in the methodological debate about MI that arises precisely from the fact that, within the framework of this debate, the word *reductionism* is often used without considering its partially polysemic nature.

Second, this chapter shows that both of these reductionist interpretations of MI are incorrect, and that they paint a caricature of this approach that does not reflect the views actually held by its most important theorists. I argue that MI has been a systemic, or structural, approach from the outset, even if, as pointed out above, there are atomistic versions of MI.

Third, this chapter critiques the theory that since MI is reductionist, it should be replaced by a new approach, sometimes called *structural individualism*, that would represent a middle ground between holism and MI (Demeulenaere, 2011; Udehn, 2002). Almost all contemporary critics of MI agree on the need to develop this third way. They mistakenly regard systemic analysis as a characteristic invention of the holistic tradition that is incompatible with MI, although they do recognize the validity of the MI's critique of sociological determinism.

2 Psychological Reductionism

The interpretation of MI in terms of psychological reductionism was developed relatively independently by thinkers from a variety of intellectual traditions. Despite some differences in terminology, they all critiqued MI along essentially the same lines of argument. This interpretation of MI was espoused by the Swedish sociologist Lars Udehn (2001, 2002), for example, who published the only history of MI available in English, and whose work has influenced analytical sociology. The British philosopher Roy Bhaskar (1979), who was the originator of critical realism, also put forward this interpretation of MI, as did two famous proponents of his epistemological approach: the economist Tony Lawson (1997, 2003) and the sociologist Margaret Archer (1995). Another well-known thinker who has supported the idea that MI is based on psychological reductionism is the Popperian philosopher Joseph Agassi (1960, 1975).

According to the thinkers above, MI is psychological reductionism in the sense that it seeks to reduce the social to the mental. From their point of view, MI denies the presence of a social structure characterized by rules, roles, and sanctions, and that exists objectively—that is, independently of what the actors think about their freedom of action in social life. In other words, MI does not take into consideration the fact that constraints must be conceived of as realities that exist outside the minds of individuals and that limit freedom of action precisely because of their external character. This critique of MI, stemming from Durkheim and Marx's work, is based on the central importance that this approach places on the comprehensive method in sociology. The fact that proponents of individualism (such as Menger, Weber, Simmel, and Schütz) have insisted on the need to seek the ultimate cause of

426   F. Di Iorio

action inside the actor, and, in particular, in his or her way of interpreting the world, is seen as proof that MI is a purely subjectivist (and therefore anti-realistic and psychologistic) theory of socio-cultural constraints. These constraints are thereby reduced to purely mental constructions, despite being objective because they exist independently of individual opinion (Di Iorio, 2016b).

Let us consider an everyday example. In many American cities it is against the law to drink alcohol in the street. Now imagine that a European tourist is unaware of the existence of this prohibition and walks through the streets of one of these cities drinking a beer. Even if the tourist thinks that he or she is free to behave in this way, there is in fact an objective constraint that exists independently of his or her subjective opinion. As a result, if a police officer sees the tourist behaving in this way, the officer will intervene. This trivial example highlights the way in which MI is considered incapable of explaining the structural constraints that limit the freedom of the actor by those who interpret it in terms of psychological reductionism.

As Udehn (2002, p. 487) points out, 'according to Weber, von Mises, [and] von Hayek... [s]ociety and culture are subjective phenomena existing only in the minds of individuals'. The difference between objective constraints and psychological beliefs about social constraints is explained by Udehn using the *exogenous/endogenous* dichotomy: *exogenous* refers to objective reality, that which is outside the human mind, while *endogenous* refers to the subjective beliefs of the actors, or to the psychological dimension (ibid.). With regard to MI, Udehn wrote that 'social institutions appear... only among the endogenous variables', meaning that they are only what individuals conceive of them as being in the light of their own interpretation of the social world and its constraints (ibid.: 448). Consequently, MI cannot truly take into account the way in which these institutions create an objective social conditioning by limiting the freedom of the actors: 'institutionalism is incompatible... with... methodological individualism' (ibid.; see also Udehn 2001, p. 355). According to Udehn (2001, p. 318), we must get rid of the original version of MI and use an approach that he calls 'structural individualism', following the Dutch sociologists Reinhard Wippler (1978) and Werner Raub (1982). This interpretation should be seen as a 'synthesis of individualism and holism' (Udehn, 2001, p. 318). The difference between structural individualism and MI is that in MI, 'no causal, or explanatory, power is attributed to social phenomena' (ibid.) in the holistic sense, whereas structural individualism recognizes that social dynamics depend partly on the beliefs and evaluations of individuals, and partly on holistic factors. In the case of structural individualism, objective institutional factors that exist independently of the individual

Methodological Individualism and Reductionism   427

conscience, such as the social roles of the actors and the legal structure of the society, are taken into account in the explanation.

A similar view to that of Udehn is shared by all those philosophers, economists, and sociologists who are inspired by the critical realism of Roy Bhaskar. As Lawson (1997, p. 137)—one of the most famous successors of the British thinker—explained, MI, as it is conceived by theorists of the comprehensive approach in the social sciences, such as Hayek, 'does not acknowledge a reality of social structures existing apart from their being conceptualised in action' (see also Lawson, 1997, p. 28ff). In other words, the problem with MI is that it cannot define these structures 'in... objective terms... but only... in terms of human beliefs' (ibid.: 138–139). This means that society consists 'ultimately, in the opinions, beliefs and attitudes of individual agents' (ibid.: 139). However, this view is mistaken since the 'structures' that govern and influence human interactions are 'irreducible to individual conceptions' (ibid.), even if they 'are dependent upon... the concepts and actions of human agents' (ibid.: 147). Consequently, MI lacks an 'adequate... ontology' (ibid.: 148) of the social world and its structural dimension. It represents an anti-realist and anti-objectivist—or *idealist* (King, 2004)—theory of this world and its constraints, which is erroneous since these result from the fact that 'social practices', 'social rules', and 'social positions' exist as objective realities (Lawson, 2003, p. 39).

Another scholar who has interpreted at least part of the individualist tradition in terms of *psychologism*—in particular, that which relates to the comprehensive sociology developed by Weber—is Joseph Agassi (1960, 1975). A disciple of Popper, he suggested getting rid of the concept of MI in favor of that of *institutional individualism*. The distinguishing feature of this institutional individualism is the fact that it assumes that 'the existing institutions constitute a part of the individual's circumstances which together with his aims determine his behavior' (Agassi, 1960, p. 247). Agassi believes that according to the traditional psychologistic version of MI, these institutions play no role (ibid.; see also Udehn 2002, p. 489). His institutional individualism, on the other hand, 'denies that individuals' aims and physical circumstances alone determine human action' (Agassi, 1975, p. 147). Within the framework of this anti-psychologistic individualism, 'people's opinions enter as a major factor in the social situation; but they enter not so much as personal opinion but rather as institutional or public opinion' (Agassi, 1960, p. 266). In addition, institutional individualism recognizes that 'institutions mould character and character transforms institutions' (ibid.: 267). As Udehn points out when analyzing the approach held by Agassi, 'in the original version of methodological individualism, social institutions are something to be explained in terms of individuals. They appear only in the explanandum or, better, the consequent of an explanation, but never in the explanans, or antecedent. In institutional individualism, on the other hand, social institutions explain and, therefore, also appear in the explanans, or antecedent of an explanation' (2002, p. 489). Institutional individualism, according to Agassi, is a happy medium between holism and individualism. It seems obvious that this new type of individualism proposed by Agassi is very close to Udehn's concept of structural individualism mentioned earlier.

3   Collective Beliefs and Their Consequences

Psychological reductionism accuses MI of reducing the social to the mental and of denying that objective sociocultural or institutional constraints influence action. This accusation is based on a misunderstanding of the nature of MI's interpretive approach (*verstehen*). It is certainly true that MI insists that it is necessary to take into account the meaning of human actions, that is, the way that individuals 'see' things (Weber, [1922] 1978, p. 13; see also Bronner & Géhin, 2017, pp. 58–128; Di Iorio, 2015, pp. 55–74; Picavet, 2015, 2018). On the other hand, it is incorrect to say that according to MI socio-cultural constraints should be conceived of in an anti-realistic or purely subjectivist way. To understand this, it is necessary to consider two points.

The first is to note that, contrary to what Udehn and the other scholars mentioned in the previous section have argued, MI does not focus on the issue of interpreting strictly personal beliefs, for example, the isolated opinion of a particular individual about whether or not he or she is free to drink on the streets of American cities (see also Bronner, 2011). In order to explain the socio-cultural constraints that influence action, MI gives little importance to this type of belief because, in the words of the British interpretive sociologist Anthony King (2004, p. 190), these 'are not the basis of social life'. In reality, MI recognizes the centrality of a principle that Alfred Schutz (1998), reinterpreting Weber in the light of Husserl, called *intersubjectivity*. To use the words of Weber ([1922] 1978), this principle precisely identifies the fact that individualist sociology is not the science of strictly subjective opinions, but, on the contrary, the science of typical ways of acting and thinking. In other words, MI applies its interpretive approach to a shared meaning or, if one prefers to use the phrase that Raymond Boudon (2001) has borrowed from Durkheim, to 'collective beliefs'. As Hayek (1952, p. 34) has pointed out, from MI's point of view, social systems must be regarded as 'the implications of many people holding certain views', that is, as 'the consequences of the fact that people perceive the world and each other through sensations and concepts which are organized in a mental structure common to all of them' (ibid.: 23).

The second point to note in order to clearly distinguish MI from any form of psychological reductionism, is simply that within MI, the interpretive conception of collective beliefs—according to which they must be explained in terms of 'good reasons'—is strictly linked to the analysis of the objective and real consequences produced by these beliefs, which are often unintentional (Di Iorio, 2015, p. 104). MI takes into account the different types of social constraints by combining the dimension of *verstehen* and that of the study of aggregate effects produced by collective beliefs (Boudon, [1977] 1979; Cherkaoui, 2006; Yoshida, 2014). For Weber, for example, the caste system in India and the powerful socio-cultural constraints it produces at the social level are the consequence of a set of shared magical and religious beliefs that can be understood through an interpretive process (Weber, 1946, pp. 396–415).

Confusing MI with a theory of psychological reductionism is mistaken, precisely because MI is an interpretive approach that considers the objective consequences of collective beliefs shared by the actors. Among these consequences that define the nature of social conditioning, there is indeed the emergence of socio-cultural or institutional structures, often formed unintentionally (see Di Iorio, 2015, p. 75ff). In the section 'Individuals and systems' below, I will follow in the footsteps of Mises and Hayek and consider the case of the spontaneous order of the competitive market. MI presupposes, on the one hand, the existence of structures that transcend the will and understanding of isolated individuals, and on the other hand, the possibility that these structures will produce a feedback effect on the actors by limiting their freedom of choice, as we will see in more detail in the subsequent pages (see Bulle, 2018; Hayek, 2011). The simple fact that, according to MI, a large proportion of these structures are unintentional renders clearly untenable the idea that this approach aims to reduce the social dimension to the mere content of individual psyches. As Popper (1963, pp. 124–125) wrote, from MI's point of view, 'the real task of the social sciences is to explain those things which nobody wants'[1] (see also Antiseri, 2004).

---

[1] For reasons that will be specified in this section, I consider the dichotomy developed by Udehn between structural individualism, which he conceives as a novel and recent invention within the framework of sociology, and traditional individualism of the psychologistic-type (for example Weber) to be erroneous. Traditional sociological individualism does not seem to me to be psychologistic or anti-structural. The expression *structural individualism* is sometimes used in a sense that is slightly different from that given by Udehn, that is, to refer to the non-atomist individualism of sociologists and to distinguish it from the atomist individualism of orthodox economics (see for example

430   F. Di Iorio

4   Semantic Reductionism

The interpretation of MI in terms of semantic reductionism was developed within analytic philosophy during the second half of the twentieth century, and today it has also become popular in sociology and economics circles. Multiple scholars have contributed to the dissemination of this interpretation of MI, among which we can mention, for example, Mandelbaum, Lukes, Kincaid, Pettit, and Sawyer. The expression *semantic reductionism* that I use to describe this way of thinking about MI is borrowed from Antonio Rainone (1990).

As is well-know, MI locates the cause of social dynamics in individuals and their way of conceiving the world instead of looking for it in holistic-type factors such as, for example, the economic structure—in Althusser's sense— or the *habitus* in the sense of culturalists (see Boudon, 2010, pp. 28, 29; Demeulenaere, 2015; Picavet, 2015). From the point of view of holism, the actors and their motivations have little importance because the individual experience is determined unconsciously and mechanically by hidden causes (Boudon & Bourricaud [1982] 2004, p. 387ff). In order to oppose this type of sociological perspective, proponents of MI have often insisted that social phenomena must be explained in terms of individuals and not in terms of holistic factors (Bronner & Géhin, 2017, pp. 9–13). Take, for example, this famous passage by Max Weber that is often cited in the literature on MI:

> If *I* have now become a sociologist (according to my documents of appointment!) it is to a large measure because I want to put an end to the whole business—which still has not been laid to rest— of working with collective concepts. In other words: sociology, too, can only be pursued by taking as one's point of departure the actions of one or more (few or many) *individuals*, that is to say, with a strictly "individualistic" method. (Weber, [1920] 2012, p. 410)

According to the interpretation of MI in terms of semantic reductionism, the idea that researchers should not explain social phenomena in terms of holistic-type collective concepts means that the social sciences should conform to a principle of semantic reduction of social predicates to individual predicates (Kincaid, 2017; Petroni, 1991; Rainone, 1990). Social phenomena should therefore be explained by presupposing that it is possible

---

Hedström & Bearman, 2009; Manzo, 2014). I have no problem with this kind of use of the concept of structural individualism and even recognize that it may have some practical utility. My analysis has simply aimed to show that, historically, anti-atomist and anti-reductionist individualism is not a recent invention.

Methodological Individualism and Reductionism

to reduce the vocabulary referring to social properties to the vocabulary referring to individual properties. In other words, MI is interpreted as a theory where the *meaning* of descriptions, concepts, and explanations used in the social sciences should be understood as always and necessarily referring to a simple sum of individual properties (see Elder-Vass, 2014; Kincaid, 1986; 1995, pp. 142–190; Kincaid & Zahle, 2019; Lukes, 1973; Mandelbaum, 1955; Ruben, 1985; Little, 1990; Pettit, 1996, pp. 165–215; Petroni, 1991; Rainone, 1990; Sawyer, 2002, 2003; Zahle & Finn, 2014). This type of interpretation of MI is linked to a reformulation, in terms of language analysis, of the old idea that the whole is more than the sum of the parts. As Steven Lukes explains, according to this interpretation of the individualist tradition, 'facts about society and social phenomena are to be explained solely in terms of facts about individuals' (1968, p. 120).

However, any approach based on a semantic reduction of social predicates to individual predicates is 'obviously implausible' (Kincaid, 1986, p. 504). It follows that MI is mistaken. For example, the phrase 'nation X is richer than nation Y' is, from a semantic point of view, irreducible to a sum of individual qualities because it does not mean that every member of nation X is richer than every member of nation Y (Di Nuoscio 2018). Those who interpret MI in terms of semantic reductionism have sometimes used very technical arguments to show the impossibility of this type of reductionism, for example, the famous 'multiple realizations' problem (Kincaid, 1986). I will not consider these technical arguments here because presenting them in detail would be pointless for this analysis, given that I readily admit that they are correct and valid and my approach is simply to explain that MI is not semantic reductionism in the sense specified above (Bulle, 2018).

In summary, interpreted as a form of semantic reductionism, MI should be rejected for two reasons: on the one hand, the analysis of social phenomena must necessarily be based on the use of semantic concepts and laws that are irreducible to individual properties—the whole being, from a semantic point of view, more than the simple sum of the parts; on the other hand, being based on a principle of semantic reduction, MI is unable to recognize the fundamental fact that explanations in the social sciences must take into account social factors (for example, the religious culture of a country) and laws (such as the law of supply and demand) that, in addition to being semantically irreducible to individual or psychological properties, create systemic constraints that influence individual actions.

432   F. Di Iorio

5   Individuals and Systems

The interpretation of MI in terms of semantic reductionism is not plausible for three reasons that are all linked to the same problem: while the analytic philosophers who have put forward this interpretation have certainly developed formally precise analyses of the necessarily systemic nature of the social sciences, they have paid little attention to carefully studying the original work of the thinkers they criticize. Philologically and historically, this interpretation of MI, like that in terms of psychological reductionism considered above, is not valid.

The first reason is simply that semantic reductionism has been explicitly attacked by several theorists who adhere to MI, for example, the sociologists Raymond Boudon and James S. Coleman, or the members of the Austrian school of economics, as well as Karl Popper and some of his followers. These scholars recognized the systemic and irreducible nature of social phenomena. According to Popper, the fact that the whole is more than the sum of the parts in any science is simply a truism:

> the triviality as well as the vagueness of the statement that the whole is more than the sum of its parts seems to be seldom realized. Even three apples on a plate are more than 'a mere sum', in so far as there must be certain relations between them (the biggest may or may not lie between the others, etc.): relations which do not follow from the fact that there are three apples, and which can be studied scientifically. (Popper, 1957, p. 82)

Carl Menger, another famous methodological individualist, also emphasized that social phenomena always present systemic characteristics, long before Popper (see also Antiseri, 2004, p. 141ff; Campagnolo, 2013, 2016). In his view, 'social structures... in respect to their parts are higher units' (Menger, [1883] 1985, p. 142). They are characterized by 'functions' that 'are vital expressions of these structures in their totality' (ibid.:139). According to Menger (ibid.:147), society is a system inasmuch as each of its parts—each individual or each subsystem (such as a family or a business)—'serves the normal function of the whole, conditions and influences it, and in turn is conditioned and influenced by it in its normal nature and its normal function' (see also Simmel, [1892] 1977). Hayek (1967, p. 70), influenced by Menger, wrote that society 'is more than the mere sum of its parts' and that it is a system characterized by the fact that its constituent parts 'are related to each other in a particular manner'. As for Boudon, he notes that there is no example of an explanation that does not refer to systemic properties in the social sciences, structural analysis being quite simply an inevitable and indispensable characteristic of any sociological research (Boudon, 1971, pp. 1–4; see also Boudon & Bourricaud, 2004, pp. 387–388; Di Iorio, 2016b, 361).

As I have already pointed out, methodological individualists have stressed the fact that social phenomena must be explained 'in terms of individuals' (Kincaid, 2017, p. 87). This does not mean that they wanted to develop explanations in terms of semantic reductionism, but that they aimed to provide a critique of sociological determinism, that is, the tendency to explain human action as simply the mechanical product of holistic factors (see Popper, 1966a, 1966b, [1977] 2003). Presupposing that the cause of social phenomena is to be sought at the individual level, and not at the level of holistic-type sociocultural factors, means, among other things, that the influence of emergent structures produced by human interactions (see the second section) is not deterministic, but mediated by the tacit or explicit interpretative capacities of the actors (Boudon, 2010, pp. 30–32; Bulle, 2018; Bronner & Géhin, 2017, pp. 53–128; Di Iorio, 2015, pp. 11–74). Herein lies the crux of the difference between holism and individualism: holism refutes the crucial importance of individual interpretive capacities and, therefore, conceives of action as a mechanical product of the context (Antiseri & Pellicani, 1995; Boudon, 2010; Bulle & Phan, 2017; Bulle 2018; Demeulenaere, 2000; Di Nuoscio, 2018). As Menger (1985, p. 133) explains, from the point of view of MI, human systems are not deterministic machines, since this type of machine is 'composed of elements which serve the function of the unit in a thoroughly mechanical way. They are the result of purely... mechanical... forces'. In contrast, human systems 'simply cannot be viewed and interpreted as the product of purely mechanical force effects. They are, rather, the result of human efforts, the efforts of thinking, feeling, acting human beings' (ibid.).

The second reason why interpreting MI in terms of semantic reductionism is untenable is related to the important role played by the concept of unintended consequences within this approach (see Antiseri, 2004; Laurent, 1994). This concept is concerned with emergent effects that cannot be reduced to predicates that describe the individuals' psychological and behavioral properties, for the simple reason that it refers to neither human will nor to the properties of individual action but to the global behavior of a system that is spontaneous or unplanned (see Boettke & Candela, 2015; Boudon [1977] 1979; Cherkaoui, 2006, pp. 1–15; Elster, 1989, pp. 91–100; Hayek, 1952, pp. 39–40; Jarvie, 1972, pp. 173–178; 2001; Menger, 1985, p. 133; Merton, 1936; Popper, 1957, pp. 157–158; Rainone, 1990). Let us consider the 'complex' version of the theory of the invisible hand of the market developed by Mises and Hayek, two famous individualists, that built upon Adam Smith's work (see Caldwell, 2003, pp. 205–231; Di Nuoscio, 2018; Dumouchel & Dupuy, 1983; Hayek, 1948, pp. 77–91; 2011; Laurent, 1994; Mises, [1922] 1981, p. 28ff; Nemo, 1988, pp. 67–106; Petitot, 2009, 2012, 2016). According to this theory, the functioning of the market is not semantically reducible to concepts and laws that concern individual properties. Indeed, the market is based on an unintentional coordination of economic activities (Di Iorio & Chen, 2019, p. 9). The theory of the invisible hand asserts that the laws governing the economic system are not psychological and strictly individual since the emergent effects prevail over and surpass the intentions of the actors. For example, the law of supply and demand, as conceived by Mises and Hayek, is linked to an aggregation effect that stems from an unintentional mechanism that is irreducible to psychological laws. This fact explains why if someone owns a car and wishes to sell it on, he or she will not be free to determine the selling price without taking into account its market value, which is an emergent phenomenon. According to this Austrian theory of the invisible hand of the market, since a price is determined through a systemic effect that depends on the combination of a very large number of individual actions and the use of distributed information, it cannot be explained using predicates that only describe human desires and individual properties (Hayek, 1948, pp. 77–91; O'Driscoll & Rizzo, 1985).

The third and final reason why interpreting MI in terms of semantic reductionism is wrong is that, as shown by Menger's systemic conception of the social mentioned above, where the whole influences the parts and vice versa, it is untrue that MI denies the existence of semantically irreducible factors capable of creating constraints that affect individuals and limit their freedom (Bouvier, 2011; Bulle, 2018; Demeulenaere, 2015; Di Nuoscio, 2016; Manzo, 2007, 2015; Rainone, 1990; Raub et al., 2011). The example of the individualistic explanation of the market in terms of invisible hand processes set out above is also useful for understanding this point. In the theory of Mises (1981, p. 28ff) and Hayek (1948, pp. 77–91; 2011)—which is partly inspired by the work of Menger—prices, which are unintentional emergent properties, allow the coordination of economic activities precisely because their variation influences the decisions of actors by limiting their freedom of choice (Di Iorio & Chen, 2019, p. 9). There is a feedback mechanism through which the price system, which is an emergent phenomenon created unintentionally by individuals, influences individuals because of their budget limitations (ibid.). This mechanism is governed by a principle of circular causality, or, by a double causality (downward and upward) that allows the adaptation from the global level to the individual level and vice versa (Bouvier, 2011; Di Iorio, 2016a; Di Iorio & Chen, 2019; Nemo, 1988, pp. 67–106; Petitot, 2016). For this reason, Hayek (1967, 1978, 2011) describes the economic order based on monetary prices as a complex self-organizing system (Caldwell, 2003, p. 363ff; Dumouchel & Dupuy, 1983; Laurent, 1994; Petitot, 2009, 2016).

Throughout the history of MI, it is possible to find countless examples of explanations in terms of circular causality between the micro and macro factors (Coleman, 1990; Raub et al., 2011). Let us consider another example taken from the work of Max Weber ([1953] 2005, p. 181ff). Weber explained that the emergence (upward causation) of capitalism (macro phenomenon) in Northern Europe was produced unintentionally by several Calvinist entrepreneurs (micro level) who all saw their success in the pursuit of wealth as a sign of divine election. As part of this analysis, Weber also pointed out that the development of capitalism has dramatically altered the way of life of individuals (downward causation from macro to micro) because it created a mechanized and highly regulated system of production that has become typical of modernized societies, and that ended up restricting certain aspects of human freedom.

6   Nominalism and Reductionism

The scholars who have developed the interpretation of MI in terms of semantic reductionism considered in the previous sections regard this approach as closely linked to a nominalist ontology. Quite often they are not opposed to this type of ontology; in fact, most of these thinkers accept and even defend it (Epstein, 2015; Little, 1990, 2016), but they criticize MI because, in their opinion, MI does not combine nominalism with a systemic and emergentist approach. There is, however, an anti-nominalist variant of the interpretation of MI in terms of semantic reductionism, that is, in terms of an atomist theory of the social that denies that the whole is more than the sum of the individual parts. According to the philosophers who developed this variant, in particular Mario Bunge (2000) and Brian Epstein (2015), the criticism of MI analyzed in the previous two sections is partially incorrect due to the impossibility of combining anti-reductionism and nominalist ontology. For them, the incompatibility between MI and the existence of irreducible global properties is strictly linked to the intrinsic limits of the type of ontology of this approach. In other words, they criticize MI as a nominalist perspective because they equate its metaphysical dimension with reductionism.

According to Bunge (2000, p. 148), MI fails to understand that 'social systems such as families, tribes, villages, business firms, armies, schools, religious congregations, informal networks, or political parties... are just as real and concrete as their individual constituents'. In contrast, 'individualists insist that all these are just collections of individuals: they underrate or even overlook structure' (ibid.). The consequence of this is that 'individualists resist the systemic approach. They insist on studying only the components of social systems, that is, individuals, while overlooking their structure or set of connections' (ibid.). According to Bunge's view, MI is therefore wrong because it is linked to a nominalist ontology, and to defend this type of ontology would be to deny the existence and the crucial importance of any systemic dimension.

Epstein (2015, p. 36ff) has proposed an analysis that is partially similar to Bunge's, but, unlike Bunge's analysis, it centers on the concept of 'supervenience'. Like Bunge, Epstein (ibid.) argues that because of its nominalist ontological assumptions, MI is unable to take into account the fact that social phenomena cannot be reduced to individual phenomena. However, unlike Bunge, he insists that nominalism entails the theory that social facts supervene on the individualistic ones (Di Iorio & Herfeld, 2018, p. 18). According to Epstein, this means that society 'is entirely composed and determined by individual properties' (Di Iorio & Herfeld, 2018, p. 18). The problem, he contends, is that social properties do not supervene only on individual properties, but also on 'physical' properties (Epstein, 2015, p. 46ff). The economic situation of a region, for example, is not only affected by human factors such as political decisions on taxation or investment, but also by material factors such as crop failures, earthquakes, or flooding. Therefore, according to Epstein, MI cannot grasp the irreducible nature of social phenomena, nor the complexity of their causes, due to a strictly ontological problem.

The first thing to note about the position of Bunge and Epstein is that, even if many individualistic scholars, such as Simmel, Mises, Hayek, and Popper, have explicitly defended a nominalist ontology of the social, others, such as Boudon and most of his followers, prefer to avoid engaging in metaphysical debates and steer clear of defining their approach in terms of nominalism (see Boudon, 1996). Boudon sometimes seems to conceive of nominalism as a reductionist theory in Bunge's sense, incapable of correctly defining the analytical presuppositions of interpretative sociology:

> Churches, political institutions, birth rates, nations, climate data, cities, traditions exist just like individuals. It seems difficult to distinguish these two types of entities on the basis of their degree of reality, if this expression means

Methodological Individualism and Reductionism 437

anything. It seems even more difficult to distinguish them on the basis of a causal or temporal priority. (Boudon, 1996, p. 378, translation)

Boudon and other contemporary individualistic thinkers' distrust of nominalism is not taken into account in the work of Bunge and Epstein, since these two philosophers simply equate MI with a form of nominalism, which is a problematic aspect of their perspective.[2]

Moreover, the line of argument they propose does not seem convincing for other reasons. Bunge and Epstein both give a caricatured view of nominalism, as it was historically conceived and defended by individualists who accepted it as the basis of their explanatory approach.

Regarding Bunge's critique, let me stress that it is incorrect to assert that nominalist thinkers such as Popper, Mises, and Hayek have outright denied the existence and influence of systemic and institutional structures. Rather, they have argued that these structures should not be viewed as Platonic substances that 'exist independently of the individuals which compose them' (Hayek, 1948, p. 6). In other words, the nominalism of these scholars is a critique of substantialism, not of emergentism, downward causation (understood in non-deterministic terms), or the systemic approach (see Antiseri 2004; Antiseri & Pellicani, 1995; Bouvier, 2020; Di Iorio, 2015, pp. 75–120; Di Nuoscio, 2018; Nadeau, 2016; Pribram, [1912] 2008). These thinkers all recognize that social interactions are characterized by irreducible global properties, but they regard these global properties as derivatives of the existence of individuals, not as sui generis entities. From their point of view, there is no contradiction between nominalism and a systemic approach, since these two concepts are two sides of the same coin. In other words, correctly conceived, ontological individualism recognizes that any social phenomenon is characterized by semantically irreducible global properties but denies that these properties create or define a new substance that exists independently of individuals (see Di Iorio & Chen, 2019, pp. 11–13). Indeed, these global properties disappear if the individuals and their interactions disappear

---

[2] That said, it is necessary to note that when he refers to collective entities, Boudon often seems to develop anti-realistic arguments accepted by ontological nominalists, for whom these are pseudo-entities. He writes, for example, that MI 'on principle refuses to treat a group as an actor that, like the individual, is endowed with an identity, a conscience and a will' (Boudon, 1988, p. 35, translation). His position on nominalism seems to me to be characterized by a certain ambiguity. The few lines he devotes to ontological problems seem to sketch a vision that is not entirely incompatible with the interpretation of nominalism defended in the work of ontological individualists and that is analyzed in the subsequent part of the section. This interpretation of nominalism is rather different from that advanced by Bunge, which seems more or less accepted by Boudon when he expresses his distrust vis-à-vis nominalism.

(Popper, 1957, p. 82). For example, from this point of view, the tactical organization of a soccer team exists precisely as an emergent property, not as a sui generis entity.

Consequently, Bunge's thesis, in which the metaphysical position of ontological individualists denies the structural and institutional dimension of social phenomena, amounts to criticizing a caricature, or at best an approximation of the point of view denounced. This passage from Mises is very instructive in this regard:

> It is uncontested that in the sphere of human action social entities have real existence. Nobody ventures to deny that nations, states, municipalities, parties, religious communities, are real factors determining the course of human events. Methodological individualism, far from contesting the significance of such collective wholes, considers it as one of its main tasks to describe and to analyze their becoming and their disappearing, their changing structures, and their operation. And it chooses the only method fitted to solve this problem satisfactorily. (Mises, [1949] 2004, p. 42)

Epstein's position is also problematic. Historically, nominalist individualists have never conceived of their ontological conception as a theory of supervenience in the sense specified above (Di Iorio & Herfeld, 2018). To maintain that collective entities are not sui generis substances, existing independently of individuals (what individualists mean by nominalism), in no way amounts to asserting that social phenomena are only determined by human factors (ibid.). The difference between these two perspectives seems to escape Epstein, who, like Bunge, does not seem to have a satisfactory command of the texts he criticizes. The mundane truth that social phenomena are also influenced by material factors is clearly reflected in the theoretical work of individualists, as well as in the empirical explanations they have offered (see Bouvier, 2015, p. 574). For example, for the members of the Austrian school of economics, who all endorse the subjectivism of value, the scarcity of a metal affects its price and, consequently, the allocation of resources and the economic structure of production (Hayek, 1948, pp. 77–91). Epstein's mistake therefore seems to be to confuse the nominalism of the individualists with a naive idea that they never defended: the *reductionist* theory according to which any aspect of social life should be explained causally in terms of individual properties. As I have pointed out, the nominalism of individualists must be seen as the idea that social explanations, including the analysis of the influence of physical factors on sociological and economic phenomena, cannot be based on the tendency to hypostatize social groups by treating them as substances. However, respecting this onto-logical rule does not prevent considering the causal role of physical properties at the social level. As the nominalist Hayek (1952, p. 25) puts it, according to MI, the social sciences deal with not only 'the relations between man and man', but also with 'the relations between men and things'. Contrary to what Epstein writes, according to this approach, the social supervenes on both individual and non-individual properties.

7 Conclusion

As Nathalie Bulle (2018, p. 1) rightly states, the 'importance of... MI... for explanation in the social sciences and the breadth of controversy surrounding it are only equaled by the misunderstandings of which it has been, and still is, the object'. Indeed, as we have seen in the previous sections, the widespread tendency to interpret MI as a reductionist theory that should be discarded is not valid. The call to replace this approach by a new systemic theory that is neither holistic (and determinist), nor individualist (in the sense of reduc-tionist) seems to be ill-suited since MI is, by its very nature and because of its history, anti-reductionist and systemic.

The confusion that persists today about MI in philosophy depends in large part on the fact that in this field, thinking about social science methodology has often become too abstract and self-referential, that is detached from the intellectual debates of the past and of the detailed analysis of the empir-ical explanations proposed by the classic social scientists. During the last few decades, the study of the reductionism/anti-reductionism dichotomy has acquired a disproportionate role within the Anglo-American analytic philos-ophy of the social sciences, but it is difficult to justify the importance it is accorded from the point of view of the concrete and historical dimen-sion of the research (Bulle, 2018). Indeed, as Raymond Boudon (1971, pp. 1–4) explained, the structural approach is quite simply indispensable and inevitable in the social sciences, given that there is no example of genuinely reductionist explanation and that this type of explanation is quite simply impossible (see also Petitot, 2009). Proponents of MI and, in particular, indi-vidualistic sociologists, have always been aware of the necessity of systemic analysis. It is true that there is an atomistic version of MI that, instead of being interested in the real world and its constraints, aimed to define abstract and simplified models of society by conceiving the actors in hyper-rationalist terms (see Hayek, 1948, pp. 1–32). However, even the atomistic thinkers of the Enlightenment, who developed social contract theories based on a strongly idealized conception of individuals and inspired by a mechanistic philosophy close to the semantic reductionism criticized by Lukes and Kincaid, have never been, strictly speaking, reductionists. They had to implicitly consider multiple systemic constraints that affect human action, such as those implied by the rules of language, since they must be respected if one wishes to discuss between co-contracting parties in order to arrive at the definition of a social contract (see Di Iorio, 2015, pp. 91, 92).

In this chapter, I have considered the relationship between MI and naive reductionism by showing that there are roughly two types of incorrect interpretations of MI. I have analyzed and criticized the interpretation of MI in terms of psychological reductionism as well as that in terms of semantic reductionism. In addition, I have clarified that the latter of these interpretations of MI has two variants: one nominalist and the other anti-nominalist.

References

Agassi, J. (1960). Methodological individualism. *The British Journal of Sociology, 11*(3), 244–270. https://doi.org/10.2307/586749

Agassi, J. (1975). Institutional individualism. *The British Journal of Sociology, 26*(2), 144–155. https://doi.org/10.2307/589585

Antiseri, D. (2004). *La Vienne de Popper*. Presses Universitaires de France.

Antiseri, D., & Pellicani, L. (1995). *L'Individualismo Metodologico: Una Polemica sul Mestiere dello Scienziato Sociale*. Franco Angeli.

Archer, M. S. (1995). *Realist social theory: The morphogenetic approach*. Cambridge University Press.

Bhaskar, R. (1979). *The possibility of naturalism*. Sussex: Harvester.

Boettke, P., & Candela, R. (2015). What is old should be new again: Methodological individualism, institutional analysis and spontaneous order. *Sociologia, 2*, 5–14.

Boudon, R. (1971). *Uses of structuralism*. Heinemann.

Boudon, R. ([1977] 1979). *Effets Pervers et Ordre Social*. Presses Universitaires de France.

Boudon, R. (1988). Individualisme ou Holisme: Un Débat Méthodologique Fondamental. In H. Mendras & M. Verret (Eds.), *Les Champs de la Sociologie Française* (pp. 31–45). Armand Colin.

Boudon, R. (1996). Risposte alla Domande di Enzo Di Nuoscio. In E. Di Nuoscio (Ed.), *Le Ragioni degli Individui. L'Individualismo Metodologico di Raymond Boudon*. Rubettino.

Boudon, R. (2001). *The origins of value: Essays in the sociology and philosophy of beliefs*. Transaction Publishers.

Boudon, R. (2010). *La Sociologie comme Science*. La Découverte.

Boudon, R., & Bourricaud, F. ([1982] 2004). *Dictionnaire Critique de la Sociologie*. Presses Universitaires de France.

Bouvier, A. (2011). Individualism, collective agency and the "micro-macro relation". In I. C. Jarvie & J. Zamora-Bonilla (Eds.), *The Sage handbook of philosophy of social science* (pp. 198–215). Sage. https://doi.org/10.4135/9781473913868.

Bouvier, A. (2015). Review of *The Ant Trap: Rebuilding the Foundations of the Social Sciences* by Brian Epstein. *Revue Philosophique de la France et de l'Étranger, 140*(4), 567–594.

Bouvier, A. (2020). Individualism versus holism. In P. A. Atkinson et al. (Eds.), *Sage research methods foundations*. Sage.

Bronner, G. (2011). *The future of collective beliefs*. Bardwell Press.

Bronner, G., & Géhin, É. (2017). *Le Danger Sociologique*. Presses Universitaires de France.

Bulle, N. (2018). Methodological individualism as anti-reductionism. *Journal of Classical Sociology, 19*(2), 161–184. https://doi.org/10.1177/1468795X1876 5536

Bulle, N., & Phan, D. (2017). Can analytical sociology do without methodological individualism? *Philosophy of the Social Sciences, 47*(6), 1–31. https://doi.org/10.1177/0048393117713982

Bunge, M. (2000). Systemism: The alternative to individualism and holism. *Journal of Socio-Economics, 29*(2), 147–157. https://doi.org/10.1016/S1053-535 7(00)00058-5

Caldwell, B. (2003). *Hayek's challenge: An intellectual biography of F. A. Hayek*. University of Chicago Press.

Campagnolo, G. (2013). *Criticisms of classical political economy: Menger, Austrian economics and the German historical school*. Routledge.

Campagnolo, G. (2016). The identity of the economic agent—Seen from a Mengerian point of view in a philosophical and historical context. *Cosmos and Taxis: Studies in Emergent Order and Organization, 3*(2–3), 64–77.

Cherkaoui, M. (2006). *Le Paradoxe des Conséquences: Essai sur une Théorie Wébérienne des Effets Inattendus et Non Voulus des Actions*. Librairie Droz.

Coleman, J. S. (1990). *Foundations of social theory*. The Belknap Press of Harvard University Press.

Demeulenaere, P. (2000). Individualism and holism: New controversies in the philosophy of social science. *Mind & Society, 1*(2), 3–16. https://doi.org/10.1007/BF02512311

Demeulenaere, P. (2011). Introduction. In P. Demeulenaere (Ed.), *Analytical sociology and social mechanisms* (pp. 1-30). Cambridge University Press.

Demeulenaere, P. (2015). Methodological individualism: Philosophical aspects (pp. 308–313). In J. D. Wright (Ed.), *International encyclopedia of the social & behavioral sciences* (2nd ed). Elsevier. https://doi.org/10.1016/B978-0-08-097 086-8.63050-7.

Di Iorio, F. (2015). *Cognitive autonomy and methodological individualism: The interpretative foundations of social life*. Springer.

Di Iorio, F. (2016a). Introduction: Methodological individualism, structural constraints and social complexity. *Cosmos and Taxis: Studies in Emergent Order and Organization, 3*(2–3), 1–8.

Di Iorio, F. (2016b). World 3 and methodological individualism in Popper's thought. *Philosophy of the Social Sciences, 46*(4), 352–374. https://doi.org/10.1177/0048393116642992

Di Iorio, F., & Chen, S.-H. (2019). On the connection between agent-based simulation and methodological individualism. *Social Science Information, 58*(2), 1–23. https://doi.org/10.1177/0539018419852526

Di Iorio, F., & Herfeld, C. (2018). Review of *The Ant Trap: Rebuilding the Foundations of the Social Sciences* by Brian Epstein. *Philosophy of the Social Sciences, 48*(1), 105–128. https://doi.org/10.1177/0048393117724757

Di Nuoscio, E. (2016). Herbert Spencer and Friedrich von Hayek: Two parallel theories. *Cosmos and Taxis: Studies in Emergent Order and Organization, 3*(2–3), 56–63.

Di Nuoscio, E. (2018). *The logic of explanation in the social sciences*. Bardwell Press.

Dumouchel, P., & Dupuy, J.-P. (Eds.). (1983). *L'Auto-Organisation de la Physique au Politique*. Le Seuil.

Elder-Vass, D. (2014). Social entities and the basis of their powers. In J. Zahle & F. Collin (Eds.), *Rethinking the individualism-holism debate: Essays in the philosophy of social science* (pp. 39–53). Springer.

Elster, J. (1982, July). Marxism, functionalism, and game theory: The case for methodological individualism. *Theory and Society, 11*(4), 453–482.

Elster, J. (1985). *Making sense of Marx*. Cambridge University Press.

Elster, J. (1989). *Nuts and Bolts for the social sciences*. Cambridge University Press.

Elster J. (2023). What's the alternative? In N. Bulle & F. Di Iorio (Eds.), *The Palgrave handbook of methodological individualism volume 1*. Palgrave Macmillan.

Epstein, B. (2015). *The ant trap: Rebuilding the foundations of the social sciences*. Oxford University Press.

Hayek, F. A. (1948). *Individualism and economic order*. University of Chicago Press.

Hayek, F. A. (1952). *The counter-revolution of science: Studies on the abuse of reason*. The Free Press.

Hayek, F. A. (1967). *Studies in philosophy, politics and economics*. University of Chicago Press.

Hayek, F. A. (1978). *New studies in philosophy, politics, economics and the history of ideas*. Routledge & Kegan Paul.

Hayek, F. A. (2011). *Droit, Législation et Liberté*. Presses Universitaires de France.

Hedström, P., & Bearman, P. (2009). *The Oxford handbook of analytical sociology*. Oxford University Press.

Jarvie, I. C. (1972). *Concepts and society*. Routledge & Kegan Paul.

Jarvie, I. C. (2001). *The republic of science: The emergence of Popper's social view of science, 1935–1945*. Rodopi.

Kincaid, H. (1986). Reduction, explanation, and individualism. *Philosophy of Science, 53*(4), 492–513.

Kincaid, H. (1995). *Philosophical foundations of the social sciences: Analyzing controversies in social research.* Cambridge University Press.

Kincaid, H. (2017). Philosophy without borders, naturally: An interview with Harold Kincaid. *Erasmus Journal for Philosophy and Economics, 10*(1). https://doi.org/10.23941/ejpe.v10i1.281.

Kincaid, H., & Zahle, J. (2019). Why be a methodological individualist? *Synthese, 196*(2), 655–675.

King, A. (2004). *The structure of social theory.* Routledge.

Laurent, A. (1994). *L'Individualisme Méthodologique.* Presses Universitaires de France.

Lawson, T. (1997). *Economics and reality.* Routledge. https://doi.org/10.4324/9780203195390

Lawson, T. (2003). *Reorienting economics.* Routledge.

Little, D. (1990). *Varieties of social explanation: An introduction to the philosophy of social science.* Westview Press. https://doi.org/10.2307/2185667

Little D. (2016). *New directions in the philosophy of social science.* Rowman & Littlefield International.

Lukes, S. (1968). Methodological individualism reconsidered. *The British Journal of Sociology, 19*(2), 119–129. https://doi.org/10.2307/588689

Lukes, S. (1973). *Individualism.* Harper Torchbooks.

Lukes, S. (2006). *Individualism: New introduction by the author.* Harper & ECPR Press.

Mandelbaum, M. (1955). Societal facts. *The British Journal of Sociology, 6*(4), 305–317. https://doi.org/10.2307/587130

Manzo, G. (2007). Comment on Andrew Abbott/2. *Sociologica, 1*(2), 1–8. https://doi.org/10.2383/24752

Manzo, G. (2014). Data, generative models, and mechanisms: More on the principles of analytical sociology. In G. Manzo (Ed.), *Analytical sociology: Actions and networks.* Wiley.

Manzo, G. (2015). Macrosociology-microsociology. In J. D. Wright (Ed.), in vol. 14 of *International encyclopedia of the social & behavioral sciences* (2nd ed., pp. 414–421). Elsevier.https://doi.org/10.1016/B978-0-08-097086-8.32086-4.

Menger, C. ([1883] 1985). *Investigations into the method of the social sciences with special reference to economics.* New York University Press.

Merton, R. K. (1936). The unanticipated consequences of purposive social action. *American Sociological Review, 1*(6), 894–904. https://doi.org/10.2307/2084615

Nadeau, R. (2016). Cultural evolution, group selection and methodological individualism: A Plea for Hayek. *Cosmos and Taxis: Studies in Emergent Order and Organization, 3*(2–3), 9–22.

Nemo, P. (1988). *La Société de Droit Selon F. A. Hayek.* Presses Universitaires de France.

O'Driscoll, G. P., & Rizzo, M. J. (1985). *The Economics of time and ignorance*. Basil Blackwell.

Petitot, J. (2009). *Per un Nuovo Illuminismo: La Conoscenza Scientifica come Valore Culturale e Civile*. Bompiani.

Petitot, J. (2012). Individualisme méthodologique et évolution culturelle. In R. De Mucci & K. R. Leube (Eds.), *Un Austriaco in Italia: Studi in Onore di Dario Antiseri*. Rubbettino.

Petitot, J. (2016). Complex methodological individualism. *Cosmos and Taxis: Studies in Emergent Order and Organization, 3*(2–3), 27–37.

Petroni, A. M. (1991). L'Individualisme Méthodologique. *Journal Des Économistes Et Des Études Humaines, 2*(1), 25–61.

Pettit, P. (1996). *The common mind: An essay on psychology, society, and politics*. Oxford University Press. https://doi.org/10.1093/0195106458.001.0001

Picavet, E. (2015). Methodological individualism in sociology. In N. J. Smelser & P. B. Baltes (Eds.), *International encyclopedia of the social & behavioral sciences* (2nd ed., pp. 9751–9755). Elsevier.https://doi.org/10.1016/B0-08-043076-7/02022-2.

Picavet, E. (2018). Rationality and interpretation in the study of social interaction. In G. Bronner & F. Di Iorio (Eds.), *The mystery of rationality: Mind, beliefs and the social sciences*. Springer.

Popper, K. R. (1957). *The poverty of historicism*. Beacon Press.

Popper, K. R. (1963). *Conjectures and refutations: The growth of scientific knowledge*. Routledge & Kegan Paul.

Popper, K. R. ([1945] 1966a). *The open society and its enemies, vol. 1: The spell of Plato*. Princeton University Press.

Popper, K. R. ([1945] 1966b). *The open society and its enemies, vol. 2: The high tide of prophecy: Hegel, Marx and the aftermath*. Princeton University Press.

Popper, K. R. ([1977] 2003). *The self and its brain: An argument for interactionism*. Routledge.

Pribram, K. ([1912] 2008). La Genesi della Filosofia Sociale Individualistica. In E. Grillo (Ed.), *L'Individualismo nelle Scienze Sociali*. Rubbettino (original in German: *Die Entstehung der individualistischen Sozialphilosophie*, Leipzig, C. L. Hirschfeld, 1912).

Rainone, A. (1990). *Filosofia Analitica e Scienze Storico-Sociali*. ETS.

Raub, W. (1982). The structural-individualistic approach towards an explanatory sociology. In W. Raub (Ed.), *Theoretical models and empirical analyses: Contributions to the explanation of individual actions and collective Phenomena*. E.S. Publications.

Raub, W., Buskens, V., & Van Assen, M. (2011). Micro-macro links and microfoundations in sociology. *The Journal of Mathematical Sociology, 35*(1–3). https://doi.org/10.1080/0022250X.2010.532263.

Ruben, D.-H. (1985). *The metaphysics of the social world*. Routledge & Kegan Paul.

Sawyer, R. K. (2002). Nonreductive individualism. Part I: Supervenience and wild disjunction. *Philosophy of the Social Sciences, 32*(4), 537–559. https://doi.org/10.1177/004839302237836.

Sawyer, R. K. (2003). Nonreductive individualism. Part II: Social causation. *Philosophy of the Social Sciences, 33*(2), 203–224. https://doi.org/10.1177/004839310 3252207.

Schutz, A. (1998). *Éléments de Sociologie Phénoménologique* (introduction and translation by Thierry Blin). L'Harmattan.

Simmel, G. ([1892] 1977). *The problems of the philosophy of history: An epistemological essay*. The Free Press.

Udehn, L. (2001). *Methodological individualism: Background, history and meaning*. Routledge.

Udehn, L. (2002). The changing face of methodological individualism. *Annual Review of Sociology, 28*, 479–507. https://doi.org/10.1146/annurev.soc.28.110 601.140938

Yoshida, K. (2014). *Rationality and Cultural Interpretivism: A Critical Assessment of Failed Solutions*. Lexington Books.

von Mises, L. ([1922] 1981). *Socialism: An economic and sociological analysis*. Liberty Found.

von Mises, L. ([1949] 2004). *Human action: A Treatise on economics*. Mises Institute.

Weber, M. ([1920] 2012). Letter to Robert Liefmann 9 Mars 1920. In H. H. Brunn (Ed.), *Max Weber: Collected methodological writings* (H. H. Brunn, Trans.). Routledge.

Weber, M. ([1922] 1978). *Economy and society*. University of California Press.

Weber, M. (1946). *From Max Weber: Essays in sociology*. In H. H. Gerth & C. Wright Mills (Eds.). Oxford University Press.

Weber, M. ([1953] 2005). *The protestant ethic and the spirit of capitalism*. Routledge.

Wippler, R. (1978). The structural-individualistic approach in Dutch sociology: Toward an explanatory social science. *The Netherlands Journal of Sociology, 14*(2), 135–155.

Zahle, J., & Collin, F. (Eds.). (2014). *Rethinking the individualism-holism debate: Essays in philosophy of social science*. Springer.

Methodological Individualism Facing Recent Criticisms from Analytic Philosophy
Artificial Reconstructions and Genuine Controversies

Alban Bouvier

Methodological individualism (MI), of which the best known contemporary representatives are probably Jon Elster (in political science and philosophy), James Coleman (in general sociology) and Raymond Boudon (*idem*), has been the object, quite recently, of a new wave of criticism within English-language analytic philosophy.[1] Carl Menger (1840–1921) is considered by historians of the social sciences as the pioneer of the MI tradition in economics, and Max Weber (1864–1920) as the one who specified the meaning of this research tradition in sociology. Weber in turn influenced the second and third generations of Austrian economists, notably von Mises (1881–1973) and Hayek (1899–1992) (Antiseri, 2005; Caldwell, 2003). Recent criticism is indirectly rooted in the specific philosophical tradition

---

[1] This text is an amended and enriched version of a first version that appeared under the title "L'individualisme méthodologique au défi des critiques de la philosophie analytique récente" published in *L'année sociologique*, 2020/2021 Vol. 70, pp. 45–67, itself a development of a paper presented in Rome in June 2019. This new version has benefited from many subsequent exchanges with Nathalie Bulle and Francesco di Iorio, who pushed me to clarify my ideas. Nevertheless, this text commits only me.

A. Bouvier (✉)

Philosophy and Social Sciences, Institut Jean Nicod, Sciences & Lettres University, Paris, France
e-mail: alban.bouvier@ens.psl.eu

Aix-Marseille University, Paris, France

© The Author(s), under exclusive license to Springer Nature Switzerland AG 2023
N. Bulle and F. Di Iorio (eds.), *The Palgrave Handbook of Methodological Individualism*,
https://doi.org/10.1007/978-3-031-41508-1_19

447

inaugurated by Friedrich Hayek's colleague and friend Karl Popper (1902–1994), during World War II, and more directly in the analytical turn (in a sense I will soon specify) that two of his disciples at the London School of Economics, John Watkins and Joseph Agassi, have introduced to it. These criticisms, as well as the defense of certain kinds of "holism" (in a sense I will also outline) that often accompany them, deserve attention. They express a concern for linguistic, conceptual, and logical rigor that is, indeed, sometimes lacking in these questions within the social sciences themselves and from which anyone interested in the progress of knowledge in the social sciences could arguably benefit. However, these criticisms suffer from a major flaw: they propose definitions of the perspectives in question (MI, holism) which, when reconstructed, often correspond very little to what is meant by the same terms in the social sciences (especially in sociology); more generally, they almost completely ignore the actual debates within these sciences. The critics of MI (as they understand it) similarly imagine the arguments that MI's proponents (in their sense) *might* put forward to defend their cause rather than examining the *genuine* arguments of the researchers who have claimed to be Methodological Individualists. What adds to the confusion is that these authors sometimes quote, here and there, statements from classical or contemporary economists and sociologists (Weber, Schumpeter, Hayek, Coleman, Elster, etc.) in order to illustrate their own theses—but since these statements are quoted out of context, their meaning is sometimes transformed to the point of being unrecognizable.

The result is a kind of sophisticated but artificial intellectual jousting that has received little notice from social scientists, except for a few economists interested in the philosophy of the social sciences in general and usually more accustomed to the analytical style than sociologists. And even those who did not feel annoyed from the start by these controversies have often remained bewildered (Hodgson, 2007; Sugden, 2016). On the face of it, such a scholastic, source-biased debate seems sterile. However, if we try to understand the origins of the considerable distortions of the actual positions of sociologists and economists that the philosophers of this tradition have caused, we may be able to identify certain equivocations in the formulation of these positions whose clarification can serve to enrich our debate.

In the first part, I characterize the specifically post-Watkinsian criticisms in the more general context of the debates on MI within the philosophy of the social sciences. I then identify two definitions of MI that are particularly representative of the analytic tradition in question, which themselves seem to stem from two major confusions, if we follow the definition of MI given by the one who introduced the term, Joseph Schumpeter, in 1909, a definition that is obviously still considered relevant by major figures in MI. In the second part, I examine the first confusion, which consists in taking for a *rejection* of *collective entities* what is only *suspicion* towards the use of *concepts of collective entities*. The third part is devoted to the dissipation of a second confusion, which consists in not clearly distinguishing *real* collective entities from purely *nominal* collective entities and, as a result, the "individualistic" method proper to MI from a method that can also be called "individualistic", but which consists in searching at the microsociological level (i.e. at the level of individuals) for the explanation of the regular correlations between certain statistical aggregates, which has nothing specifically "Schumpeterian" about it.

1 The Post-Popperian and More Precisely Post-Watkinsian Tradition in the Analytic Philosophy of the Social Sciences

Popper's publications on MI, in particular *The Poverty of Historicism* (1957 [1944–1945]), have had a continuation in very contrasting traditions within the philosophy of social sciences. One of them, inspired by Wittgenstein's later work, initiated by Peter Winch (1958) and continued in France by Vincent Descombes (2014) [1996], took a critical stance from the start. A second one, parallel and completely independent of the first one, but initially very favorable to the Popperian IM, emerged at the same time, through a series of articles by John Watkins (notably 1994 [1957]), a disciple of Popper, and, to a lesser extent, by Joseph Agassi (1975), also a Popperian, but already critical of Watkins. Both of them wanted to *support* the Popperian understanding of MI more effectively by giving it a more analytical turn and thus by breaking it down into several logically independent theses, then by drawing up an inventory of the arguments in support of the different theses thus identified and the counter-arguments that could be used against them (they did the same for what they called "holism"). This post-Watkinsian tradition has resurfaced today, but this time in the context of a *critique* of MI, thanks to the re-emergence, on still different grounds, neither Popperian nor Wittgensteinian, of subtle forms of holism, which are no longer incompatible with MI (Gilbert, 1989, 1994; Gold & Sugden, 2007; Pettit, 2003, 2014; Sellars,

450   A. Bouvier

1974; Tuomela, 1995).[2] The debate is formally very rigorous, but it suffers from the major flaw I pointed out in the introduction: its detachment from actual debates in the social sciences.[3]

A recent collective work (Zahle & Collin, 2014a) attempts to take stock of these issues in the same way that John O'Neill (1973) sought to do forty years earlier, in a volume that included contributions from Watkins and Agassi. The new volume includes chapters by a number of authors from the post-Watkinsian tradition (notably Epstein, 2014, 2015; Ylikoski, 2014, 2017; Zahle, 2016; Zahle & Collin, 2014b), but also authors located at the confluence of other traditions, notably Marxist (Kincaid, 2008, 2014; Little, 2014), and finally authors from traditions outside the anti-Popperian perspectives, neo-holist in particular (Pettit, 2003, 2014). In this abundant literature, an article by Julie Zahle and Harold Kincaid (2019), who have both written extensively on the subject, is particularly challenging because of the direct questioning form of its title: "Why to be methodological individualist?" The authors take as their starting point a definition they borrow from Christian List and Kai Spiekermann (2013), who did not present themselves as advocates of MI, but who aimed, like many others, to go beyond the opposition between MI and holism: a laudable undertaking, but one whose outcome is somewhat compromised by the fact that they take the terms in question in the Watkinsian sense, as we shall soon verify:

> Crudely put, methodological individualism is the thesis that good social scientific explanations should refer *solely* to facts about individuals and their interactions. (List & Spiekermann, 2013, p. 629)

I have emphasized the word "solely" because the main debate for these authors still seems to concern the question of whether or not social scientific explanation can or should refer only to individuals (and possibly their interactions)—which would, according to these authors, characterize MI—or whether it can, or even should, introduce other elements, which would

---

[2] Thus there are at least *three* different, largely autonomous traditions dealing with the relationship between MI and holism within the analytic philosophy of social sciences. I will mention others in the conclusion. It should be noted that the proponents of a neo-holism (usually focused on the issue of *collective intentionality* or "we-intentions") sometimes recall that etymologically the best spelling should be "wholism" (from "whole", all, itself from the Greek *olos*, all). It was Popper (1956 [1944–1945], p. 15) who introduced the term "holism" (initially translated very correctly into French as "*totalisme*", from the Latin *totus*, all), by explicit borrowing from biology, as more generic than "organicism", used at the time of Menger, Weber or Schumpeter. Symptomatically, the relation to the idea of totality, although crucial, often takes a back seat in the post-Watkinsian tradition.

[3] For analyses that instead examine these, see, for example, Di Iorio (2015, 2020), Bulle (2020) and Bulle and Phan (2017).

Methodological Individualism Facing Recent ...   451

supposedly characterize holism. Zahle and Kincaid do not give an explicit definition of holism in their article, but it is reasonable to look for one in an article written by one of these authors from the same period. Thus, Zahle (2016) writes:

> Methodological holists engaged in this debate defend the view that explanations that invoke social phenomena (e.g., institutions, social structures or cultures) should be offered within the social sciences: their use is indispensable. Explanations of this sort are variously referred to as holist, collectivist, *social (-level), or macro (-level)* explanations [italics mine]

This definition seems quite congruent with the definition of MI given in the article co-authored with Kincaid in the sense that it seems to be symmetrical. However, Zahle (2016) immediately adds two examples of "holistic" explanation[4]: "They are exemplified by claims such as 'the unions protested because the government wanted to lower the national minimum wage', or 'the rise in unemployment led to a higher crime rate'". While the first example fits the proposed definition—the unions and the government are institutions—the second example does not: the rise in unemployment or the rise in the crime rate (however related or unrelated these social phenomena may be) are as such *neither* institutions *nor* social structures, even if it is conceivable that they may be related to them (e.g., the government may have different economic and social policies). The relationship of the examples to the definition seems to be due only to the ambiguity of the terms "social level" or "macro level", which can be used, in effect, to designate ontological levels (first example) or epistemological levels (second example).[5] In her introduction to the collective volume, Zahle & Collin (2014b) equates the two kinds of "social phenomena" (and others such as norms and culture) even though she intends to distinguish them: "Within the individualism-holism debate more generally, it is common to distinguish between various kinds of social phenomena. Some of the most frequently mentioned ones are: (a) social organizations, as exemplified by a nation, a firm, and a university; (b) statistical properties like the literacy or suicide rate of a group of individuals [...]" (ibid., p. 2).

---

[4] Zahle uses"collectivist" as well as "holist". This is not a problem. The two terms are in fact, in these debates, generally interchangeable. Hayek (2010 [1942–1944]), for example, preferred "collectivism", likewise O'Neill (1973). Later, Elster (1985, p. 6) also used "collectivism".

[5] On the very great equivocality of the notions of "macro" and "micro" in the social sciences, see, e.g., Walliser and Prou (1988, pp. 107–123), and Bouvier (2011, 2020).

452   A. Bouvier

That this position is not peculiar to these two authors, but is a legacy of Watkins, is verified in one of the articles that seems to be in the background: "Sociological holism means that some superhuman agents *or factors* are supposed to be at work in history" [my italics] (Watkins, 1994 [1957], p. 442). Watkins did not specify what he meant by "superhuman agents" here, but what would be the closest to it, if one assumes that Watkins is thinking of the very beginnings of the individualist tradition (notably Weber), is probably Hegel's "World Spirit" (1965 [1822 and 1828]), that is to say, Divine Providence, or the "spirit of the people" (*Volksgeist*), the "spirit of the language" (*l'esprit et le caractère d'une langue*), expressions that have been widespread since Herder, Fichte, Schlegel, or de Staël, all of which refer to entities overhanging social actors.[6] But the most problematic element lies rather in the "or factors", which is not highlighted in Watkins' paper. The meaning of this a priori anodyne precision is however explained by Watkins: "An example of such a superhuman, sociological factor is the alleged long-term cyclical wave in economic life which is supposed to be self-propelling, uncontrollable, and inexplicable in terms of human activity". That an economic curve, which only appears as such on a graph, can exert a constraint on individuals in the same way as God or even a social group (people or ethnic group) is difficult to understand. Such an idea seems simply to be inherited from the ancient meaning of the word "law", which referred to the idea that it was God who, like a monarch, decided the laws that would govern the universe.

Zahle and Kincaid further specify what they want to discuss in their article, namely the respective relevance of two variants of MI (they are noted as MI1 and MI2, respectively), in their relation to methodological holism as the authors seem to conceive it (Zahle & Kincaid, 2019, §2: "Some preliminaries"):

MI1 *Individualists explanation <u>alone</u> should be put forward within the social sciences*; they are indispensable. Holist explanations may, and should, be dispensed with [the emphasis on "alone" is mine].
MI2 Purely holist explanations may never stand on their own; they should <u>always</u> be supplemented by accounts of the underlying individual-level microfoundations [the emphasis on "always" is mine].

What I have emphasized in the previous quotations is what the authors focus on throughout their article (as do List and Spiekermann in the initial definition chosen by the authors). The two variants of MI, according to Zahle

---

[6] See also Popper (1960, §31, p. 149): "ideas of 'spirits' – of an age, of a nation, of an army".

Methodological Individualism Facing Recent ...   453

and Kincaid, can be rephrased as follows: "individualistic" explanations are *necessary* and *sufficient*, and "holistic" explanations are superfluous (the first variant of MI); "individualistic" explanations are *necessary* even if "holistic" explanations can be relevant (the second variant of MI). We are dealing, in a way, with a radical and a moderate versions of MI. The authors then consider the arguments in support of these two theses. But a preliminary question is whether the two versions of MI mean the same thing, depending on whether one understands "holism" as an explanation that introduces reference to institutions, as in the first example (unions and government), or as an explanation that introduces reference to statistical considerations, as in the second example (the unemployment rate and the crime rate); and, if not, whether it would be necessary, by crossing the criteria, to distinguish at least two versions of MI1 (MI1a and MI1b) and two versions of MI2 (MI2a and MI2b), thus providing *four* variants of MI—assuming that MI1 and MI2 make sense—the two distinctions crossing each other in a kind of orthogonal way (see table below).

In the rest of this chapter, I will limit myself to considering successively (in the two following parts) two variants of what is generally called the "individualistic method" in the social sciences: one (MIa) corresponding to the example of unions and government and to which alone, from my viewpoint, one should reserve the term "methodological individualism", in the Schumpeterian sense; and the other (MIb), corresponding to the examples concerning the suicide rate in relation to the unemployment rate or concerning the cyclical character of the economic crises, and which has nothing specifically in conformity with MI as defined by Schumpeter. Thus, in the social sciences, we find researchers who follow the first method without following the second, and vice versa; and researchers who follow both. I will give some examples.

"Holism"/ "Methodological Individualism"	Radical versions of "MI"	Moderate versions of "MI"
"Holism" in the institutional sense; "macro/micro" in the ontological sense E.g. unions, governments, firms, universities (=institutions, therefore *real groups*)	MI 1a *Individualist explanation alone should be put forward within the social sciences*; they are indispensable. **Holist** explanations may, and should, be dispensed with	MI 2a *Purely* **holist** *explanations* may never stand on their own; they *should* *always* be *supplemented by accounts of the underlying individual-level* **microfoundations**

(continued)

454   A. Bouvier

(continued)

"Holism"/ "Methodological Individualism"	Radical versions of "MI"	Moderate versions of "MI"
"Holism" in the statistical sense; "macro/micro" in the epistemological sense E.g. unemployment rate, crime rate, suicide rate (= correlations between statistical aggregates, therefore between **nominal groups**)	MI 1b *Individualist explanation alone should be put forward within the social sciences*; they are indispensable. **Holist** explanations may, and should, be dispensed with	MI 2b *Purely* **holist** *explanations* may never stand on their own; they *should always* be *supplemented by accounts of the underlying individual-level* **microfoundations**

2 The *Central* Intention of the MI Tradition in Its Logically *Primitive* form (the Critical Vigilance Against the "Fallacy of Misplaced Concreteness") and in Its *Derivative* Form (the Individualizing Method)

Let's first look at the place that institutions should or should not take in MI according to the sociologists and economists who claim to be methodological individualists, taking Schumpeter's definition as our starting point. We will thus see that there is a first distortion in the definitions of MI given by Zahle and Kincaid (or before them, in a more surprising way since he is supposed to support MI, Watkins): the "radical" version of MI that *would exclude* taking institutions into consideration *does not*, in fact, exist, in the work of sociologists and economists from Carl Menger to Jon Elster.

Among recent authors, Kincaid, Zahle, Epstein, Ylikoski, and List and Spiekermann (as well as G. M. Hodgson) all mention that it was Schumpeter who first introduced the term "Methodological Individualism" in a 1909 article. Some, like Ylikoski, even note that the German equivalent "Methodologische Individualismus" was introduced a little earlier in a book (Schumpeter, 1998 [1908]). But they generally say little more. Geoffrey M. Hodgson (2007), for example, merely states on this historical issue, which is already valuable, that the recent characterizations of MI in analytic philosophy are often remote from Schumpeter's, and he seems to think that the Schumpeterian definition no longer corresponds to any meaning currently used. I believe, on the contrary, that a "Schumpeterian" thread can be high-lighted, and even unwound, both upstream towards Carl Menger and Max Weber and downstream towards Friedrich Hayek, Raymond Boudon, Jon Elster and, more allusively, James Coleman. To do so, I will not follow the chronological order, as some retrospective insights are valuable. Moreover, I will go through the strictly economic tradition and the sociological tradition successively, as they are largely autonomous, even if they sometimes intersect intimately.

We thus find formulated in these various authors what they hold to be the central intention of MI. This fundamental intention has, it seems, if we read them closely, two forms: (1) the use of the individualistic or "individu-alizing" method proper; however (2) this form is itself only *derived*, logically speaking, from a more *primitive* methodological attitude, which consists in an acute vigilance towards the fallacy of the *abusive reification of concepts (= hypostasis)*—and in particular (but not only) of abusive reification of *collective* concepts. This specific vigilance means that what is challenged is not the recourse to collective *entities*, such as institutions or social structures, but the careless use of collective *concepts* (and of concepts in general) and the consid-erable risk then incurred of introducing *pseudo-entities* into the explanation. This is a considerable difference in orientation in the characterization of MI compared to the post-Watkinsian literature.

(a) *MI from the standpoint of the purely economic tradition*

The first part of Chapter VI, entitled "Der Methodologische Individualis-mus" of *Das Wesen und der Hauptinhalt der theoretischen Nationalökonomie* (Schumpeter, 1908; 1998 [1908]), distinguishes methodological individu-alism from political individualism—this is the only part that is retained by the few philosophers who cite this chapter (e.g. Ylikoski, 2017). Yet it is the second part that exposes the "essence" or, more prosaically, "nature" of MI ("Wesen des methodologischen Individualismus"). For Schumpeter, the essence of MI is the critique of collective *concepts* - Schumpeter speaks of "concepts" (*Begriffen*) and not of collective *entities*. Schumpeter even distin-guishes two groups of collective concepts (*die beiden Gruppen von "sozialen" Begriffen in der Theorie*) (Schumpeter, 1908, p. 97; 1998, p. 7) that deserve criticism, and he gives examples in each case. The first list includes the concepts of national income (*Volkseinkommen*), national wealth (*Volksver-mögen*) and social capital (*Sozialkapital*); the second is the concept of social value (*sozialer Wert*). Schumpeter then refers to the article to be published in the 1909 issue of the *Quarterly Journal of Economics* as developing more

456   A. Bouvier

specifically the critique of the concept of "*social value*" (translation of *sozialer Wert*).

The more precise analysis of the example of the concept of "social value" proposed in the *Quarterly* article is very enlightening on some of the contextual issues, sometimes internal to the Austrian school—for example, on how to conceptualize socialism as an alternative to communism (Antiseri, 2005; Boettke, 1994a;Littlechild, 1990): for Schumpeter, it was notably a question of showing that the concept of "social value" would not make any sense in economics except in a country that was communist (in the sense that there would be, in this country, collectivization of production), because the State is not a *unit of action* in economics—except in a communist country; the units of action are necessarily only individuals (or quasi-individuals, as are, in principle, households and firms), except in a communist country, and one can therefore only speak of the economic value that a good has for individuals (or quasi-individuals), not for the State or the nation, and therefore only speak of "individual values". States, on the other hand, are units of *political* action; from this we can infer that there would be no objection to speaking of the properly *political* values of a well-circumscribed state. The criterion is thus this one: a collective concept is legitimate when and *only* when it refers to a unit of action, which can obviously, according to the proposed criterion, still be possibly broken down into sub-units.[7]

What is particularly interesting is that Hayek—in the short preface he gave to the very late English translation of this Chapter IV—saw in it "*the* classical exposition" [my emphasis] and "one of the most explicit expositions of the Austrian School's Methodological Individualism" (von Hayek, 1980, p. 160; commented on by Antiseri, 2005, p. 254).[8] And it is in von Hayek (2010 [1942–1944]) himself that we find an even clearer explication of the logically primitive form of MI's central intention, of which Schumpeter (1998 [1908]) drew the general framework and gave a particular example (1909). Hayek then wrote, this time taking the example of concepts shared by laymen: "Is it the ideas which the popular mind has formed about such collectives as

---

[7] Schumpeter is not very explicit on these various points; but Coleman, probably the most accomplished representative of MI in American sociology, is much more explicit in a note that is almost half a page long (Coleman, 1990, p. 5, "A note on Methodological Individualism").

[8] There might be some excessive emphasis and some inaccuracy in this formulation from Hayek's viewpoint itself. On the basis of what Hayek wrote earlier (von Hayek, 1948), the Austrian MI *research tradition* is founded on *two principles*: (methodological) individualism strictly understood (and defined quite correctly and with great subtlety by Schumpeter with its two forms: primitive and derivative) *plus* subjectivism (see Boettke, 1994b, p. 4). Thus, Methodological Individualism as a principle is opposed to Holism (von Hayek, 1948, chapter IV) while Subjectivism as a principle (that is taking into account beliefs, motivations, intentions, etc.) is opposed to Objectivism (taking only external data into account) (Chapters 3 and 5).Schumpeter is only concerned in 1908–1909 with MI as the first *principle* of MI as a *research tradition*.

society or the economic system, capitalism or imperialism, and other such collective entities,[9] which the social scientist must regard as no more than provisional theories, popular abstractions, and which he must not mistake for facts? That he consistently refrains from treating these pseudo-entities as facts, and that he systematically starts from the concepts which guide individuals in their actions and not from the results of their theorising about their actions, is the characteristic feature of that methodological individualism [...]." (von Hayek, 2010 [1942–1944], p. 100). Later, in the next chapter, he writes "The naïve realism [...] uncritically assumes that where there are commonly used concepts there must also be definite 'given' things which they describe" (p. 118). A sentence before, Hayek wrote, "They thus become, when they least suspect it, the victims of the fallacy of 'conceptual realism' (made familiar by A. N. Whitehead as the 'fallacy of misplaced concreteness')" (p. 118). In a note (n.3, p. 102), Hayek makes the link with the individualist method of Menger (1883), quoting a passage from the *Untersuchungen* "[...] In the exact social sciences [...] the human *individuals* and their *efforts*, the final elements of our analysis, are of empirical nature, and thus the exact theoretical social sciences have a great advantage over the exact natural sciences", whose "ultimate elements", such as "atoms" and "force", are "of theoretical nature" (Hayek's emphases). Although Menger's global context is the opposition between empirical and theoretical concepts depending on sciences (which is another issue), Hayek's emphasis is clearly on the relevance of the individualistic method in the social sciences..[10]

(b) *MI from the standpoint of the sociological tradition*

Raymond Boudon plays, later on, a role similar to that played by Hayek with respect to Schumpeter and Menger. At the very beginning of the dictionary he co-authored with François Bourricaud (Boudon & Bourricaud, 2003,

---

[9] It is crucial to notice that Hayek does not take examples such as "nations, states, municipalities, parties, religious communities" (see footnote on Popper and Mises below), which can be action units in one domain or another.

[10] Schumpeter is sometimes introduced strangely (and wrongly) as a former student of Weber (e.g. Heath, (2020 [2005]); Hodgson, 2007) explaining Weber's method (instead of Menger's). Thus, according to the historians of Austrian economics (e.g. Antiseri, 2005; Boettke, 1994a, 1994b; Bouvier, 2007; Littlechild, 1990), Schumpeter (1883–1950) studied constantly at the University of Vienna, where he took his Ph. D. in 1906 after attending the seminar of Eugen Böhm-Bawerk, one of the three co-founders—along with Carl Menger and Friedrich von Wieser—of the Austrian School. As for Weber (1864–1920), he was first appointed in 1894 at Freiburg, then at Heidelberg in 1896. He suffered from depression from 1899 onwards and was unable to teach regularly to the point that he eventually retired from his chair as soon as 1903 (Bendix, 1977; Radkau, 2011) until 1919, when he briefly taught in Vienna (for the first time), then in Munich before dying of the Spanish flu in 1920.

p. 11), a famous letter from Max Weber to Robert Liefman in 1920 is quoted, in which Weber writes: "If I have finally become a sociologist [...] it was mainly so as to bring to a definite conclusion these essays based on collective concepts [*Kollectivbegriffen*] whose specter still prowls. In other words: sociology [...] can *only* [my emphasis] come from the actions of one, or several, or a number of separate individuals. *This is why* [my emphasis] it is bound to adopt methods which are strictly 'individualistic' [*individualistisch in der Methode*]" (Weber, 2012a [1920]).[11] This letter is crucial[12] not only because it attests once again that what the economists or sociologists of MI oppose, from Schumpeter to Hayek and from Weber to Boudon, are not collective *entities as such* (various institutions or communities such as State, municipalities or companies, churches, universities or armies)— but rather the *uncontrolled use of collective concepts* because these can refer to pseudo-collective entities (capitalism, imperialism, communism, social system, etc.)—but also because this letter expresses very clearly the *logical* relation of the derived principle (the individualist method) to the primitive principle (the rejection of "exercises based on collective concepts").[13]

In the passage from *Economy and Society*, where he returns to this question, Weber refers more specifically to the fact that the concept of State does not necessarily refer to the same thing (or: to the same entities) in lay and scholarly usage, nor to the same thing in law and sociology, so that the entity to which the concept of State, as long as it is not specified, refers is a chimera (Weber, 1978 [1922], vol. 1. A. Methodological Foundations, §9: 13–5). In *Über einige Kategorien der verstehen Soziologie* he writes, "[...] the peculiar nature [...] of our thinking [...] makes the concepts by which we grasp an activity appear as a durable reality, a 'reified' [*dinghaften*] structure, or a 'personified' ['*personenhaften*'] structure having an autonomous existence" (my translation) (Weber, 1981 [1913], p. 158; 1985 [1913], §3. p. 439). This is followed by examples of the concepts of "state", "association" and "feudalism". What Weber criticizes here is the "cognitive bias" at the source

---

[11] It is hardly necessary to add that Weber assumes not only the principle of methodological individualism but also, like Menger, von Mises, Hayek and so on, the principle of "subjectivism" (see footnote above). Weber (1981 [1913]) will call it "interpretive sociology" or "interpretive explanation" (*verstehendes Erklärung*) as well as "explanatory understanding" (*erklärendes Verständnis*).

[12] Unfortunately, only a few short passages of this letter have been translated into English (Weber, 2012b) and access to this letter was difficult even in German before the publication of the *Max Weber-Gesamtausgabe* (Weber, 2012a).

[13] The word "only" in Weber's quotation above and passages of the same kind found in Max Weber (1985 [1913]; 1992 [1922]), taken out of context, might have been misinterpreted by Watkins and the post-Watkinsian tradition.

of the *fallacy of conceptual realism or misplaced concreteness.*[14] Weber's actual practice in *The Protestant Ethic and the Spirit of Capitalism*, therefore as early as 1904, is itself exactly along these lines. Weber in fact gave an analysis of what he ironically called, using quotation marks in the first edition, the "spirit" of capitalism (to which Hayek's previously mentioned work (1953 [1952]) seems to refer, so to speak, since it was targeted in particular at the concept of capitalism); to do this, Weber meticulously analyzed the opuscules written by Benjamin Franklin, considered to be a typical *individual* way of thinking shared by thousands or millions of people, the true "action units" of "capitalism". This did not prevent Weber from reintroducing, in *Economy and Society*, the concept of capitalism itself, thus a collective concept, nor from referring to collective entities, such as the State, or the Monetary System, but by making a *narrowly circumscribed use of* these concepts (see, for example, Weber, 1978 [1922], vol.1, ch. 2: "Sociological Categories of Economic Action 32. "The Monetary *System of the Modern State* and the Different Kinds of Money: Currency Money" [my emphasis]).

Boudon is also particularly close to Schumpeter and Hayek on this point when he writes, at the end of *L'idéologie*: "As for holistic methods, when they are set up as an ideal, they tend to erase [the actor] purely and simply, by reducing him to being, according to a memorable formula, only a 'support of structure'" (Boudon, 1986, p. 288). Boudon certainly has in mind, given the body of his work, the role accorded by Pierre Bourdieu and Jean-Claude Passeron to "social structures" which often, if not always, seem to be conceived as the true agents of social processes (Bourdieu & Passeron, 1977 [1970]). However, Boudon's most direct and emblematic opponents in this passage, even if they are not explicitly named, are obviously Louis Althusser, and his heir in sociology Nicos Poulantzas, for they are the ones who introduced and used the "memorable" formula that Boudon refers to or similar formulas. For example, here is a passage from Althusser: "[…] All that [Marx] has told us puts us on the way to conceiving that these [juridico-political and ideological] relations also treat concrete human individuals as 'carriers' of relations ('porteurs' de rapports), 'supports' of functions (*'supports' de fonctions*), in which men are only involved because they are caught up in them" (Althusser, 1976, p. 168, my translation).[15] This does not, of course, prevent

---

[14] One almost thinks one is reading Vilfredo Pareto (1935 [1916], "Personifications", §1070–1085) who, as a member of the Lausanne School founded by Léon Walras, did not claim to follow Menger or Schumpeter, but nevertheless also defended MI (in the strict sense). So it can be said that Elie Halévy (1904) was right in characterizing Pareto's method as "individualisme méthodologique" (using the same term as Schumpeter four years earlier, as mentioned by Frobert (1996), but in a paper that has fallen into oblivion).

[15] Also see Poulantzas (1968, p. 66).

460   A. Bouvier

Boudon from formulating statements elsewhere about collective entities such as, for example, the French university, because these are then circumscribed and without imposing himself to make an inventory of all the *individual* decisions that produced the French university, precisely because it is a clearly identifiable entity (see Boudon, 1982 [1977], chapters III and IV).

I do not have enough space to develop the case of Jon Elster, although it is all the more interesting since Elster claimed to be a Marxist as much as Althusser, Poulantzas—or Kincaid (see in particular Elster, 1989 [1985]).[16] I merely indicate the concepts (which are not all "collective concepts") targeted by Elster: that of "social classes", which does refer to collectives, but also those of "productive forces" and "relations of production", which no longer refer to them, at least not directly (Elster, 1986a), but are also often hypostasized. In parallel analyses, Elster (1983) attacks above all the "structures" in so far as they are also hypostasized by Bourdieu or by Foucault.

So if we briefly go back to MI1 and MI2 (in the sense of Zahle and Kincaid) interpreted as concerning the place that should or should not be reserved to institutions in the sociological explanation (thus MI1a and MI2a), it appears that MI1a has *no* relevance (this position is not defended by *anyone*): the "holistic" explanations in the sense that one would designate by this word explanations introducing institutions are a priori not the object of *any* discredit on the part of economists and sociologists who claim to be MI. And as far as MI2 is concerned, even reinterpreted as MI2a, the formulation *lacks* relevance, because, as soon as collective entities are well circumscribed (which are, in principle, States and governments as well as parties and unions, churches and armies) and that one can thus in principle go back to the individuals who make decisions in them, a statement like the one Zahle and Kincaid take as an example *does not* pose any particular problem in the framework of MI (as originally defined by Schumpeter).[17]

---

[16] Although Elster's understanding of MI is entirely consistent with Schumpeter's characterization, Elster never situates himself in the continuity of either the Austrian school or Weber. But he shows convincingly that Marx himself practiced MI (as later formulated by Schumpeter) in some crucial passages of his work (Elster, 1985). See also Boudon (1991 [1984], Chapter 5).

[17] One reason for the many controversies on this question, since Winch (1958, pp. 126–128), may lie in the enigmatic formulations of Popper (1960 [1944–1945]. Thus, Popper writes, in §29: "For most of the objects of social science […] are abstract concepts; they are *theoretical* constructions. Even 'the war' or 'the army' are abstract concepts, strange as they may sound to some. What is concrete is the many who killed; or the men and women in uniform, etc." (p. 135, Popper's emphasis). This statement, taken out of context, could be understood as if Popper were entirely denying the existence of institutions and, more generally, of collective entities, and recognizing only the existence of individual human beings. But this passage and the whole of §29 must be read in conjunction with §10 and §31. On this basis, it seems that Popper used a rather reckless shortcut to address at least *two* distinct issues in a brief sentence: (1) The first one is the relationship between *universal concepts* ("Universals") such as, in the social sciences: *the* Army (p. 135), *the* War (p. 135), *the* Military Commander, *the* Soldier, or in the natural sciences: *the* celestial body, on the one hand and,

3 Another Intention Mixed up with the Central Intention of MI: The Search for the Micro-sociological Basis of Statistical Correlations Observed at the Aggregate Level

The second confusion mentioned in the introduction is a confusion, or at least an absence of sufficiently careful distinction between, *real* (e.g. institutionalized) *groups* and mere *nominal groups*. The equivocation of the terms "macro" and "micro", which can have either an ontological meaning (which I considered in the previous section) or an epistemological one (which I consider here), may have played a harmful role.[18] Some representatives of MI have insisted with great clarity on the difference between these problems, either theoretically, like Hayek, or practically, like Boudon. However,

---

on the other hand, *singular items*, such as, in the social sciences, the French and the Russian armies at Borodino in 1812 (§31, pp. 148–149), the First World War (§10, p. 27), Alexander the Great (§10, p. 27), Napoleon and Kutuzov (§31, p. 148), "men and women in uniform" (p. 135) or in astronomy, Halley's comet (§10); (2) The second one is the relationship between singular *collective entities* (such as singular armies [§29, §31] and other singular social institutions [§31], singular wars [§10, 29, 31] and other singular social movements [§31]), on the one hand and, on the other hand, *individuals* in the sense of *singular human beings* (military commanders and soldiers or "men and women in uniform" [§29]). In *The Poverty of Historicism*, Popper seems to recognize the existence of singular things or items—including individual persons as well as singular armies, singular institutions, singular wars, etc. although he does not elaborate metaphysically much on the problem of knowing in what sense (singular) institutions and other (singular) collective entities can be said to exist in relation to (singular) human beings. He just writes, "we must try to understand all collective phenomena *as due to* [my emphasis] the actions, interactions, aims, hopes, and thoughts of individual men, and *as due to* [idem] traditions created and preserved by individual men' (pp. 157–158). Then, Popper (1974, 1978) will recognize the possibly autonomous existence of three kinds of "Worlds": the World of material objects (World I), the world of beliefs and mental states in general (World II) and even the possibly autonomous existence of a world of concepts (World III). This latter World, on which Popper wrote at length, obviously includes the conceptual *models* of an army, of a war, of a state, of a celestial body, etc. (although not the "*essence*" of the State or the "essence" of an army or the "essence" of anything [§10]). But, contrary to what many authors have claimed (typically Uedehn, 2001), it seems that the problem of the existence of World III (or of what medieval philosophers called "Universals") has nothing *specific* to do with collective entities: these are two quite distinct issues (abstract concepts *versus* concrete items; collective entities *versus* individual human beings).

Be that as it may, the main point here is that no *economist* or *sociologist*—whether or not they claim to be Methodological Individualists (Alpert, 1961, p. 147, note 116)—has ever argued that institutions such as armies and states (or companies, churches, etc.) or social movements do *not* exist *at all* and do not have be taken into account as such in explanations. Ludwig von Mises (1949), clearer than Popper on this issue, wrote as follows: "It is uncontested that in the sphere of human action social entities have real existence. Nobody ventures to deny that nations, states, municipalities, parties, religious communities, are real factors determining the course of human events. Methodological individualism, far from contesting the significance of such collective wholes, considers it as one of its main tasks to describe and to analyze their becoming and their disappearing, their changing structures, and their operation" (p. 42).

[18] It should be added that, in *The Poverty of Historicism*, Popper is far less explicit than Hayek on this point, which Popper does not tackle directly.

462   A. Bouvier

it must be admitted that even when the distinction is made clearly in certain key passages, among those who claim to be Methodological Individualists, the distinction between the two problems is sometimes obscured in other passages, either by comments that veil the recourse to statistics as the starting point of analysis (e.g. Coleman on Weber), or by a decidedly over-inclusive use of the term "MI" (e.g. Boudon on Durkheim).

Hayek, for example, made the diffeHayekrence between these two kinds of problems very clear by noting that the notion of "collective beings" is sometimes used, but wrongly, in connection with simple statistical sets (von Hayek, 2010). In particular, he wrote: "The collectives of statistics […] are thus emphatically not wholes in the sense in which we describe social structures as wholes" (p. 124). But Durkheim's case is more complicated. On the one hand, Durkheim is sometimes concerned with quite well circumscribed real "collective beings", e.g. this or that Arunta clan in Australia (Durkheim, 1964 [1912]), so that we can reconstruct the interactions between their members (Alpert, 1961); on the other hand, he also frequently—in all his works (Alpert, 1961, pp. 131–162)—makes unguarded use of other collective concepts (e.g. "collective consciousness", "suicidogenetic current," etc.), and thus frequently *violates* the principle of MI as characterized by Schumpeter. However, in *Le Suicide*, despite these unfortunate violations, it is with yet other "totalities" that Durkheim (1951 [1897]) is essentially concerned: statistical aggregates. This latter work establishes correlations between the suicide rate in a given country and other rates (single/married people; married people with children/without children; Protestants/Catholics, etc.). But Durkheim is not content, contrary to the method set out in the *Rules* (Durkheim, 1982 [1895]), to explain "social facts" by "social facts"; he also seeks the basis of these regularities by distinguishing different types of suicide (egoistic, anomic, etc.) according to the possible types of psychological functioning of *individuals* in their social life (need for integration, need for norms, etc.). This has been remarkably shown by Boudon and Bourricaud (1982, s.v. "Individualisme", p. 306), but only up to a certain point, to which I will return.

Examining statistical correlations is also what Weber undertakes at the very beginning of *The Protestant Ethic and the Spirit of Capitalism* (Weber, 2001 [1904–1905], pp. 3–12 and the corresponding footnotes: 131–138) by noting the correlations between, on the one hand, religious affiliation and, on the other hand, entrepreneurship, commerce and banking; although he does this kind of exercise with much less care than Durkheim (ibid., pp. 6–10),[19]

---

[19] See H. Trevor-Roper (1967) on this point.

and so quickly ("a glance", ibid., p. 3) that these pages tend to be forgotten. Weber also looks for the microfoundations of these statistical regularities (not, however, by reconstructing very general and simple plausible motives of Protestants like Durkheim for suicides, but by meticulously reconstructing the history of the genesis of very particular motivations specific to Protestantism). What is properly methodologically individualistic, in the sense of the logically primitive intention of MI (again as defined by Schumpeter), in this work of Weber, is the critical dimension with regard to the *concept* of the spirit of capitalism. But the *statistical* analysis which, in *The Protestant Ethic,* is barely sketched, is obviously not especially typical of the MI tradition; it is even more characteristic of Durkheim's method, which is incomparably more accomplished on this level.

However, it must be acknowledged that Coleman himself veils Weber's statistical concerns when he reinterprets—in the very first pages of the *Foundations,* just after reformulating the holism/IM debate (Coleman, 1990, pp. 1–6)[20]—the beginning of *The Protestant Ethic* without making *any* reference to Weber's examination of statistical regularities and speaking, instead, of the pressure that would be exerted on individuals by pre-existing Protestant values (i.e., some sort of collective representation). What Weber is supposed to have started from, according to Coleman (1990, p. 6), is a "macro-social proposition" containing collective entities (which can be found, incidentally, poorly circumscribed) of the kind: "Capitalism encourages Protestantism" (ibid., p. 6). These pages, among the most read and therefore the most influential of the *Foundations,* because they contain the now famous diagram (under the name of "Coleman's boat") of the different levels of analysis (macro *versus* micro) and their relations, besides considerably distorting Weber's approach, contribute to veiling the difference between the ontological and the epistemological meaning of the notion of levels of analysis, which is nevertheless at the heart of the problem I am raising.

For their part, Boudon and Bourricaud (1982) go so far as to characterize Durkheim's method in *Le Suicide* as a *typical illustration* of "an *individualistic* explanation (in the methodological sense) [italics by the authors]" (Boudon & Bourricaud, 1982, s.v. "Individualisme", p. 306). The comparison is striking (for it is a reference to Weber that most readers, in particular sociologists, expect at this point), but it leads, at the same time, to a veiling, in a manner symmetrical to Coleman's, of the very specificity of the individualizing method of MI as a means of rejecting collective pseudo-entities, since it is clear that Durkheim often abusively reifies concepts (notably collective

---

[20] Typically, Coleman wrote: "No assumption is made that the explanation of systemic behavior consists of *nothing more* than individual actions and orientations […]" (my emphasis), p. 5.

ones), in a way that could not be more typical. It therefore seems more reasonable to distinguish the individualistic - or microfoundationalist - method of seeking the explanation of statistical correlations at the level of individuals from that of MI.[21]

It should be noted, moreover, that the method used by Raymond Boudon (1985 [1973]) in *L'inégalité des chances*, his masterpiece, is in conformity with Schumpeterian (and Weberian) requirements: not only does Boudon not introduce pseudo-entities, but he even tracks them down when they are presented as endowed with an agentivity of their own (and thus with an autonomous existence). And it is Pitirim Sorokin who is then incriminated: "[with him] the social structures are described as having the control of the game" (ibid., p. 89). But Bourdieu and Passeron, discussed just after, are in no way incriminated (which would have been justified, by the way) for having committed, like Sorokin, the *fallacy of misplaced concreteness*. They are only criticized insofar as their explanation by the *habitus* - an *individual* psychological disposition even if it proceeds from the internalization of social structures—could only account for a complete social immobility (and thus for the mechanisms of reproduction of the structures); whereas, even if social mobility is weak, it is *statistically* attested (ibid.). This explanation is considered "more convenient than convincing", but not rejected as such. Bourdieu and Passeron are even credited by Boudon with "the merit of having posed a problem that Sorokin had hardly tackled", namely that of the foundations at the "microsociological" level of the data obtained by "macrosociological analysis" (ibid., pp. 90, 91). So, to sum up, it seems that Durkheim, Bourdieu and Passeron (but not Sorokin) do practice an individualistic method (or the search for microfoundations), but not the one that is characteristic of MI, since they violate (like Sorokin) the very principle of MI (as finely characterized by Schumpeter). And it is important to distinguish between the two; otherwise it is not surprising that there is confusion in the MI/holism debate.[22]

To come back precisely to the two variants of MI, MI1 and MI2, defined by Zahle and Kincaid, concerning *the role, in the sociological or economic explanation, of the search for the microfoundations of statistical correlations*,

---

[21] Boudon's excessive "ecumenism", making Durkheim himself a practitioner of MI, has been frequently pointed out (see for example the questions of Massimo Borlandi: Boudon, 1998).

[22] Unfortunately, for our purpose, *Education, Opportunity and Social Inequality* is, according to Boudon (1974) himself, "a new English version rather than a translation of [his] earlier work in the French language, *L'inégalité des chances* […]" (p. XVI). The discussion of Sorokin's and Bourdieu and Passeron's theories was not repeated in detail. The theoretical refinement of the previous analysis and, as a consequence, the recognition of Bourdieu and Passeron's specificity, however crucial in this context, have disappeared.

it is clear that MI2a simply does not make any sense: no sociologist, no economist, seriously challenges the use of statistical methods as such, even when they highlight their limits and in particular their bad use. As for MI2a, i.e. the thesis according to which analyses highlighting statistical correlations are *never* sufficient and that they *always* require microfoundations, this proposition does make sense and it is indeed bitterly debated.[23] But it is not at all part of MI strictly understood.

4 Conclusion

In this chapter, I have focused on two very important kinds of confusion, which seem obvious, on analysis, in the article by Zahle and Kincaid (2019), and which are, it seems to me, widespread throughout the post-Watkinsian tradition: on the one hand, the confusion between *vigilance towards concepts of collective entities* (and, more generally, towards any concept) and *rejection of collective entities* (and, more generally, the rejection of "misplaced concreteness"); on the other hand, the non-distinction between *real groups* (or real collective entities) and *purely nominal groups*. I have also tried to understand the reason for these confusions: even the authors who are the most aware of them sometimes produce statements that contribute to spreading them, hence the importance of an analysis of the language of the social sciences. But there are still other confusions or hasty assimilations that are present either in the article in question or, more generally, in this post-Watkinsian (and fundamentally post-Popperian) tradition which I have not examined. I merely point out a few of them. Thus the individualizing method—in the sense in which it was originally understood by Schumpeter and that was particularly highlighted by Hayek—is still only a particular expression of a more general method, the *analytical method,* which can be adopted not only to eliminate pseudo-entities (such as *Volksgeist*, capitalism, socialism, protestantism, social structures, etc.), but also to break down any entity that is expected to be divisible into more elementary components. Furthermore, this method does not necessarily aim to go down to the level of individual actors. One can thus analyze institutions, states, churches, companies, unions, soccer teams, with their rules of operation (Coleman 1990, Part III: Corporate Action, chap. XIII to XVIII, pp. 325–502); and one can even—one must sometimes (Coleman, 1990, chap. XIX, pp. 503–530; Elster, 1986a, 1986b)—descend

---

[23] Boudon himself sometimes supports a position as strong or combative as this one (1992), but sometimes retreats to a weaker or more conciliatory position (the "in general" replaces the "never" and the "always": Boudon, 1986).

466   A. Bouvier

to the infra-individual level. This is the whole program of analytical sociology (Hedström & Bearman 2009) which, in this sense, is broader than that of MI. As an *analytical method*, one can certainly say that the individualizing method is opposed to "methodological holism" (Boudon, 1986)[24] or to "methodological collectivism" (Elster, 1985), but it would be more reasonable, in order to avoid confusion between the problems, to oppose here the analytical method and the global method or "methodological *analyticism*" (of which MI is a simple species) and "methodological *globalism*" (Walliser & Prou, 1988, p. 109). Nor have I addressed the question of whether the criticism of the excessive emphasis on *social internalization processes* by authors such as Durkheim or Bourdieu is inherent in the individualist method. Zahle and Kincaid do not discuss this in their article, but Watkins, who does discuss this question, answers it positively. Nor have I addressed the question of the connection between the individualist method and the search for the unintended effects of human actions. Certainly, these quite fruitful perspectives are present in the MI *tradition*,[25] but their connection to the individualizing method does not seem to be based on a *logical link*.

This chapter therefore leaves in the shade a certain number of important questions raised by the Zahle-Kincaid article we have considered here (and by the post-Watkinsian tradition)—and of which I have only examined the preliminaries. It does not say anything about the questions that this same tradition formulates by borrowing its instruments from contiguous domains, such as the philosophy of cognitive sciences or the philosophy of mind. But, regarding a very important example, I have wanted to implement a conception of the philosophy of the social sciences that is *internal to* the social sciences and that perceives them as being *in themselves* spaces of debate by using a style that is both analytical and historical. From this point of view, it seems obvious to me that Watkins, as well as Agassi, and those who follow them in this direction, have hardly taken advantage of the paths opened in this sense a long time ago by Thomas Kuhn (1970) and other authors following him, leading to the examination of effective debates and thus of exchanges of effective arguments, which are themselves restituted, when this can be enlightening, in their context of production and reception. But Agassi (2014, p. 53), rather indifferent, like Watkins, to the actual debates and to the context of these debates, seemed at least well aware of his difference in

---

[24] See, in particular, the very last chapter (Chapter 10) of this book.

[25] On the crucial role of this search in the social sciences, which dates back to Adam Smith (1977 [1776]), and Adam Ferguson (1995 [1767]), see Hayek in particular (and Di Iorio's comments [2015]). Also see Boudon (1982 [1977]). Pareto (1935 [1916]) identified these specific causal chains as well (as a class of "non-logical actions" or paradoxical actions).

style from Kuhn's since he saw the latter as a (friendly) rival. Agassi, of course, might have had in mind Kuhn's relativist views above all and rightly wanted to reject them, but one does not have to assume them when one focuses on real historical debates.

References

Agassi, J. (1975). Institutional individualism. *The British Journal of Sociology, 26*(2), 144–155.

Agassi, J. (2014). *Popper and his popular critics: Thomas Kuhn, Paul Feyerabend and Imre Lakatos.* Springer.

Alpert, H. (1961). *Emile Durkheim and his sociology.* Russell & Russell.

Althusser, L. (1976). *Positions.* Éditions sociales.

Antiseri, D. (Ed.). (2005). *Epistemologia dell'economia nel "marginalismo" austriaco.* Rubbettino.

Bendix, R. (1977). *Max Weber: An intellectual portrait.* University of California Press.

Boettke, P. J. (Ed.). (1994a). *The Elgar companion to Austrian economics.* Edward Elgar.

Boettke, P. J. (1994b). "Introduction" to Boettke P. J. (1994). *The Elgar companion to Austrian economics* (pp. 1–6.). Edward Elgar.

Boudon, R. (1974). *Education, opportunity and social inequality: Changing prospects in western society.* Wiley.

Boudon, R. (1982 [1977]). *The unintended consequences of social action.* Macmillan.

Boudon, R. (1985 [1973]). *L'Inégalité des chances. La mobilité sociale dans les sociétés industrielles.* Armand Colin.

Boudon, R. (1991 [1984]). *Theories of social change. A critical appraisal.* Polity Press.

Boudon, R. (1986). *L'Idéologie ou l'origine des idées reçues.* Fayard.

Boudon, R. (dir.), (1992). *Traité de sociologie.* Presses Universitaires de France.

Boudon, R. (1998). Should one still read Durkheim's rules after one hundred years? Raymond Boudon interviewed by Massimo Borlandi. In R. Boudon (Ed.), *Études sur les sociologues classiques* (pp. 137–163). PUF.

Boudon, R., & Bourricaud, F. (1982). *Dictionnaire critique de la sociologie.* Presses Universitaires de France.

Boudon, R., & Bourricaud, F. (2003). *A critical dictionary of sociology* (partial trans. of Boudon & Bourricaud, 1982, by Peter Hamilton). Routledge.

Bourdieu, P., & Passeron, J.-C. (1977 [1970]). *Reproduction in education, society and culture.* SAGE.

Bouvier, A. (2007). Qu'est-ce qu'un 'engagement de groupe' en sciences sociales? L'exemple de l'école autrichienne en économie (de Carl Menger à Murray Rothbard). In A. Bouvier et B. Conein (dir.), *L'épistémologie sociale. Une théorie sociale de la connaissance.* (pp. 255–294). Ed. de l'EHESS.

Bouvier, A. (2011). Individualism, collective agency and the 'micro-macro relation'. In I. Jarvie & J. Zamora (Eds.), *Handbook of philosophy of social science* (chap. 8: pp. 198–215). Sage.

Bouvier, A. (2020). Individualism versus holism. In P. A. Atkinson, S. Delamont, M. Hardy & M. Williams (Eds.), *Sage research methods foundations*. Sage.

Boyer, A. (1999). Le tout et ses individus ou d'une querelle à l'autre. *Revue philosophique de la France et de l'étranger* , 435–465.

Bulle, N. (2020). Trois versions de l'individualisme méthodologique à l'aune de l'épistémologie. *Année sociologique, 70*, 97–128.

Bulle, N., & Phan, D. (2017). Can analytical sociology do without methodological individualism? *Philosophy of the Social Sciences, 47*(6), 1–31. https://doi.org/10.1177/0048393117713982

Caldwell, B. (2003). *Hayek' challenge: An intellectual biography of F.A. Hayek.* Chicago University Press.

Coleman, J. S. (1990). *Foundations of social theory.* Harvard University Press.

Descombes, V. (2014 [1996]). *The institutions of meaning. A defense of anthropological holism* (S. A. Schwartz, Trans.). Harvard University Press.

Di Iorio, F. (2015). *Cognitive autonomy and methodological individualism. The interpretative foundations of social life.* Springer.

Di Iorio, F. (2020). Individualisme méthodologique et réductionnisme. *Année sociologique, 70*, 19–44.

Durkheim, É. (1951 [1897]). *Suicide: A study in sociology* (J. A. Spaulding & G. Simpson, Trans.). The Free Press.

Durkheim, É. (1964 [1912]). *The elementary forms of the religious life* (G. W. Swain, Trans.). George Allen & Unwin.

Durkheim, É. (1982 [1895]). *The rules of the sociological method* (W. D. Halls, Trans.). The Free Press.

Elster, J. (1983). *Sour grapes. Studies in the subversion of rationality.* Cambridge University Press.

Elster, J. (1985). *Making sense of Marx.* Cambridge University Press.

Elster, J. (1986a). Marxisme and individualisme méthodologique. In P. Birnbaum & J. Leca (Eds.), *Sur l'individualisme. Théories et méthodes.* Fondation nationale des sciences politiques.

Elster, J. (Ed.). (1986b). *The multiple self* . Cambridge University Press.

Epstein, B. (2014). What is individualism in social ontology? Ontological individualism vs. Anchor individualism. In J. Zahle & F. Collin (Eds.), *Rethinking the individualism-holism debate. Essays in philosophy of social science.* Springer.

Epstein, B. (2015). *The ant trap: Rebuilding the foundations of the social sciences.* Oxford University Press. https://doi.org/10.1093/acprof:oso/9780199381104.001.0001

Ferguson, A. (1995 [1767]). *An essay on the history of civil society.* Transaction Publishers.

Frobert, L. (1996). La controverse Halévy-Pareto au IIe congrès international de philosophie, Genève 1904. *Revue européenne des sciences sociales, 34*(106), 51–67.

Gilbert, M. (1989). *On social facts.* Princeton University Press.

Gilbert, M. (1994). Durkheim and social facts. In W. Pickering & H. Martins (Eds.), *Debating Durkheim.* Routledge.

Gold, N., & Sugden, R. (2007). Collective intentions and team agency. *The Journal of Philosophy, 104*(3), 109–137.

Halévy, E. (1904). (Report on the) General Session (of the 11th Congress of Philosophy—Geneva). *Revue de Métaphysique et de Morale, 12*(6), 1103–1113.

Heath, J. (2020 [2005]). Methodological individualism. In E. N. Zalta (Ed.), *The Stanford encyclopedia of philosophy.* https://plato.stanford.edu/archives/sum2020/entries/methodological-individualism/

Hedström, P., & Bearman, P. (2009). *The Oxford handbook of analytical sociology.* Oxford University Press.

Hegel, G. W. F. (2012 [1822 and 1828]). *Lectures on the philosophy of world history. Introduction, reason in history* (H. B. Nisbet, Trans.). Cambridge University Press.

Hodgson, G. M. (2007). Meanings of methodological individualism. *Journal of Economic Methodology, 14*(2), 211–226. https://doi.org/10.1080/135017807013 94094

Kincaid, H. (2008). Individualism versus holism. In S. Durlauf & L. E. Blume (Eds.), *The new Palgrave dictionary of economics.* Palgrave Macmillan.

Kincaid, H. (2014). Dead ends and live issues in the individualism-holism debate. In J. Zahle & F. Collin (Eds.), *Rethinking the individualism-holism debate. Essays in philosophy of social science* (pp. 139–152). Springer.

Kuhn, T. (1970 [1962]). *The structure of scientific revolutions.* University of Chicago Press.

List, Chr, & Spiekermann, K. (2013). Methodological individualism and holism in political science: A reconciliation. *American Political Science Review, 107*(4), 629–643. https://doi.org/10.1017/S0003055413000373

Littlechild, S. (1990). *Austrian economics I.* Edward Elgar.

Little, D. (2014). Methodological localism and actor-centered sociology. In J. Zahle & F. Collin (Eds.), *Rethinking the individualism-holism debate. Essays in philosophy of social science.* Springer.

Menger, C. (1883). *Untersuchungen über die Methode der Sozialwissenschaften, und der politischen Oekonomie insbesondere.* Dunker & Humblot.

O'Neill, J. (Ed.). (1973). *Modes of individualism and collectivism.* Heinemann Educational.

Pareto, V. (1935 [1916]). *The mind and society* (Ed. A. Livingston). Harcourt, Brace.

Pettit, P. (2003). Groups with minds of their own. In F. Schmitt (Ed.) *Socializing metaphysics* (pp. 167–93). Rowman & Littlefield.

Pettit, P. (2014). Three issues in social ontology. In J. Zahle & F. Collin (Eds.), *Rethinking the individualism-holism debate. Essays in philosophy of social science* (pp.77–96). Springer.

Popper, K. (1960 [1944–1945]). *The poverty of historicism.* Routledge.

Popper, K. (1974). *Unended quest. An intellectual autobiography.* Routledge.

Popper, K. (1978). *Three worlds. The Tanner lecture on human values.* University of Michigan.

Poulantzas, N. (1968). *Pouvoir politique et classes sociales.* Maspero.

Radkau, J. (2011). *Max Weber: A biography* (P. Camiller, Trans.). Polity Press.

Schumpeter, J. A. (1908). *Das Wesen und der Hauptinhalt der theoretischen Nationalökonomie.* Duncker & Humblot.

Schumpeter, J. A. (1909). On the concept of social value. *Quarterly Journal of Economics, 23*(2), 213–232. https://doi.org/10.2307/1882798

Schumpeter, J. A. (1998 [1908]). *Methodological individualism.* Trans. of the chapter 6 of J. A. Schumpeter, 1908 by M. van Notten. Institutum Europaeum.

Sellars, W. (1974). *Essays in philosophy and its history.* Reidl.

Smith, A. (1977 [1776]). *An inquiry into the nature and causes of the wealth of nations.* University of Chicago Press.

Sugden, R. (2016). Ontology, methodological individualism, and the foundations of the social sciences. *Journal of Economic Literature, 54*(4), 1377–1389.

Trevor, R. (1967). *The Crisis of the Seventeenth Century: Religion, the Reformation, and Social Change, and Other Essays.* Liberty Funds.

Trevor Roper, H. (1967). *The crisis of the seventeenth century: Religion, the reformation, and social change, and other essays.* Macmillan.

Tuomela, R. (1995). *The importance of us. A philosophical study of basic social notions.* Stanford University Press.

Uedehn, L. (2001). *Methodological individualism: Background, history and meaning.* Routledge.

Walliser, B., & Prou, C. (1988). *La Science économique.* Le Seuil.

Watkins, J. (1994 [1957]). Historical explanation in the social sciences. In M. Martin & L. McIntyre (Eds.), *Readings in the philosophy of social science.* MIT Press.

Weber, M. (1978 [1922]). *Economy and society* (Eds. G. Roth & C. Wittich). University of California Press.

Weber, M. (1981 [1913]). Some categories of interpretive sociology. Trans by Edith E. Graber. *The Sociological Quaterly, 22,* 151–80.

Weber, M. (1985 [1913]). Über einige Kategorien der verstehenden Soziologie. In M. Weber (Ed.), *Gesammelte Aufsätze zur Wissenschaftslehre* (pp. 427–473). Johannes Winckelmann.

Weber, M. (2001 [1904–1905]). *The protestant ethic and the spirit of capitalism* (T. Parsons, Trans.). Routledge .

Weber, M. (2012a [1920]). Brief an Robert Liefmann (9.03.20). In *Max Weber-Gesamtausgabe*, II, 10. Morh-Siebeck.

Weber, M. (2012b). *Collected methodological writings* (Eds. H. Henrik Bruun & S. Whimster). Routledge.

Winch, P. (1958). *The idea of a social science and its relation to philosophy.* Routledge & Kegan Paul.

von Hayek, F. (1948). *Individualism and economic order. A critical analysis of socialist economics and a plea for the preservation of "true individualism".* Routledge.

von Hayek, F. (1980). Preface to J. A. Schumpeter, 1998 [1908]. *Methodological Individualism.*

von Hayek, F. (2010 [1942–1944]).. Scientism and the study of society Hayek. In F. A. Hayek (Ed.), *Studies on the abuse and decline of reason: Text and documents* (ed. Bruce Caldwell). University of Chicago Press.

von Mises, L. (1949). *Human action. A treatise on economics.* Yale University Press.

Ylikoski, P. (2014). Rethinking micro-macro relations. In J. Zahle & F. Collin (eds.), *Rethinking the individualism-holism debate. Essays in philosophy of social science* (pp.117–135). Springer.

Ylikoski, P. (2017). Methodological individualism. In L. McIntyre & A. Rozenberg (Eds.), *The Routledge companion to philosophy of social science* (pp. 135–146). Routledge.

Zahle, J. (2016). Methodological holism in the social sciences. In E. N. Zalta (Ed.), *The Stanford encyclopedia of philosophy.* https://plato.stanford.edu/archives/win2016/entries/holism-social/.

Zahle, J., & Collin, F. (Eds.). (2014a). *Rethinking the indivdualism-holism debate. Essays in the philosophy of social science.* Springer.

Zahle, J. & Collin, F. (2014b). Introduction. In J. Zahle & F. Collin (Eds.), *Rethinking the individualism-holism debate. Essays in philosophy of social science* (pp. 1–14). Springer.

Zahle, J., & Kincaid, H. (2019). Why be a methodological individualist? *Synthese, 196*(2), 655–675.

Individualism-Holism Debate in the Social Sciences: Political Implications and Disciplinary Politics

Branko Mitrović

1 Introduction

The debate between individualism and holism in the social sciences is often marred by political assumptions, alleged ideological connections and (as we shall see) even unpleasant political insinuations. It is particularly common to associate individualist positions with libertarian political views and neo-liberal economics. Not rarely, advocates of social holism state or imply such associations as their motivation or even as direct arguments against individualist views. In this paper I will start by analyzing the impact of such assumptions on the quality of the debate in general and then proceed to show that such assumptions cannot be theoretically defended. I will then argue that numerous social policies in favor of disadvantaged sections of society can be introduced only if one relies on individualist assumptions about social groups. I will also argue that holist assumptions about social entities really favor the interests of administrators and owners of corporate capitalist entities. Finally, I will discuss the consequences of disciplinary politics and especially the concerns that individualism entails the reduction of the social sciences to psychology.

B. Mitrović (✉)
Faculty of Architecture and Design, Norwegian University of Science and Technology, Trondheim, Norway
e-mail: branko.mitrovic@ntnu.no

© The Author(s), under exclusive license to Springer Nature Switzerland AG 2023
N. Bulle and F. Di Iorio (eds.), *The Palgrave Handbook of Methodological Individualism*,
https://doi.org/10.1007/978-3-031-41508-1_20

The distinction between methodological and ontological individualism happens to be of secondary significance when it comes to the political implications of the individualist understanding of society. The same political arguments are often used against both methodological and ontological individualism and generally have the same (limited) value and relevance. Consequently, in this paper I am discussing the political claims made in debates about individualism in general—it would be pointless to limit discussion to methodological individualism, since the scope of arguments often pertains to ontological individualism as well. For the same reason I will not address here various ways to define methodological individualism and differentiate it from ontological individualism—I assume this has been covered in other papers in this volume.

2 Form of the Debate

It is necessary to start here by analyzing the general form of the debate between individualism and holism in the social sciences, otherwise it will be impossible to understand the impact that political claims and assumptions make on these discussions. The discussion between individualist and holist perspectives on society has been going on for more than a century, and it has been quite lively at least since the 1960s. One aspect of the form of the debate that is particularly important for our analysis here is the disproportionally high number of articles and books that seek to refute individualism, ontological or methodological. One rarely reads articles that argue in favor of individualism. At the same time, the arguments against individualism presented in these publications are often marked by remarkably poor quality. As we shall see from the examples presented below, one can talk about an actual avalanche of poor and unconvincing anti-individualist arguments that have passed reviewers and editors and have been published in top-ranking journals or in books produced by prestigious publishers. Not rarely, the same weak arguments are further repeated in later publications, even though they have been refuted in the meantime. All this suggests that strong extra-theoretical motivation plays a major role in the debate. The phenomenon should not escape our attention especially since political views are often stated to motivate the debate between individualism and holism. In order to provide the wider perspective that is necessary in order to explain the politicization of the theoretical dilemma it is necessary to start here by describing the phenomenon.

Consider, for instance, the anti-individualist claim that social entities cannot be understood as individuals and their interactions because in order to identify these individuals one has to refer to other social phenomena. In other words, it is argued that one cannot say that educational or penal systems are sets of individuals and their interactions, because some of these individuals have to be teachers or inmates—while without social items such as schools or prisons there can be no teachers or inmates. The obvious individualist response is going to be that schools or prisons, being social entities, are themselves nothing more than sets of individuals and their interactions. The argument assumes that they are not, while this is something that it is only meant to prove. In other words, the argument seeks to prove that social institutions cannot be individuals and their interactions by assuming that social institutions are not individuals and their interactions. Although it relies on a classic logical fallacy, the argument has had extremely wide circulation in the literature. So far I have managed to trace its history, it was mentioned for the first time in an article by Mandelbaum (1955, p. 309). Subsequently it was repeated in articles by Lukes (1968, p. 122) and Weldes (1989, p. 362) and in an article and a number of times in a book by Kincaid (1986, p. 499; 1997, pp. 23, 34, 35, 51). Giddens (1979, pp. 94–95) seems to imply it as well.[1] Similarly, Zahle and Kincaid (2019, p. 658) de facto endorsed the holist view that describing an individual as a CEO presupposes the existence of corporations and corporate structures, without noting the logical error.[2] For the analysis of the politics of the debate later in this article, the significant point is not merely that all these authors made the same elementary logical error, but that their texts that contain that same error received institutional endorsement by the editors and reviewers of *The British Journal of Sociology*, *Theory and Society*, *Philosophy of Science*, *Synthese* and The Macmillan Press who published them. In fact, reviewers of leading journals sometimes endorse invalid arguments even after they had been repeatedly refuted and rejected in earlier publications. Ritchie (2013, p. 266) in an article published in *Philosophical Studies* argued that teams, committees, clubs or courts cannot be sets of individuals because such 'groups' (as she calls them, Ritchie, 2013, p. 257) can change members, whereas sets cannot. The simple individualist

---

[1] Giddens (1979, pp. 94–95) is somewhat unclear, but I assume that he relies on this argument when he says that social institutions 'are the outcome of action only in so far as they are also involved recursively as the medium of its production.'

[2] Since Zahle and Kincaid (2019, p. 658) refer to Kincaid's own book *Philosophical Foundations of the Social Sciences* for this argument, one can assume that they endorse it, though the situation is unclear because they do not state the page. Their statement later on the same page 'explanations invoking roles in organizations are the most social' confirms that they assume that individualist understanding of organizations is impossible. There is no reason why individualist explanations could not invoke roles in organizations.

476   B. Mitrović

response is that the claim that institutions ('groups') are sets of individuals and their interactions does not mean that every institution is a *single* set of individuals and their interactions. An institution can be various sets of individuals and their interactions at various times. For instance, one can conceive of the Supreme Court of the USA as different sets of justices and their interactions at different times—or, alternatively, as a set of *all* justices who have ever served or will serve on the Court, whereby different subsets of these justices make decisions for the Court at different times. These two responses were presented in well-known papers by Uzquiano (2004, p. 139) and Ruben (1982, p. 301), before the publication of Ritchie's paper. Again, the interesting point is not merely that Ritchie presented an invalid argument, but that reviewers failed to notice the fallacy that had been already well established in the existing literature. They also failed to notice the absurdity of another argument she presented in the same paper, that because sets are 'normally taken to be abstract' they cannot be in space: 'Sets are not, and groups are in space,' she said (Ritchie, 2013, pp. 266–267). The idea of the argument is that 'set' is a mathematical concept, and such concepts are abstract entities that exist outside space and time. Since human individuals are in space and time, it is impossible to talk about 'sets of individuals' or assume that they participate in social institutions (which are also in space and time). This is like arguing that there cannot be seven cars on a parking lot because the word 'seven' refers to a number and numbers are mathematical abstract entities outside space–time. Since cars are physical objects, it would follow that they can never be counted. Following the same reasoning, any application of mathematics to the physical world would be impossible. The simple response is that sets of human individuals or any other physical objects are modeled on mathematical sets. Many mathematical concepts were originally derived from our everyday experience. The terms used to refer to such experience do not cease to be applicable to our everyday experience once mathematicians start theorizing about them. As for sets, the founder of the set theory, Georg Cantor, actually did not conceive of them as abstract Platonic non-mental entities.[3]

Looking at the wider picture, it is sometimes hard to resist the impression that *any* claim can be endorsed by reviewers and published in well-reputed journals, insofar as it purports to refute individualism. Consider the following claim presented by Bunge (2000, p. 394) in an article published in *Philosophy of Social Sciences*:

---

[3] As he put it, he regarded them as 'Objecten [sic] unserer Anschauung oder unseres Denkens' (Cantor, 1869, p. 481).

Individualism-Holism Debate in the Social Sciences ... 477

> If methodological individualism were adequate, to know a triangle it should suffice to know its sides regardless of its relations, namely, the inner angles—which is not even true in the exceptional case of equilateral triangles.

In other words, Bunge is actually arguing that:

> If methodological individualism were adequate, then knowing the sides of a triangle should suffice to know its inner angles.
> But knowing the sides of a triangle does not suffice to know its inner angles (not even in the case of equilateral triangles).
> Therefore, methodological individualism in the social sciences is inadequate.

The claim is nonsense: the dilemma between individualism and holism in the social sciences has nothing to do with the relationship between the sides and the angles of a triangle. To make things worse, it is directly embarrassing that the author would state, the reviewers endorse and a reputed journal would publish a false claim about high-school-level trigonometry as a supporting thesis for the argument. If all sides of a triangle are known, it is easy to calculate the angles on the basis of the cosine theorem. In the case of an equilateral triangle, all angles are 60° and no calculation is necessary.

The impression that anything goes when theorists of the social sciences want to refute individualism is further strengthened by the widespread presence of straw-man refutations in the literature. The most common tactics is to disregard the emphasis that advocates of individualism have traditionally placed on the role of interactions in social phenomena, redefine individualism as the claim that social phenomena consist of passive individuals who do not interact, and then easily refute it. It should be obvious that interactions have to play a central role in all individualist attempts to explain social phenomena. They are standardly mentioned in the existing literature when it comes to definitions of individualism. Agassi (1987, p. 150) actually described the view 'that social phenomena are but interactions between individuals' as 'classical individualistic idea.' The understanding that individualist understanding of social phenomena must refer to interactions goes back at least to Simmel (1908, pp. 3–5; 1917, pp. 6–15), who emphasized the role of interactions and joint actions early in the last century (he used the terms *Wechselwirken* and *Zusammenwirken*). It is also not hard to find references to interactions in various definitions of individualism proposed by modern authors. Elster (1992, p. 13) thus defined methodological individualism as the view that explanations of social institutions show how they arise 'as the result of the action and interaction of individuals.' Sawyer (2005, p. 6) identified methodological individualism as the approach that seeks to explain how

478   B. Mitrović

macro social phenomena, such as institutions, social movements, norms and role structures result from 'individual actions and dyadic interactions.' List and Spiekermann defined methodological individualism as the thesis that good social-scientific explanations should refer solely to facts about individuals and their interactions (List & Spiekermann, 2013, p. 629). One may agree or disagree with the theoretical views of these authors, and one may oppose or endorse their definitions—but in any case, it is hard to deny that there exists a long and established tradition to include interactions in definitions of individualism. This is certainly reasonable. A soccer match is not a set of individuals who belong to two teams, it is what these individuals do on the field. Very often, the interactions in which individuals engage define the nature of the social entity. One and the same set of bank employees can also constitute a rugby team, and when they play rugby they are not a bank, they are a rugby team. They are a bank when they perform bank-specific interactions. Also, it is natural to assume that they can interact with the material environment (e.g. computers) as well as non-members of the set (e.g. clients of the bank). Finally, when individualists talk about individuals, it is natural to assume that the properties (intrinsic and extrinsic) of these individuals are included as well—it would be absurd to talk about individuals without properties.[4]

Classic straw-man arguments against individualism simply omit any reference to interactions from the definition of individualism and then seek to refute the position that they have thus defined as individualism. In the article cited above Bunge points out that 'social facts can only be understood by embedding individual behavior in its social matrix and by studying interactions among individuals' (Bunge, 2000, p. 394). Contrary to standard definitions in the literature, he claims that individualism does not take individuals' interactions into account and then proceeds to refute it by pointing out that social facts cannot be understood without taking interactions into account. Hodgson (2007, p. 220) in an article published in the *Journal of Economic Methodology* accurately notes that attempts to explain social phenomena in terms of individuals alone have never been successful and that '[i]n modern social theory, structures are typically defined as sets of

---

[4] An intrinsic property is the property of the object (Paul is six feet tall) whereas an extrinsic property is the property the object has in relation to other objects (Paul is taller than Peter). (For an explanation of this distinction see Marshall & Weatherson, 2018.) Extrinsic properties are also sometimes referred to as 'relations.' The fact that a person has a certain mental state that instantiates a mental content is an intrinsic property of that person. Sometimes extrinsic properties may be accounted of in terms of interactions, as Weatherson and Marshall put it, 'We have other [extrinsic] properties in virtue of the way we interact with the world.'

interactive relations between individuals.' He then decides that it is unwarranted to use the term 'methodological individualism' for the position that takes individuals' interactions into account and calls this refutation based on such re-definition of individualism 'devastating.' As we have seen, such re-definition is in obvious collision with the common practice to include interactions in definitions of individualism. Hodgson actually recognizes that 'many advocates of methodological individualism' as he says 'fail' to specify individualism in terms of individuals alone—i.e. the way he needs them to define their position so that his refutation would be valid. De facto this recognition amounts to an open admission that his is a straw-man refutation. The impression is that Hodgeson felt compelled to reject something called 'individualism' and then adjusted the definition in order to fit his agenda. The only thing that is really devastating is the thought that reviewers would pass and the *Journal of Economic Methodology* (note the word 'methodology' in the title) would publish a paper whose author almost openly admits that he is presenting a straw-man refutation.

A similar, but more elaborate straw-man refutation of individualism was presented by Epstein (2009). Epstein's main target is ontological individualism, but this kind of straw-man argument could be seen as directed against methodological individualism as well. His account of individualism completely omits to mention interactions in which individuals engage. As he presents it, 'Ontological individualism is committed to the claim that individual people are the ultimate constituents of the social world in which they reside. ... [It] holds that once the individualistic properties of people and relations among them are appropriately understood, those properties and relations suffice to determine the social facts' (Epstein, 2009, p. 208). In his view (ontological) individualists 'mistakenly commit themselves to limiting the determinants of social properties to the properties of individuals' (Epstein, 2009, p. 209). The understanding that for individualists interactions in which individuals engage often play an important role in the constitution of social entities, events or phenomena has been suppressed.[5] It is hard not to wonder

---

[5] One may be tempted to try to interpret his concept of 'individualistic properties' (i.e. the properties of individuals) to include interactions in which individuals engage. Similarly, the definition of ontological individualism in the Abstract of his article says that '[o]ntological individualism is the thesis that facts about individuals exhaustively determine social facts' and one may think that 'facts about individuals' include facts about interactions in which individuals engage, including their interactions with the physical environment. However, these interpretations cannot be correct, considering the point that Epstein seeks to make. His line of argument is to point out that the physical environment contributes to the social environment—which in his view is meant to show that individuals and their properties are not sufficient to explain the properties of the social environment. Obviously, the impact of the physical environment on the social environment is something that individualists would seek to explain by referring to the interactions of individuals with the physical environment. Since he leaves this line of response unanswered, one can thus infer that by 'individualistic properties' he

480   B. Mitrović

about the motivation of the *Synthese*'s reviewers who overlooked such an omission—especially since Epstein's 'refutation' of ontological individualism is precisely based on that omission. The line of argument then leads him to claim:

> It is difficult even to conceive of any satisfactory characterization of explanation of a social phenomenon such as dance or an orchestral performance or a riot, without incorporating physical factors as well as psychological ones. Likewise, physical factors are involved in the determination of membership in groups as well. It is not only the dance and the orchestra and the riot that involve physical factors, but also the holding of the properties being a dancer, being a cellist and being a rioter. If there were no cellos, then regardless of Yo-Yo Ma and the rest of us thought and did, there would be no cellists. (Epstein, 2009, p. 202)

However, rioters with their rioting equipment or musicians with musical instruments are not enough for a riot or an orchestral performance; they must also *do* something. They must interact with that physical equipment for something to happen. In other words, cellos and musicians could exist, but if musicians did not interact with cellos (i.e. played them), there would be no concerts. Yo Yo Ma would not be a cellist if he did not interact with cellos. Since definitions of individualism commonly include a reference to the inter-actions, and interactions are often with the physical objects, one cannot refute individualism by merely pointing out that the physical environment plays an important role in social phenomena. That is a platitude. When definitions of individualism mention interactions this necessarily includes the physical environment as something that is interacted with. In order to refute individ-ualism, one would have to show that the impact of the physical environment on social phenomena can be independent of interactions between individuals and the physical environment.[6]

---

does not mean the interactions in which individuals engage. He actually cites J. W. N. Watkins's statement that physical causes in society 'operate either by affecting people, or through people's ideas about them' and then claims that according to Watkins 'the social facts themselves are not depen-dent of physical factors at all' (Epstein, 2009, p. 202). Here too, it is significant to note that such openly self-contradictory interpretation of Watkins's view could have passed unnoticed the reviewers of *Synthese*.

[6] In a later book Epstein presented an equivalent straw-man argument in the form of a thought experiment about Starbucks that also suppressed the consideration of interactions between individuals and the physical environment (Epstein, 2015: 46). Epstein there imagined that a late night power spike causes serious damage to a great number of Starbucks stores and the company became insolvent. He claimed that '[i]n this example, the transition to insolvency involves property and equipment, not individuals.' However, insolvency is impossible if no human individuals are involved. For insolvency to occur some human individuals would have to discover the broken equipment, employees had to be prevented from doing their work, and customers would have to stop buying coffee from Starbucks.

Arguments about coincidence are another group of straw-man anti-individualist arguments that systematically disregard the role of interactions in the individualist understanding of social reality. Gabriel Uzquiano's article published in *Nous* presents three fine examples of this strategy. (Uzquiano, 2004) It is enough to present one of them here in order to illustrate how it works (for the remaining two see the footnote).[7] Uzquiano's arguments pertain to the imaginary situation in which the Senate of the USA appoints all and only justices of the Supreme Court to a Committee on Judicial Ethics. According to Uzquiano, it is a problem that the set of individuals that make up the Committee on Judicial Ethics will be identical to the set of individuals serving as Supreme Court justices. For instance, the Supreme Court may be in session at a certain time—and yet, he points out, that does not mean that the Committee on Judicial Ethics is in session at the same time. The argument is a mere trick with words. Uzquiano uses the same phrase 'to be in session' to refer to different types of interactions that characterize and define these two institutional bodies. Institutions, as mentioned earlier, are often defined on the basis of the interactions that the participating individuals engage in. The Supreme Court is in session when the justices interact in accordance with their duties as Supreme Court justices. When the Committee on Judicial Ethics is in session, these same individuals are performing another kind of interactions. It is perfectly normal that sets of individuals can perform different tasks at different times and that they are classified as different social entities depending on the tasks they perform. The argument is therefore invalid in a very trivial sense—merely because it fails

---

If the spike merely affected equipment that was not in use, and its destruction passed completely unnoticed, insolvency would not have happened. This thought experiment is thus not suitable to show that individualism cannot deal with the impact of the physical on the social environment. In order to show this, one would have to construct an example in which the physical environment can affect the social environment in a way that would be independent of its interactions with the human individuals who participate in that social environment.

[7] The argument overlooks the role of interactions in the constitution of social entities. If the Senate appoints 'individuals who perform interaction X' this does not mean that the Senate appoints 'individuals who perform interaction Y'—even though these happen to be the same individuals. In his third argument, Uzquiano imagines a situation in which the Committee on Judicial Ethics, but not the Supreme Court, joins the Committee of Ethics Committees. (As mentioned, the assumption is that the Committee on Judicial ethics and the Supreme Court are the same sets of individuals but they perform different interactions.) If membership in the Committee of Ethics Committees were merely a relation between a set and its elements, Uzquiano reasons, then both the Supreme Court and Special Committee on Judicial Ethics would be members of the Committee of Ethics Committees— and this was not meant to be the case. Once again, the response will be that the phrase 'Supreme Court' refers to individual justices together with the interactions they perform as justices, whereas the Committee on Judicial Ethics is individual members of the Committee together with the interactions that they perform as the members of the Committee on Judicial Ethics. Consequently, it does not follow that the Supreme Court joins the Committee of Ethics Committees if the Committee on Judicial Ethics does so.

482   B. Mitrović

to take into account the well-established role of interactions in definitions of institutions. One certainly has the right to expect that the reviewers who evaluated the article for *Nous* should have recognized the argument as a mere straw-man refutation—and one has the right to wonder why they failed to do so.

The examples presented here are merely the tip of the iceberg. The limited size of this article makes it impossible to list more of them. The examples should be however sufficient to point out that this widespread proliferation of bad arguments, and that their endorsement by reviewers and editors of leading journals constitutes a remarkable phenomenon in its own right. We are talking about an environment in which even false claims about high school-level trigonometry can pass referees of prestigious journals, insofar as they are presented as arguments against individualism. It becomes hard to avoid the unfortunate impression that anything goes and that reviewing practices in the social sciences—at least when it comes to the polemic against individualism—are merely an instance of what Chomsky (2002, p. 338) described as 'the desperate attempt of the social and behavioral sciences to imitate the surface features of sciences that really have significant intellectual content.' The phenomenon is so pervasive and systematic that it cannot be explained if one does not assume the existence of widespread and unstated motives that inspire it. Considering the nature of the topic, it is reasonable to expect that the motivation has to be a political one.

3   Political Values and (as) Argumentation

The tone of the debate, indeed, often suggests that the participants have strong, politically motivated, feelings about the dilemma between individualism and holism. Commonly, individualist perspectives are associated with libertarian political views, neo-liberal economics and opposition to social programs. This kind of association often reminds of Margaret Thatcher's statement that there is no society but only individuals. Consider Kincaid's formulation of this view:

> Individualism also supports distinctive views about the place of government policy and the nature of social justice. Holists locate the cause of poverty, crime, and the distribution of income in large social forces; individualists look for causes in the traits and preferences of individuals. Quite naturally, individualists are skeptical that governments can solve what they see as individual problems, while holists are likely to demur. Individualists are likewise naturally

Individualism-Holism Debate in the Social Sciences ... 483

attracted to accounts of justice emphasizing individual preferences, contributions, merits, and rights; giving social structure an essential place raises doubts about that program. (Kincaid, 1997, p. 2)

Similarly, in the article that he and Zahle wrote together they say:

Individualism is consistent with the view that social outcomes are entirely the responsibility of individual decisions. This means social circumstances and society are down played for social injustices such as poverty, inequality, racism and the like.[8]

The two paragraphs are a remarkable collection of non sequitur conclusions inferred from false claims, based on a misrepresentation of individualist views about the nature of social phenomena. Contrary to what Kincaid and Zahle suggest, there is no reason why individualists cannot see the causes of poverty, crime, and the distribution of income in large social forces. Rather, individualists differ from holists because they assume that these large social forces are themselves individuals and their interactions and not something else. It is true that individualists do not understand social institutions such as governments as something over and above individuals that make them up and their interactions—but that does not mean that individualists cannot advocate welfare programs, free healthcare and education and so on. The assumption that governments as social institutions are sets of individuals (e.g. ministers) and their interactions is not contradictory to the expectation that governments should organize welfare programs and provide for pensions, free healthcare and education. Saying that poverty is always instantiated in individuals does not mean that governments cannot be expected to help overcome problems with mass poverty. It is true that individualists assume that decisions made by social entities result from decisions and interactions of individuals. Social entities such as classes, governments, boards and so on do not have mental capacities over and above the mental capacities of participating individuals that would enable them to make decisions. In that sense all decisions made in a society do depend on the decisions made by interacting individuals.

---

[8] Zahle and Kincaid (2019, p. 672). The claim is not only inaccurate but also remarkably badly phrased. 'Individual decisions' could be even made by holistically understood social entities in the sense that decisions are countable and thus individual (a holist social entity could thus make its first, second, third individual decision). Also, *decisions* are not human beings and they cannot have responsibility so it is meaningless to talk about 'the responsibility of individual decisions.' The authors are likely to have meant 'decisions made by individuals.' Such details are significant for the argument made in this article, since they show that in the case of a paper that argues an anti-individualist position even sloppy, anything-goes formulations can pass the reviewers and the editors of *Synthese*.

484   B. Mitrović

It follows that social entities indeed cannot have responsibilities independently of the responsibilities of participating individuals who make decisions for them. (This is an important point, because we shall see that the opposite holist view absolves from any responsibility the protagonists of the most rapacious practices of corporate capitalism.) But the view that individuals are responsible for the state of society does not entail that 'social circumstances and society are down played for social injustices such as poverty, inequality, racism and the like' as Kincaid and Zahle assert in the paragraph cited above. Rather, it means that social circumstances that generate poverty, inequality or racial problems are understood to result from interactions between individuals. For an individualist, there is nothing else they could result from. From the individualist point of view, when a government recklessly reduces spending for the healthcare system and causes it to malfunction in order to decrease taxes, the responsibility for the consequences is not with some abstract entity called 'the government'—rather, the responsibility is with the ministers who voted for the decision. Similarly, in order to prove their position, holists cannot merely point out that, for instance, widespread poverty causes an increase in crimes. (Not many people would deny this, but this is not the question.) They need to show that poverty is something more than a property instantiated in individuals.

The association of ontological or methodological individualism with political libertarianism is yet another non sequitur. First, there are no logical reasons why a political libertarian could not be a holist in matters of social theory. An American libertarian may, for instance, believe that the USA is a supra-individual holistic entity that arranges, or ought to arrange, social relations within the country on libertarian principles. There is no reason why an American political libertarian may not believe that the USA as a holistic political entity is over and above American citizens and their interactions. Libertarians seek to limit the rights of governments to intervene in the society and in interactions between individuals—but this political program need not commit them to a specific theory about the nature of society or social institutions. Also, left-wing individualism has a long tradition and possibly even goes back to the early writings of Karl Marx and Friedrich Engels.[9] It has

---

[9] The attribution of individualist views to Marx and Engels may sound unusual because many people associate their views with various Communist traditions and ultimately a holist worldview. Nevertheless, individualist views are particularly strong in their early writings such as the *German Ideology* and *Holy Family* where they associate individualism with materialism and correlate it to their rejection of Hegelian idealism. (See especially the first part of their *German Ideoloogy*, Marx, Engels, 1953, pp. 9–78). There exists an extensive discussion about their views when it comes to individualist perspectives on social phenomena. See for instance Israel (1971); Elster (1982, 1985), Dumont (1977, pp. 113, 125, 136–137), Weldes (1989), Wolff (1990), Tarrit (2006), Kumar (2008), and Levine et al. (1987).

Individualism-Holism Debate in the Social Sciences ...

been pointed out long time ago that it is impossible to make theoretical links between methodological individualism and libertarian economics or politics. Important observations in that sense go back to Joseph Schumpeter's work early in the twentieth century (Schumpeter, 1998 [1908], pp. 88–98). Schumpeter insisted that it is important to differentiate between political and methodological individualism. As he put it, there is 'no connection between individualist research on society and political individualism' and 'the [individualist social] theory provides no arguments in favor or against political individualism' (Schumpeter, 1998 [1908], p. 90).

There are two significant problems when it comes to the association of ontological or methodological individualism with neo-liberal economics and the opposition to social programs. The first is that economic needs and social oppression exist only at the level of (sets of) individuals and can only be addressed at that level. The second is that social programs can only be formulated and developed if one relies on the individualist understanding of human collectives targeted by these programs.

In order to illustrate the first point, consider Burman's (1979, p. 368) description of the factors that enter in the plight of a black child in an American inner-city school of the 1970s:

> ...a home life with little conventional intellectual tutoring; unemployment of parents leading to apathy and hostility towards the society at large; white teachers who have a low opinion of blacks and verbally deficient students; a streaming process in school that perpetuates the poor expectations of the child's performance; absence of role models with which the child can identify; a school program that is geared in its discipline and curriculum to verbal, white, middle-class children; a social system that prizes academic ability, offering few rewards for those who are inept; few job prospects once schooling is over; a socio-economic system that offers all its riches to winners and nothing but contempt, discrimination and the dole to its losers; vested interests who do not want any substantial government support for inner-city schools.... (Burman, 1979, p. 368)

The list is convincing, but his account is marred by his further claim that individualists are ill-equipped to make an effective diagnosis of these problems. In fact, *many elements of the list*—home life with little intellectual tutoring, unemployed parents, absence of role models and so on—can *only* be understood and described from the individualist perspective. 'Home life with little intellectual tutoring,' 'unemployed parents,' and 'absence of role models' can only refer to types of interactions between individuals. There are no supraindividual social forces or holist social entities called 'home life with little intellectual tutoring' or 'absence of role models.' 'Unemployed parents' are human individuals who lack employment, not holist entities. Also, a 'school system,' 'socio-economic system,' 'government support' and similar can affect a child only through that child's interactions with other individuals, and these other individuals can be affected by such systems (represent it, act on its behalf, etc.) only as a result of their interactions with other individuals within the system. Only some concepts that Burman's account relies on can have holist interpretations at all ('school program' or 'school system'). At the same time, they can have individualist interpretations as well.[10] Speaking in general, it is hard to imagine how there could be oppression that somehow affects human collectives understood holistically—i.e. a form of oppression that would not be oppression of human individuals. If this is so, then only the individualist understanding of society can provide the understanding of the specific forms of oppression that is necessary in order to oppose it.

Social welfare programs are often possible only if one assumes rigidly individualist understanding of society. By 'rigidly individualist' I mean an understanding of social groups that assumes that they are not only sets of individuals, but sets of *identifiable* individuals. Many kinds of social programs become impossible if one cannot identify all the individuals that constitute the social group that these programs are directed to. One need not always know the personal data of the recipient, but recipients need to be identifiable as recipients—otherwise the social program becomes impossible. Social welfare programs cannot target social groups *in abstracto.* They do not target a social group as a holist entity over and above the individuals that make it up. If a social group is something else and not the individuals that make it up, if it is something over and above these individuals, then it will be impossible to provide free healthcare for the group because it will be impossible to identify the individuals who should be entitled to free healthcare. If individuals who make up a social group cannot be identified as the set of individuals who would be the recipients of social programs, then social programs cannot be applied to the group. Holist perspectives on social groups thus make social welfare programs impossible.

---

[10] The school system is those individuals who work in education and their education-related interactions. There is no need to postulate it as a holistic social entity. School programmes are widely shared mental contents (among teachers, school administrations etc.) that describe the material that needs to be thought in schools. They are instantiated in the minds of individuals and can be codified (e.g. by means of texts)—but there is no need to postulate them as immaterial Platonic entities, over and above mental states of individuals.

4 Social Holism as the Ideology of Corporate Capitalism

So far I have argued that methodological or ontological individualism in the social sciences do not entail political individualism, libertarianism or neo-liberal economics. Nevertheless, such claims have been repeatedly made, and it is hard to avoid the impression that they belong to the culture of poor arguments that, we have seen, dominates polemics against individualism. The coin however, has its other side. If we consider the political implications of holism, one of its significant implications is that the holist understanding of social entities absolves government officials, army officers, various CEOs and so on from any responsibility for the actions of these social entities, bureaucracies, military units or companies that they lead. I will concentrate here on the implications of social holism in the context of corporate capitalism. Legal systems in various countries make it easier or more difficult to prosecute, for instance, a CEO of a medical company whose drugs kill people or a CEO of a car company that produces cars that pollute more than this is allowed. If companies and corporations are, however, consistently understood as holistic entities, then this should not be possible—an individual could not be held responsible for the actions of a collective body such as a corporation even if this individual ordered these actions. Insofar as it understands social entities as entities *sui generis*, as entities of their own kind, as something over and above individuals and their interactions, social holism entails the impunity from prosecution of various CEOs, military officers, high government officials and bureaucrats for the illegal or criminal actions of the corporate bodies that they lead.

For our discussion here New Zealand's leaky buildings disaster is particularly suitable example. In the late 1980s and the early 1990s the New Zealand government passed a series of laws that deregularized building industry and opened it to market forces.[11] Up to that time New Zealand had a strong system of building codes and regulations that ensured good technical quality of residential architecture. Problems with waterproofing were rare. All this changed within a decade after deregulation. By 2008 there were over 89,000 dwelling failures and in 2011 the total damage was estimated to be NZ$23 billion (or about 12% of the country's yearly GDP). Most issues pertained to waterproofing—precisely the kind of problem that would arise once building codes have been abolished. The problems were not limited to the extremely high cost of repairs. Humidity in buildings caused a wide range of health

---

[11] There exists a massive literature about New Zealand problems with leaking buildings. See Dyer (2012).

488   B. Mitrović

problems. Many apartment owners were forced to file for bankruptcy because they were unable to pay for the necessary repairs and some committed suicides.

The disaster illustrates two important points. First, critics of neo-liberal economic principles may rightly point out that such deregulation is a perfect example of the negative consequences of economic neo-liberalism. It would be hard to deny that the New Zealand government was motivated by neo-liberal ideas during the era. The system of strict building codes and regulations was arguably rigid and it slowed down the introduction of new (untried) building technologies. The government may have believed in the power of market to regulate the quality of work in building industry. It may have assumed that if some developers and building companies sold bad apartments, their reputation would make it impossible for them to survive on the market. All this admitted, there is nothing about ontological or methodological individualism that would make them incompatible with the idea of building codes and regulations. Individualists are not obliged to oppose the idea of building codes, regulations or legislations introduced by governments. For instance, an individualist may think that rules, regulations and codes are contents of mental states instantiated in individuals that specify how things should (not) be done. Individuals learn about them through interactions, verbal or written, from other individuals. There is nothing in the individualist perspective that would make rules and regulations impossible.

At the same time, there is another side of New Zealand's leaking buildings scandal, that illustrates the consequences of the holists' incapacity to individualize the responsibility for the actions of social entities, as discussed earlier. We have seen that for individualists decisions of social entities result from decisions and interactions of participating individuals. Consequently, the individualist view is that responsibility for the decisions made by social entities falls on those individuals who contributed to decisions. (Obviously, one also has to take into account that some decisions may have unintended consequences.) Contrary to this view, the holist position has to be that decisions of social entities do not result from decisions and interactions of individuals. It follows that no responsibility can be assigned to the individuals who participated in the decision-making process. The New Zealand's leaking buildings scandal finely illustrates the perils of such holist understanding of social institutions. Implicit[12] holist assumptions about social institutions were actually

---

[12] By 'implicit' I mean that I am unaware that the social-theoretical understanding of building companies as defined by the law has ever been discussed in relation to the scandal, but the background legal assumption that enabled the disaster was clearly that building companies were entities on own and of their own kind, unidentifiable with individuals (such as their owners or share holders) and their interactions.

an important contributing factor to the leaky buildings disaster, since they removed the responsibility from individual developers and the owners of building firms that built leaky buildings. This made it possible for developers to build a shoddy buildings, sell the apartments, take the profits and close down their companies. By the time the problems with the building were discovered and the owners faced repair bills, the building companies had ceased to exist and could not be held responsible or sued. The developer could not be held responsible because (in line with the holist understanding of social institutions) the financial responsibilities of the company are separated from those of the company owners who derive financial gain from the profits of the company. The company was seen as an entity on its own and its responsibilities could not be identified with those of the owner. Individualists thus do have a good case to argue that social holism supports some of the worst and most dishonest practices of capitalist economy. From the holist point of view, it is acceptable that capitalists, developers, various managers, CEOs and so on earn profits and salaries by making decisions without being responsible for the outcomes of these decisions; the responsibility is located in the corporate body. To make things worse, problems with the holist incapacity to assign responsibility to individuals for actions of the social entities that they manage are not limited to absolving the officials of corporate capitalism from the responsibility for their decisions. The same kind of reasoning easily expands to numerous other social issues pertaining to the responsibility for actions of various other social entities. Following similar logic, for instance, it becomes unclear how one could hold responsible and prosecute military officers who order their units to commit war crimes or government officials whose decisions actively violate human rights.

5 Disciplinary Politics

It is, at the same time, hard to believe that enthusiasm for corporate capitalism, government bureaucracies and military institutions or the desire to advocate the impunity of their managers and officials could motivate the vehemence of anti-individualist polemics in the social sciences. More likely, many social scientists may be motivated by their desires to change the society for better or they may consider it ethically proper to promote theories that in their views can help the disadvantaged sections of society. They may wrongly believe that by promoting social holism they are indeed acting in accordance with such intentions. There is, however, another kind of political explanation of the vehemence of the anti-individualist polemic in the social sciences, that pertains to disciplinary politics and not to political issues in society at large. To some social scientists, individualism may seem to endanger the very possibility of the social sciences (economics excluded because it operates mostly on individualist assumptions), because in their view it entails the reduction of social sciences to psychology. Speaking in general, the argument that seems to disturb many social scientists is that if actions of social entities result from the actions of individuals, and if actions of individuals are ultimately caused by their decisions and psychology, then the social sciences should be reducible to psychology. In that case, the social sciences, it may seem, would lose the very purpose for their existence. Udehn (2001, p. 3) describes this concern explicitly:

> Many social scientists, especially sociologists, have, no doubt, rejected methodological individualism, because they believed that it implies psychologism, or the reduction of sociology to psychology. If so, methodological individualism would rob sociologists of their discipline. For all these reasons, the debate between methodological individualists and their critics has been more confused than usual in social science and philosophy.

Quite appropriately, Udehn (2001, p. 3) himself qualified this motivation as 'crass.' One should also note that the background concern about the reduction of the social sciences on psychology is baseless on two levels. First, it is not clear how such reduction could be possible. Social phenomena do not result only from what individuals think but also from unintended consequences of their actions—and unintended consequences are not mental properties, so they could not be described or explained by psychology (see for instance Di Iorio, 2013, p. 27). Second, the idea that individualism could entail such reduction could only be based on the misunderstanding of the individualist position and especially the individualist emphasis on interactions. By definition, 'interactions' have to include what individuals do and not merely what they think—so they cannot be exclusively mental, nor fully described in psychological terms. Third, even if such reduction of sociology to psychology were possible, this would not deprive sociologists of their discipline. Consider the case of chemistry. Chemical phenomena ultimately result from the interaction between physical particles and physics explains the behavior of these particles. This, however, does not mean that chemistry is not a legitimate science. Laws of chemistry result from the physical behavior of particles, but that does not mean that there are no laws of chemistry.

All this admitted, Udehn may nevertheless be right in his view that fears from the reduction of the social sciences to psychology in many cases motivate anti-individualist polemics. A comparison with the historical humanities may be taken to confirm his view. The dilemma about the nature of social items such as states, armies, cultures or battles is certainly very important for historians as well. However, the disciplinary possibility of history does not depend on the outcome of the polemic against individualism, and historians are much less worried about the problem. In the early decades of the twentieth-century holist tendencies were dominant in German-speaking historical scholarship (though much less outside Germany).[13] In modern English-speaking historiography or the philosophy of history the debate between individualism and holism is not nearly so often addressed as in the social sciences. As Tucker (2004, p. 212) pointed out, historians generally accept that ontological individualism is true, and assume that methodological individualism (the way they understand it) is false. For practicing historians the decisive argument against (what they mean by) methodological individualism, as described by Førland (2017, p. 140), is that it is often impossible to identify all individuals who made up social entities in history because the necessary information is not to be found in archives. For instance, it is a methodological constraint that the existing archives do not enable us to identify all soldiers of Napoleon's *Grande Armée*, but as regards ontology, that does not mean that the *Grande Armée* was anything more, over and above individuals and their interactions. Obviously, this is a perspective of practicing historians, who are predominantly interested in what they can prove through archival work. Now, it seems reasonable to assume that social scientists and scholars in the historical humanities are equally prone to subscribe to theoretical positions that support social change for better and tend to favor the disadvantaged sections of society. There is no reason to think that one group has different or stronger political commitments than another. This would then suggest that the vehemence of anti-individualist polemic among social scientists does not derive from such wider social commitments. The remaining explanation would be that it derives from disciplinary politics and concerns about the integrity of the social sciences—the kind of concerns that historians simply do not have. This impression is further strengthened by the fact that the historians' reason for the rejection of methodological individualism—that through archival work one often cannot identify all members of a social entity—never gets mentioned by the social scientists who write against methodological individualism (at least in the literature I have surveyed). The fact that the individuals and the interactions that make up social phenomena often cannot be identified is not a relevant argument for social scientists because it is not sufficient

---

[13] This particularly pertains to so-called historicist tradition. For surveys of German historicism see Iggers (1968), Beiser (2011); for the role of individualism-holism debate within historicist theories see Mitrović (2015).

492   B. Mitrović

in order to claim that social entities have causal capacities that derive from something else and not from the psychologies of the participating individuals. It cannot be used in order to defend the possibility of the social sciences, such as sociology, as a field independent of other fields of research.

6   Conclusion

There is something profoundly wrong with the idea that positions in the social sciences can be defended or dismissed on the basis of their political consequences and not on the basis of their theoretical validity, arguments and evidence. One would expect social scientists to be motivated by the pursuit of truth—and yet, we have seen that political arguments commonly play a role in the polemics against individualism, and possibly even motivate them. In an environment motivated by genuine search for truth a researcher who is motivated by political inclinations would be ashamed to admit it, let alone present political arguments as relevant in arguing his or her position. The fact that these political arguments, that should not have been used in the first place, are poorly argued or based on invalid assumptions, makes things even worse. The claims that identify individualism in the social sciences with libertarian political views or neo-liberal economics have never been properly justified, and it is hard to see how they could be because a credible logical link is simply not to be found. Such claims have been repeatedly made, and the lack of credible arguments to support them suggests that they are part and parcel of a wider phenomenon—the fact that virtually any claim may pass reviewers and be published in leading journals if it is said to support arguments against ontological or methodological, individualism. When a well-reputed philosophy journal publishes baseless political insinuations about individualism such editorial practices belong to the same phenomenon as the publication of a non sequitur 'refutation' of individualism based on false claims about elementary trigonometry. In principle, anything goes when it comes to such 'refutations' of individualism. Disciplinary politics in the social sciences thus enables and allows for the systematic publication of material that would be quite unusual in most other fields.

The suggestion that concerns about the integrity of the social sciences and fears from their reduction to psychology motivate many social scientists in their opposition to individualism needs to be taken seriously. This suggestion is particularly devastating, insofar as it may be taken to suggest that many social scientists parade wider social concerns in order to dismiss the views that they actually regard as a threat to their discipline—that they are ready to sacrifice their intellectual integrity in order to defend the integrity of the discipline. The credibility of Udehn's claim that 'many social scientists, especially sociologists' reject individualism because they fear the reduction of their disciplines to psychology is further confirmed by the fact that the debate between individualism and holism in the philosophy of history has none of the vitriolic tone characteristic of the philosophy of the social sciences. In their work, historians deal with social items as much as social scientists, but they hardly worry about the reduction of the historical humanities to psychology. In fact, the reduction of social sciences to psychology is improbable because of the argument about unintended consequences—but even if it were possible, it is not clear that it would be particularly devastating for the social sciences since it would merely mean that they depend on psychology the way chemistry depends on physics. (In fact, much less, because of the arguments about unintended consequences and interactions that I mentioned above.) All this admitted, the belief that if social sciences are going to be legitimate sciences, then social entities must be entities of their own special kind, '*sui generis*,' has been with the social sciences at least since the time of Emil Durkheim.[14] If Udehn is right, then for many social scientists the very idea of social science is inseparable from the holist understanding of society—and when they attack individualism they are ferociously defending the credibility of the field they work in, the way they understand it. The belief that social items are *sui generis* seems to be a firmly ingrained fundamental disciplinary assumption—so firmly ingrained that questioning it could cause the vitriolic reactions that we have seen. How and why such a belief came about, and how and why it has been sustained, is a question that should be addressed by historians of the social sciences. It is not a topic for this paper.

---

[14] The classic formulation is in Durkheim (1982 [1895]: 39). See also Udehn (2001, pp. 35, 181).

Bibliography

Agassi, J. (1987). Methodological individualism and institutional individualism. In J. Agassi & I. C. Jarvie (Eds.), *Rationality: The critical view* (pp. 119–150). Martinus Nijhoff Publishers.

Beiser, F. (2011). *The German historicist tradition*. Oxford University Press.

Bhargava, R. (1992). *Individualism in social science: Forms and limits of a methodology*. Clarendon Press.

Bulle, N. (2019). Methodological individualism as anti-reductionism. *Journal of Classical Sociology, 19*(2), 161–184.

Bunge, M. (2000). Ten modes of individualism—None of which works—And their alternatives. *Philosophy of Social Sciences, 30*, 384–406.

Burman, P. (1979). Variations on a dialectical theme. *Philosophy of Social Science, 9*, 357–375.

Cantor, G. (1869). Beiträge zur Begründung der transfiniten Mengenlehre. *Mathematische Analen*, 481–512.

Chomsky, N. (2002). *American power*. The New Press.

Di Iorio, F. (2013). *Nominalism and systemism: On the non-reductionism nature of methodological individualism*. Duke University: Center for the History of Political Economy. https://papers.ssrn.com/sol3/papers.cfm?abstract_id=2289318. Downloaded on 7 May 2022.

Dumont, L. (1977). *From Mandeville to Marx*. The University of Chicago Press.

Durkheim, E. (1982 [1895]). *The rules of sociological method*. The Free Press.

Dyer, P. (2012). A very homegrown disaster. *North and South, 311*, 42–51.

Elster, J. (1982). A case for methodological individualism. *Theory and Society, 11*, 453–482.

Elster, J. (1985). *Making sense of Marx*. Cambridge University Press.

Elster, J. (1992). *Nuts and bolts for the social sciences*. Cambridge University Press.

Epstein, B. (2009). Ontological individualism reconsidered. *Synthese, 166*, 187–213.

Epstein, B. (2015). *The ant trap*. Oxford University Press.

Førland, T. E. (2008). Mentality as social emergent: Can the Zeitgeist have explanatory power? *History and Theory, 47*, 44–56.

Førland, T. E. (2017). *Values, objectivity and explanation in historiography*. Routledge.

Giddens, A. (1979). *Central problems in social theory: Action, structure and contradiction in social analysis*. The Macmillan Press.

Hodgson, G. (2007). Meanings of methodological individualism. *Journal of Economic Methodology, 14*, 211–226.

Iggers, G. (1968). *The German conception of history. The national tradition of historical thought from Herder to present*. Wesleyan University Press.

Israel, J. (1971). The principle of methodological individualism and Marxian epistemology. *Acta Sociologica, 14*, 145–150.

Jones, T. (1996). Methodological individualism in proper perspective. *Behavior and Philosophy, 24*, 119–128.

Kincaid, H. (1986). Reduction, explanation and individualism. *Philosophy of Science, 53*, 492–513.

Kincaid, H. (1997). *Individualism and the unity of science: Essays on reduction explanation and the special sciences.* Rowman & Littlefield Publishers.

Kumar, C. (2008). A pragmatist spin on analytical Marxism and methodological individualism. *Philosophical Papers, 37*, 185–211.

Levine, A., Sober, E., & Wright, E. O. (1987). Marxism and methodological individualism. *New Left Review*, 67–84.

List, C., & Spiekermann, K. (2013). Methodological individualism and holism in political science. *American Political Science Review, 107*, 629–643.

Lukes, S. (1968). Methodological individualism reconsidered. *The British Journal of Sociology, 19*, 119–129.

MacDonald, G. (1986). Modified methodological individualism. *Proceedings of the Aristotelian Society, 86*, 199–211.

Mandelbaum, M. (1955). Societal facts. *The British Journal of Sociology, 6*, 305–317.

Marshall, D., & Weatherson, B. (2018). Intrinsic vs. extrinsic properties. In E. N. Zalta (Ed.), *The Stanford encyclopedia of philosophy.* https://plato.stanford.edu/arc hives/spr2018/entries/intrinsic-extrinsic.

Marx, K., & Engels, F. (1953 [1932]). *Die deutsche Ideologie.* Dietz. English translation: *The German ideology.* Prometheus Books, 1998.

Mitrović, B. (2015). *Rage and denials: Collectivist philosophy, politics and art historiography 1890–1947.* Penn State University Press.

Ritchie, K. (2013). What are groups? *Philosophical Studies, 166*, 257–272.

Ruben, D. H. (1982). The existence of social entities. *The Philosophical Quarterly, 32*, 295–310.

Sawyer, K. (2002). Nonreductive individualism: Part I—Supervenience and wild disjunction. *Philosophy of the Social Sciences, 32*, 537–559.

Sawyer, K. (2005). *Social emergence.* Cambridge University Press.

Schumpeter, J. (1998 [1908]). *Das Wesen und der Hauptinhalt der theoretischen Nationalökonimie.* Duncker & Humboldt. English translation: *The nature and essence of economic theory.* Routledge, 2010.

Simmel, G. (1908). *Soziologie. Untersuchungen über die Formen der Verge-selschaftlichung.* Duncker & Humboldt. English translation: *Sociology: Inquiries into the construction of social forms.* Brill, 2009.

Simmel, G. (1917). *Grundfragen der Soziologie* [Individuum und Geselschaft]. Göschen'sche Verlagshandlung.

Steel, D. (2006). Methodological individualism, explanation and invariance. *Philosophy of Social Sciences, 36*, 440–463.

Tarrit, F. (2006). A brief history, scope and peculiarities of 'analytic Marxism.' *Review of Radical Political Economics, 38*, 595–618.

Tucker, A. (2004). *Our knowledge of the past: A philosophy of historiography.* Cambridge University Press.

Udehn, L. (2001). *Methodological individualism.* Routledge.

Uzquiano, G. (2004). The supreme court and the supreme court justices: A metaphysical puzzle. *Nous, 38*, 135–153.

Webster, M. (1973). Psychological reductionism, methodological individualism, and large-scale problems. *American Sociological Review, 38*, 258–273.

Weldes, J. (1989). Marxism and methodological individualism. *Theory and Society, 18*, 353–386.

Wolff, R. (1990). Methodological individualism and Marx: Some remarks on Jon Elster, game theory and other things. *Canadian Journal of Philosophy, 20*, 469–486.

Zahle, J., & Kincaid, H. (2019). Why be a methodological individualist. *Synthese, 196*, 655–675.

Economics: A Methodological Individualism in Search of Its Own Incompleteness

Olivier Favereau

The question of the practice of methodological individualism (MI) in economic theory comes up against an obstacle that is well known to economists, but of a rather paradoxical nature: on the one hand, it is almost a pleonasm to speak of methodological individualism of the "dominant" economic theory. The split between micro and macroeconomics, together with the concern to found the latter on the former, has been a quasi-universal fact of life in all economics departments since the 1950s, both in teaching and in research. On the other hand, the pleonasm ceases to be a redundancy as soon as we try to specify what we mean by *mainstream* economics. This term, more neutral in English than in French ("*courant dominant*"), must be taken in a quasi-sociological sense. By this, we mean the most widespread academic works, in scientific journals of international rank (and consequently, in English). However, we must recognize the diversity of content of this dominant theory over time, and even the increase in this diversity over the last thirty or forty years. The paradox is clear: maximum diversity among economists of the same generation contradicts the very idea of a "dominant" current.

A previous version of this article was originally published in French in *L'Année sociologique* 70 (1) 2020, 231–259: "cindividualisme méthodologique à la recherche de sa propre incomplétude."

O. Favereau (✉)
Economics, University of Paris-Nanterre, Nanterre, France
e-mail: ofavereau76@gmail.com

© The Author(s), under exclusive license to Springer Nature Switzerland AG 2023
N. Bulle and F. Di Iorio (eds.), *The Palgrave Handbook of Methodological Individualism*,
https://doi.org/10.1007/978-3-031-41508-1_21

The objective of this paper is to resolve this paradox by shifting it: it is indeed the relationship to MI that is the key to the coherence of *mainstream* economics, but this relationship is not one of conformity. On the contrary, it is the *impossibility* of conforming to its own requirements, those which it has formulated with absolute rigor, that will structure the present dominant economy, and give meaning to its—relative—variety.

After having characterized what the practice of MI in economics consists of, since the advent of neoclassical marginalism against classical/Marxist thought, we will show in the first part that it took on an axiomatic form in the 1970s with the theory of general equilibrium, but that this very form revealed *a double phenomenon of incompleteness*, the first in the modeling of individual rationality (R-incompleteness), the second in that of inter-individual/market coordination (M-incompleteness). The two fundamental tools of theoretical economics have been shown to be of only limited use in casting a net around the reality they seek to capture: the coordination of optimizing economic agents by the market alone. This negative result—indeed a fundamental one—bears the stamp of MI that we will call "closed."

The second part will be devoted to showing how the dominant economy, from the 1970s to the present day, has reacted to this double negative discovery. We will distinguish three types of reaction that define as many different regions in contemporary mainstream economics: the majority reaction (contract and incentive theory) has been the treatment of M-incompleteness without R-incompleteness; an important but minority reaction (evolutionary and behavioral economics) has been the opposite: R-incompleteness treatment without M-incompleteness; a third reaction (located in macroeconomic modeling: Real Business Cycles, RBC, and Dynamic Stochastic General Equilibria, DSGE) has been in denial of the two incompletenesses. There remains the fourth logically possible region: the simultaneous treatment of the two incompletenesses, which would mark a radical break with the neoclassical tradition. Cumulatively, the two incompletenesses show a positive side, source of a MI, rich in elements that the "closed" MI would have qualified as holistic or structuralist. We will therefore speak of an "open" MI.

1 Incompleteness as the Endpoint of a "Closed" Individualism

MI and *Mainstream*: A Co-construction 1870–1970

The 1870s saw the break-up of the so-called "classical" tradition (Smith, Ricardo, Stuart-Mill), which combined contradictory elements, into a so-called "neoclassical" current of individualist methodology, with the first marginalists (Jevons, Menger, Walras), and a methodologically holistic, or structuralist, current with Marx. We shall see below that the individualist *mainstream* underwent a deep change in the 1970s. But we will venture to embrace the whole period, to project ourselves to the present period, and to deduce from it a characterization as neutral and consensual as possible of the MI that characterizes the professional practice of *mainstream* economists.

What is MI for *Mainstream* Economics?

The MI of *mainstream* economists can be condensed into two obviously related but quite distinct propositions:

> MI-1: Individual behaviors only become intelligible to the economist once they are embodied in a general model of individual rationality
> MI-2: For an economist, explaining a macroeconomic phenomenon means deducing it from a composition of individual behaviors, understood according to MI-1

These two propositions have a polemical history. They are opposed to the positions that would define Methodological Holism (MH), represented at the end of the nineteenth century essentially by Marxist theory, as a result of the classical model.

Proposition 1 has a better-defined and better-known status than Proposition 2. As history develops, it will lead to what is now called Rational Choice Theory (RCT), which has received a strict formulation in the discipline of economics (through the criterion of maximum expected utility, with subjective probabilities, when the decision-maker has no objective probabilities). We shall examine it very closely, since its incompleteness is the keystone of our argument.

Proposition 2 perhaps calls for more introductory comments. We have used the generic term "composition" to suggest that it can take various forms. We immediately encounter two problems.

500   O. Favereau

The first arises from the simplest and seemingly most natural form of "composition": aggregation by addition. The fundamental result is negative. The perfect aggregation theorem[1] states that only linear relationships of equal slope can lead to an aggregated relationship of the same nature, allowing an immediate reading in terms of individual behavior. In all other cases, the aggregate observations combine individual behaviors and the statistical distribution of these behaviors. In other words, the hope of providing macroeconomics with a rigorous microeconomic foundation in full generality is close to zero.

The second problem lies in the interweaving of the two propositions. What is the origin of this "composition?" If it is institutional (for example, participation in a well-organized market, where supply and demand are aggregated[2]), we must make explicit the rationality of the agents who operate this institution (and/or who created it), unless we admit that a part of the macro-phenomena does not fall under an explanation that conforms to MI: a structural element is decisive. Once again, the hope of a rigorous articulation between micro and macro fades away.

Despite this, *mainstream* economics has always claimed a strict methodological individualism. Starting from individuals, postulated to be rational in an increasingly precise sense, in order to account for macroeconomic phenomena, it became a dogmatic, normative, quasi-moral position between the last third of the nineteenth century and the present period (with a significant hardening from the 1970s onwards), not devoid of hypocrisy or casuistry.

The typical example of this casuistry is the adoption of a representative agent hypothesis in order to switch through a simple change of scale from micro to macroeconomics, which cleverly bypasses the restrictive conditions of perfect aggregation, as well as a large part of the institutional mechanism of the meeting between agents, which becomes artificially bilateral.

With these definitions, we can move from questions of method to those of substance. This will be all the easier since *MI's two propositions correspond exactly to the two axes of the so-called "neo-classical" tradition, in its canonical form of general equilibrium theory.*

---

[1] See Malinvaud (1991) (Chapter 6, and in particular pp.195–200).

[2] In this context, the economist comes up against the negative theorems propounded by Sonnenschein, Mantel, and Debreu between 1972 and 1974. They amount to showing (let us put it metaphorically) a certain reciprocal independence between the properties of individual behavior and those of their sum on the market.

What Was *Mainstream* Economics (at the Threshold of the 1970s)?

In the 1870s, through the marginalist revolution and more particularly the work of Walras, what would become known as "General Equilibrium Theory" (GET) appeared. It combined, on the MI-1 side, a "strong" individual rationality (individuals maximize their utility, firms maximize their profit), and on the MI-2 side, a market-like coordination: the economy was thought of as a system of interdependent markets. Walras made a bold assumption: he generalized the image of a well-organized market, such as the Paris stock exchange, to all goods and all factors of production. On such a market, all offers and demands (of securities) are aggregated, and transactions only take place at the equilibrium price, computed by the representative of the institution. Note that this characterization of economic activity exclusively in terms of supply and demand corresponds to a case where MI-2 is respected in an apparently absolute manner. Each economic agent proposes to the relevant market a certain quantity (either offered or demanded) that is supposed to be aggregable with all the others. We are therefore in a vision that is consistent with a quasi-perfect MI,[3] in both its elements.

In 1954, the mathematician Leonard Savage published *Foundations of Statistics*, in which the almost definitive axiomatics of RCT were formulated, while the economists Kenneth Arrow and Gérard Debreu gave in the journal *Econometrica* the first mathematically correct proof (by means of a fixed point theorem) of the conditions for the existence of a vector of prices that ensured general equilibrium on all markets. The following years were devoted to the consolidation and extension (to the uncertain case) of this rigorously individualistic general model of economic functioning.

The decade of the 1970s began with the publication by Kenneth Arrow and Frank Hahn, in 1971, of their great work *General Competitive Analysis*, which can be viewed as the canonical expression of the GET, initiated by Walras almost exactly one century earlier. Since a similar judgment can be made of Savage's work, which brings to its conclusion an even more venerable tradition of building a model of individual rationality,[4] one conclusion is obvious. *At the beginning of the 1970s, the discipline of economics had, thanks to a seemingly flawless MI, foundations of exceptional rigor in the scientific field, all disciplines taken together. Its two "pillars" (individual rationality, coordination*

---

[3] Except for the fact that the computation of equilibrium prices is left to the institutional mechanism of the market (the auctioneer, as Walras would say), whose origin nobody knows. See also note 1.

[4] Ian Hacking traces this tradition back to Pascal, with the invention of numerical probabilities, and to his "wager" presented as the first problem in decision theory (Hacking, 1975).

by the market) had been axiomatized. And these two axiomatizations reinforced each other. Arrow and Hahn (1971, pp. 122–128), after having built an existence proof of a general equilibrium, without uncertainty, would buy the Savagian model of maximizing subjective expected utility to show that the certainty formalism "remains valid," on condition that "each commodity now must be interpreted as a contingent claim, with the double index, *i.e.*, a promise to supply one unit of commodity i if state s occurs, and nothing otherwise" (p. 124).

Now, it is this MI that will reveal its own—irreducible—limitations.

The Discovery of the Two Incompletenesses

The First Incompleteness

Arrow and Hahn's work is the subject of the most laudatory reviews, but often not without scepticism concerning the empirical relevance of this framework of analysis when applied to actual capitalist economies. One example is the Hungarian economist Janos Kornaï, in his book *Anti-Equilibrium*, published shortly after *General Competitive Analysis*, who led our two authors to react separately but identically—and this reaction is the first sign of an essential shift in the dominant economic theory. The common part of their reaction consisted in vigorously refuting the idea (a priori plausible!) that this economic theory of general equilibrium would constitute an apology for market capitalist economies. In fact, their enterprise turns out to be a *critical* work on these economies. Since Adam Smith, the metaphor of the "invisible hand" has served as proof that the interplay of market coordination and individual utilitarian rationality was sufficient to produce, if not a permanent equilibrium, at least a permanent tendency to return to equilibrium. However, according to Arrow and Hahn, one only has to look at the list of axioms to understand that it is not fulfilled, in particular the Savagian axiom that requires the existence of as many futures markets as there are predictable states of nature from now until the end of time.

We see here the strategic importance of the reference to the axiomatics of individual rationality—at the heart of the axiomatics of general equilibrium (i.e., of market coordination). Obviously, this *sine qua non*-condition is not verified in practice. Market economies do *not* have—far from it—all the markets that would be necessary, according to the theory, for them to coordinate perfectly, with a single mode of coordination: markets—all the more so as we want only well-organized markets, where prices always balance supply and demand, like on the Stock Exchange. In the end, axiomatization appears to be an operation of strengthening a research program, but one that is not without cost: *by making explicit the foundations of its validity,*[5] *it also sets its limits.* We shall therefore call this incompleteness of coordination by the market according to the axiomatics of general equilibrium "M-incompleteness." It is an *empirical* phenomenon, even if in order to be able to "see" it, one needs a precise theoretical framework, which will indeed allow one to see "what is missing" in the actual world: a full set of futures markets.

The Second Incompleteness

We have been able to measure, through M-incompleteness, the advantages and disadvantages of axiomatization for a scientific research program. The advantage is the complete explicitness of the conditions of validity; the disadvantage is the strict delimitation of a domain of validity and thus, outside it, of a domain of non-validity.

We have just experienced this rule on the "market coordination" axis of the GET. The paradox that will inspire the rest of our approach is that the same operation must be conducted on the "individual rationality" axis. Arrow was also one of the great neoclassical authors who contributed to the construction of the RCT, particularly in a context of uncertainty (see, for example, his 1963 *Essays on risk-bearing*). But within this general orientation, he curiously distinguished himself very early on by focusing on extreme situations known as "ignorance" where the decision-maker does not have enough information to give any particular weight to one state of nature rather than another. As early as 1953, he had simplified the theorem obtained in 1951 by Hurwicz, who proposed as a criterion of optimality in such a situation to maximize a weighted average of the best and worst expected results, the weighting reflecting the degree of optimism or pessimism of the decision-maker. The publication in 1954 of Savage's axiomatics has since led all authors to aim at the same level of generality. Thus, in 1972, Arrow and Hurwicz resumed their joint work to redefine the criterion of individual rationality, outside the "now more standard framework of subjective probabilities by not presupposing a fixed list of states of nature" (1972, p. 1).

Arrow and Hurwicz will therefore define four "properties" that an optimality criterion should respect, when the decision-maker has no reliable reference point on the list of possible states of nature. For example, an optimal action in a certain formulation of a decision problem must remain optimal

---

[5] Validity of its formal tools, which does not exhaust its conditions of empirical validity. The reader who would like a deep reflection on the axiomatic method, in relation with the economic discipline, should consult Philippe Mongin (2003).

in a reformulation of actions and states of nature, isomorphic to the first one. They then show that the necessary and sufficient condition for an optimality criterion to respect these invariance properties is that it takes the following form: an action is optimal in a certain decision problem if its pair of consequences, minimal and maximal, is superior or equivalent to the similar pair of any other possible action in the same decision problem. In other words, we must abandon any idea of a utility function, whose mathematical expectation must be maximized with numerical probabilities, either objective (Von Neumann & Morgenstern, 1953) or subjective (Savage, 1954).[6]

The result was generalized at the end of the 1970s by two French mathematicians, Michèle Cohen and Jean-Yves Jaffray. Its epistemological importance for MI in economics, behind its technical disguise, did not escape these two authors. When the logic of research comes to defining what it means to be "rational" for an individual, there is an unbridgeable boundary, a "discontinuity" (Cohen & Jaffray, 1980, p. 1296), between two universes, the one where the individual "considers only a fixed set of states of nature" (Arrow & Hurwicz, 1972, p. 3) and the one where he cannot presuppose "a fixed list of states of nature" (ibid., p.1). In the former, rationality has a quasi-intrinsic link, elucidated by a tradition of three centuries, with the mathematical tool of optimization (and a treatment of the future in the form of numerical probabilities). In the second, there is no more utility, no more probability, *a fortiori* no more expected utility, and naturally *no more optimization.*

We can now integrate the second incompleteness, the R-incompleteness, that which concerns the standard model of individual rationality—it proves to be inappropriate for all contexts other than those where the decision-maker is confronted with a determined, constant, exogenous list of possible states of nature (between which the future will choose—we are still in an uncertain future). It is only valid for a world without "surprises."

Thus, the great neoclassical tradition, leaning on a double axiomatics, that of the general equilibrium and that of rational choice, was confronted in the 1970s with its quite involuntary discovery of a double phenomenon of incompleteness, and this, in the name of the same requirement of methodological individualism:

– The M-incompleteness, with the demonstration of the incapacity of concrete market economies to know full coordination of the behaviors of rational economic agents, by the only means of organized markets.

---

[6] For a pedagogical synthesis, see John C. Harsanyi (1977).

– R-incompleteness, with the demonstration of the incapacity of optimizing
  individual rationality to deal with any situation likely to bring about an
  "unforeseeable contingency."[7]

Let us add that these two incompletenesses rest on the same basis: the
practical impossibility of having a conception of the future with an exhaustive list of possible scenarios (and thus, with the corresponding futures
markets). From this point of view, R-incompleteness is in some ways "more"
fundamental than M-incompleteness.

How will *mainstream* economics absorb the shock of this double discovery?
This is the subject of the second part.

2 Incompleteness as a Starting Point for "Open" Individualism

If the conclusion of the first part is correct, then the only truly scientific
approach for a *mainstream* economist would be to work from now on in an
analytical space admitting *both* R-incompleteness and M-incompleteness. But
this would mean facing a double unknown, by managing a double exit, one
methodological outside the familiar MI, the other epistemological, outside
the dominant economy.

What we have observed, for the last thirty or forty years, is a dispersion of
economists around the four logically possible positions: dealing with one of
the two incompletenesses; dealing with neither; and dealing with both. But
there is nothing random about this dispersion. It has a structure. This is what
we shall try to identify in this second part, in order to account simultaneously
for the continued existence[8] of a *mainstream* economy[9] (and therefore necessarily of a heterodox, non-standard economy) and of a relative variety within
it. The key to this structure is precisely the relation to this double incompleteness: do we accept it or do we refuse it? If one refuses it (perhaps, because one
does not want to "see" it), what form does this refusal take (refusal of both
incompletenesses, or of only one of them)? Finally, is there a (chrono)logical
link between the various positions?

The answer lies in distinguishing three moments, corresponding to three
re-readings of MI: first, the thoughtful and majority choice of the treatment of the M-incompleteness alone; then (and partly, as a reaction to this

---

[7] As David Kreps (1990) puts it.

[8] One can even speak of reinforcement (Orléan [ed.], 2015; and see note 10).

[9] Of which we have given a "sociological" characterization in the introduction.

majority choice) the choice of the other two paths by reference (positive or critical) to the general equilibrium tradition; and finally, the pioneering work, almost always heterodox, of dealing with the two incompletenesses, which is beginning to be assumed as such.

The Treatment of M-incompleteness by Hypertrophy of the "Closed" MI

In the account we have given, Arrow and Hahn show themselves to be very lucid about the existence of this "empirical" phenomenon of M-incompleteness. For them, as for all *mainstream* economists who read them, the question arises of how to deal with it.

The Extension of the Field of Optimization

Arrow immediately outlined a new research program, destined for a great future, since its implementation, as progressive as it was systematic, would completely reconfigure the *mainstream*, with a renewed ambition for MI (and for *mainstream* economics). In 1974, in a small book, oddly entitled *The limits of organization* (since it is more about the limits of the market), he observed that according to GET, formulated properly, market economies should encounter insuperable coordination problems. But this is not the case, even if unemployment, poverty, and inflation forbid idealizing the results of market economies. *The most reasonable scientific hypothesis is therefore that other modes of coordination than the market are in action.* On the basis of this hypothesis, he lists three avenues to be explored (thus three additional modes of coordination):

– rules of law (sanctioned by public force)
– private contracts (especially the relationship of authority instituted by the employment contract)
– and "the principles of ethics and morality" (p. 26).

A new orientation of the dominant economic theory then takes shape, which could be called "Arrow's program," although there was never anything deliberate in this sense: its "coordination by the market" pillar has failed; what remains is the "individual rationality" pillar. This one, Arrow will admit, is more solid, empirically and analytically—especially in view of the requirements of MI. This is the moment to remind the reader why we have systematically indicated above that GET is "almost" perfectly in line with the MI requirement. All modern authors of GET know that there is only one individual whose behavior escapes the axiomatics of individual rationality, and it is not an ordinary one: it is the person whom Walras calls the "market secretary" and who is kind enough to compute the equilibrium price(s), i.e., to find the fixed point, the solution to the system of equations describing the behaviors of supply and demand on all markets. This point had long embarrassed Arrow. From then on, the natural strategy was to *extend*[10] RCT to the choice of coordination modes, starting with legal rules and private contracts. The theory of contracts and incentives was to become the new dominant figure in economics, with a "law and economics" component, developing rapidly from the principles laid down by Coase, and with the reinforcement of the theory of human capital from the mid-1960s. Any inter-individual arrangement, any institutional rule, which claims to be "robust," must be compatible with individual incentives ("incentive-compatibility"), in the sense that it must be deduced from an optimal behavior such that none of the individuals concerned has an interest in deviating from it.

Forty years after the first intuitions, this theoretical framework, supported by the progress of game theory (since every contract is a relationship between actors), represents almost all of the teaching of microeconomics throughout the world.[11] It has even spread to business schools, which marks a huge difference from the traditional *mainstream*, that of General Equilibrium Theory, which had nothing to say to managers. As far as research is concerned, it has been crowned with a dozen "Nobel" prizes—twenty if we add game theory.

It is true that Arrow started from *economic* considerations. But in the end, the perfectly legitimate desire to tackle the problem of M-incompleteness head-on led the dominant economics to extend the sphere of application of economists' MI to virtually *any social phenomenon*, even if the cost of this is probably, for reasons of insuperable complexity, the renunciation of a perspective of general interdependence.

The current MI of the post-1970 *mainstream* economists is thus, in its applications, even more ambitious and demanding than that of their predecessors—but infinitely more fragile in its foundations, since it denies the limits it has itself defined.

---

[10] For this reason, if the General Equilibrium Theory constitutes the "Standard Theory," we can call this new central figure of mainstream economics the "Extended Standard Theory," using a classic terminology from the philosophy of science. Arrow (1985, p. 38) was perfectly aware of this.

[11] From an immense literature, let us only quote the manuals of Jean Tirole (1988, 2006, 2018) as well as Bernard Salanié (2005), Patrick Bolton and Mathias Dewatripont (2005) or for "Law and economics" the synthesis of Steven Shavell (2004) or the collection of articles by Alain Marciano (2009).

508   O. Favereau

The Consequences of a Hypertrophied MI

(i) The refusal to take into account the axiomatic foundations of its main tool, optimizing rationality, has first of all *internal* consequences for the academic field. We will wait for p. 516 to gather all the observations on the functioning of the academic community of economists. Here, we will only mobilize the fact that the extension of the field of optimization, together with the abandonment of reflection on its foundations, has nourished the development of an imperialist attitude toward the Social and Human Sciences (SHS) close to economics (an attitude profoundly foreign to the generations of the pre-1970s, for whom the SHS studied the exogenous of the GET). Indeed, this mathematical tooling, with a little ingenuity, allows economists to approach almost any social fact, with, in the background, the idea that the SHS produce a lot of "facts," which they have difficulty in treating "scientifically," that is to say, in reducing them to "rational" behavior.

The treatment of M-incompleteness, without a correlative treatment of R-incompleteness, has above all *external* consequences of an importance that goes far beyond the community of academic economists.

It is no exaggeration to say that this "economics" has provided and still provides most—not all, as we shall see in the next section—of the intellectual infrastructure, founding its scholarly legitimacy, of *neo-liberalism*, that project of modernizing classical liberalism by generalizing *homo economicus*. For lack of being able to deploy here all the necessary argumentation, we will insist on this simple key of understanding: *for classical liberalism, the individual is the measure of all things; for neo-liberalism,*[12] *homo economicus is the measure of all things.*[13] A quote from the economist Michael Piore will allow us to get to the point quickly: "The term [neo-liberal] is a shorthand for a framework of thought which has two key components. First and most prominent is the notion of the competitive market as understood in standard economic theory as the template for social and economic organization. The second key component is an understanding of human behavior as motivated by narrow self-interests which can generally be captured by the maximization of monetary gains" (2010, p. 2).

---

[12] From an overabundant literature, let us extract the small work of the political scientist Wendy Brown (2007) and the sums of Pierre Dardot and Christian Laval (2009), without forgetting the prophetic seminars of Michel Foucault at the Collège de France in 1978–1979. See also the interesting list of components drawn up by Wendy Brown (2007, p. 58).

[13] Two references stand out: Pierre Demeulenaere (1996) and Christian Laval (2007).

These two components are clearly found in the logic of the new *mainstream* that we have chosen to characterize by the *extension* of the domain of optimization (beyond traditional economic behaviors) to any contractual, organizational, institutional, conventional, or even cultural arrangement, with the nuance that the second component is more emphasized than the first, which continues to privilege competition through the market. This nuance will find its place in the elucidation, in the next section, of complementary approaches to neoliberalism.

(ii) In order to *assess* all of these consequences, both internal and external, we must now go deeper into the technique of modeling relationships. Any recurrent relational scheme is thus endogenized in the form of an agreement (possibly implicit) between rational agents (i.e., maximizing their expected utility), in a context of probabilistic uncertainty, with an additional characteristic that will prove decisive: the information available to the agents is unevenly distributed. The asymmetry of information may relate to a behavioral variable (for example, the industriousness of the employee, who knows that he is not permanently controlled by his employer); in this case, we speak of "moral hazard." If the asymmetry relates to an exogenous variable (such as the quality of a used car), this is called "adverse selection."

The common issue in all these arrangements, whatever they may be, is to neutralize the negative consequences of these two types of asymmetric information. The risk is that one of the agents involved in the arrangement will take advantage of this asymmetry to adopt opportunistic behavior, and will not respect the prescribed behavior, insofar as the deviation is not observable.

In such a framework, it is obvious that R-incompleteness is a fundamentally "negative" phenomenon, which totally undermines the relevance of the theory. If it is true that the only purpose of an institutional arrangement for the economist is to prevent the agreed coordination from giving rise to fraud or deception, then the impossibility—more than frequent, if not general—of not having considered in advance all possible states of nature, condemns without appeal this approach to institutions *lato* sensu. And the economist is even completely disarmed in the face of all cases of contractual relations that are manifestly schematic or incomplete (such as the labor contract, to give just one example). How can it be that rational agents agree to enter into such dangerous relationships?

The non-economist reader may be surprised by the extravagant role played by the possibility of fraud in the design of institutional rules. That this possibility should be taken into account is common sense—but that it should be the *only* element taken into account becomes a normative position that is unaware of itself, under the guise of scientific impartiality.

Now, we are able to achieve our attempt at assessment. It is instructive to return to the enumeration of modes of coordination "other than the market" (and its price system) envisaged by Arrow (1974b), which we quoted at the beginning of this p. 506. Arrow referred to three great institutions of market economies in the contemporary Western world: the rule of law, the capitalist enterprise, and "those invisible institutions: the principles of ethics and morality" (p. 26).

After forty-four years of neo-liberalism, or, to be more sober, of integration into the new *mainstream* economic theory, let us recapitulate what the systematic application of the strict model *of homo economicus* has produced from top to bottom of these three "institutions":

- *The firm* becomes a network of contracts, with shareholders (as principals) inducing managers (as their agents) to control the work of employees (agents of the managers), making it a quasi-financial asset owned by shareholders, who demand that financial returns be given priority over any other evaluation criteria (on behalf of "shareholder value"). The financialization that followed from the 1980s onwards caused a "great deformation" of the creative collective that is the business firm (Favereau, 2014).
- *The State*, where civil servants and officials, from top to bottom, behave just like any other *homo economicus*, has the sole responsibility of ensuring the proper functioning of the rules of competition in the broadest possible part of the economy (for example, internalizing externalities), and must be subject, in its internal organization and in its handling of legal constraints, to the strictest supervision by the rest of the economy and society. The "phobia of the State" (Foucault, 2004, p. 77) is going to find itself formidably stimulated.
- *Ethical and moral values* are completely outside the framework of homo economicus. Nothing to say, therefore? Certainly, at first sight. But on closer inspection, we notice a strange shift in perspective: *homo economicus* becomes a normative model—a citizen responsible for the economic coherence of his choices and responsive to "good incentives." This is certainly not what Arrow wanted or anticipated (it is less clear-cut for the preceding points), but here, unfortunately, we are confronted with an absolutely direct consequence of the maintenance of optimizing rationality. Indeed, Savage's axiomatics, which justifies, it was conceived for games "against nature." But in the new *mainstream*, it is extended to situations of interaction between humans—as if this were equivalent from a normative point of view! The Weberian principle of scientific neutrality is here turned into its opposite. Cheating, when possible, becomes rational; not cheating, out

of honesty (which is what Arrow was certainly thinking of), obliges one to leave standard rationality, since it is contradicted head-on.

The hypertrophy of MI is the principal mode of reconfiguration of the dominant economics—but it is not the exclusive one. By situating the minority modes in relation to the latter, we will be able to draw the overall map of the current theoretical positions in economics, *by reference to MI*. See Fig. 1.

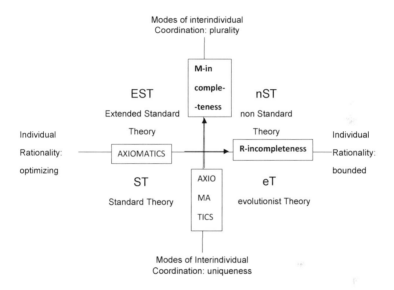

Fig. 1 Map of contemporary economic theories of individualist methodology

Other (Non)treatments of Incompleteness: An Atrophied MI

Since the beginning of this chapter, we have implicitly operated with a two-axes orientation table: one horizontal axis dealing with individual rationality, the other vertical axis dealing with interindividual coordination.

The two axiomatizations mentioned in Part I define, if we now cross these two axes, a region (South-West, if you like), which is the canonical region of the GET. Because of its historical importance, we can speak of the "Standard Theory." It combines optimization (RCT) and uniqueness of coordination (via the market). The theory of contracts and incentives, which we have just reviewed at length, delimits a new zone (North-West, according to our metaphor), which continues to use the standard rationality model to address an indefinite plurality of coordination modes. Because of the extension of the use of optimization, we can speak of "Extended Standard Theory." In this area, the *mainstream* deals with the M-incompleteness problem, leaving the R-incompleteness problem intact.

The impact of this new *mainstream* has exceeded its own limits, and in many respects has given rise, by reaction, to two subsidiary forms, very different, but which we shall be entitled to bring together because of a similar relationship to MI, *mutatis mutandis*. All these postures will appear to us to be indicative of a MI that is *no longer hypertrophied but atrophied*.

The Atrophy of the MI and Macroeconomic Theory (RBC, DSGE)

The new *mainstream* associated with the theory of contracts and incentives has revolutionized microeconomics, in the sense of an almost inexhaustible formal enrichment, since we now have the tools to tackle any institutional arrangement. The downside of taking into account the potentially infinite diversity of these arrangements is that the connection between micro and macroeconomics (i.e., the addition of MI 2 to MI 1 in our initial characterization of MI), which was already excessively difficult, has become virtually unaffordable within an explicit model. To put it simply, one does not know how to think about the general equilibrium of an economy with the sophisticated contracts studied at length in microeconomics. Many observers have remarked grumpily that the new *mainstream* embodies a return to the partial equilibrium ... of the 1930s, before the revival of interest in macroeconomics and general equilibrium brought about by the Keynesian revolution and the complete mathematization of GET.

This macroeconomic blindness of the new *mainstream* has given rise (nature abhorring a vacuum) to the development of either small simplified models, inspired by monetarism, or, on the contrary, large models using the technological advances made in the modeling of stochastic processes. Concerning the latter, the Real Business Cycles (RBC) approach, which originated in two articles by Finn Kydland and Edward Prescott in 1977 and 1982 (together they received the 2004 "Nobel Prize" in economics), will be prolonged and operationalized with the Dynamic Stochastic General Equilibrium (DSGE) approach, following an article by John Long and Charles Plosser in 1983. This theoretical lineage[14] directly inspires a number of macroeconomic forecasting models used by government agencies or academic research centers in France.

In both cases, there is a regression in the quality of MI compared to that of the new *mainstream*. In the first case, it is obvious. The second deserves more attention. The modeling technique consists of endowing economic agents with sophisticated optimizing rationality, but limiting itself to supply or demand behaviors on organized markets, assumed to be in equilibrium at each moment, but subject to stochastic shocks calibrated so that the sequence of temporary equilibria reproduces the observed macroeconomic series. The reader will have noted the apparent similarity with GET: optimization, on the one hand; exclusive coordination by the market, on the other. This calls for three comments:

– this region is one of explicit refusal to deal with the two incompletenesses.
– Although (or may be because) agents are sophisticated optimizers, this region is characterized by a weakening of the MI requirement. Indeed, the markets are always cleared (as with the Walrasian market secretary) and, above all, the stochastic shocks that will decide the macroeconomic dynamics are totally exogenous and inexplicable in minimally intuitive economic terms.[15] All this has little to do with an even tolerant MI. Similarly, the systematic use of representative agents.
– The use of flawless optimization behaviors, the high technicality of stochastic modeling, and the reference to market mechanisms mean that, so to speak, by capillarity, the academic work in this region, far from

---

[14] The interested reader can refer to the collective work directed by David Colander, which offers both a pedagogical synthesis and an enlightened critique (Colander [dir.], 2006).

[15] For an accessible technical discussion, we refer the interested reader to Romer (2016). The latter is a co-recipient of the 2018 "Nobel" Prize in Economics. See also the earlier critique by Lawrence Summers (1991).

any heterodox affiliation, is considered part of *mainstream* economics, according to the "sociological" criterion given in the introduction.

What to think now about the South-East region?

Evolutionary/Behavioral Theory: Another Atrophy?

This region (South-East) is the symmetrical opposite of the heart of the new *mainstream* (the North-West region), since this time R-incompleteness is fully assumed, but in an analytical framework that excludes M-incompleteness, by recognizing only one mode of coordination.

This calls for two clarifications, one per axis.

- First, on the horizontal axis: the works we are going to collect do not use our vocabulary and therefore do not propose to deal with R-incompleteness. But the question is indeed one of rejecting perfect computational rationality, in favor of a more realistic model of "bounded rationality" in the face of radical uncertainty and/or finite capacities of memory, analysis, and computation.
- Then, on the vertical axis, the analysis firmly retains the principle of coordination by the market alone, but thought of here as a general space of competition, rather than as a specific market organization (such as the stock exchange, which had inspired the Walrasian notion of the market, and which was somehow "born again" in the RBC and DSGE models).

It draws much of its inspiration from evolutionism, and its central figure is the thought of Hayek, who, it should be remembered, developed a fierce critique of the abstraction of GET, while sharing the same requirement of methodological individualism.[16] While Arrow broke out of this framework by exploring all sectors of the economy and society by means of the model of optimizing rationality, Hayek did the same, *mutatis mutandis*, by generalizing a logic of competition that makes individuals more intelligent collectively than they are individually; thanks to the cognitive sciences, of which Hayek was more than an enlightened amateur, it has become impossible to ignore the natural limits of human computing capacity.

Behavioral economics, under the impulse of Kahneman and Tversky, starting (once again) in the 1970s, came to feed this area, completely independently, with the building blocks of a *microeconomics of bounded rationality*.

---

[16] All references can be found in Gilles Dostaler's excellent synthesis (2001).

The possibility of experimentation, and the collaboration with other HSS facilitated by the renunciation of optimization, have given a very great vitality to this region.[17]

The essential problem of this region is the transition to a macro level, as shown by the collections of articles up to the turn of the year 2000, with an extreme imbalance between micro advances and macro reflections integrating the latter (moreover, limited to the labor market or the financial market: see, for example, Camerer et al. [ed.], 2004).

The macro level has long been monopolized, in a non-formalized reflexive mode, by the thought of Hayek, within an easily identifiable political philosophy that is that of neo-liberalism, but a neo-liberalism distinct from and complementary to the one(s) we have already attached to the North-West and South-West regions. By mobilizing Piore's synthetic quotation again, we see that here the "maximization" aspect (in the second component) has receded, but the "market" aspect (in the first component) has advanced. As the "market" has become the generic term for a general competitive dynamic, the vision of neo-liberalism is no longer so constructivist (or even no longer, in Hayek's case) and becomes more and more evolutionist.

But since the turn of the year 2000, the transition to the macro level[18] has begun to be made by a multitude of models, mixing evolutionism and behaviorism in various ways, either in a post-DSGE framework with standard markets populated by agents with non-standard behavior, or in a non-DSGE framework with ad hoc connections between non-standard agents.[19] The techniques of simulation and multi-agent modeling (which allow us to free ourselves from the bondage of the representative agent) increase the space of possibilities for the theorist. This multiplication will explain the prudence of our judgment, with regard to our reflection on the relationship to MI. We are perhaps at a crossroads. On the one hand, these works mark, as announced in the subtitle of this section, a regression with respect to the demands of the "closed" MI. Indeed, the question of coordination increasingly exceeds the power of intelligibility displayed by the MI. The *homo economicus* of bounded rationality seems even more disarmed in the face of coordination problems, which should nevertheless proliferate, because of these limits of rationality. In the end, all these models ask questions that they seem structurally incapable not only of solving *but even of formulating*.

---

[17] See Daniel Kahneman et al. (1982), Daniel Kahneman (2011) and Gerd Gigerenzer and Reinhard Selten (dir.)(2001), and more recently Richard Thaler and Cass Sunstein (2008).

[18] With the intermediate levels of firms, markets, and institutions (see Jacques Lesourne et al., [2002] from an evolutionary perspective, and Samuel Bowles [2004] from a behaviorist perspective).

[19] See David Colander et al. (2008), for a brief overview, and Mauro Napoletano et al. (2012) for a concrete example.

516   O. Favereau

One comes to deplore the fact that the Arrovian discovery of the multiplicity of modes of coordination has been "forgotten." But on the other hand, these works can be considered as an exploration whose horizon is the North-East region, the one where both M-incompleteness and R-incompleteness would be faced—at last.

Preliminary Reflections on the Joint Treatment of the Two Incompletenesses

The field being fundamentally to be constructed, we will start the process by taking a retrospective and synthetic look at this scientific attitude which consists in claiming a "closed" MI, falsely demanding since it turns out to be incapable of respecting the limits of validity that it has itself discovered by giving itself axiomatic foundations. Then, we will question the change of epistemological posture of a MI that would assume these limits—and what does "assuming" these limits mean, if not *no longer considering them as limits*, but as invitations to a new realism, expected, hoped for, required of a renewed MI?

Voluntary Blindness to the Foundations Leads to a Flight to Empiricism

First, we must be aware of the scientific damage caused by the collective decision—more or less intentional—of the dominant economy not to take into account the limits of the foundations that a previous generation, particularly demanding in its MI, had itself identified.

The first is the reinforcement of the instrumentalist, or rather anti-realist, bias of mainstream economics. A second is the theoretical disinvestment of younger generations of mainstream economists. And the two are linked.

Let us detail them.

Seen from civil society, the academic community of economists is perceived as embodying a conception of scientificity centered on the use of certain scholarly tools while refusing a debate on the scope and meaning of these tools,[20] which would make possible a critical exchange with civil society. The only thing that matters to economists is the conformity of these tools to

---

[20] It would be more accurate to say that the debate is carefully circumscribed to specialized segments, such as experimental economics, behavioral economics, institutionalist theory, even neuro-economics, etc., without ever developing a collective will to revise all levels of the scientific representation of economics, based on what is discovered in these specialized segments.

the methodological canon in force. The instrumentalism of the discipline, skilfully defended by Friedman[21] (under the guise of adherence to Popperian falsificationism), is exacerbated to the point of producing a kind of anti-realism. The relevance of forecasting is much too risky to be put forward. It gives way at best to performance in the replication (rather than the explanation) of past statistical data. What remains of theoretical modeling, by dint of fleeing from reflection on the foundations, gives rise to a malaise in the young generations of economists, who are subjected to a paradoxical injunction, like architects asked to innovate by continuously building additional floors without touching the foundations, whereas the most consequential innovations would require other foundations. Hence the refocusing of these young generations of economists on econometric work, which has the double advantage of demonstrating a high degree of technicality and avoiding entering into properly theoretical debates, which are considered perilous in all respects (professionally, intellectually, politically)[22]; hence also the intensification of internal competition between economists, who are encouraged to all place themselves on the same level of these empirical works (the technicality of the tools used thus becoming the criterion of vertical differentiation of qualities[23]).

The end of this process, if we are not careful, may be the establishment of a society for which the real economy is reduced to *more and more data*, which it is *less and less able to interpret*. It is therefore not only of academic interest to want to deal jointly with the two incompletenesses.

The Limits of a Closed MI Are the Levers of an Open MI

We will indicate three series of reference points, for a research program in the direction of this North-East region, by underlining as much as possible their logical interdependence.

1. The experience of MI over the last fifty years in economics suggests that the neoclassical tradition can overcome (not without drawbacks) either R-incompleteness or M-incompleteness, but not both, for that would be to overcome itself. What is the element common to both incompletenesses

---

[21] See Milton Friedman (1953)—and his legacy, rather hardened by his successors, which I have studied in Favereau (1995, 2013).

[22] The controversial book by Pierre Cahuc and André Zylberberg (2016) offers a caricatural version of this, while the monograph by Angus Deaton and Nancy Cartwright (2016) places the debate at the appropriate level.

[23] See Geoffard's lucid chronicle (2019).

that makes the neoclassical tradition, with its two axiomatized pillars, unable to bear to give up, without disintegrating? Put in simple terms, it is the fact that *time can bring something new.* We have seen with Arrow and Hurwicz that this forbids translating the axioms of rationality into terms of maximizing a utility function, with the uncertainty of the future reduced to numerical probabilities; and with Arrow and Hahn that this forbids instituting all the insurance or option markets, which would allow rational agents to make their supply or demand decisions, leading to a general equilibrium in an economy regulated exclusively with organized markets. But if time brings new things, this means that the individual can … learn. *Learning is the flip side of incompleteness.*[24] *So there is a positive side to R-incompleteness. But this is no less true of M-incompleteness.* The market with its prices is not enough—this is the negative side of M-incompleteness. The positive side was immediately spotted by Arrow: there are other modes of coordination. Arrow listed three. But the list is not closed: we can … learn new ways of coordinating.[25] The universe of the two incompletenesses, because of these learning possibilities, is perhaps more complex but it is richer and more interesting than the universe of the closed MI which, *by excluding incompleteness, excluded creativity*, at the *individual* and *collective* levels.[26] This introduces the following benchmark.

2. This creativity, via the learning of new modes of coordination, will naturally produce *collective entities*, which the closed MI had by construction the greatest difficulty in accounting for. The new MI will start to manipulate—this time without denying itself—collective entities. These collective entities will retroact on the definition of individual rationality, depending on whether the individual considers himself as a member, or not, of such or such entity. This is a major change in the definition of the individual: on the one hand, he has a direct interest in coordination, which contrasts completely with the *homo economicus* of the closed MI; on the other hand, he must ask himself the question of his *identity*—and this identity, depending on whether it is "personal," "social," or "human,"[27] modifies his perception and his evaluation of the costs and benefits of his choices. Conversely, this mutation of the individual will have an impact on

---

[24] "[In the face of genuine uncertainty] Learning, in the form of reacting to perceived consequences, is the dominant mode in which rationality manifests itself." Herbert Simon (1978, p. 8).

[25] It can be shown that incompleteness can paradoxically make cooperation easier, contrary to the whole theory of contracts (Favereau, 1995).

[26] See recent work on modeling "creative" rationality by Pascal Lemasson et al. (2017).

[27] See n.31 below.

the question of coordination, by putting the agent in a space crossed by multiple collective memberships—in short, by establishing *a third order of reality*[28]: *intersubjective*, which is added to the two orders—subjective, objective—to which the ontological equipment of the closed MI was limited. And this ontological enrichment[29] definitively forces individual rationality to no longer be satisfied with computational skills—and to complete them with *interpretative* skills.[30]

3. This to-and-fro between mutations of rationality and coordination, typical of the approach of joint treatment of the two incompletenesses, should help the reader to understand to what extent the North-East region will differ analytically from the others. An illustration will be helpful. Akerlof (see the bibliography of his work in the 2010 book with Kranton) used, as we did (implicitly) in the previous point, the "theory of social identity" (which comes from social psychology),[31] but he did so by *adding* social membership variables in individual utility functions to be maximized, and then immersing his individuals in *standard* game theory or market exchange schemes. This is doubly incoherent: *a social membership variable requires an interpretive and not just a computational rationality; the coordination, and therefore equilibrium, patterns should be modified.* Two specific properties of this North-East region thus emerge from this illustration: (i) interdisciplinary relations change in nature (far from any imperialism, economics must comply with the canons of the discipline from which it borrows an essential resource) and (ii) the increasing role of collective variables no longer justifies a watertight partition between MI and MH, as with the "closed" MI.

* * *

To explore this further would be the subject of another chapter. There is no need to go back over the panorama of the different "stations" of "closed" MI in the dominant economy, and its avatars up to the present time. It is up to

---

[28] Let us quote Charles Taylor (1971), Karl Popper (1979) with worlds One, Two, Three, and especially Vincent Descombes (1996). Conventions are the paradigmatic example of intersubjective entities, neither subjective nor objective.

[29] The first economist to have defended this thesis is Tony Lawson (2003). See also Edward Fullbrook's collection (2002).

[30] Formally, this implies a change of logic from standard axiomatics: no longer extensional, but intensional (see Favereau (2003). This will be the basis of our critique of Akerlof in the next paragraph.

[31] See John C. Turner (1987). We show all the affinities with the research program of the economics of convention (Bessis et al., 2006).

the reader to judge whether the content announced in the introduction corresponds to the path taken. On the other hand, by way of conclusion, we will provide a final illustration, symbolizing and condensing the main elements of the "open" MI whose advent this chapter conjectures (with a mixture of prediction and wishful thinking). What becomes of the firm, in the North-East region, where R-incompleteness and M-incompleteness become the very spring of the argument?

The firm is an individual agent in the South-West and South-East regions; it is a mode of coordination in the North-West region; *it is both* in the North-East region, for a reason that has to do with the law (here a new form of interdisciplinarity appears): the most usual enterprise rests on two legal vehicles: labor contract and the statutes convened by the initial shareholders creating a "company" as a legal *person*—just as there are *physical* persons. These concepts are paradigmatic examples of incomplete contractual relationships, precluding not only the *possibility* but even the *interest* of specifying what precisely will be done respectively with labor, and with owners' stock, in every forthcoming state of nature. A business firm is a dispositive of collective creation[32]—or it is doomed to fail. Moreover, the emergence of powerful, although fictitious, legal "persons" ends destabilizing the "closed" MI of the dominant economy. The law itself turns a collective (the business firm) into an individual decision-maker—this seems to blur the boundary between MI and MH. At the same time, no legal entity, either private or public, could act without natural persons to animate it. That immediately reopens an indefinite new space for MI. Nothing could better represent what we call an "open" MI than this question of the relationship between physical persons and legal persons (private and public). Now, this question is probably the most important one in contemporary economics, society, and politics.

So, our tentative conclusion is a three-layer proposition: first, economics has been under the control of MI for one century and a half; second, the king is dead; third, long live the king.

References

Akerlof, G. A., & Kranton, R. E. (2010). *Identity economics: How our identities shape our work, wages and well-being*. Princeton University Press.

Arrow, K. (1974a). Limited knowledge and economic analysis. *American Economic Review, 64*(1), 1–10.

Arrow, K. (1974b). *The limits of organization*. Norton.

---

[32] For a development, see Favereau (2019, pp. 40–48).

Arrow, K. (1985). The economics of agency. In J. W. Pratt & R. J. Zeckhauser (Eds.), *Principals and agents: The structure of business* (pp. 37–51). Harvard Business School Press.

Arrow, K., & Debreu, G. (1954). Existence of equilibrium for a competitive economy. *Econometrica, 27*, 265–290.

Arrow, K., & Hahn, F. (1971). *General competitive analysis.* Holden-Day.

Arrow, K., & Hurwicz, L. (1972). An optimality criterion for decision making under ignorance. In C. F. Carter & J. L. Ford (Eds), *Uncertainty and expectations in economics: Essays in honor of G. L. S. Shackle* (pp. 1–11). Basil Blackwell.

Bessis, F., Chaserant C., Favereau, O., & Thévenon, O. (2006). L'identité sociale de l'homo conventionalis [Social identity of homo conventionalis]. In F. Eymard-Duvernay (Ed.), *L'économie des conventions: méthodes et résultats* (pp. 181–195). La Découverte.

Bolton, P., & Dewatripont, M. (2005). *Contract theory.* MIT Press.

Bowles, S. (2004). *Microeconomics: Behavior, institutions and evolution.* Princeton University Press.

Brown, W. (2007). *Les habits neufs de la politique mondiale: néo-libéralisme et néo-conservatisme* [The new clothes of world politics: neo-liberalism and neo-conservatism]. Les Prairies ordinaires.

Cahuc, P., & Zylberberg, A. (2016). *Le négationnisme économique et comment s'en débarrasser* [Economic negationism and how to get rid of it]. Flammarion.

Camerer, C., Loewenstein, G., & Rabin, M. (Eds.). (2004). *Advances in behavioral economics.* Princeton University Press.

Coase, R. H. (1960). The problem of social cost. *The Journal of Law and Economics, 3*, 1–44.

Cohen, M., & Jaffray, J. Y. (1980). Rational behavior under complete ignorance. *Econometrica, 48*(5), 1281–1299.

Colander, D. (Ed.). (2006). *Post-walrasian macroeconomics: Beyond the dynamic stochastic general equilibrium model.* Cambridge University Press.

Colander, D., Howitt, P., Kirman, A., Leijonhufvud, A., & Mehrling, P. (2008). Beyond DSGE models: Toward an empirically based macroeconomics. *American Economic Review, 98*(2), 236–240.

Dardot, P., & Laval, C. (2009). *La nouvelle raison du monde: essai sur la société néo-libérale* [The new reason for the world: An essay on neo-liberal society]. La Découverte.

Deaton, A., & Cartwright, N. (2016). *Understanding and misunderstanding randomized controlled trials* (National Bureau of Economic Research, Working Paper). Cambridge.

Demeulenaere, P. (1996). *Homo oeconomicus: Enquête sur la constitution d'un paradigme.* Presses Universitaires de France.

Descombes, V. (1996). *Les institutions du sens* [Institutions of meaning]. Les Éditions de Minuit.

Dostaler, G. (2001). *Le libéralisme de Hayek* [Hayek's liberalism]. La Découverte.

Favereau, O. (1989). Marchés internes, marchés externes [Internal markets, external markets]. *Revue Économique*, special issue: l'économie des conventions, *40*(2), 273–328.

Favereau, O. (1995). La science économique et ses modèles. In A. D'Autume & J. Cartelier J. (Eds.), *L'économie devient-elle une science dure?* (pp.130–153). Economica [Translated in Favereau, O. (1997). Economics and its models. In: A. D'Autume & J. Cartelier (Eds.), *Is economics becoming a hard science?* (pp.120–146). Edwar Elgar].

Favereau, O. (1997). L'incomplétude n'est pas le problème, c'est la solution [Incompleteness is not the problem, it is the solution]. In B. Reynaud (Ed.), *Les limites de la rationalité, tome 2: Les figures du collectif* (pp. 219–233). La Découverte.

Favereau, O. (2003). La pièce manquante de la sociologie du choix rationnel [The missing piece of the sociology of rational choice]. *Revue française de sociologie, 44*(2), 275–295. [Translated by Amy Jacobs in: The missing piece in rational choice theory. An annual English selection from *Revue française de sociologie, 46*, supplement, 2005, 103–122].

Favereau, O. (2010). La place du marché [The market place]. In A. Hatchuel, O. Favereau, & F. Aggeri (Eds.), *L'activité marchande sans le marché* [Commercial activity without the market]? (pp.111–131). Presses des Mines.

Favereau, O. (2013). Keynes after the economics of convention. *Evolutionary and Institutional Economics Review, 10*(2), 179–196.

Favereau, O. (2014). *Entreprises: la grande déformation.* Collège des Bernardins-Parole et Silence.

Favereau, O. (2019). The economics of convention: From the practice of economics to the economics of practice. In O. Favereau & R. Diaz-Bone (Eds.), *Markets, organizations and law.* Special Issue, *Historical Social Research, 44*(1), 25–51.

Foucault, M. (2004). *Naissance de la biopolitique: cours au Collège de France 1978–1979* [Birth of biopolitics: Lectures at the Collège de France 1978–1979]. Gallimard/Seuil.

Friedman, M. (1953). The methodology of positive economics. *Essays in positive economics* (pp. 3–43). The University of Chicago Press.

Fullbrook, E. (Ed.). (2002). *Intersubjectivity in economics: Agents and structures.* Routledge.

Gauthier, D. (1986). *Morals by agreement.* Clarendon Press.

Geoffard, P.-Y. (2019, September 17). Chronique "économiques": L'économie est-elle un sport de combat ["Economic" column: Is the economy a combat sport]? *Libération.*

Gigerenzer, G., & Selten, R. (Eds.). (2001). *Bounded rationality: The adaptive toolbox.* MIT Press.

Hacking, I. (1975). *The emergence of probability.* Cambridge University Press.

Hahn, F. (1974). The winter of our discontent. *Economica, 40*(159), 322–330.

Harsanyi, J. C. (1977). *Rational behavior and bargaining equilibrium in games and social situations.* Cambridge University Press.

Kahneman, D. (2011). *Thinking, fast and slow.* Farrar, Straus and Giroux.

Kahneman, D., Slovic, P., & Tversky, A. (1982). *Judgment under uncertainty: Heuristics and biases.* Cambridge University Press.

Kornaï, J. (1971). *Anti-equilibrium: On economic systems theory and the tasks of research.* North-Holland Publishing Company.

Kreps, D. (1990). Corporate culture and economic theory. In J. E. Alt & K. A. Shepsle (Eds.), *Perspectives on positive political economy* (pp. 90–143). Cambridge University Press.

Laval, C. (2007). *L'homme économique: essai sur les racines du néo-libéralisme* [The economic man: An essay on the roots of neo-liberalism]. Gallimard.

Laval, C. (2018). *Foucault, Bourdieu et la question néo-libérale* [Foucault, Bourdieu and the neo-liberal question]. La Découverte.

Lawson, T. (2003). *Reorienting economics.* Routledge.

Lemasson, P., Weil, B., & Hatchuel, A. (2017). *Design theory: Methods and organization for innovation.* Springer.

Lesourne, J., Orléan, A. & Walliser, B. (2002). *Leçons de microéconomie évolutionniste* [Lessons in evolutionary microeconomics]. Editions Odile Jacob.

Malinvaud, E. (1991). *Voies de la recherche macroéconomique* [Ways of macroeconomic research]. Éditions Odile Jacob.

Marciano, A (2009). *Law and Economics, A reader.* Routledge.

Mariotti, M. (1995). Is bayesian rationality compatible with strategic rationality? *Economic Journal, 105*(432), 1099–1109.

Mongin, P. (2003). L'axiomatisation et les théories économiques [Axiomatization and economic theories]. *Revue Economique, 54*(1), 99–138.

Napoletano, M., Dosi, G., Fagiolo, G., & Roventini, A. (2012). Wage formation, investment behavior and growth regimes: An agent-based analysis. *Revue de l'observatoire Français des Conjonctures Économiques, 124*(5), 235–261.

North, D. (1990). *Institutions, institutional change and economic performance.* Cambridge University Press.

Orléan, A. (Ed.). (2015). *A quoi servent les économistes s'ils pensent tous la même chose* [What good are economists if they all think the same thing]? Les Liens qui Libèrent.

Piore, M. (2010). *A turning point in American politics and in economics and social policy: The lessons of the conventional school.* Paper presented at: Colloque de Cerisy «Conventions: l'intersubjectif et le normatif» (3 Sept. 2009).

Popper, K. (1979). *Objective knowledge.* Oxford University Press.

Romer, P. (2016). *The trouble with macroeconomics.* Sage.

Salanié, B. (2005). *The economics of contracts: A primer* (2nd ed.). MIT Press.

Savage, L. (1954). *Foundations of statistics* (2nd ed., 1972). Wiley.

Shavell, S. (2004). *Foundations of economic analysis of law.* The Belknap Press of Harvard University Press.

Simon, H. (1978). Rationality as process and as product of thought. *American Economic Review, 68*(2), 1–16.

Summers, L. (1991). The scientific illusion in empirical macroeconomics. *Scandinavian Journal of Economics, 93*(2), 129–148.

Taylor, C. (1971). Interpretation and the sciences of man. *Review of Metaphysics, 25*(1), 3–51.

Thaler, R., & Sunstein, C. (2008). *Nudge: Improving decisions about health, wealth and happiness.* Penguin Books.

Tirole, J. (1988). *The theory of industrial organization.* MIT Press.

Tirole, J. (2006). *The theory of corporate finance.* Princeton University Press.

Tirole, J. (2018). *Economie du Bien* [Commun economics of the common good]. Presses Universitaires de France.

Turner, J. C. (1987). *Rediscovering the social group: A self-categorization theory.* Basil Blackwell.

Von Neumann, J. & Morgenstern, O. (1953). *Theory of games and economic behavior* (3rd ed.) [1st ed.: 1944]. Princeton University Press.

Williamson, O. E. (1996). *The mechanisms of governance.* Oxford University Press.

Methodological Individualism in Terms of Rational-Choice and Frame-Selection Theory: A Critical Appraisal

Richard Muench

1 Introduction

For knowing what methodological individualism (MI) means in historical sociological explanation, Max Weber is still the first address to consult. He was central to the so-called first methods dispute at the end of the nineteenth and the beginning of the twentieth centuries. Friedrich von Hayek and Karl Popper were most important for the revival of methodological individualism after WW II. For contemporary sociology, it is the marriage of methodological individualism and rational-choice theory (RCT) or value-expectancy theory (VET), which dominates the understanding of methodological individualism (cf. Heath, 2005; Udehn, 2002). While MI in the Weberian sense is widely practiced implicitly without spelling it out as a doctrine, RCT understands MI as a doctrine, and this is a very specific and much more narrow way than Weberian MI. Therefore, a discussion of this kind of MI is most important for the methodology of sociological analysis. It is this kind of rational-choice methodological individualism (RC-MI), which I want to discuss in this chapter with regard to its understanding of historical–sociological explanation, exactly that kind of sociological analysis for which Max

---

R. Muench (✉)
Zeppelin University, Friedrichshafen, Germany
e-mail: richard.muench@zu.de

Otto Friedrich University, Bamberg, Germany

© The Author(s), under exclusive license to Springer Nature
Switzerland AG 2023
N. Bulle and F. Di Iorio (eds.), *The Palgrave Handbook of Methodological Individualism*,
https://doi.org/10.1007/978-3-031-41508-1_22

Weber's work stands. I take two leading contributions to RC-MI by demonstrating its special features and for explicating its explanatory potential and shortcomings. These are James Coleman's (1990) rational-choice theory and Hartmut Esser's (2005) extension of rational choice theory by a frame selection theory turning RC-MI into FS-MI literally. Both together constitute what we might call RC/FS-MI. Because of their direct link to one another, we can also take RC-MI as the more general concept which includes FS-MI as an extension.

2 What is Methodological Individualism?

With Max Weber, I share the understanding of social phenomena as constituted by the actions and expectations of individuals related to one another. A social order or an institution exists in as far as we can expect in certain situations regularly specific actions of individuals who assume the binding validity of certain norms and respective sanctions applied by authorized people—possibly any third party—or special authorities in the case of their violation. The authorities must be believed as legitimated for doing so by all actors who can exert influence on the application of sanctions. MI in this Weberian sense is focused on how we understand social phenomena like social relations, social groups, social orders, associations, or the state as defined in Weber's introductory chapter to *Economy and Society* on the basic categories of sociology (Weber, 1976 [1922], part I, ch. 1, pp. 1–30). I totally subscribe to this Weberian MI. In addition to institutionalized norms and regularly expectable action, we may take social structures like class structures, constellations and configurations of power, constellations, and configurations of actors as macro phenomena. In these cases too, we speak of possessions and chances to act of individual actors which may be put together in classes or groups. And whenever the social phenomenon to be explained on the macro level entails actors as described so far, we need to involve actors in the explanation of such social phenomena, even in the extreme case that only physical features are addressed in a partial causal role. An example maybe the scarcity of water on a certain territory, which causes more frequent and fiercer distribution struggles on that territory than on territories where such scarcity does not exist. The frequency, violence, and outcome of these struggles are determined by how the actors living on this territory perceive this scarcity, how they are willing to share water, and how much power they dispose on to enforce their interests.

This Weberian type of MI includes definitely the position of Raymond Boudon (1987, 2001), which has been forcefully renewed by Nathalie Bulle (2019) who pleas for an anti-reductionist MI, which acknowledges that social structures are entities of their own and as such play a crucial role in the explanation of social phenomena. However, as Bulle argues, the impact of these structures on social phenomena is mediated through the subjective meaning attributed to these structures by individual actors; and attached subjective meaning is the driving causal force to be recognized in the causal analysis (pp. 2, 8–13). Hence, social structure is a condition for action, which is transformed into a causal force through the attribution of subjective meaning to it by individual actors. However, we should distinguish two kinds of linking social structure to individual action, an objective and a subjective one. Just take the class structure of a society according to the distribution of economic, social, and cultural capital as macro-level feature and its reproduction or transformation from time t_0 to time t_1. First, the distribution of capital at t_0 directly determines the means available to pupils to achieve at school, independent of the subjective meaning they attribute to their available capital in relation to the capital of other pupils. It is an opportunity structure for action. Second, the subjective meaning they attribute to the distribution of capital determines among other perceptions and motives how far they trust in their own capabilities to achieve at school. Both together—objective and subjective class position—determine how far pupils achieve at school and how far the class structure is reproduced or transformed in this way. Here, a feature of a society—its class structure at t_0—determines via achievement of pupils at school involving objective and subjective class position another feature of this society—its class structure at t_1.

However, for reasons of parsimony, we sometimes may ignore the variable of subjective perception and motivation to act and simply focus on the variable of objective opportunity structure. For example, the higher inequality of educational achievement of pupils in one society compared to another society may be simply explained by higher inequality in the distribution of economic, social, and cultural capital among their parents whatever the perception of this situation and the motivation of the pupils and their parents is and whatever they do. For example, a study by Godin and Hindriks (2018), using data from the OECD's Programme for International Student Assessment (PISA), reveals at the test waves from 2003 to 2015 that mobility of pupils in terms of the decile of their socioeconomic family background and the decile of their PISA score is lower in countries with higher social inequality. Studying data from the Progress in International Reading Literacy Study (PIRLS) and PISA, Chmielewski and Reardon (2016) also demonstrate that countries of higher social inequality produce higher inequality in educational achievement than countries of lower social inequality. These explanations of more reproduction of social inequality in countries showing higher social inequality than in countries showing lower social inequality holds independent of the perceptions and motivations of the pupils. Whatever they do, the class structure determines to a large extent their educational achievement (cf. Muench & Wieczorek, 2022; Muench et al., 2022). Therefore, we do not always need to step down to individual perceptions and motivations when explaining macro phenomena, just for reasons of parsimony. The class structure of a society is a social fact in Durkheim's sense that nobody can escape. At least, it has to be taken as real by everybody affected and by everybody who wants to change it. In this respect, Durkheim's most elementary unit of sociological analysis—the social fact—is as valuable as Weber's most elementary unit of sociological analysis—social action—in other respects (Durkheim, 1982, ch. 1; Weber, 1976 [1922], part I, ch. 1).

Durkheim's (2005) study on suicide is definitely only focused on explaining higher or lower suicide rates of various communities or at times of stability or crisis as macro-level phenomena and is not interested in individual suicides. Regarding what he called egoistic suicide, he found that suicide rates are always higher among Protestant communities than Catholic communities, independent of being in a majority or minority position in a certain region. His explanation was the difference in the availability of help in situations of mental distress between these two religious communities. He assumed that there was more of such help by the Catholic church because it administered the mental life of its members while the Protestant had to cope with distress self-responsibly by him/herself. Interestingly, this is very close to Weber's characterization of Protestantism and Catholicism. For Durkheim, this difference between Protestantism and Catholicism explained higher suicide rates among Protestants than Catholics, for Weber it explained higher rates of higher education and entrepreneurship among Protestants than Catholics. In both cases, a macro-level feature of religious communities determines another macro-level feature of the same communities. In both cases, we can step down to the micro level to find that a self-responsible individual has a greater chance and willingness to achieve in education and occupation but has also a greater risk to fail in his/her endeavors and then lacks help and has to cope him/herself with this situation. Hence, there is a greater risk of finding no other solution than ending one's life in cases of mental distress. Stepping down to the micro level helps here to "understand" the actions which are aggregated to constitute the explaining and the explained features of the two religious communities. It is, however, crucial for explaining suicide or entrepreneurship rates as macro-level features to also address a macro-level feature on the explanatory side. And this feature is not the individual availability of administered help or individual self-responsibility but the frequency and intensity of administered help or the frequency and intensity of being in situations in which one has to act self-responsibly and without administered help. These are clearly macro-level features, which make a crucial difference between the two religious communities.

Another example may further illustrate the argument: We want to explain why in France the number and length of unauthorized strikes were much higher than in Germany in the period from 1950 to 1990. This is a feature of these two societies on the macro level. For explaining this feature, we look for other features on the macro level which make a difference between these two societies and may exert an effect on the frequency and length of unauthorized strikes. First, we may observe that income inequality, economic up- and downswings, periods of worker dissatisfaction with their wages, and overall situation are not very different between these two societies. What is, however, very different is the organization of workers in unions. In France, unions are fragmented in terms of political ideology, in Germany, workers are organized in unitary unions of industrial sectors. This is on the one hand an objective or opportunity structure and a subjective structure represented in the subjective perception of workers on the other. Hence, in situations of dissatisfaction, which motivate workers to act for improving their situation the objective or opportunity structure of a unified union system turns protest directly into negotiations with also unified employer federations so that the protest is immediately calmed down. The fragmented union system does not offer this opportunity so that protest is not directly calmed down in negotiations. This objective situation is complemented by the subjective perception of German workers that their unions will successfully negotiate a new collective agreement so that they do not have to do anything by themselves. In contrast, the French workers' perception tells them that direct negotiations will not take place and are complicated because of fragmented unions, so that they have to go on strike directly in order to improve their situation. This link of different objective and subjective macro structures in these two countries explains the different frequencies of unauthorized strikes in these two countries in the period from 1950 to 1990. The explained phenomenon is on the macro level, that is the frequency and length of unauthorized strikes in a society. The explaining variable is on the macro level too, namely a feature of the union system (objective opportunity structure) and the number of workers who trust or do not trust that their unions will improve their situation (subjective perception of the situation). That is, even the subjective perception counts here first in its frequency as a feature of society on the macro level. Only secondly entailed in this macro phenomenon is the perception and motivation of individual actors to act in certain ways. What is definitely necessary for explaining the macro phenomenon of the frequency of unauthorized strikes is the aggregation of individual perceptions and motivations so that they are a feature of society and not only of individuals. Otherwise, we could not explain the macro phenomenon of the frequency and length of unauthorized strikes in a society. Hence, for explaining macro-level features, we first of all have to look for possible explanatory macro-level phenomena. Otherwise, we do not find an explanation. Stepping down to the micro level is helpful for understanding the causal link between macro-level features, but for reasons of parsimony, very often, it is only implicitly addressed when speaking of macro-level phenomena.

Contemporary RC-MI conceives methodological individualism in a much more rigorous way than MI in the tradition of Max Weber. For adherents of RC, MI is not simply what is done implicitly in most sociological research but a doctrine that demarcates "scientific" sociology from "non-scientific" sociology. In this way, first, individual action is conceived in a much too narrow way, second, there is so much emphasis on stepping down to the individual level that the specific features of the macro level in their explanatory role are missed. This is different in Boudon's (1987, 2001) approach of MI, which explicitly attributes the configurations on the macro level, a crucial role for explaining social phenomena. I just want to remind on this kind of MI in a critical assessment of RC-MI. The latter turns historical–sociological explanation into a procedure, which is not covered by Weber's practice in his own historical-sociological explanations. The difference between Weber's MI and RC-MI begins with the understanding of individual action. While Weber distinguished four types of action—purposeful, value-rational, traditional, and affective action—RC-MI knows only one, the purposeful utility-maximizing action. This difference is crucial for how historical–sociological explanation proceeds. A second major difference is the requirement of RC-MI that any macro-sociological analysis has to step down explicitly to the micro level of individual action and climb back up from there to the macro level in order to explain macro phenomena by some causal laws working on the micro level. And these laws are limited to the conception of individuals as utility-maximizing actors. I will show in this chapter that in this way RC-MI does not meet the requirements of historical–sociological explanation as we find them in the work of Max Weber himself.

For RC-MI, a clear-cut differentiation of the object of sociological analysis into the micro and macro levels is fundamental. The requirement is to link macro-sociological analysis always to micro-sociological analysis, otherwise, it would be incomplete and unsubstantiated. In sociology, not only the differentiation of its object of analysis into these two levels, but its differentiation into three levels is commonplace, that is the differentiation into the macro, meso, and micro levels. For example, regarding the comparison of Protestantism and Catholicism, the macro level is how dominant is one of these religious communities in a country or a certain region of that country, the meso level is the church as a more or less bureaucratic organization, the micro level is the belief and behavior of the individual Protestant or Catholic. In discussions of RC-MI, this three-level differentiation is mostly not addressed because the meso level is already a social phenomenon above the micro level the analysis of which has to be founded in micro-level analysis. Therefore, the logic of meso analysis is the same as the logic of macro analysis, why it does not need special attention from the methodological point of view.

That the differentiation of the macro, meso, and micro levels of sociological analysis is commonplace in sociology does not include that all kinds of sociology share the requirement of RC-MI to always link macro-sociological analysis to micro-sociological foundations. This is also true for historical–sociological explanations of historical phenomena. Ronald Jepperson and John W. Meyer (2011) demonstrate with the example of Max Weber's study of ascetic Protestantism and the spirit of capitalism that not only the macro and micro levels but all three levels of sociological analysis are at issue in it, and that the organizational meso level and the institutional macro-level make independent contributions to explaining the emergence of modern capitalism that cannot be reduced to the micro-level of individual motives and individual behavior in the sense of RC-MI. They show that the widely received Coleman bathtub and its application to Weber's study on ascetic Protestantism and the spirit of capitalism and the related plea for RC-MI offer a considerably truncated account of this study and need revision (cf. Coleman, 1990, pp. 6–10). And they argue for a multilevel model of sociological explanation that assigns specific qualities to all three levels that require their own analysis (cf. generally Alexander et al., 1987; Greve et al., 2008; Heintz, 2004). Following Jepperson and Meyer's thrust, I argue that the program of RC-MI is inadequate for historical–sociological explanation and needs to be replaced by a multilevel analysis as it was practiced by Max Weber himself and which is in agreement with Weber's MI (cf. Münch, 1973, 1983, 1987; Münch & Smelser, 1987). The features of a good historical–sociological analysis of social reality will be clarified by means of two examples and related to programmatic parts of RC-MI. The first example is—as in Jepperson and Meyer's analysis—Max Weber's (1972a [1920]) study of ascetic Protestantism and the spirit of capitalism as a historical–sociological explanation of a social fact; the second example is Diego Gambetta's (1993) sociological case study of the Sicilian Mafia.

3 Historical–Sociological Explanation of a Social Fact: Max Weber's Study on Ascetic Protestantism and the Spirit of Capitalism

For sociology, since its inception, the program has been to unite both natural scientific and hermeneutic methodology under one roof. The central authority for this unique character of sociology is Max Weber. For him, "sociology…, is a science which seeks to understand social action interpretatively and thereby explain it causally in its course and effects" (Weber, 1976 [1922], p. 1, translated by the author). Every sociological investigation is, therefore, to be examined with regard to its causal adequacy and with regard to its meaning adequacy (Weber, 1976 [1922], pp. 1–11). "Causal adequacy" means that a social phenomenon is proved to be a cause of another social phenomenon, "meaning adequacy" that a specific meaning of a text, a religious belief system, or an action is proved. A sociology that focuses on only one of these two sides can only be an incomplete sociology that cannot recognize, understand, and explain essential parts of social reality. For Max Weber (1973 [1904]), sociology is a science of reality (Wirklichkeitswissenschaft) so that it is not sufficient to discover basic and abstract laws of social action but needs to understand and explain what happens in concrete social reality. This is what is meant by "interpretative sociology" ("verstehende Soziologie"). Since RC-MI has a clear preference for an approximation to natural science methodology, its research program stands and falls with the ability to also grasp culture, contexts of meaning, and institutions in a way that is appropriate for the subject matter. If this is not the case, it remains blind in one eye. An example of considerable effort and advancement in this direction is the extension of rational choice theory by the model of frame selection and the model of sociological explanation by Hartmut Esser (2001, 2005, 2006, 2010). It forms an important bridge from sociology focused on causal analysis to sociology proceeding in the hermeneutic tradition of understanding the meaning of social phenomena (cf. Albert, 2005, 2007; Greshoff & Schimank, 2012). Esser follows Coleman's (1990, pp. 6–10) bathtub model of RC-MI and extends it with a special focus on the frame within which an action takes place so that rational-choice theory is complemented by a frame-selection theory. In the following, I will use this example to discuss the potential of RC/FS-MI with regard to addressing culture and contexts of meaning as well as institutions as macro phenomena. A complementation of rational-choice theory to address this side of social reality appropriately has succeeded insofar as the concept of frame serves as a tool for explaining norm-guided action without having to derive it from a consciously executed calculation of utility. The frame is so firmly anchored in the subconsciousness of an actor that it immediately triggers action. To remain in the language of rational-choice theory or value-expectancy theory, this is conceptualized as the selection of a frame. As far as patterns of perception, thought, and action, acquired through habitualization, match the perception of an action situation, they directly trigger a certain action. The less this is the case, the more the action to be selected has to be "calculated." The selected frame is then also the premise for all further decisions, which are accordingly carried out in "bounded" rationality in the sense of March (1978) and Simon (1979).

For rational choice theory, this is a considerable advancement. But does this progress render superfluous the continuation of sociological traditions that specialize in conceiving culture and meaning contexts as macro phenomena and have developed their own methods for doing so? This question, for all merits of bridge-building between the two sociologies of Esser's model of sociological explanation, must nevertheless, in my estimation, be answered in the negative. With the theory of frame selection, this side is incorporated in an adapted form into the program of RC-MI, but this is done without fully engaging with the perspectives and methods of the other side. The theory of frame selection is advancing rational-choice theory, but it does not fully represent the idiographic side of sociology and fully capture cultural meaning and institutions as macro phenomena. Take Max Weber's (1972a [1920]) study on ascetic Protestantism and the spirit of capitalism as a classic example. It combines the test of causal adequacy with the test of meaning adequacy of the analysis. Weber (1972a [1920], pp. 17–30) sees causal adequacy confirmed by the disproportionate representation of Protestants in entrepreneurship and in the higher educational strata and by the primary and most pervasive industrialization of Protestant regions. Becker and Woessmann (2009) have similarly found that in Prussia in the late nineteenth century, the level of economic development of districts, starting from the Protestant center of Wittenberg, increased with the spread of Protestantism. After controlling for the level of education, this effect disappears, so that the level of education turns out to be the crucial explanatory factor.

But Protestantism, in turn, had a strong effect on raising educational attainment, which then does assign it an important role in promoting economic development. In the end, however, we know that such correlations cannot be taken as proof of causation, because there might be simply a parallel development of the two phenomena, a causation in one or the other direction or a mutual impact of one phenomenon on the other.

For interpreting the correlation between two phenomena including gaining knowledge about a possible causal relationship, searching for meaning adequacy in understanding this correlation is crucial for Weber. By far the largest part of his study on ascetic Protestantism and the spirit of capitalism is indeed concerned with the adequacy of meaning (see Schluchter & Graf, 2005). And this is not limited to this study but has been extended by him in comparative analysis to all major world religions (Weber, 1971 [1920], 1972b [1920]), though in the case of Islam only in sections in the sociology of religion, law, and rule in *Economy and Society* (Weber, 1976 [1922], e.g., pp. 346–347, 375–376, 459–460, 629–630, 739–740). A focused comparison is made specifically in the juxtaposition of Confucianism and Puritanism (Weber, 1972a [1920], pp. 512–536). In the perspective of Esser's FS-MI, here would be the starting point for frame selection. To understand the Protestant ethic as a frame within which a rationalism of world domination and restless occupational work becomes the only consistent way of life or habitus (Bourdieu, 1984, pp. 169–225) is, in the examination of the adequacy of meaning, the very last step in explaining this way of life, a step that has a logically true character because an act of restless occupational work follows logically from the ethical expectation of the Protestant to prove himself in occupational work before God (cf. Coleman, 1990, pp. 6–10). As long as there is a match between the ethical program or habitus and the perceived situation, the Protestant acts according to the ethical program (or habitus) (Esser, 2005, pp. 11–13). The less this is the case, the more the Protestant has to find a solution to the problem that is not already given. He must reduce dissonance (Festinger, 1957). However, this is not necessarily a matter of calculating utility, but also a matter of the cognitive processing of dissonance, the management of which is a matter of "true/false" in perception, "appropriate/inappropriate" in orientation to custom, and effective/ineffective in expected problem-solving. An example may illustrate this process: According to Esser's model of sociological explanation and the corresponding bathtub model, it is to be proceeded in three steps: Macro level to micro level: logic of the situation → micro level: logic of selection → micro level to macro level: logic of aggregation (cf. Esser, 1999, pp. 91–102). We start with an actor whose actions need to be explained. It is a middle-class entrepreneur and member of a Protestant religious community believing in the interpretation of the doctrine of predestination by Calvin's successors. The logic of the situation is determined by the fit/non-fit of a religious and a civic frame:

1. Religious frame: how can I know that I am one of the elects? Answer: By your exemplary conduct of life.
2. Civic frame: What does exemplary conduct of life consist of? Answer: discipline, diligence, fidelity to contract, and humility.
3. Perception of the action situation: My costumer expects the timely delivery of goods of the best quality, but I am behind schedule.

According to the logic of selection, the entrepreneur infers from the fit of frames and situation:

4. Therefore, I have to invest more working time in order to deliver on time.

Conclusion (4) follows logically from the meaning of premises (1) and (2) and the perception of the action situation (3) alone. A missing fit of frame and situation would be given, e.g., if the increase of working time for the purpose of contract fulfillment would only be possible at the expense of late delivery to another customer and the frame did not provide any specifications for this dilemma. In this case, the entrepreneur could be guided by the amount of the compensation payment due for late delivery and, for maximizing the benefit, deliver first to the customer who would receive the higher compensation payment. But this also takes the form of a logical conclusion.

According to the logic of aggregation, a collective state can be inferred from the interaction of individual actors according to a transformation rule. The transformation rule here might say that membership in the Protestant religious community suggests a reciprocal expectation of contract fidelity which is affirmed with each individual act of contract fidelity and gradually becomes an institution. In terms of Weber's MI, this institution means that certain actions can be expected in certain situations.

However, Weber's explanation of the action of the middle-class entrepreneur according to the logic of the situation and the logic of selection in the perspective of RC/FS-MI is preceded by many other steps in which the meaning of the Protestant ethic and its formation by carrier classes and institutions from the Judeo-Christian tradition and its development are investigated in a historical-comparative way (cf. Münch, 1986, vol. 1, pp. 61–179; 2001, pp. 34–117). Without this historical-comparative reconstruction of the meaning of the Protestant ethic as a cultural macro phenomenon, I do not know what proving oneself in the religious sense of the Protestant in restless occupational work means, nor can I explain why he acts in this way. The question for the actors is not "What is the benefit?" but "What makes sense?" And this is not done unconsciously as frame selection, but in a completely conscious construction of meaning. But the construction of meaning falls outside the rational-choice model of utility-maximizing action and it is also more than unconscious behavior within an action frame according to frame-selection theory. Central actors on the macro level of analysis are widely respected intellectuals as constructors of meaning. For the sociological observer, questions such as the following are relevant: "What is the class position of the main members of a religious community?" "How much power of definition do they have, how much power of definition do their competitors have for the religious construction of meaning?" "What is the relationship between orthodoxy and heterodoxy?" This hermeneutical reconstruction of the meaning of the Protestant ethic as a cultural macro phenomenon is by no means freehand and speculative, but rather based on the testing of a series of hypotheses (cf. Gadamer, 1960). It is first assumed that certain religious compendia are representative of the culturally dominant ethic of the ascetic Protestant lifestyle at a certain time and in a geographically determined area (England in the seventeenth to eighteenth centuries as the area of origin of modern capitalism). Then hypotheses are formulated such that certain parts of a work are representative of the meaning of the entire text and similar texts. Further, assertions are made that certain sentences in a text have a certain meaning. And it is said that this meaning demands specific action in specific situations of action. In Weber's study of Protestantism, such assumptions are made throughout, and the author's work is to make sure, as best as he can, that the assumptions stand up to possible objections. Such objections may be anticipated by the author himself. But the critical examination of his text does not end here. Rather, it continues after the text has been published, with critics raising objections that may or may not require corrections. Indeed, this has happened extensively (cf. Schluchter & Graf, 2005; Tyrell, 1990).

Weber also invests a great deal of research in the logic of aggregation according to the Coleman/Esser bathtub model. The transformation rule is not simply aggregating individual acts of aspiration and contract fidelity. As already mentioned, membership in the Protestant religious community plays an important role. It acts as social capital in the form of mutual obligations and expectations (Coleman, 1990, pp. 306–310). But this is by no means the end of the story. There is also the fact that restless occupational work captures only the aspect of bourgeois lifestyle in modern "rational" capitalism. A whole series of other factors was necessary to bring about modern rational capitalism in all its material, institutional, and cultural features, such as the European town of burghers as a model of a modern legal community overcoming family particularism, the shaping of rational law by an independent legal profession (practicing lawyers and judges in England, university lawyers in resumption of Roman law on the European continent), a predictable bureaucratic administration, rational experimentation as a method of modern science, technical inventions such as the steam engine and the mechanical loom, and material conditions such as raw material deposits, transportation routes and means, and the commercialization of agriculture in England (Weber, 1924; 1976 [1922], pp. 456–467, 551–579, 741–757). To pick up just one factor: For Weber, a constitutive feature of economic intercourse under "rational" capitalism was that far-flung business relationships could be maintained between "strangers" beyond the closest ties of family and friends. This presupposes trust in business partners beyond these ties. In his eyes, overcoming family particularism was crucial for this feature of modern business. And he identifies in the civic and legal community of the medieval city of Europe and in the Protestant religious community crucial prerequisites for the expansion of economic transactions characterized by trust beyond the family (Weber, 1976 [1922], pp. 226–230, 741–757). In contrast, in China, with the purely administrative city of the empire, the primacy of loyalty to the family in the village of one's own origin and the maintenance of the family cult of ancestors, the people living and doing their business in the city did not build a community of citizens sharing a common law. Therefore, crucial prerequisites for the expansion of trustful business relationships beyond family and friends were lacking in China (Weber, 1972a [1920], pp. 290–298, 378–390, 518; 1976 [1922], pp. 293, 756). In Weber's eyes, the lack of these prerequisites explains why rational capitalism did not emerge there, although China had been technologically superior to Europe until the fifteenth century due to numerous inventions, such as paper, printing, gunpowder, the magnetic compass, porcelain, mechanical clockwork, iron foundry, stirrup, and other techniques (see Needham, 2013). Here, there is a direct cause–effect relationship at the macro level such that a structural feature in the comparison of two economic regions—Europe vs. China in the fifteenth century—is seen as a prerequisite for the ability to develop trust beyond family ties, and this trust in turn is seen as a lubricant for extensive economic transactions in a given area. Of course, these are individual actors who trust or do not trust and they establish the link between cause and effect on the macro level. Crucial for explaining the macro-level feature of family-transcending trust in economic transactions is, however, that there is not only the trust of single individuals and their ability to engage in economic exchange with any other person on the micro level but the emergence of cross-family communities in the city and of the explicitly universalistic Christian religious community that transcended all particularistic group boundaries, i.e., a macro-level structural feature of the society. And the effect is not only the action of single individuals, but the range and number of actors involved in economic transactions and the frequency of willingness shown therein to engage in business relations with strangers. The corresponding hypothesis on the macro level would read as follows:

> The larger the number of family-transcending associations and the larger the number of people being members of these associations in a social unit, the larger the number of people with sufficient trust to engage in economic transactions reaching beyond the ties to family and friends in this social unit.

Both, cause and effect in this hypothesis are features of a social unit and in this sense macro phenomena. In the ontological sense, however, both macro phenomena are constituted and linked together by the aggregation of individual acts. Nevertheless, it is a feature of the social unit that determines how the feature of the social unit to be explained looks like. That there is not only one trustful economic transaction beyond the ties of family and friends results from the number of family-transcending voluntary associations and the number of their members. Hence, a macro feature of a social unit serves as transformation rule which turns individual acts into regular behavior. Whoever enters the market can expect such behavior, and in any case of deviation from this expectation, one can expect commonly shared sanctions aiming at confirming what is considered as regular behavior. In this sense, expecting a certain behavior in a specific situation is "institutionalized" among a circle of people involved in economic transactions, and this institutionalized character of expectations is a macro phenomenon.

The same holds for the feature of restless occupational work. Modern rational capitalism as a firmly institutionalized economic form does not simply emerge from the aggregation of a multitude of like-directed individual acts of restless occupational labor, but from the interaction of a multitude of macro factors, each of which independently produces a specific aspect of the complex phenomenon at the macro level. Frame selection and rational choice of action in a decision situation as well as individual acts of restless occupational work are only the individual representations of a macro-level relationship. If one really wants to, one could give the enumerated factors the status of transformation rules that ensure that restless and reliable occupational work is embedded in a context of material, institutional, and cultural structures through which it is sustained and can only contribute a factor of lifestyle to the much more complex phenomenon of modern rational capitalism. Looked at closely, even the connection between the ascetic Protestant ethic and restless and reliable occupational work involves a specific macro-level connection, namely, that the association of ascetic Protestantism with the bourgeois way of life created, first in England in the seventeenth to eighteenth centuries, a bourgeois carrier class that provided a basis of legitimacy for the capitalist mode of production and facilitated its spread in the form of the prevalence of capitalist-working enterprises. The corresponding hypothesis reads as follows:

> The more the capitalist mode of production experiences religious or cultural legitimation in connection with a rising economic carrier class at a certain time and in a certain area, the more it spreads at that time and in that area in the shape of a growing number of capitalist-working enterprises.

Accordingly, the crucial factor is the connection between ascetic Protestantism and the bourgeoisie, the "elective affinity" between the ethic of ascetic Protestantism and the lifestyle of the rising bourgeoisie and its enforcement against the aristocratic way of life and production on the macro level. It is a macro phenomenon—the connection between ascetic Protestantism and the rising bourgeoisie—which makes the bourgeois lifestyle legitimate and binding so that restless and reliable occupational behavior is not a matter of individual acts only but a collective and thus macro phenomenon. This constellation, unique in historical comparison in England in the seventeenth–eighteenth centuries as a structural feature on the macro level, explains the legitimacy and therefore prevalence of capitalist-working enterprises as a feature of the English economy at that time likewise on the macro level. The behavior of the individual Protestant and also the aggregation of such individual behavior are only effective in connection with the macro-level configuration in causally determining the macro phenomenon of the legitimacy and high incidence of capitalist-working enterprises in England in the late eighteenth century eventually by historical and comparative standards. Crucial is the close connection or elective affinity of ascetic Protestantism and the bourgeois lifestyle likewise as a macro phenomenon. Without this macro-level relationship, the restless and reliable professional work of an individual Protestant would make no contribution to the assertion of the capitalist way of life and mode of production against the aristocratic way of life and mode of production. If you wish so, this structural dynamic is exactly what the protagonists of RC/FS-MI seek as a transformational rule of aggregation of individual behaviors into a macro phenomenon. Here, the transformation rule turns individual actions into a macro phenomenon through a structural dynamic on the macro level that pulls individual actions into a macro link between two phenomena.

The connection of ascetic Protestantism with the rising bourgeoisie makes restless and reliable work in business an institutionalized binding norm. Therefore, people in business expect this behavior from each other and find support from the legal community in the case of deviation from this expectation. That means, it can be expected with good reason that this kind of behavior does not occur accidentally but regularly. Exactly this regular behavior is the macro phenomenon determined by the connection between ascetic Protestantism and the rising bourgeoisie. This is the transformation rule which turns individual acts of restless and reliable behavior on the micro level into a collective phenomenon on the macro level. It is the link between two macro phenomena that serves here as transformation rule. And it is exactly this link that we need to discover in explaining a historical macro phenomenon. Stepping down from the macro situation to the selection of an individual action on the micro level and climbing up to the macro level again helps to understanding the macro phenomenon, but it does not in itself provide the crucial insights we need for a convincing historical–sociological explanation of the institutionalization of the capitalist mode of production as a macro phenomenon.

4 Sociological Case Analysis: Diego Gambetta's Study of the Sicilian Mafia

I would like to further illustrate the loss of sociological knowledge that would result from an epistemological monopoly of RC/FS-MI that is too narrowly conceived and proceeds mostly in a positivistically deductive-nomological manner according to Hempel/Oppenheim (1948) by means of Diego Gambetta's case study of the Sicilian Mafia.

Briefly summarized, the story reads as follows: In Sicily, the Mafia fills a gap in the protection of legal and illegal businesses from fraud and competition, because historically, due to the principle of "divide et impera" practiced by the Spanish occupation, there is no trust reaching beyond family ties and because the state is too weak to provide this protection. In the case of illegal business, the state is not even willing to do so. And the Mafia gets paid for this protection (Gambetta, 1993, ch. I). For Gambetta, the demand for protection in the absence of trust reaching beyond family ties does not yet give rise to an offer of protection in the form of the Mafia. He notes that the Mafia did not spread where the ruling class of big landowners was united and strong enough to keep order. Better conditions for the Mafia's emergence existed where this was not the case, where economic exchange grew, as in the coastal cities in particular, and where there were people striving for economic success who saw a profitable business in the supply of protection and could also find enough job-seeking service providers for their trade in the coastal cities. However, he also points out that the Mafia was not a purely urban phenomenon but flourished especially in the nexus of town and countryside (Gambetta, 1993, ch. IV). We are here throughout in descriptions of crucial macro-level factors for the existence of the Mafia as a macro phenomenon, that means, features of the Sicilian economy on both parts of the causal relationship.

The peculiarity of Mafia protection is, again, that it cannot rely on a complete and undisputed monopoly in the use of force, as in state protection, but must compete. Several Mafia families are always active within a given terrain or market, so that the struggle for drawing boundaries is omnipresent, and even more so the more Mafia families are active. At the time of Gambetta's study, for example, there were five in New York City but many more in Palermo.

So, the Mafia families have their hands full trying to stay well in business. To ensure that their clients in a given territory or market indeed use their service of protection against fraud and competition and do not switch to competitors or protect themselves, they need reputation. This reputation they gain as owners of information about the market in general and about competitors, buyers, sellers, authorities, and the police in particular. They also need to be able to guarantee protection. To do so, they must protect a client's interest in the event of conflict, using force if necessary. And they must control their territory or market in such a way that other Mafia families cannot be called upon for protection. They must defend their territory or market against competing Mafia families, if necessary, by means of violence which must be used all the more the more contested a territory is and the less secure this makes the reputation of a particular Mafia family among its clients and its competitors. Frequent use of violence is always a sign of unsafe conditions in the demarcation of protected areas. In individual cases, it is a means of ensuring rule over an area and gaining or maintaining the corresponding reputation in return. Binding agreements and cartels are other means of administering a protected area and thus also securing reputation. Furthermore, Mafia families can enhance their reputation by contributing to festive events. Thus, a large part of the study is devoted to identifying the means that help secure reputation as a prerequisite for a Mafia family to remain in the business of supplying protection. Another part is to identify the structural conditions that contribute to clients seeking protection. These are a fundamental lack of trust across family boundaries, inadequate state protection or explicit denial of state protection in the case of illegal transactions, and self-produced mistrust in the system because dishonest transactions by protected sellers with unprotected buyers are not punished, all of which are macro phenomena for explaining the existence of the Mafia as a macro phenomenon (Gambetta, 1993, ch. I).

Of course, reference is always made to the actions of individuals. However, this is always done in connection with spelling out the configurations on the macro level, which lead rational actors in a certain direction in the sense of Weber's MI. The individual actions do contribute to the macro effect by adding an element to the counted frequency of that behavior in the social unit, and in the sense of Weber's MI they are constitutive elements of the macro phenomenon. The frequency of the behavior as a feature of the social unit is determined by the preceding structural feature of the social unit in connection with the motivation of individual action. Without this structural feature, an individual action does not yet give rise to another structural feature of the social unit that is to be explained. Accordingly, we can identify individual actions at the micro level that do contribute to a structural feature of a social unit at the macro level, but only because a specific relationship exists on the macro level. Without this relationship, individual action would not make any contribution to the specific structural feature of the social unit under study. Accordingly, the sociological explanation of a macro phenomenon requires knowledge of such direct structural dynamics on the macro level. They form an emergent ontological level of their own, which one cannot get into view based on the motivational situation on the individual level alone. Individual actions are involved in the constitution of macro phenomena, but they contribute to it only under the condition of cause–effect relations on the macro level. From a purely ontological point of view, the structural feature of the frequency of a certain action in a social unit would not exist without individual actions. But the individual action would not bring about a macro-level feature without the link between two configurations on the macro level. We can see this directly from a number of hypotheses about macro connections that underlie Gambetta's Mafia study. Some of the hypotheses implicitly used by Gambetta can be formulated as follows:

1. The less trust reaching beyond family ties exists and the less the state or a coordinated and assertive ruling class bridges the trust gap in economic

transactions by means of effectively enforced laws, and the more economic trade grows, especially between urban and rural areas, the more the demand for privately granted protection increases.

2. The supply of privately provided protection is more likely to emerge the more individuals are in search of profitable business and recognize the supply of protection as a business field, and the more individuals seeking employment are available to them as service providers.

3. The more fragmented privately provided protection is, the more frequently there is unpunished fraud and consequently further demand for protection.

4. The greater the number of private protection providers that compete for control of a protected area, the more frequently conflicts and the use of force occur.

5. The more monopolies or oligopolies of protection are formed as a result of elimination battles, or the more private providers of protection fix clear boundaries through agreements or cartels, the less frequently conflicts and use of force occur.

6. The more comprehensive and precise the information that protection providers have about their protected area and the private and state actors in that area, and the more securely they can provide protection, including the use of force, the greater their reputation and, as a result, the greater the demand for their protection.

7. The more firmly the Mafia members' vow of silence is institutionalized, the more difficult it is for the state to investigate the Mafia, and the more the Mafia keeps the business of providing protection in its own hands.

In all of these hypotheses, a relationship between two macro phenomena is asserted. Both phenomena describe characteristics of a social unit rather than characteristics of individuals. For example, the point of hypothesis 4 is that a greater number of Mafia families in city A than in city B will also result in a greater number of violent acts in A than in B as a macro phenomenon. The number of Mafia families in the same area determines the greater or lesser frequency of violent acts in that area, certainly in connection with the motivation of the actors involved. This frequency as a characteristic of a city as a macro phenomenon is of first interest here, the individual motivational situation underlying the individual acts of violence contributes to the link. It is, however, only of complementary interest. Applying Esser's FS model, the motivational situation looks like this for the Mafioso using violence:

1. Frame: unrestricted rule over an area is the prerequisite for permanently secured protection business.
2. Fact finding: Competitor X endangers my rule over my territory in the long run.
3. Conclusion: Competitor X must be eliminated.

This would be a microfoundation of hypothesis 4. A microfoundation of hypotheses 1 and 6 would read as follows:

1. Frame: To make secure profits from my business, I need to be protected against fraud.
2. Fact finding 1: I cannot trust everybody in my business.
3. Fact finding 2: The state administration does not guarantee protection from fraud.
4. Fact finding 3: There is a reputed private security agency offering effective protection against fraud.
5. Utility calculation: It is more profitable to buy protection from this agency than trusting state protection or buying alternative private protection or doing it by oneself.
6. Conclusion: I need to buy protection from this agency.

Such implicit microfoundations always run along with Gambetta's historical–sociological explanation of Mafia violence and of buying Mafia protection in business but need not be specifically addressed, because they are constitutive elements of the macro phenomena in the sense of Weber's MI and are immediately present to any understanding reader, especially since it is a logical conclusion. The point is that a particular characteristic of a social unit (the number of Mafia families or the presence of no effective state protection but instead an effective private provider of protection against fraud and competition in business in a city) determines another characteristic of the same social unit (the frequency of violent acts or the frequency of buying private protection in business in that city). Of course, the statement holds only ceteris paribus. The emergence of a cartel—organized by a commission—according to hypothesis 5 may bring the frequency of violent acts closer to the frequency in a city with only 5 Mafia families, even in a city with many more Mafia families, and effective state control may replace private protection in business. As in the case of Weber's study on ascetic Protestantism and the spirit of modern capitalism, a link between two phenomena on the macro level serves as a transformation rule, which turns individual acts of violence or of buying private protection in doing one's business into a macro phenomenon in the sense of an attribute of a social unit. Many families competing for profit from supplying protection simply makes the risk of being ousted from the market omnipresent in a social unit so that the frequency of violent acts as a strategy of coping with this risk rises in this social unit ceteris paribus. Lacking administrative control along with prevailing mistrust in transactions reaching beyond family and friends increases demand for private protection in a social unit which together with the existence of respective supply increases the frequency of buying private protection in this social unit.

Gambetta's study of the Sicilian Mafia is sociology as a science of reality by means of a case analysis in which the focus is on grasping the case in the first place, describing it in sociological terms, and explaining various aspects of the case by means of sociological hypotheses at the macro level, which are not tested or explicitly micro-founded but presupposed in their validity. This is not a deductive-nomological, hypothesis-testing approach, nor is it a sociological explanation based on RC/FS-MI, but rather a historical–sociological narrative of an individual case, well-structured by sociological knowledge and based on a Weberian type of MI. One can use parts of this narrative to study similar organized crime phenomena in other countries. However, they must be adapted to the particular case, as in analyses of the Russian Mafia or the Chinese Mafia (see Varese, 1994, 2001; Shvarts, 2002; Wang, 2011). Hence, there is something to be learned from Gambetta's study for other cases. But the general, transferable knowledge is cross-case sociological knowledge, which is not generated or tested at all in this case study, but simply used. We already know substantial parts of it from Machiavelli's (2014) analysis of the acquisition and maintenance of power in his study *The Prince*. Many good sociological studies show this character. They do not proceed deductive-nomologically to "test" hypotheses, nor do they proceed in the way of RC/FS-MI to explain social phenomena; rather, they use macro-sociological analysis to make visible a social reality that we know only insufficiently and tell an interesting story about it that is well supported by sociological knowledge, in the best case a story with cultural significance in the sense of Max Weber (1973 [1904]). Sociological research of this kind advances the discipline, is societally relevant, and is interesting for interdisciplinary work bordering on other disciplines. It is macro-sociological research in the tradition of a Weberian MI with staying power that cannot be replaced by short-winded, hypothesis-testing RC/FS-MI research according to highly standardized procedures, which certainly has its merits in itself.

5   Conclusion

There is a place for methodological individualism in historical–sociological explanation in Weber's sense and in an anti-reductionist sense (Bulle, 2019). However, there must be enough attention paid to the configurations on the macro level for explaining macro-level phenomena, certainly in connection with the motivation of individual action. In Boudon's outline of the "individualist tradition" in sociology, this is very clear (Boudon, 1987, 2001). My argument for special attention to the macro level is less of a doctrinal kind, but more of a practical one, aiming not at constructing an ideal methodological world but at real sociological research. This is why I made extensive use of two sociological studies to demonstrate what is good sociology. Too much preoccupation with looking for microfoundations of macro-sociological analyzes may occupy so much time and energy that there is not enough time and energy left for studying the crucial features of social units on the macro level. There is a danger that historical-sociological explanation following the model of Max Weber's study on ascetic Protestantism and the spirit of capitalism and sociological case studies following the model of Diego Gambetta's study of the Sicilian Mafia—both situated at the macro level, but in the tradition of a Weberian MI—will fall victim of a too narrowly understood RC/FS-MI if this kind of MI alone were to dominate the field. The essentially deductive-nomological, positivist-behavioral program of this very special kind of methodological individualism has insufficient sensitivity and space for the essential features of historical–sociological explanation, particularly attention to culture, social structure, and institutions as macro phenomena in themselves. Within the framework of such a program, it is not even possible to acquire the necessary competencies to conduct a historical–sociological investigation in Weber's sense or a case analysis in the manner of Gambetta's Mafia study. In addition, major paradigmatic traditions of sociology would at best be considered as a source of variables in the frame of reference of RC/FS-MI, but by far not in their own quality. This would not be the academic milieu in which sociology can flourish as a discipline, which provides rich insights in the working, conflicts, and changes of society.

References

Alexander, J. C., Giesen, B., Münch, R., & Smelser, N. J. (Eds.). (1987). *The micro-macro link*. University of California Press.

Albert, G. (2005). Moderater methodologischer Holismus. *Kölner Zeitschrift für Soziologie und Sozialpsychologie, 57*(3), 387–413.

Albert, G. (2007). Keines für alle! *Kölner Zeitschrift für Soziologie und Sozialpsychologie, 59*(2), 340–349.

Becker, S. O., & Woessmann, L. (2009). Was Weber wrong? A human capital theory of Protestant economic history. *The Quarterly Journal of Economics, 124*(2), 531–596.

Boudon, R. (1987). The individualistic tradition in sociology. In J. C. Alexander, B. Giesen, R. Münch, & N. J. Smelser (Eds.), *The micro-macro link* (pp. 45–70). University of California Press.

Boudon, R. (2001). Which rational action theory for future mainstream sociology: Methodological individualism or rational choice theory? *European Sociological Review, 17*(4), 451–457.

Bourdieu, P. (1984). *Distinction: A social critique of the judgement of taste.* Harvard University Press.

Bulle, N. (2019). Methodological individualism as anti-reductionism. *Journal of Classical Sociology, 19*(2), 161–184.

Chmielewski, A. K., & Reardon, S. F. (2016). Patterns of cross-national variation in the association between income and academic achievement. *AERA Open, 2*(3).

Coleman, J. S. (1990). *Foundations of social theory.* The Belknap Press of Harvard University Press.

Durkheim, E. (1982). *The rules of sociological method* (W. D. Halls, Trans., 1938). French original 1895. Free Press.

Durkheim, E. (2005). *Suicide: A study in sociology.* French original 1897. Routledge.

Esser, H. (1999). *Soziologie. Allgemeine Grundlagen.* Campus.

Esser, H. (2001). *Soziologie. Spezielle Grundlagen,* Vol. 6: *Sinn und Kultur.* Campus.

Esser, H. (2005). *Rationalität und Bindung. Das Modell der Frame-Selektion und die Erklärung normativen Handelns* (SFB 504, No. 05-16). https://ub-madoc.bib.uni-mannheim.de/2660/1/dp05_16.pdf. Accessed on 21 December 2021.

Esser, H. (2006). Eines für alle(s)? *Kölner Zeitschrift für Soziologie und Sozialpsychologie, 58*(2), 352–363.

Esser, H. (2010). Sinn, Kultur, Verstehen und das Modell der soziologischen Erklärung. In M. Wohlrab-Sahr (Ed.), *Kultursoziologie. Paradigmen – Methoden – Fragestellungen* (pp. 309–335). Verlag für Sozialwissenschaften.

Festinger, L. (1957). *A theory of cognitive dissonance.* Stanford University Press.

Gadamer, H.-G. (1960). *Wahrheit und Methode. Grundzüge einer philosophischen Hermeneutik.* Mohr Siebeck. (Gadamer, H. G. (2013 [1975]). *Truth and method* [J. Weinsheimer & D. G. Marshall, Eds.]. Bloomsbury).

Gambetta, D. (1993). *The Sicilian Mafia: The business of private protection.* Harvard University Press.

Godin, M., & Hindriks, J. (2018). An international comparison of school systems based on social mobility. *Economie Et Statistique/Economics and Statistics, 499,* 61–78.

Greshoff, R., & Schimank, U. (2012). Hartmut Essers integrative Sozialtheorie–Erklärungs-und Verstehenspotenziale. In W. Hansmann (Ed.), *Professionalisierung und Diagnosekompetenz* (Vol. II.2, pp. 1–26). Universität Marburg. http://archiv.ub.uni-marburg.de/es/2012/0003/pdf/II.2_Greshoff_Schimank_Esser.pdf. Accessed on 16 July 2018.

Greve, J., Schnabel, A., & Schützeichel, R. (Eds.). (2008). *Das Mikro-Makro-Modell der soziologischen Erklärung*. VS Verlag für Sozialwissenschaften.

Heath, J. (2005). Methodological individualism. In *Stanford encyclopedia of philosophy*. Stanford University, Department of Philosophy. https://plato.stanford.edu/entries/methodological-individualism/. Accessed 13 July 2022.

Heintz, B. (2004). Emergenz und Reduktion: neue Perspektiven auf das Mikro-Makro-Problem. *Kölner Zeitschrift für Soziologie und Sozialpsychologie, 56*(1), 1–31.

Hempel, C., & Oppenheim, P. (1948). Studies in the logic of explanation. *Philosophy of Science, 15*(2), 135–175.

Jepperson, R., & Meyer, J. W. (2011). Multiple levels of analysis and the limitations of methodological individualisms. *Sociological Theory, 29*(1), 54–73.

Machiavelli, N. (2014). *The prince and other writings*. Italian original 1513. Simon and Schuster.

March, J. G. (1978). Bounded rationality, ambiguity, and the engineering of choice. *The Bell Journal of Economics, 9*(2), 587–608.

Münch, R. (1973). Soziologische Theorie und historische Erklärung/Sociological Theory and Historical Explanation. *Zeitschrift für Soziologie, 2*(2), 163–181.

Münch, R. (1983). From pure methodological individualism to poor sociological utilitarianism: A critique of an avoidable alliance. *Canadian Journal of Sociology/cahiers Canadiens De Sociologie, 8*(1), 45–77.

Münch, R. (1986). *Die Kultur der Moderne* (2 Vols.). Suhrkamp.

Münch, R. (1987). The interpenetration of microinteraction and macrostructures in a complex and contingent institutional order. In J. C. Alexander, B. Giesen, R. Münch, & N. J. Smelser (Eds.), *The micro- macro link* (pp. 319–336). University of California Press.

Münch, R. (2001). *The ethics of modernity*. Rowman & Littlefield.

Münch, R., & Smelser, N. J. (1987). Relating the micro and macro. In J. C. Alexander, B. Giesen, R. Münch, & N. J. Smelser (Eds.), *The micro- macro link* (pp. 356–388). University of California Press.

Muench, R., & Wieczorek, O. (2022). In search of quality and equity: The United Kingdom and Germany in the struggle for PISA scores. *International Journal of Educational Research Open, 3*, 100165.

Muench, R., Wieczorek, O., & Dressler, J. (2022). Equity lost: Sweden and Finland in the struggle for PISA scores. *European Educational Research Journal, 22*(3), 413–432.

Needham, J. (2013). *The grand titration: Science and society in East and West*. Routledge.

Schluchter, W., & Graf, F. W. (Eds.). (2005). *Asketischer Protestantismus und der "Geist" des modernen Kapitalismus: Max Weber und Ernst Troeltsch.* Mohr Siebeck.

Simon, H. A. (1979). Rational decision making in business organizations. *The American Economic Review, 69*(4), 493–513.

Shvarts, A. (2002). Russian Mafia: The explanatory power of rational choice theory. *International Review of Modern Sociology, 30*(1/2), 69–113.

Tyrell, H. (1990). Worum geht es in der, Protestantischen Ethik'? Ein Versuch zum besseren Verständnis Max Webers. *Saeculum, 41*(2), 130–178.

Udehn, L. (2002). The changing face of methodological individualism. *Annual Review of Sociology, 28*(1), 479–507.

Varese, F. (1994). Is Sicily the future of Russia? Private protection and the rise of the Russian Mafia. *European Journal of Sociology, 35*(2), 224–258.

Varese, F. (2001). *The Russian Mafia: Private protection in a new market economy.* Oxford University Press.

Wang, P. (2011). The Chinese Mafia: Private protection in a socialist market economy. *Global Crime, 12*(4), 290–311.

Weber, M. (1924). *Wirtschaftsgeschichte. Aus den nachgelassenen Vorlesungen* (S. Hellmann & M. Palyi, Eds.). Duncker & Humblot. (Weber, M. (2017 [1981]). *General economic history* [I. J. Cohen & F. H. Knight, Eds.]. Routledge).

Weber, M. (1971 [1920]). *Gesammelte Aufsätze zur Religionssoziologie* (Vol. 3). Mohr Siebeck. (Weber, M. (2010 [1952]). *Ancient Judaism* [H. H. Gerth & D. Martindale, Eds.]. Simon and Schuster).

Weber, M. (1972a [1920]). *Gesammelte Aufsätze zur Religionssoziologie* (Vol. 1). Mohr Siebeck. (Weber, M. (2013 [2001]). *The Protestant ethic and the spirit of capitalism* [S. Kalberg, Ed.]. Routledge. Weber, M. (1951). *The religion of China* [H. H. Gerth, Ed.]. The Free Press).

Weber, M. (1972b [1920]). *Gesammelte Aufsätze zur Religionssoziologie* (Vol. 2). Mohr Siebeck. (Weber, M. (1967). *The religion of India: The sociology of Hinduism and Buddhism* [H. H. Gerth, Ed.]. Free Press).

Weber, M. (1973 [1904]). Die ,Objektivität' sozialwissenschaftlicher und sozialpolitischer Erkenntnis. In M. Weber, *Gesammelte Aufsätze zur Wissenschaftslehre* (pp. 146–214). Mohr Siebeck. (Weber, M. (2011). "Objectivity" in social science and social policy. In M. Weber, *Methodology of social sciences* [E. A. Shils & H. A. Finch, Eds.]. Routledge).

Weber, M. (1976 [1922]). *Wirtschaft und Gesellschaft.* Mohr Siebeck. (Weber, M. (1978). *Economy and society: An outline of interpretive sociology* [G. Roth & C. Wittich]. University of California Press).

Interactionism and Methodological Individualism: Affinities and Critical Issues

Natalia Ruiz-Junco, Daniel R. Morrison, and Patrick J. W. McGinty

1 Introduction

Once a challenge to dominant approaches in the discipline, Interactionism has been a major perspective in sociology for several decades, having achieved "triumphant" status in the early 1990s as key interactionist ideas became accepted by the larger discipline (Fine, 1993). Today, interactionism is a vibrant and multifaceted perspective, with several approaches for social research (vom Lehn et al., 2021). Despite this success, it has received criticism for not addressing major concerns of mainstream sociological practice (Maines, 2001), especially for providing a microsociological perspective that ignores the macro level of analysis.

Similarly, Methodological Individualism (MI) has been a strong perspective in sociology for decades (see, e.g., Agassi, 1960; Coleman, 1986), and this continues to be the case, as this Handbook's contributions demonstrate. However, like Interactionism, it has also received criticism for its focus

N. Ruiz-Junco (✉)
Sociology, Auburn University, Auburn, AL, USA
e-mail: ncr0007@auburn.edu

D. R. Morrison
Sociology, University of Alabama in Huntsville, Huntsville, AL, USA

P. J. W. McGinty
Sociology, Western Illinois University, Macomb, IL, USA

© The Author(s), under exclusive license to Springer Nature Switzerland AG 2023
N. Bulle and F. Di Iorio (eds.), *The Palgrave Handbook of Methodological Individualism*,
https://doi.org/10.1007/978-3-031-41508-1_23

552   N. Ruiz-Junco et al.

on the micro level, and for overlooking the macrosociological dimensions that constitute the major preoccupation of mainstream sociology. In fact, the *common* criticism that both perspectives have received is that they are "microsociological reductionists."

Our chapter's purpose is to address this common criticism that both MI and Interactionism, two seemingly different perspectives, have received. Our argument is organized in three parts. First, we discuss the affinities between MI and Interactionism; in particular, we consider the classic formulation provided by Blumer, and its connection to the ideas of two founders of MI—Max Weber and Georg Simmel. We argue that both perspectives have similarities that commentators frequently overlook, despite the common criticism launched against both perspectives.

Second, we tackle this common criticism by engaging Interactionism because it is our own research tradition. We trace the origin of this controversy to Blumer's rendition of Mead's theory.[1] We contend that the interactionist tradition has relied heavily on the intellectual blueprint that Blumer created for the perspective. This has had the consequence of making Interactionism particularly vulnerable to the charge of microsociological reductionism. While we believe that this charge against Interactionism is unfair and unfounded, we suggest that Interactionism can address the common criticism by fully engaging with Pragmatism, a kin perspective, and the original theoretical inspiration for Blumer's Symbolic Interactionism.

In the last part of the chapter, we present three core ideas that begin to address the common criticism within Interactionism; in doing so, we propose a preliminary framework we call *Pragmatic Interactionism* (PI). Pragmatic Interactionism builds on past attempts at pragmatizing Interactionism, but also on new scholarship rediscovering the relevance of Pragmatism for today's social theory (Côté, 2015; Huebner, 2014; Joas, 1985; Joas & Huebner, 2016; McPhail & Rexroat, 1979; Shalin, 1986; Snow, 2001). PI responds to the common criticism, because: (1) it is interactionist, and representative of a sensitizing orientation to social phenomena (Blumer, 1969); (2) provides three core ideas—problem-solving activity; human agency as creative constraint; and multidimensional sociality—rooted in several robust areas of research within the interactionist tradition (see McGinty, 2014); and (3)

---

[1] The thought of G. H. Mead is the main foundational inspiration for this chapter, but it is not the only conceivable one. In later phases of theoretical development, we plan to incorporate in our framework the oeuvres of important classic pragmatists, who have been uneven sources for theoretical inspiration in interactionism, and even the work of other classic social theorists, such as Charles H. Cooley, whose relations to interactionism and pragmatism have been recently the object of scholarly interest (Brossard & Ruiz-Junco, 2020; Ruiz-Junco & Brossard, 2019).

moves Interactionism forward by mediating the space between our tradition and pragmatist approaches.

2 Interactionism and Methodological Individualism

Interactionism and Methodological Individualism are often presented as opposite perspectives. In Sociology, they tend to be perceived to be on opposing sides of the debate on the philosophy of knowledge; many identify Methodological Individualism with positivist approaches, and Interactionism with interpretive ones. However, this perception can lead to incorrect conclusions about the relationship between the two. MI and Interactionism have similarities that are not frequently commented on. In what follows, we outline the similarities between MI and Interactionism through a discussion of the relationship between two founders of MI—Georg Simmel and Max Weber—and the interactionist tradition.

To begin, Simmel and Weber are influences for Interactionism, but they are not influences in the same way. While the direct influence that Simmel has had, and continues to have, on Interactionism is well-known, the same cannot be said about Weber's. We first consider Simmel's influence.

Simmel's shadow has loomed large throughout Interactionism's history. The leader of the Chicago School, Robert Park, was highly influenced by Simmel (see Lengermann, 1988). Park discovered Simmel when he visited Germany. During his visit, he attended Simmel's classes; for Park, the impact of this intellectual interaction with Simmel lasted for decades. Park was drawn to Simmel's analytic focus to study social life. In fact, as it is frequently commented on, one only needs to browse the "green bible" of the Chicago School, the famous *Introduction to the Science of Society* (1921), which Robert Park edited alongside Ernest Burgess, to recognize the strength of Simmel's influence on Park.

However, Simmel's influence on Interactionism is not restricted to his direct influence on Park. Simmel's inspiration runs deeper than that. Simmel's proposal to study social forms, and social types, aligned well not only with the Chicago School's interest in the microsociological dynamics of human life, but also with the larger interactionist agenda. Simmel's analytic focus on the structural features of social processes and social interactions has attracted generations of interactionists. Indeed, interactionists continue to forge ideas on the "shoulders of Simmel" (see, e.g., Ruiz-Junco, 2017; Zerubavel, 2007, 2021).

Weber's influence on Interactionism is another story. Weber's theory constitutes an important inspiration for Interactionism; however, Weberian influences in the interactionist tradition have not been frequently acknowledged by commentators in the past. This is puzzling, since any observer can quickly attest to the similarities that exist between some of Weber's ideas and interactionist perspectives. To name a few, Weber called his own sociology "interpretive sociology," being a forceful defender of *Verstehen* as a method. He made the concept of action central to his sociology, which he defined as "a science concerning itself with the interpretive understanding of social action…" adding to his definition of social action the central component of subjective meaning (Weber, 1978, p. 4). "Action is 'social' insofar as its subjective meaning takes account of the behavior of others and is thereby oriented in its course" (Weber, 1978, p. 4).

Even though these similarities are undeniably present, scholars have debated the question of Weber's influence on Interactionism for decades (see, e.g., Kivisto & Swatos, 1990; Segre, 2014). This scholarship affirms a certain Weberian influence in Interactionism. This influence has been thought to be somewhat negligible during the First Chicago School, but with the Second Chicago School onward, there is an understanding that things start to change; for instance, Howard Becker was deeply influenced by Weber, and engaged his ideas unmistakably (Kivisto & Swatos, 1990). Despite this, some argue that Weber's reception among interactionists has been "selective and inadequate" (Segre, 2014, p. 474).

The similarities between the founders of MI and Interactionism suggest that the affinities may have been deep on a conceptual level, even if they were not always consciously recognized by authors, or explicitly acknowledged on the page, in the form of citations, homages, and other forms of intellectual lineage. Let us now move on to consider how the main premises of symbolic interactionism, the classic formulation of Interactionism, resemble some ideas from Simmel and Weber.

Herbert Blumer (1969, p. 2) defined his famous three premises of symbolic interactionism as follows:

> The first premise is that human beings act toward things on the basis of the meanings that the things have for them (…) The second premise is that the meaning of such things is derived from or arises out of, the social interaction that one has with one's fellows. The third premise is that these meanings are handled in, and modified through, an interpretive process used by the person in dealing with the things he [sic] encounters.

As can be gleaned from the classic premises, Blumer had several ideas in common with the founders of MI. Let us discuss this premise by premise. The first premise reflects a deep affinity with both founders of MI. For the symbolic interactionist, meaning guides action. Here, we find a parallel with Weber's interpretive sociology, and its emphasis on the meaning from the point of view of the actor within a cultural context in which meanings are embedded in social relationships and connected to action orientations. Later interactionist scholarship that revolves around group life, and the creation of meaning, has much in common with this Weberian focus. Examples can easily be found in the work of major interactionists such as Fine (see, e.g., Fine, 2001, 2007).

The second premise allows Blumer to identify the processes by which "meaning" is created. Like Weber, he wasn't interested in meaning as a philosophical issue, but rather, as a sociological one. Drawing on pragmatism, Blumer construed the meaning of something as an indication that something has a purpose in the social world. The purpose of a thing can only be learned by interacting with others. Thus, without human communication, there is no meaning. More importantly, without meaning, there is no complex group life, since people need meaning to be able to join lines of action. Joined lines of action, in turn, are the core of more complex patterns of group life. In other words, Blumer, like other interactionists, takes this premise to mean both that the source of meaning is socialization through social interaction, and that this meaning-making is recreated in successive social interactions.

In addition, the second premise aligns with Simmel's sociology. Like Simmel, interactionists study meaning in relation to ongoing interactions, which they view as situations in which people find themselves as they go about the making of their lives. The parallel between Simmel and Interactionism on this point is hard to miss when one considers Simmel's interest in exploring what he called "forms of social life," and the impact that "the mere *number* [emphasis in original] of sociated individuals has upon these forms of social life" (Simmel, 1950 [1908], p. 87). The Simmelian argument that the number of individuals in a particular type of group shapes the possibilities for meaning-making within the group is not lost on interactionists. Interactionists recognize that group size matters for the meaning-making process in social interaction. Indeed, in a two-person social unit, or dyad, the two members of the group adopt identities and roles that are influenced by the smallness of the group.

More importantly, as we look forward to the state of contemporary sociology, today's interactionists, as well as Methodological Individualists inspired by Simmel and Weber find common ground when they concern themselves with the organized patterns of action that ongoing negotiations at the interactional level lead up to. This is how Segre (2014, p. 479) put it: "For Weber, as for Symbolic Interactionism, society is viewed accordingly as a negotiated order that results from relations between individuals and between groups who pursue their interest and have distinct social statuses and identities."

Finally, the third premise invites us to incorporate the issue of agency and reflexivity into social analysis. This premise is one of the major achievements of the interactionist tradition, even if it appears in embryonic form in Blumer's original formulation. Rather than positing a deterministic outlook in social life, the premise narrows the distance between the emphases of macrosociologists and microsociologists. It is perhaps through the extension of this premise that interactionists may complement the best Simmelian and Weberian-inspired analyzes as they work to uncover the processes underlying small-scale patterns of social life.

In sum, Interactionism and MI have common implicit and explicit "forebears." While MI embraces them, Interactionism has had a more complicated lineage relationship with them, especially with Weber. Nevertheless, we have shown that there are clear *conceptual* affinities between the perspectives of the MI founders and Interactionism. These are important, we contend, because they help us address the *common* criticism earlier identified.

Even though in the following sections we tackle this criticism within the tradition of Interactionism, and we do not consider how this criticism specifically applies to MI, we believe that the way in which we respond to this criticism is useful for MI as well, as the common charge affects both perspectives. After all, the solutions that Interactionism finds for this controversy are solutions that could potentially be adapted by proponents of MI.

3   Tackling a Common Criticism

In this section, we begin to address the common criticism from the perspective of Interactionism. We choose to respond to it taking into consideration Interactionism, because this is the tradition we belong to, and we do so through an engagement with Pragmatism, which is a kin perspective. As it is well-known, Herbert Blumer coined the term "symbolic interactionism" in 1937, and founded the perspective based on his own reading of the pragmatist philosopher G. H. Mead's social theory. In other words, he created symbolic interactionism "on the shoulders of Mead."

For the purposes of understanding this criticism, it is important to begin by considering the context in which Herbert Blumer positioned SI within the field of sociology. The "positioning" of a theory refers to the processes by which theorists conceive of their intellectual field, and, in turn, how they define themselves, and their work, in relation to other theorists and perspectives in that field; this is a crucial factor in deciding the trajectories of intellectual ideas and their larger intellectual movements (Baert, 2015).

Blumer positioned SI against Parsonian functionalism, which he interpreted as a perspective that occupied itself with macrosociological and abstract social phenomena conceived to be overdetermining individual action. Thus, Blumer's positioning of SI as a perspective concerned with individuals' lines of action, rather than with macro-sociological structures or an "oversocialized" (Wrong, 1961) view of the actor, had an enormous impact not only on the launching of the common criticism earlier discussed, but also on the framing of Pragmatism and Interactionism within the larger discipline of sociology.

Our argument is that Blumer's positioning of SI in this intellectual context facilitated the charge of microsociological reductionism against the perspective. Moreover, while the premises have some undeniable qualities, which we have previously stated here, it is surprising that they have not changed, nor been significantly altered, even though many have called for their alteration throughout the years (Fine & Tavory, 2019; McPhail & Rexroat, 1979; Snow, 2001).

Part of the explanation for this is that the premises have come to act as symbols of Interactionism endowed with a significant load of what we might call the perspective's *representational capital*. That is, the premises came to represent Interactionism in the intellectual field, and due to the weight of their representational capital, they were more likely to be used in research and publication than other components of interactionist theory. We hypothesize that, by employing the premises for representational purposes, scholars signaled their belonging to Interactionism as a theoretical group but did not necessarily produce substantive engagement with the theory (see Brossard & Ruiz-Junco, 2020).

Moreover, the unintended consequence of the overuse of Blumer's three famous premises, has been that the premises, given their symbolic nature, have come to limit Interactionism's possibilities for development, narrowing it not only to its dominant version (Blumerian SI), but also to its reified version (to the detriment of theoretical innovation). That is, as the representational capital of Blumer's premises rose, the premises came to "mark" symbolic interactionism, and did so in powerful ways; this process in turn relegated other possibilities for interactionist development to the arena of the unremarkable (Zerubavel, 2018).

Indeed, the story of the "unremarkable" currents of SI is extensive. That Blumer was selective in his interpretation of G. H. Mead, and in using Mead to build the interactionist project, has been perhaps the most discussed, most divisive, most commonly accepted—yet until recently unexplored—truism in contemporary Interactionism. Blumer's positioning vis-à-vis "structural" sociological approaches is also one of the topics in the larger debate within the discipline about Interactionism's presumed astructural bias (Maines, 1989; McGinty, 2016; Ruiz-Junco, 2016).

We take Blumer's selective act of interpreting Mead (and marking certain concepts or features) as being born out of the search for a solution for a particular problem in a particular context, as Neil Gross (2018) notes when comparing Blumer to Parsons. Using the language of positioning theory previously discussed, Blumer utilized Mead's theory to create an alternative to the dominant perspective at the time, Parsonian functionalism; however, this alternative was conditioned and influenced by the rejection of parsonianism. This and other limitations and conditionings of Blumer's interpretation of Mead, as evinced by the reception of the Blumerian premises, have been examined in the literature, and they will not be discussed extensively in this chapter.

Several authors have worked on extracting Mead from the historical influence of SI. Côté (2015), Huebner (2014), and other recent contributors (e.g., Joas & Huebner, 2016) demonstrate that Mead's work provides a dynamic and yet consistent base from which to promote contemporary theorizing. Together, these authors show that recasting, transforming, and reconstructing Interactionism on the basis of an engagement with its pragmatist roots is possible.

Our work in this area follows the direction by one of these scholars, Dmitri Shalin (1986), who made a call to better understand, articulate, and strengthen the pragmatist roots of Interactionism. Shalin (1986) suggested that scholars should incorporate the diverse elements of Mead's pragmatist thinking which were obscured by Blumer's canonical writing. Shalin argued (1986, p. 26):

> interactionists need to reexamine the origins of interactionist sociology, its metatheoretical foundation - pragmatist philosophy, and particularly the notion of the indeterminancy of the situation. If interactionists are serious about tackling the issues of power, class, and inequality, they will also have to reclaim the political commitments of their pragmatist predecessors.

Interactionism and Methodological Individualism ...   559

Shalin's early agenda has only been partly fulfilled (Côté, 2015; Joas, 1993, 1996, 1997; Joas & Huebner, 2016; Mead, 2015). Our framework, Pragmatic Interactionism, constitutes a further step in responding to this call for the renewal of Interactionism.

4   The Core Ideas of Pragmatic Interactionism

In this section, we address the *common* criticism surrounding both MI and Interactionism by outlining three core ideas of PI that begin to address these critical issues. In our elaboration of these ideas, we acknowledge our relationship to the pragmatist literature that extracts Mead from the limiting influence of Blumer's adoption of his ideas (Côté, 2015; Huebner, 2014; Joas & Huebner, 2016). In addition, our core ideas draw on other interactionist literature, including: the New Iowa School of Symbolic Interaction, Straussian concerns regarding action and interaction, and the study of social organization within the interactionist tradition.[2]

Before we introduce the new core ideas, it is important to note that Pragmatic Interactionism does not deny the importance and significance of symbols and meaning. Instead, we assert that the processually organized, embodied, practical act—which includes meaning—is our primary focus of attention. That is, PI extends Blumer's legacy (1969) and draws from several interactionist research traditions. In support of our argument, we offer brief definitions and elaborations of the implications of the core ideas for the study of social action. In this elaboration, we begin by emphasizing the need to theorize individual action in conjunction with its social interactional elements.

---

[2] The core ideas guiding our perspective follow from our analysis of three traditions in interactionism: the New Iowa School, Anselm Strauss' work on action, and the study of social organization. Here, we offer a brief and necessarily incomplete review of the main "take-aways" from these traditions. From the New Iowa School, we take a focus on social processes in context (Buban, 1986; Couch, 1989; Littrell, 1997; Miller & Hintz, 1997; Miller et al., 1975; Travisano, 1975). From Strauss' *Continual Permutations of Action* (1993), we take that interactionism should be a theory of action/interaction (see also Strauss, 1991). From the social organization paradigm, we take the negotiated order concept, as well as this tradition's focus on the mesodomain (Hall, 1987, 1995; Hall & McGinty, 1997, 2002; Maines, 1977, 1982), meta-power (Burns & Hall, 2013; Burns et al., 2013; Hall, 1997) and the need to highlight the processual nature of social action in organizational contexts (Hallett, 2010, Hallett & Ventresca, 2006).

First Core Idea

> Social action typically involves *problem-solving* activity, developed in and beyond social interaction.

The essential Blumerian take on the act—an idea he adapted from Mead—notes that humans act toward things on the basis of the meanings ascribed to them, that meanings arise out of social interaction, and that those meanings are modified through an interpretive process. Blumer's perspective is improved here by recognizing that social action, as oriented by consciousness, involves the definition of problems, and the potential for their solution. This does not mean that all human action is conscious nor that all conscious action is oriented toward the definition of problems.

The idea that social action is typically "problem-solving" activity is an addition to the interactionist understanding of the act. We use the "problem-solving" expression, characteristic of pragmatist theory, to signify not individually defined "problems" but rather a processual and sometimes organizational articulation of potential forms of acting defined within the interactional and extra-interactional context. In recasting this process, our framework contends that humans act toward things on the basis of the meanings of those things couched within the actor's (ever-changing) intentionality, or intended use. While this may seem provocative, our view of the act is derived from Hans Joas, who noted, "… we find our purposes or ends in the world, and we are practically embedded into the world in every intentional act" (as quoted in Shalin, 1986, p. 13). Dmitri Shalin (1986, fn. 1, p. 13) qualified this statement, reminding us that the intentional act is dialectically related to the situation. That is, based on social interaction with others, intended uses might be individualized, but they could be, or become, collective, consensual, and coordinated, or just as easily conflictual.

Thus, the meanings of objects emerge from ways of acting toward, handling, sensing, and receiving anticipated responses from things. Meaning can therefore be understood as a telescoping act (Strauss, 1993)—extending outward where responses from others/objects can change and alter meaning. In this way, objects become multivocal—their meaning and significance varies by use and the intentionalities and perspectives of other actors within situational contexts that are always necessarily temporally and spatially defined. Furthermore, meanings also depend upon responses by the objects—the completion of the act, resistance, surprise, or some other form of the unexpected. Indeed, the unexpected is to be understood as a common experiential factor that motivates actors' adaptations, and creative responses to situations.

Interactionism and Methodological Individualism ... 561

More importantly, this reconceptualization of the act, leads us to suggest a broader notion of *social action as action-interaction*, the core idea behind *any and all forms of Interactionism*. This notion is inspired by Anselm Strauss and his pragmatist concept of action/interaction. For Strauss, a theory of action is best understood as the theorizing of *acting* (1993). Such a theory, Strauss writes, "… can help guide us to informed observations and reflections about action, whether individual or collective" (1993, p. 49).[3]

From a PI standpoint, interaction is not a narrowly defined social-psychological concept, but rather a multidimensional idea, the culmination of a long tradition of reflection on the interconnection between self and society. Following from this insight, humans are *interactional actors*: *active, engaged*, pursuing their projects through interaction, stymied by the unexpected, called upon to reflect upon and assess the effects of their actions, frequently forgetting that they are the result of interactions, revising and developing ongoing projects as a result. Of course, routine activities, organizational practices, or habitual behavior are frequently disrupted, thus forcing people to reflect on their interactions (as well as their intentions) and chart new individual and collective paths and new intentions.

This is not to suggest that actors do not experience systemic forms of social inequality. Our pragmatic interactionist framework takes as a starting point that existing social inequalities prevent people from fully experiencing the potential inscribed in their capacity for creative action, and from accessing needed resources in their everyday lives to tackle oppressive conditions of living. The organizing processes that individual and collective actors engage in to mobilize resources (of different types) to accomplish their purposes are imagined (Cooley, 1992 [1902]) and real, as are their effects. However, they are not a thing apart from the conditions—interactions—consequences chain; they are merely rooted in places and times not always immediately available to the impacted actor(s). As a result, while the effects of interactions, including the effects that the actions of others have, can be felt by the actors, the capacity to act in ways to overcome oppressive organizing processes is sometimes achieved, sometimes not.

---

[3] Discussing trajectory, work, embodiment, thought processes, symbolizing, representation(s), and social worlds, Strauss follows Chicago pragmatists in rejecting false dichotomies and advancing ideas that help us make sense of social life across levels of analysis. For a recent discussion of the Straussian framework, see Clarke (2021).

Empirical Example

Consider an office worker whose routine includes packing herself a lunch. Before work, she rises, going through the typical workday tasks of readying herself for work and her children for school. As always, she prepares lunch and stores it in several plastic tubs, which go in her grown-up lunch box. She packs for the kids as well. Then, she rushes to drop off children at daycare and begins the twenty-five-mile commute from her suburban home to her downtown office building. On the way, she realizes that she has failed to move her lunch from the kitchen countertop to her car. As an hourly worker, she cannot afford to return home before clocking-in, and has a standing morning meeting with her supervisor in any case. She also cannot afford a long lunch break to return home, rush through lunch, and return on time.

As an embodied creature in our culture, however, she still needs to eat. How will she solve this problem? Finding a solution involves coordinating a number of human and nonhuman actors, including the money system. The constraints she encounters extend to the options available for purchasing lunch, such as the dining options within easy proximity to the workplace. Money that might have been spent on an afternoon coffee or treat is suddenly more significant. Will it be enough for a lunchtime meal? Are any food items easily accessible, left over from past meals, or otherwise available, such as from a personal "snack drawer?" If the worker does have some disposable income at hand, the problem can usually be solved easily enough. Yet, as recent supply chain disruptions and shortages show us, contingency, interruption, and surprise are constant features of social action that is interaction. This empirical example thus illustrates how social action involves problem-solving activity, which is developed through action-interaction, i.e., in social interaction with others (with the store or restaurant worker, with the office mate, etc.) and beyond immediate social interaction (through employment structures and regulations and through economic exchanges involving a wide variety of actors).

To conclude, our theoretical framework proposes the claim that social action as action-interaction is socially processual, embedded in unfinished, discontinued, and sometimes completed activities, involving different actors—whether these actors are present in the consciousness of the other actors or not—and forms of action-interaction across space and time. The way in which this happens—where and when it happens, how it happens, who makes it happen, and why—is suggested by the next two core ideas.

Second Core Idea

> The spatial/physical and social organization of the world we inhabit is constrained by the expression of human agency toward a projected future, which itself is constrained by the shaping of opportunities for agentic expression in the present by the past.

Human activity always takes place within more or less perceptually bounded settings (temporally, spatially, and physically). Such settings influence both the definition of the situation as well as the manner in which actors experience each setting. As a result, particular experiences are shaped by these settings. These settings, including both "natural" and socially designed environments, are interacted with relative to their intentional use; importantly, they are also socially organized, for instance, designed by planners, built by construction workers, used and inhabited by human and even nonhuman actors. Thus, settings for human interaction are socially conditioned spaces and are part of the conditions for action-interaction.

Following from the former idea, we contend that our sociality is premised on our "working out" of our experience in these settings. The creative, active, dynamic, sensemaking human actor is acting their way through the practical accomplishments of life, but always in a world that is socially conditioned. Based on Mead's understanding of the act, we posit that "the act" sometimes reaches a culminating climax, but recognize that social life is full of instances in which acts are only imagined, impaired, disrupted by social factors that are beyond the actor's capacity to act in desired ways.

This core idea is partly based on, but also extends, mesodomain analysis and the study of social organization (Burns & Hall, 2013; Burns et al., 2013; Hall & McGinty, 2002), the New Iowa School,[4] and the "inhabited institutions" approach.[5] The mesodomain approach features five concepts: collective activity, network, conventions-practices, processuality-temporality, and grounding (Hall, 1987). For Hall, mesodomain analysis helps us understand, "… how past and contemporary social forces… shape situated activity"

---

[4] The New Iowa School elaborates Mead's concern with the temporality of human action that involves the coordination of implied and projected pasts and futures (Mead, 1932; Couch, 1989). In line with this, Steven Buban suggests, "… society in all its component parts is conceptualized not as a thing but as an intertwined set of events, an ongoing accomplishment of persons acting in concert" (1986, p. 35). Other scholars in this tradition identify the structure and temporality of social interaction and the conditions for the possibility of momentary or ongoing interaction (Miller & Hintz, 1997).

[5] The "inhabited institutions" (Hallett, 2010; Hallett & Ventresca, 2006; Hallett et al., 2009) approach demonstrates how processual action and generic social processes work in relation to organizational/institutional rituals and practices.

564 N. Ruiz-Junco et al.

and "how social organization emerges from collective activity/social action, how it extends across space and time, and how the units and levels fit together" (1995, p. 399).

Empirical Example

Recent work from Hirsch and Khan (2020) demonstrates the significance of the built environment on college campuses, as well as university-controlled policies and procedures, in organizing sexual assault. For example, the residence hall room. Often, such spaces are sparsely furnished, including one or more beds, one or more desks with desk chairs, and perhaps a small refrigerator. These authors argue that when residents invite others to their rooms, whether for friendly conversation or as a part of relationship formation, they face a dilemma built into the way the room was designed by past agents. The room and its furnishings have been designed to enable a certain set of activities: sleeping and studying, not "hanging out." They must decide whether to sit on a business-like chair associated with academic work, or on a bed. Sitting on the bed, while it is likely more comfortable, also enables a certain kind of intimacy associated with sharing a sleeping space, and thus generates the potential for unwanted sexual activity.

Consistent with our second tenet, Hirsch and Khan take a social ecological approach to campus design, arguing that spatial arrangement and campus policies create opportunities for fun, sexual pleasure, and also abuse. The social organization of the Columbia campus and its role in facilitating rape culture was dramatically illustrated by Emma Sulkowicz, whose senior art thesis, "Carry That Weight" indicted the university's flawed handling of her rape, which occurred at a Columbia University residence hall room (Izadi, 2015). When our core idea is combined as part of a pragmatic interactionist framework, the organizational aspect described in this particular example needs to be placed in relation to the first core idea. That is, the actors who interact in the campus environment not only are conditioned by the organizational aspects of their built environment, but also, and significantly, by how they engage in the definition of their action as part of "problem-solving" activities: in this case, a dichotomous understanding of the act as "sex" or "conversation" could be an object of negotiation or it could not; due to the processual nature of action-interaction, it is frequently not negotiated, because human actors exist in a system of power structures that condition and structure, but also enable them to act in ways that may or may not align with the interpretations of other actors involved. When there is no alignment in this type of situation in question, and no successful negotiation of the definition of the situation, there can be misunderstanding, abuse and the use of force.

To summarize, our pragmatic interactionist framework recognizes that the actions of people and groups are conditioned by, and responsive to, understandings of time, history, and subjective experience. In our view, each person holds a stock of both explicit, tacit, and thoroughly social knowledge of their life trajectory and an action repertoire built from the experienced past, which constitutes the basis for their expression of human agency. Furthermore, we agree with Eviatar Zerubavel (see, e.g., 1985, 2009) that our understanding of time itself is a social product and subject to change across time and place.

Third Core Idea

The human actor is never alone; emotion, affect, cognition and language all imply a sociality of past, present, and future that is available through embodied experience.

The existence of human emotion, affect, cognition and language means that the human actor is never alone. While Blumer understands this important point, he focuses on language as the symbolic root of the act. We instead assert that this is a limited understanding of the act. In what follows, we outline why.

First, and consistent with interactionist theory generally, we recognize the significance of human language as one significant symbolic system. Indeed, written, spoken, and enacted language (even body language) are essential carriers of meaning for individuals and social groups. Most of our sociality emerges from the structure of language in its many forms. Human language is important in extreme cases of social isolation (say, locked-in syndrome) and in mundane cases (say, communication with a partner). Clearly, in many social situations, the desire to use, receive, and enjoy the use of language as a communicative resource is activated, and the practice accomplished. However, language is not the only way in which we communicate with others. Language is not equivalent to the extent of human experience. As the example of human infants and nonhuman animals shows, human language is not the only way by which we develop a sense of self, and communicate with others (Irvine, 2021).

Human language functions as a tool (even if a highly sophisticated one) to aid in certain past, present, and future interactions. As a tool to aid in interaction, language expresses thinking, imagining, memory in ways previously unrecognized by human actors; language promotes communication, cooperation, conflict, and community.

More importantly, the development of language and symbolic systems require embodiment (Hall, 2016). Or, as Peter Hall and Dee Ann Spencer-Hall (1983, p. 253) note, through embodied action:

> human beings can reach out, grasp, take apart, build upon, reassemble, act against, and overcome resistance. Such action can contribute to perception and conception, new meanings and solutions, and at the same time carry out intended actions… On the other hand, blockage, interruption, ambiguity, and resistance to physical and social activity leads to problem-solving thinking activity and the use of language and mind.

Thus, we arrive at the embodied, meaningful act—where meaning does not merely reside in symbols, but is also grounded in the intentionality (Hall & McGinty, 1997) of action-interaction, object use, and sense-making suffused with emotion and affect. In other words, *the body is a source and a means of the sustaining of society. Embodiment* entails the engagement of cognition, emotion, and affect, as inescapable parts of the human experience. Thus, this third core idea helps us envision not only the symbolic aspects of human selfhood, but also the sensory, embodied ones—ourselves and our physical forms as vehicles of expression and of activity, and as objects of action-interaction. This is perhaps the simplest of the core ideas, but its implications are sociologically illuminating.

Empirical Example

Actors on stage and screen, comedians, musicians, and other entertainers know that the body is an essential component in their efforts. This is not simply because the body is the instrument through which they create their art. The body *becomes* part of the art that is created through the work of performance. The embodiment of emotions and affect, whether "for play," "for show," or "for real" are always and already objects of action-interaction. Most importantly, even in the most mundane of cases, where performance is less scripted, we can find an illustration of our last tenet.

As new college graduates walk across the stage and join the ranks of the full-time employed, they often move to cities and find modest apartments. Those with a decent income and a taste for low prices and functionality may choose to patronize an IKEA store. Regardless of their style and value,

IKEA furniture comes with some assembly required. Correctly assembling such furniture is, as many know, a full-body experience, as sofas, bookshelves, chairs, tables, kitchen cabinets, and many other housewares and furnishings feature lengthy assembly instructions with complex diagrams and little explanatory text. Many express frustration and stress, and are often left wondering about the "extra" screws and other assembly bits left behind after a piece of furniture is "assembled" (Householdquotes.co.uk, 2022). Human emotions—frustration and anger—are expressed and communicated as part of the furniture assembly experience.

Applying our framework to this particular example, we combine the third core idea with the other two. A person assembling IKEA furniture feels a set of emotions, i.e., understanding that these are social labels that we apply to states of sensation and feeling. When someone starts to feel in their body a sensation of discomfort, and even a mild sense of heat in their face, or a higher pitch in their voice as they speak about the situation, they may realize they are shocked, frustrated, angry, and they come to reflect on their feelings as part of their embodiment of emotion. As our first core idea suggests, the person realizes that the activity is not "problem-solving," and thus, needs to be modified in order to solve the problem at hand. Once we incorporate the possibility of affective states being activated, and intersecting in this situation (for example, someone suddenly becomes affected by the "furniture assembly rage" that has overcome the roommate building the IKEA furniture), we understand the undeniable sociality of a seemingly individual act. Moreover, the person who is angry at the process comes into contact not only with an object (IKEA furniture), but with a whole series of potential actors, who take the person away from immediate social interaction, and place them into action-interaction. In the spatial/physical context of this person's activity, we find both a sense of agency expressed through the action of solving a problem, and a sense of constraint, determined by a context of production and consumption based on the Do-It-Yourself concept, which removes the final stage of furniture assembly from the production sphere and moves it into the consumption sphere. This development in capitalist consumption, depending on the availability of time and skills on the part of consumers, is the product of past conditionings in the capitalist market.

5 Conclusion

This chapter has brought closer two perspectives, MI and Interactionism, that are usually assumed to be opposites in significant ways. We argue that there are similarities between these perspectives, partly emerging from the conceptual affinities that exist between the founders of MI and the interactionist perspective. We also contend that both perspectives are victims to the same common criticism, the charge that they are microsociological reductionists.

The chapter goes on to show how this charge can be neutralized, in the case of Interactionism, through a deeper engagement with Pragmatism. For the past several decades, the tenor and tone of conversations about Interactionism have assumed SI to be reducible to Blumer's famous premises. While we affirm the significance of Blumer and his three tenets, our work expands the conversation beyond them by including concepts at the intersection of recent work on Mead and the contextually situated interactionisms just sketched: social processes, human action-interaction, and social organization. By highlighting the pragmatist influences of interactionism more generally, we have arrived at a set of key ideas that refocus the interactionist project on a set of concerns less centrally social psychological, and more decisively sociological. Our work thus responds to the common criticism that both MI and Interactionism have fallen victims to.

To recap, building on insights from Interactionism, this chapter begun to respond to the controversy of microsociological reductionism, by proposing *Pragmatic Interactionism.* This framework offers three core ideas that support processually oriented empirical work across the multiplicity of sites, phases, and locales of human activity. Thus, the perspective extends the efforts by Fine and Tavory (2019) to renew Interactionism.

We think that Pragmatic Interactionism provides its future practitioners with an interactionist, pragmatist sociology. Our perspective draws on key traditions within Interactionism generally and develops them beyond the microsociological level. As interactionists deepen their study of social inequalities concerning class, gender, race, sexuality, (dis)ability, nation, and of power structures more generally, through a more pragmatic lens, a new conversation about the potential of pragmatism for renewing interactionist theory is forthcoming.

Acknowledgements We thank the careful and thoughtful feedback provided by the editors, Nathalie Bulle and Francesco Di Iorio, and by Ran Keren, on an earlier version of this work. We especially appreciate the thoughts and insights on Pragmatic Interactionism that the late Peter Hall shared with us.

References

Agassi, J. (1960). Methodological individualism. *The British Journal of Sociology, 11*, 244–270.

Baert, P. (2015). *The existentialist moment: The rise of sartre as a public intellectual.* Polity Press.

Blumer, H. (1969). *Symbolic interactionism: Perspective and method*. University of California.

Brossard, B., & Ruiz-Junco, N. (2020). On the shoulders of citers: Notes on the organization of intellectual deference. *The Sociological Quarterly, 61*, 567–587.

Buban, S. L. (1986). Studying social process: The Chicago and Iowa schools revisited. *Studies in Symbolic Interaction* (Suppl. 2) (Part A), 25–38.

Burns, T. R., & Hall, P. M. (Eds.). (2013). *The meta-power paradigm: Impacts and transformations of agents, institutions, and social systems.* Peter Lang.

Burns, T. R., Hall, P. M., & McGinty, P. J. W. (2013). Conceptualizing power and meta-power: Causalities, mechanisms, and constructions. In T. R. Burns & P. M. Hall (Eds.), *The meta-power paradigm: Impacts and transformations of agents, institutions, and social systems* (pp. 19–82). Peter Lang.

Clarke, A. (2021). Straussian negotiated order theory c.1960–present. In D. vom Lehn, N. Ruiz-Junco, & W. Gibson (Eds.), *The Routledge international handbook of interactionism* (pp. 47–58). Routledge.

Coleman, J. S. (1986). Social theory, social research, and a theory of action. *American Journal of Sociology, 91*, 1309–1335.

Cooley, C. H. (1992 [1902]). *Human Nature and the Social Order*. Transaction.

Côté, J.-F. (2015). *George Herbert Mead's concept of society: A critical reconstruction.* Paradigm.

Couch, C. J. (1989). *Social processes and relationships: A formal approach.* General Hall.

Fine, G. A. (1993). The sad demise, mysterious disappearance, and glorious triumph of symbolic interactionism. *Annual Review of Sociology, 19*, 61–87.

Fine, G. A. (2001). *Gifted tongues: High school debate and adolescent culture.* Princeton University Press.

Fine, G. A. (2007). *Authors of the storm: Meteorologists and the culture of prediction.* The University of Chicago Press.

Fine, G. A., & Tavory, I. (2019). Interactionism in the twenty-first century: A letter on being-in-a-meaningful-world. *Symbolic Interaction*. Online first. https://doi.org/10.1002/symb.430

Gross, N. (2018). Pragmatism and the study of large-scale social phenomena. *Theory and Society, 47*, 87–111.

Hall, P. M. (1987). Interactionism and the study of social organization. *The Sociological Quarterly, 28*, 1–22.

Hall, P. M. (1995). The consequences of qualitative analysis for sociological theory: Beyond the microlevel. *The Sociological Quarterly, 36*, 397–415.

Hall, P. M. (1997). Meta-power, social organization, and the shaping of social action. *Symbolic Interaction, 20*, 397–418.

Hall, P. M. (2016). *There is more to reality than meets the eye: Reaching out to grasp the hand*. Paper presented at the Annual Meetings of the Society for the Study of Symbolic Interaction, Seattle, WA.

Hall, P. M., & Spencer Hall, D. A. (1983). The Handshake as interaction. *Semiotica, 45*, 249–264.

Hall, P. M., & McGinty, P. J. W. (1997). Policy as the transformation of intentions: Producing program from statute. *The Sociological Quarterly, 38*, 439–467.

Hall, P. M., & McGinty, P. J. W. (2002). Social organization across space and time: The policy process, mesodomain analysis, and breadth of perspective. In S. C. Chew & D. Knottnerus (Eds.), *Structure, culture, and history: Recent issues in social theory* (pp. 303–322). Rowman and Littlefield.

Hallett, T. (2010). The myth incarnate: Recoupling processes, turmoil, and inhabited institutions in an urban elementary school. *American Sociological Review, 75*, 52–74.

Hallett, T., & Ventresca, M. (2006). Inhabited institutions: Social interactions and organizational forms in Gouldner's *Patterns of Industrial Bureaucracy. Theory and Society, 35*(2), 213–236.

Hallett, T., Shulman, D., & Fine, G. A. (2009). Peopling Organizations: The promise of classic symbolic interactionism for an inhabited institutionalism. In P. S. Adler (Ed.), *The Oxford handbook of organization studies: Classical Foundations* (pp. 486–509). Oxford University Press.

Hirsch, J. S., & Khan, S. (2020). *Sexual citizens: A landmark study of sex, power, and assault on campus*. W. W. Norton & Company.

Householdquotes.co.uk. (2022). *Where does assembling IKEA furniture cause the most stress?* Retrieved March, 24, 2022. https://householdquotes.co.uk/ikea-stress/

Huebner, D. R. (2014). *Becoming Mead: The social process of academic knowledge*. University of Chicago.

Irvine, L. (2021). Animal selfhood. In D. vom Lehn, N. Ruiz-Junco, & W. Gibson (Eds.), *The Routledge international handbook of interactionism*. Routledge.

Izadi, E. (2015). Columbia student protesting campus rape carries mattress during graduation. *Washington Post*. Retrieved March 24, 2022. https://www.washingtonpost.com/news/grade-point/wp/2015/05/19/columbia-student-protesting-campus-rape-carries-mattress-during-commencement/

Joas, H. (1985). *G.H. Mead: A contemporary re-examination of his thought*. MIT Press.

Joas, H. (1993). *Pragmatism and social theory*. University of Chicago.

Joas, H. (1996). *The creativity of action*. University of Chicago.

Joas, H. (1997 [1985]). *G. H. Mead: A contemporary re-examination of his thought*. MIT Press.

Joas, H., & Huebner, D. R. (Eds.). (2016). *The timeliness of George Herbert Mead*. University of Chicago Press.

Kivisto, P., & Swatos, W. H. (1990). Weber and interpretive sociology in America. *The Sociological Quarterly, 31*, 149–163.

Lengermann, P. M. (1988). Robert E. Park and the theoretical content of Chicago sociology: 1920–1940. *Sociological Inquiry, 58*, 361–377.

Littrell, B. (1997). Carl Couch and pragmatism: Naturalism, temporality, and authority. *Studies in Symbolic Interaction* (Suppl. 3), 3–20.

Maines, D. R. (1977). Social organization and social structure in symbolic interactionist thought. *Annual Review of Sociology, 3*, 235–259.

Maines, D. R. (1982). In search of mesostructure: Studies in negotiated order. *Urban Life, 11*, 267–279.

Maines, D. R. (1989). Repackaging Blumer: The myth of Herbert Blumer's astructural bias. *Studies in Symbolic Interaction, 10*, 383–413.

Maines, D. R. (2001). *The faultline of consciousness: A view of interactionism in sociology*. Aldine de Gruyter.

McGinty, P. J. W. (2014). Divided and drifting: Interactionism and the neglect of social organizational analyses in organization studies. *Symbolic Interaction, 37*, 155–186.

McGinty, P. J. W. (2016). The astructural bias in symbolic interactionism. *Studies in Symbolic Interaction, 46*, 19–56.

McGinty, P. J. W. (2017). *Pragmatic interactionism: Organizing attention to the power of and possibilities for a 21st century interactionism*. Paper presented as the Peter M. Hall Lecture at the Annual Meetings of the Midwest Sociological Society, Milwaukee, WI.

McPhail, C., & Rexroat, C. (1979). Mead vs. Blumer: The divergent methodological perspectives of social behaviorism and symbolic interactionism. *American Sociological Review, 44*, 449–467.

Mead, G. H. (1932). *The philosophy of the present*. Prometheus Books.

Mead, G. H. (2015 [1934]). *Mind, self, and society: The definitive edition* (C. W. Morris, Ed., Annotated by D. R. Huebner and Hans Joas). University of Chicago Press.

Miller, D. E., Hintz, R. A., & Couch, C. J. (1975). The elements and structure of openings. *The Sociological Quarterly, 16*, 479–499.

Miller, D. E., & Hintz, R. A. (1997). The structure of social interaction. *Studies in Symbolic Interaction* (Suppl. 3), 87–107.

Rochberg-Halton, E. (1986). *Meaning and modernity: Social theory in the pragmatic attitude*. University of Chicago.

Rochberg-Halton, E. (1987). Why pragmatism now? *Sociological Theory, 5*, 194–200.

Rose, A. M. (1962). *Human behavior and social processes*. Houghton Mifflin.

Ruiz-Junco, N. (2016). The persistence of the power deficit? Advancing power premises in contemporary interactionist theory. *Studies in Symbolic Interaction, 46*, 145–165.

Ruiz-Junco, N. (2017). Advancing the sociology of empathy: A proposal. *Symbolic Interaction, 40*, 414–435.

Ruiz-Junco, N., & Brossard, B. (2019). *Updating Charles H. Cooley: Contemporary perspectives on a sociological classic*. Routledge.

Segre, S. (2014). A note on Max Weber's reception on the part of symbolic interactionism, and its theoretical consequences. *The American Sociologist, 45*, 474–482.

Shalin, D. (1986). Pragmatism and social interactionism. *American Sociological Review, 51*(1), 9–29.

Simmel, G. (1950 [1908]). *The sociology of Georg Simmel*. The Free Press.

Snow, D. A. (2001). Extending and broadening Blumer's conceptualization of symbolic interactionism. *Symbolic Interaction, 24*(3), 367–377.

Strauss, A. L. (1991). *Creating sociological awareness: Collective images and symbolic representations*. Transaction.

Strauss, A. L. (1993). *Continual permutations of action*. Aldine de Gruyter.

Travisano, R. V. (1975). Comments on 'A research paradigm for symbolic interaction.' In C. J. Couch & R. A. Hintz (Eds.), *Constructing social life* (pp. 263–271). Stipes.

vom Lehn, D., Ruiz-Junco, N., & Gibson, W. (2021). Introduction. *The Routledge international handbook of interactionism* (pp. 3–21). Routledge.

Weber, M. (1978). *Economy and Society*. Volume 1. University of California Press.

Wrong, D. H. (1961). The oversocialized conception of man in modern sociology. *American Sociological Review, 26*, 183–193.

Zerubavel, E. (1985). *Hidden rhythms: Schedules and calendars in social life*. University of California Press.

Zerubavel, E. (2007). Generally speaking: The logic and mechanics of social pattern analysis. *Sociological Forum, 22*, 131–145.

Zerubavel, E. (2009). *Social mindscapes: An invitation to cognitive sociology*. Harvard University Press.

Zerubavel, E. (2018). *Taken for granted: The remarkable power of the unremarkable*. Princeton University Press.

Zerubavel, E. (2021). *Generally speaking: An invitation to concept-driven sociology*. Oxford University Press.

Explaining Social Action by Embodied Cognition: From *Methodological Cognitivism* to *Embodied Individualism*

Riccardo Viale

1   Starting Points

Since the late 1980s, I started publishing articles on the cognitive basis of social action and rationality. The starting points were the psychologism of John Stuart Mill and Herbert Simon's theory of Bounded Rationality. Some years ago I published two books "Methodological Cognitivism. Mind, Rationality, and Society" (2012) and "Methodological Cognitivism. Cognition, Science and Innovation" (2013) that were a collection of my articles and papers published since the end of the 80s until 2011. In these articles, I tried to elaborate a new theory in the methodology of the social sciences, dubbed *Metodological Cognitivism* (*MC*) that overcame some philosophical weaknesses of *Methodological Individualism* (*MI*) and that was able to exploit the new discoveries in the cognitive sciences.

When I began the development of *MC* in the late 1980s, the social sciences and economics were not open to an approach to social rationality that was not of an axiomatic type. Human reason had to be analyzed according to a priori principles and in the case of economics according to the formal rationality of the utility theory of Von Neumann and Morgenstern (1944) and of the subjective expected utility of Jim Savage (1954).

R. Viale (✉)
Behavioral Sciences and Cognitive Economics, University of Milan Bicocca, Milan, Italy
e-mail: viale.riccardo2@gmail.com

© The Author(s), under exclusive license to Springer Nature Switzerland AG 2023
N. Bulle and F. Di Iorio (eds.), *The Palgrave Handbook of Methodological Individualism*,
https://doi.org/10.1007/978-3-031-41508-1_24

574   R. Viale

One of the most important confirmations of *MC* comes with the development and affirmation, starting from the 80s, of behavioral economics and, starting from the first decade of 2000, of the behavioral approach to public policy. Herbert Simon had already, starting from 1947, with his book "Administrative Behavior," opened the way to overcome the a priori dimension of economic rationality and of the related methodological individualism. The introduction of the concept of Bounded Rationality had the objective of building a normative theory of choice on a descriptive basis that was able to deal with the uncertainty and complexity of most human decisions (Boudon &Viale, 2000; Viale, 2011b). Since the 60s, epistemology and philosophy of science were supporting the descriptive and empirical approach to knowledge and rationality. The following are some examples.

According to Quine (1969), among the most fundamental questions that epistemology has sought to answer are the following: (1) How ought we to arrive at our beliefs? (2) How do we arrive at our beliefs? (3) Are the processes by which we do arrive at our beliefs the ones by which we ought to arrive at our beliefs? Question 1 cannot be answered independently of Question 2. The question of how we actually arrive at our beliefs is therefore relevant to the question of how we ought to arrive at our beliefs. This position is well summed up by the following passage from Quine:

> Epistemology becomes as a chapter of psychology and hence of natural science. It studies a natural phenomenon, viz. a physical human subject. This human subject is accorded a certain experimentally controlled input – certain patterns of irradiation in assorted frequencies, for instance – and with a little of time the subject delivers as output a description of the three-dimensional external world and its history. The relation between the meagre input and the torrential output is a relation that we are prompted to study for somewhat the same reasons that always prompted epistemology; namely, in order to see how evidence relates to theory, and in what ways one's theory of nature transcends any available evidence for it. (Quine, 1985, p. 24)

In the article "Can Human Irrationality Be Experimentally Demonstrated?" Jonathan Cohen (1981) points out that the presence of fallacies in reasoning and decision-making is wrongly evaluated by referring to external normative criteria which ultimately derive their own credentials from a systematization of the intuitions that agree with them. These normative criteria cannot be taken, as some have suggested, to constitute a part of natural science, nor can they be established by metamathematical proof. Since a theory of competence has to predict the very same intuitions, it must ascribe rationality to ordinary people.

Finally, according to Goldman (1986), cognitive science can tell us which aims and what inferential procedures are clearly outside the scope of human cognitive capacities. This modality, dubbed by Goldman as "feasibility principle" (1986, 1993) is the cognitive variant of Laudan's "anti-utopistic" criterion (1984) in philosophy of science. The theory of scientific rationality can not be founded on cognitive abilities that are unrealistic and beyond the inferential skills of scientists (Viale, 2011b, 2013).

2 Overcoming the Weakness of Methodological Individualism: The Methodological Cognitivism

MC is connected to the previous naturalizing epistemological program. MC can be regarded as a naturalized psychologistic evolution of methodological individualism since it appears to neutralize some of its epistemological and methodological difficulties and is more firmly rooted in the fabric of scientific knowledge, which is now more widely accredited in the study of human behavior.

In *Methodological Cognitivism* (Viale, 2012, 2013), the incipit starts from a fundamental question of John Stuart Mill: "Are human actions subject to invariable laws like all other natural events? Are they really ruled by the constant of causality that underlies every scientific theory of successive phenomena?" (Mill, 1843, 1st edition; 1956, 8th edition, p. 827). Mill answers that empirical generalizations are possible about society. They are generalizations of some aspects of social life. However, they derive their truth from causal laws, of which they are the consequence. If we are familiar with those laws, we know the limits of the derived generalizations; instead, if we have not yet justified the empirical generalization—if it is based on observation alone—then there is no guarantee in applying it beyond the limits of time, place, and circumstances in which the observations were made. Causal laws that can justify empirical generalizations must refer to the human mind. In other words, the laws of ethology, derived deductively from the laws of psychology, should allow us to explain the different characters of social or national contexts in the presence of different starting conditions. But are the fundamental laws of psychology that constitute the causal barycenter of social explanation?

Mill's position was either neglected or criticized by later contributions to scientific methodology. These were marked by a generalized antipsychologism

576   R. Viale

expressed by authors whose theses were radically divergent on other essential methodological questions. Marx, Weber, Menger, L. von Mises, Popper, von Hayek, Watkins, (and to a certain extent Boudon and Elster) all share a stringent criticism of Mill's psychologism and, more in general, of the thesis concerning the reduction of social action to causal mechanisms of the human mind. It is significant that Popper takes Marx as one of the main objects of his critical analysis, but he finds himself in complete agreement with his antipsychologism and with Marx's famous maxim: "It is not men's conscience that determines their being, but, on the contrary, their social being that determines their conscience" (Marx, 1859). According to Popper "The error of psychologism consists in claiming that methodological individualism in the field of social sciences entails the need to reduce all social phenomena and all social regularities to psychological phenomena and psychological laws" (Popper, 1966, 5th edition, vol. II, p. 131).

What kind of methodological individualism is justified by psychologism? The reasons for the criticism of psychologism raised by Popper and many other authors can be traced back to methodological individualism. The structure and complexity of this school of methodology of social science is often not taken sufficiently into account (Antiseri, 1996). The contraposition of methodological individualism (MI) to the various holistic theses often overshadows the fact that it contains incompatible and contradictory positions. The first reason for this confusion is the lack of differentiation between the different philosophical dimensions present within this approach (Bhargava, 1992).

In the first place, a distinction should be drawn between at least three forms within MI: explanatory individualism (EI), ontological individualism (OI), and semantic individualism (SI). The first can be summed up in the thesis that all social phenomena can, in principle, be explained using methods that refer to individuals and their properties. The second aims to demonstrate that because only individuals and their properties exist, then social phenomena must be identified with them. Semantic individualism is based on a different thesis from the first two, although it can be derived from them. Every noun and attribute referring to social attributes can be reduced to nouns and attributes referring to individual entities.

The structure of psychologistic or antipsychologistic theses belongs, mainly to the dimension of OI, even if, according to the position expressed in EI, consequences can be derived in favor of one or the other position.

(a) Ontological Individualism (OI) deals with the domain of the application of individualist theses. What characteristics and properties of the individual are crucial and determinant at an explanatory level? In the first place, the most likely individual attributes are those of a physical and psychic kind. Psychologism can be interpreted as a form of psychological atomism that intends to trace social events back only to the individual's mental properties. This is the aspect that most clearly distinguishes psychologism from antipsychologism. As pointed out by Bhargava (1992, pp. 42–45), psychological atomism can be regarded as the version of OI that supports the following theses:

- The only acceptable explanations of mental states are those accomplished by empirical science, by knowledge on human psychology. Only causal laws of the mind rather than the philosophical conceptual and hermeneutical analyzes can be used for the explanatory purposes of EI. The antipsychologism of Weber (1978, p. 19) and Dilthey (1989) is aimed precisely at challenging this position.
- The only relevant facts for individualistic explanation are the mental ones. This form of psychologism is disputed by various authors like Popper (1966, Italian translation, 1974, pp. 75–76), Agassi (1973, pp. 187–188), von Hayek (1973, p. 40), and Elster (1983). In addition to those illustrated when describing Popper's antipsychologism (the historicist accusation; the subalternity of mental causes to situational logic; the psychological irreducibility of unintentional consequences: see Viale, 2012, pp. 56–64), the reasons for their opposition also include the refusal to exclude, as psychologism requests, the physical causes and the role of the action.

(b) According to EI, what can be explained is only the particular social event (that is considered linguistic fiction without any real content[1]) and the explanation is based on laws and the starting conditions of individuals and their properties. EI has no substantive ontological implications relating to what we propose as laws on individuals and their properties. The possibility of social laws is implicitly denied because the ontological existence of social phenomena is rejected. It is not assumed, however, which type of entity and individual properties are important for the explanation. The entities might be everything related to individual

---

[1] The EI might have a conventional approach to the use of social concepts. They may be judged pragmatically useful tools for social explanation and prediction even if they do not correspond to real entities. There are other similar examples in science and philosophy. For example, according to the monist approach in the philosophy of mind mental phenomena and concepts such as pain, hunger, and fear are useful fictions for research and communication, but the only real entity is the neural structure. The neurocomputational approach (Anderson & Rosenfeld, 1988; Churchland & Sejnowski, 1992) has the goal to substitute the mental language with neural language but until then, the mental language may be used among neurocognitivist scientists.

action. However contrary to ontological individualism psychologism does not include physical atomism, which seeks to explain social events on the basis of the physical and behavioral properties of single individuals. On the contrary, psychologism can be interpreted as a form of psychological atomism that intends to trace social events back to the individual's mental properties. Psychological atomism supports that only acceptable explanations of mental states are those accomplished by empirical science, by knowledge on human psychology, and that the only relevant facts for individualistic explanation are the mental ones. However, psychologism does not neglect that the social events correspond to the interaction of two or more individual actions and to their aggregate consequences and effects. Nevertheless, it emphasizes that the explanation of action should be found at the cognitive level of the causal mechanisms of action.

While the mainstream of MI was antipsychologist until the recent past, from the 70s onward, the Millian tradition started to put forward new interpreters. For example, Homans' thesis (1970) is that the explanation of social events by psychological propositions cannot be proved philosophically. It is a matter of empirical investigation and analysis. He states that all social phenomena can be analyzed without residue into the actions of individuals. And since methodological individualism entails psychologism, all sociological facts can be explained by the use of psychological propositions.

MC starts from the previous remarks. It has three main philosophical components:

1. The causal explanatory model
2. The mind–body identity theory
3. The cognitive models of social action

The following remarks are a snapshot of *Methodological Cognitivism* (Viale, 2012, 2013):

1. The *first component* concerns the preferred type of explanation. In order to respond to the "social interrogatives," various models of explanation have been put forward over the past years. In short, two key problems have emerged regarding explanation: one pragmatic, the other causal. The first was examined in an interesting article by van Fraassen (1980), above all concerning the logic of answers to the questions "why," using what he

also called erothetic logic.[2] The pragmatic context K, which includes all contextual factors, ranging from knowledge of the phenomenon to our philosophical points of view and our interests, is what makes us select the reference class of the explanans when we try to provide an explanation. This factor is particularly valid when we are dealing with several causal chains, whose choice and selection cannot be justified solely on methodological grounds, or when we have to decide the level of aggregration at which to stop the explanation. However, factor K cannot rule out a spurious explanation or strong regularities without explanatory value.

This difficulty leads us to the second problem, the causal one. What matters in an explanation is not the formal subsumption among assertions,[3] but the physical subsumption among facts. This is based on the nomological—or probabilistic-type causal relation that describes the physical connection—either constitutive (the micro intimate causes of the macro) or non-constitutive (the causal relation between phenomena at the same level of aggregation)—between facts. It can be identified for its productive or propagative typology, through the empirical control of conjunctive and counterfactual conditionals on the effects of the perturbation or modification of the causal chain.[4] Therefore, it is possible to avoid the false steps of earlier models only according to causal modeling.

The causal problem, that is the explanatory need to identify the causal chain that "constitutes" the social phenomena, brings us to the heart of the debate on methodological individualism. Since the epistemic purpose is a causal explanation, the study of social phenomena must use an individualistic type approach. But what kind of individualism is this?

The candidates for this role are a form of reductionist individualism, which attempts to reduce social laws and theories to those of the actor through the use of "bridge" laws or correlation.[5] This microreduction allows the social macro properties to be identified with individual micro properties, which may

---

[2] While the main task of logic is to define the consistency (or inconsistency) of ideas (sentences) and the definition of inference in the erothetic logic, the definition of questions, and rules have to ascertain whether a sentence can be conceived as an answer to a given question.

[3] Because assertions are linguistic entities whereas causal explanation focuses on real facts that may or not linguistically represented.

[4] In other words by empirically testing the results of changes of a fact (the cause) on the other connected fact (the effect).

[5] An example of bridge law between social laws and psychological laws are: imitation heuristics drive the individual to imitate the crowd; the imitation of the crowd generates social phenomena as flock behavior in stock market; flock behavior in the stock market causes financial bubbles.

have an unlimited number of specific combinations.[6] This type of individualism, therefore, offers causal and constitutive type explanations that seem to satisfy our epistemic desiderata. At first sight, the reductionist version and above all the microreductionist version seem to be closer to psychologism. By asserting the identity between social phenomena, on the one hand, and individuals and their psychological properties on the other, the epistemological barycentre of social analysis shifts towards the reality of individual minds.[7] Obviously, when MC accounts for macro-level phenomena such as financial bubbles or political votes in terms of concrete individuals, it does not refer to their biographical characteristics, but to ideal–typical psychological phenomena and mechanisms[8] and consequent typical patterns of social interaction. Moreover, the MC represents social action as the behavioral effects of individual causal cognitive mechanisms and their interactions with other individual behaviors. Then social action corresponds to the recursive interaction of different individual behaviors, caused by individual cognitive mechanisms. Any interaction generates a behavioral change that produces a further different interaction.

2. The *second component* refers to the identity theory in philosophy of mind. If we identify the mind with brain because of our goal to represent the causal properties of mind, how we can link the representational features of mind to the causal structure of the brain? There appear to be two main options. One solution is given by a sophisticated functionalist theory (Fodor, 1990, p. 22): the idea is of a mental syntax that is realized in the causal structure of the brain. Using the comparison with the computer, the causal properties of its processors (for example, outputting current when and only when it receives electric currents from both inputs) realize the syntactic properties of the symbol (in this case, the formal properties of the logical conjunction) that realizes the semantic properties of the symbol (in this case truth-preserving, using the rule of conduction). According to this analogy, the AND-GATE processor (that of conjunction) can correspond

---

[6] Indeed, the identity is not token-token or type-token, but type-type (Bhargava, 1992, pp. 68–78). Furthermore, this microreductionist individualism offers fundamental, rock-bottom explanations, whose essential nature is decided—by convention—according to the pragmatic Factor K mentioned earlier with reference to van Fraassen.

[7] To do this, however, we must assume the existence of social laws and social properties, the object of the reduction. If this type of assumption corresponds to an affirmation of the epistemological reality of social laws, this would not be accepted by psychologism because, according to it, only the laws of the human mind are real. On the contrary, according to the non-reductionist option, what can be explained is only the particular social event and the explanation is based on laws and the starting conditions of individuals and their properties.

[8] For example the imitation heuristics.

Explaining Social Action by Embodied Cognition ...   **581**

to a given neural connection in the brain. Therefore, the goal of cognitive science is to find the formal syntactic rules that process the representations and are realized by given neural connections, analogs of computer processors. Using this program, it would be possible to connect mental states with the causal properties of the neural connections of the brain via the mental syntax.

The second possible solution starts from a critique of the unbiological features of the previous functionalist description. The computer model of mind aims at a level of description of the mind that abstracts away from the biological realizations of the cognitive structures. As far as the computer model goes, it does not matter whether our gates are realized in grey matter, switches, or other substances. But if the mind is identified with the causal structure of the brain, an answer to our questions, therefore, lies in how the neural structure works. Computational neuroscience (Anderson & Rosenfeld, 1988; Churchland & Sejnowski, 1992) and its artificial counterpart, connectionism, are the programs that pursue this goal. According to computational neuroscience, the basic unit of occurrent cognition is apparently not the sentence-like state, but rather the high dimensional neuronal activation vector (that is a pattern processing is apparently not an inference from sentence to sentence, but rather a synapse-induced transformation of large activation vectors into other such vectors. In this way, speaking of representations of reality means speaking of prototype vectors in the higher populations of cortical neurons. Strong identity theory and our causal desiderata seem to incline towards this second eliminative option, computational neuroscience.

3. The scientific content is the *third component* of methodological cognitivism. If the ultimate aim, the motive for our research, is to explain and if the best form of explanation is causal and constitutive, at what level of aggregation should we position our fundamental explanations? I define the proposed methodology as *methodological cognitivism* precisely because I feel that the conventional base level of our attempts of explanation must be the cognitive. I use the cognitive attribute in a narrow sense compared to most sociologists, but in a broader sense than cognitive scientists, to include all the psychological mechanisms responsible for the decision, and therefore, the action. Therefore, not only the superior psychological processes, like memory, learning, reasoning, but also emotion, instinct, and perception.

582   R. Viale

Explanation based on cognitive mechanisms cannot always be accomplished. In some cases, the state of our knowledge of the subject will make us raise the level of aggregation to the concepts of folk psychology, or commonsense psychology. However, these rock-bottom explanations must be sought, in macro–micro explanation at the cognitive and, in prospect, at the neurocognitive level, as indicated by the program of Patricia Churchland (1986) and Paul Churchland (1989, 1998).

3   The Applied Successes of Methodological Cognitivism: Behavioral Economics and Nudge Theory

Since I started focusing on the theory of rationality and social action some years ago, something has never ceased to puzzle me: the blind eye that the economic and legal sciences have turned on the real nature of the human mind (Viale, 2022a). I have always wondered how a discipline that deals with explaining and predicting economic behavior (economics), or one that aims to identify norms to guide the behavior of citizens toward achieving goals in the public interest (the law) could deliberately not take into account the way in which individuals reason and make decisions, that is, the natural mechanisms of behavior. If economics aspires to be capable of explaining and predicting economic phenomena, it should base its models on the psychology that underlies reasoning and decision-making, i.e., on how the individual estimates the probability of an event, how the utility of a choice is evaluated, how the alternatives are pondered, etc. In short, on the mechanisms behind the individual's economic behavior. Similarly, if the law aspires to be effective in its normative activity, it should take into account how the norm is represented mentally by the citizens, which effects the norm has on their existing beliefs, and how this effect determines their subsequent behavior. In other words, how the individual reacts psychologically to the norm, in terms of aligning to or distancing from its goals, according to the interpretation given and to the corresponding cognitive and emotional effects on their behavior.

Surprisingly, both economics and the law have not seemed particularly concerned with understanding human behavior and putting to good use the scientific knowledge available about it. Economics has operated according to a fictitious and unreal of economic model individual (*homo oeconomicus*) whose "mind" constitutes a set of optimization algorithms and whose behavior is driven only by positive or negative monetary incentives. The law, in turn, has created an even more general *homo juridicus*, that is, an unsophisticated model individual guided, in Pavlovian conditioning,[9] chiefly by the need to avoid sanctions deriving from not respecting the norm.

The path opened by Herbert Simon in 1947 found partial application in the development of the "Heuristics and Biases" program launched by Amos Tversky and Daniel Kahneman (1974). Starting in the 1970s, Kahneman and—up to his premature death - Tversky, along with a growing number of collaborators and colleagues, have investigated various aspects of human cognition linked to the processes of judgment and decision-making.

These contributions, from Herbert Simon to Kahneman and Tversky, draw a picture of the human being as somewhat limited in their rational capacity, guided by decisional automatisms, falling into traps and suffering from cognitive illusions. This is an individual that is more inert than active, more interested in "seizing the day" than investing in the future, etc. At the start of the new millennium Richard Thaler, a behavioral economist from Chicago Booth and a collaborator of Tversky and Kahneman, and Cass Sunstein, a Harvard lawyer with a keen interest in the behavioral and economic aspects of the law, began to wonder about which serious contribution psychology could make to public policy. In fact, Thaler had started in 1994 with a short article on savings policies (1994), which was followed a few years later, after a series of experiments, by the successful behavioral program "Save More Tomorrow" (Thaler & Benartzi, 2004). This program showed how, based on a series of natural cognitive propensities, important social and economic objectives could be achieved, such as increasing the propensity to set money aside. Soon thereafter, the behavioral approach was extended to a whole set of other public policies. Thaler, Sunstein, and a number of young colleagues, including Hersh Shefrin, Shlomo Benartzi, Brigitte Madrian, and Eric Johnson, began experimenting with the introduction of behavioral solutions in various areas of environmental, health, tax compliance, and credit policy. What characterized these proposals was not only their sophisticated behavioral design. Economic incentives or legal prohibitions too rest on behavioral logic, albeit often an inadequate one. Their most innovative feature was their *paternalistic* dimension, i.e., their

---

[9] Classical conditioning (also known as Pavlovian conditioning) refers to a learning procedure in which a biologically potent stimulus (e.g., food) is paired with a previously neutral stimulus (e.g., a bell). It also refers to the learning process that results from this pairing, through, which the neutral stimulus comes to elicit a response (e.g., salivation) that is usually similar to the one elicited by the potent stimulus. Classical conditioning is distinct from operant conditioning (also called instrumental conditioning), through which the strength of voluntary behavior is modified by reinforcement or punishment. However, classical conditioning can affect operant conditioning in various ways; notably, classically conditioned stimuli may serve to reinforce operant responses. Classical conditioning was first studied in detail by Ivan Pavlov, who conducted experiments with dogs and published his findings in 1897.

effort to nudge citizens into making choices that would improve their well-being, combined with a need for *libertarian* protection of their autonomous choice.[10]

The public and academic success of Heuristic and Biases program and of nudge theory do not let us forget the critical remarks that can be done to both the programs. In Viale (2018, 2022a, forthcoming), I introduce some of the main critical remarks to the normative theory of rationality and the experimental method of Heuristic and Biases program and to the contradictory libertarian tenet of the Nudge Theory. However, both programs are the results of the application of the cognitive theory of social action to economics and political sciences. They are the exemplaries of the MC approach.

4 The Limiting Features of Representationalist Cognitivism

In the second section, I proposed a snapshot of the arguments in favor of MC referring to *Methodological Cognitivism* (Viale, 2012, 2013). One of them is that MC refers to the psychology of decision-making as the main set of models to explain social action. The empirical study of human behavior characterizing the psychology of decision-making, from the postwar period to today, developed above all by looking at the normative model of decision theory and in particular at Subjective Expected Utility (SEU) decision-making, with the formal theory of utility of Von Neumann and Morgersten (1944), and the subjective expected utility theory of Savage (1954) as reference. Actually, this conceptual reference constitutes an epistemological weakness of the psychology of decision-making (for a detailed analysis of the arguments about the negative influence of SEU model to the psychology of decision-making see Viale, forthcoming).

The cognitive psychology of decision-making precisely reflects the conceptual structure of formal decision theory. In relation to this structure and the normative component derived from it, empirical research in the cognitive psychology of decision-making has been developing since the 1950s. As highlighted in Weiss and Shanteau (2021), it was in the 1950s that Ward Edwards,

---

[10] Economic incentives can be both paternalistic and libertarian. The difference with nudging lies in the fact that they rely on an abstract model of rationality that does not correspond to the real human decision-maker. Moreover, paradoxically the ideal aim of economic incentives is to find the exact amount of gains or losses that would lead the rational citizen to choose the option that is preferable for the government. From this perspective, the libertarian element seems rather weak.

the founder of the psychology of decision-making, began to carry out laboratory experiments on how people decide. His experiments, which became the reference of subsequent generations and in particular of Daniel Kahneman and Amos Tversky's Heuristics and Biases program, have two fundamental characteristics: firstly, the provisions of the SEU are set as a normative reference, and the experimental work has the aim of evaluating when and how the human decision-maker deviates from the requirements of the SEU.

The highly artificial experimental protocols of the Heuristics and Biases program are based on one-shot situations. They do not correspond to how people learn and decide in a step-by-step manner, thus adapting to the demands of the environment. There is no room for people to observe, correct and craft their responses as experience accumulates. There is no space for feedback, repetition, or opportunities to change. Consequently, conclusions about the irrationality of the human mind have been based on artificial experimental protocols.

From the post-war period to today, the study of reasoning, judgment, and decision processes has mainly been carried out within the classic cognitive model, which can also be defined as "Information Processing Psychology." It has three characteristics: first, thought is computation and takes place as a form of computation. Every mental activity is performed by algorithms similar to the computer's machine language. Cognition derives from computational procedures that are carried out on abstract symbolic structures. This idea is not new. The concept of reasoning as a form of calculus dates back to Aristotelian philosophy and subsequently to Thomas Hobbes (for whom reasoning was comparable to arithmetic calculus), Leibniz (for whom thought was a symbolic process of combining signs), and Descartes (for whom the ideal of reasoning was the deductive chains of geometry).

Second, thought has as its objects mental images that are representations of external reality, derived from perceptual activity or its elaborations. It is a postulation very similar to that of Descartes, according to which every thought has as its object images projected into an internal space called the mind. When we perceive something from the outside, the stimulus is translated into these amodal representations that become the real content of the experience and cognitive activity. These mental entities—which may be in the form of images, propositions, or a mix of the two—are the content of the inferential and decision-making activity.

Third, the psychological activity takes place independently or can be explained independently of the physical substrate that carries it out. The study of how we reason on a logical or probabilistic level, how we represent external reality, how we store and retrieve data from memory, and how

586   R. Viale

we decide on the basis of our knowledge and preferences does not require an understanding of the central nervous system, let alone of the peripheral one or the rest of the body. Research and hypotheses on the thinking and decision-making processes of the human mind can be compared to the research activity of a computer scientist who develops software programs. This can happen relatively independently of how the reference hardware is structured. According to the computer metaphor, the mind is to the software as the brain is to the hardware.[11] Linked to this philosophical position, cognitivism has developed a series of theories that hypothesize different levels of analysis of cognitive reality. Among the most popular theories about a multi-level ontology is David Marr's (1982) tripartite level between a computational level (e.g., the result of a computation), an algorithmic level (e.g., the ways in which the computation is done) and one of physical implementation (e.g., the type of material structure that executes it). Cognitivism was primarily interested in studying the first two levels, that is, how mental representations are generated and computed after the sensory organs transfer data from the outside.[12]

The classic cognitivist approach just described represents the central theoretical pillar of the psychology of decision-making and of the empirical study of the Heuristics and Biases program. According to this approach, the main limits of rationality are derived from the three structural characteristics of the cognitivist approach: 1) the mind decides through computations that use 2) internal representations and acts at a 3) level separate from the body and the environment. These characteristics have meant that research in the psychology of decision-making has been focused on the computational "benchmark" between normative SEU models and decision-making activities, highlighting the pathological aspects of human rationality and failing to recognize the adaptive and bodily dimensions in problem-solving activity and action.

Now, we will see how this representationalist approach of classical cognitivism is not acceptable, both because rationality is constrained by the

---

[11] The theoretical support for this model is varied. In the philosophy of mind, according to the functionalist approach (Fodor, 1975; Putnam, 1975), mental states are functional to the brain in the same way that the states generated in the computer by the software are functional to its hardware. This is the position of "Dualism of Properties" (in its version of "anomalous monism" see Davidson, 1970; in its version of "biological naturalism" see Searle, 1983) which supports the theory that mental activities described by cognitive psychology are different from those of any kind of hardware, for example, from those of the brain, capable of implementing them. This is the theory of the multirealization of mental properties by various types of physical substrate inspired by Hilary Putnam and Jerry Fodor.

[12] In another metaphor, the "mental sandwich" introduced by Susan Hurley (1998), the cognitive processes are the tasty, protein-laden internal part of the sandwich, while the sensory and motor parts are the two tasteless external slices of bread.

individual ability to adapt to the environment, and because this adaptation occurs through the active coupling and coping with the environment by brain and bodily action.

5   Embodied Cognition

It is commonly held that our most precious gift, the central nervous system, developed over the course of evolution from the most primitive animals to ourselves in order to enhance our thinking ability. In nature, however, this dogma does not always seem to be confirmed. For example, consider sea squirts, a type of marine sponge. They have a very peculiar life cycle. When they are born, the larvae have a nervous system consisting of a brain, spinal cord, and sensory organs that are sensitive to light. Sea squirts are able to move and do so immediately as soon as they are born in order to find a place to settle. This occurs a few hours after birth and generally, they find asylum on some rock or submarine wreck. At this point, an amazing process takes place: as soon as they are implanted on the rock and no longer need to move, they begin to ingest and reabsorb their brain. That is, they appear to regress to a lower stage of phylogenetic development. This strange metamorphosis has a functional explanation: if the sea squirts no longer need to move, the nervous system that gave them perception and guided them in movement becomes an unnecessary and costly handicap from the perspective of energy use. Colombian neuroscientist Rodolfo Llinas is very clear in generalizing the principle underlying this phenomenon: the development of the nervous system is primarily necessary to carry out actions and not to implement cognitive processes, that is, to think (Llinas, 2002). This is a Copernican reversal of the relationship between brain and body. It is no longer the body that is at the service of the brain, as cognitivism claims, but the opposite, that is, the brain is the tool that allows the individual to physically interact with the environment. The very center of gravity of the decision-making process at this point is no longer located in the cognitive computational part, but shifted to the pragmatic part of the possible actions that the environment allows.

Cognitivism has portrayed the mind that thinks and decides as if it were in a vat, separated from the body and the environment. The mind is "disembodied" from the body that carries it and "detached" from the environment in which it interacts. The new perspective introduced by neuroscience reveals instead an "embodied" and "embedded" cognition: that is, a cognition integrated with the body through action and shaped by the environment with which the body interacts and where it is located.

There is no dualism between mind and body, perception and action, reason and emotion. There are no vertical hierarchies between high and low, cortical cognition and subcortical emotion. Between the perceptive, cognitive, and motor processes, there is no relationship of temporal "sequentiality" for which we first perceive an event, then we think about what to do and then we act. On the contrary, the processes are intertwined and integrated. We can speak of "circularity" instead of sequentiality, because the action influences both perception and thought, and is also influenced by them. Cognitive processes are the result of various factors, first and foremost the sensorimotor dynamics between agent and environment. Perception and action are interpenetrated. Indeed, perception is a particular type of exploratory activity. Visual perception, for example, is not comparable to the static snapshot of a camera. The photograph of our visual apparatus is always "moved," as the continuous movement of the eye integrates with the sensorimotor and cognitive apparatus and in this way, the image can be interpreted by the brain.

The acting body must no longer be understood as a mere physical tool guided by the mind, as if it were the physical structure of a robot guided by its software. Rather, the body is one with the cognitive activity which, jointly, interacts with the environment. In this interaction, we acquire motor and perceptive experiences that comewith or are later reactivated by cognition. Body states are, therefore, necessary for cognition and for the simulation of perceptual and motor experiences, of sensorimotor models ("patterns"), extrapolated from their motor function and exploited in cognitive processes different from those for which they are created. Think, for example, of the importance of simulation in our social activity, of interacting with others, of deciding what to do in aroup work or within a market where several parties find themselves operating. In these cases, we act after reading the minds of others through the simulation in our body of their possible actions and the consequent affective and emotional results that prompt our action. These simulations are based on the reactivation of sensorimotor experiences previously acquired by the individual in similar contexts. This type of simulation experience demonstrates the falsity of another assumption of cognitivism: thinking is not based only on the use of a-modal mental representations[13] that are derived from the translation of external perceptive stimuli into

---

[13] In the sense that mental representations are the product of a translation from a language of sensory modalities to one independent of the senses, therefore a modal.

neutral mental objects or from their mnemonic recovery. The representations can also be in body and modal format. They can be motor, sensory, and affective simulations based on past experiences or sensorimotor routines introduced in evolution for one function and used today for another. In this last case, it is a pre-adaptation process whereby a biological structure evolved for a certain function takes on a new one, without losing its original function.[14] Many cognitive processes such as intersubjectivity, social action, and moral judgment are guided, shaped, and realized by the structure of pre-adapted sensorimotor mechanisms.

How does this use of the pre-adapted sensor-motor mechanisms take place? Motor programs can be disconnected from the original motor output thanks to inhibitory mechanisms exploited by higher cognitive systems.[15] In this way, the body simulation of a motor pattern takes place without activating the actual motor action.

We have seen that the brain and related bodily activity is the tool that allows the individual to physically interact with the environment. This bodily interaction with the environment shapes and models the same cognitive activity. The center of gravity of the action is therefore no longer located in the computational and cognitive part, but it shifts to the pragmatic one, that is, to the possible actions that the body–environment interaction allows. This position, which places the constraints of the rational activity of choice and decision not so much in the computational possibilities of the human mind as in the mind–body–environment interaction, represents a further development of Herbert Simon's theory of Bounded Rationality. The environment cannot be analyzed only as a structure of the task through its computational variables.[16] The physical and social environment also generates sensory and motor constraints that influence reasoning and action. And in determining a choice, the possible bodily action and the simulated one have an influence in shaping the range of possible options and the value

---

[14] A classic example of pre-adaptation is constituted by the feathers of birds, evolved from dinosaurs presumably for thermal insulation purposes and which then proved to be very useful for flight; or the primitive lung that evolved from the swim bladder of fish. In the human species, the laryngeal folds, which appeared to prevent the regurgitation of food from entering the lungs during vomiting, were subsequently co-opted to produce sounds and transformed into the vocal cords, while maintaining their original function.

[15] Neuroscientists have baptized this reuse of motor programs disconnected from motor output with various names such as "neural exploitation," "neural reuse," and "neural recycling."

[16] For example, refer to the characteristics of the structure of the environment introduced by Gigerenzer and colleagues (Gigerenzer & Gassmaier, 2011) such as uncertainty, redundancy, variability, number of alternatives, and sample size. These characteristics derive from symbolically deconstructed empirical phenomena that are manipulated as cues with statistical meaning (tallied, weighted, sequenced, and ordered).

590   R. Viale

attributed to them. Based on these considerations, it is appropriate to extend the concept of bounded rationality to ecological, as Gigerenzer does, and add the "embodied" attribute.[17]

6   The Embodied Dimension of Social Action[18]

In Viale (2012, 2013), I dubbed the new theory in the methodology of social sciences as *methodological cognitivism* because the conventional base level of our explanations of social phenomena should be the cognitive one. I used the cognitive attribute in a narrow sense compared to most sociologists, but in a broader sense than cognitive scientists, to include all the psychological mechanisms responsible for the decision and therefore the action. Therefore, not only the superior psychological processes, like memory, learning, reasoning, but also emotion, instinct, perception. In the 2012–2013 version, the focus was on classical representationalist cognitivism. We have seen how the computational approach of symbol manipulation and information processing to the study of human cognition is limiting, abstract and unrealistic. If it is true that the focus on human cognition remains central to the explanation of social action, this must be done by characterizing it in its embodied dimension and interaction with the external environment. Let's now see how this articulation of embodied cognition has developed and how it becomes the new backbone of the theory of social action.

The embodied cognition (EC) underlying bounded rationality is a heterogeneous and controversial concept. It presents philosophical and neurocognitive differences that have consequences on the way in which we have to reinterpret problem-solving. The main difference is between models of embodied cognition that accept or do not accept a representational account of embodied cognition in its relationship with the environment. There are several ways of characterizing these differences, and I will do this by following Gallagher (2017).

Weak EC attributes the significant role to what are variously called body-formatted *neural* representations.[19] *Strong* EC argues a significant role for the *non-neural* body cognition. Another way to differentiate EC is between

---

[17] The term "embodied rationality" was introduced by Spellman and Schnall (2009). The term "embodied bounded rationality" was introduced by Viale (2019, 2020) and Gallese et al. (2021).

[18] The present section is based on Viale (forthcoming).

[19] *Weak* EC issues from the Classic Computational model of cognitivism. Goldman and Vignemont (2009) believe that the body plays an important role through the brain representations of its states. Every body representation is formatted in the brain ruling out any role of anatomy and body activity (actions and postures). Such B-formats are representations and processes that represent or respond to

narrow EC, which focuses only on body representation,[20] and *wide* EC, which extends the cognition beyond the embodiment to the dynamic inter-action with the environmental affordances (Enacting), to its external support to cognition (Extended), to its situated dimension (Embedded). This is also known as the "four Es" cognition.

Strong and *wide* EC is based on a series of considerations that demon-strate how the psychological experience does not always correspond to the central computational cognitive functions but to those of the body. Perceptual experience and thought are caused also by peripheral and affective processes. Perceptual stimuli that change the posture of the whole body are interpreted as changes in the surrounding environment.[21] The extraneural structural features of the body shape our cognitive experiences. The way the body is made, its binocular vision, its vertical position, its peculiar rotation of the back and neck, and its manual and movement skills are all characteristics that determine the perceptual and cognitive style. The visceral stimuli derived from the intestine, aptly known as "second brain" (Gershon, 1999) due to its numerous nerve endings, act on an emotional, sensorial, perceptual, and cognitive level. The phenomenon of gut feeling (Gigerenzer, 2007) or the concept of somatic marker (Damasio, 1994) represents this role of visceral stimuli in human action. Hormonal and biochemical changes, for example, blood sugar, have an evident and dramatic effect on human perception and thinking. The body regulates the brain as much as the brain regulates the body. Parts of the brain operate on homeostatic principles via mutual influ-ences between parts of the endocrine system and signals from the autonomic system (Gallagher, 2017, p. 39). These phenomena confirm a *strong* dimen-sion of embodied cognition that does not correspond to a representational account of cognition. Moreover, EC extends beyond the limits of body repre-sentations alone. According to Clark's Extended Mind (2008), the body as well as objects of the environment can function as non-neural vehicles of cognitive processes, performing a function similar to neural processes. For Clark, the body is part of an extended system that starts from the brain and includes the body and the surrounding environment. Just as we store

---

the body, such as perception of a bodily movement, and they are representations and processes that affect the body, such as motor commands.

[20] Goldman and Vignemont (2009) believe that the B-formats are purely internal to the brain and they have no interest in understanding the body interacting with and embedded in the environment. From this point of view, their *weak* EC is also *narrow.*

[21] For example, in a recent interview with *Corriere della Sera* on 17 April 2022, the Italian downhill champion Sofia Goggia stated how the vibrations she felt on the sole of her foot during the descent were essential for her to understand both the slope of the track itself and the speed and fluidity of her skiing. This information generated postural adjustments that had the aim of increasing the speed and fluidity of sliding skis.

information in our brains, we can also do this in external objects and our body has the function of perceiving this external information and allowing the computational activity of the brain on it.

Merleau-Ponty (1962) offers a nonrepresentational account of the way the body and the world are coupled that also points to the wide dimension of strong embodied cognition. According to Merleau-Ponty, as an agent acquires skills, those skills are "stored" not as representations in the mind but as a bodily readiness to respond to the solicitations of situations in the world. If the situation does not clearly solicit a single response or if the response does not produce a satisfactory result, the learner is led to further refine their discriminations which, in turn, solicit more refined responses. Merleau-Ponty calls this feedback loop between the embodied agent and the perceptual world the *intentional arc*. He writes: "Cognitive life, the life of desire or perceptual life, is subtended by an 'intentional arc' which projects round about us our past, our future, [and] our human setting" (Mearleau-Ponty, 1962). Describing the phenomenon of everyday coping as being "geared into" the world and moving toward "equilibrium" suggests a *dynamic* relation between the coper and the environment. Timothy van Gelder (1997) calls this dynamic relation *coupling*. He explains the importance of coupling as follows:

> The fundamental mode of interaction with the environment is not to represent it, or even to exchange inputs and outputs with it; rather, the relation is better understood via the technical notion of coupling... The post-Cartesian agent manages to cope with the world without necessarily representing it. A dynamical approach suggests how this might be possible by showing how the internal operation of a system interacting with an external world can be so subtle and complex as to defy description in representational terms -- how, in other words, cognition can transcend representation. (Van Gelder, 1997)

Van Gelder shares with Brooks (1991) the idea that thought is grounded in a more basic relation of agent and world. Even when abstract thought such as mathematics or theoretical philosophy is involved, it relies on a basic dimension of being in the world. As Van Gelder puts it: "Cognition can, in sophisticated cases [such as breakdown, problem solving and abstract thought], involve representation and sequential processing; but such phenomena are best understood as emerging from [i.e., requiring] a dynamical substrate, rather than as constituting the basic level of cognitive performance" (Van Gelder, 1997).

The situation of the individual is also an important point for differentiating between how human beings and computers relate to the world.[22] The computational approach considers the context irrelevant to the data with which a computer operates. For a computer to be able to respond to a problem, it needs to operate with a set of determined data to which it should assign a set of determined values. Unlike the computer, the human being has the possibility of processing the information in the world holistically, operating even with undetermined and uncertain data. This is possible because of the intentional arc and embodied skills by means of which we have direct access both to what is going on in the world and to the adaptable available solutions to solve our tasks. The environment in which the human being operates is very different from that of games, in which AI is so effective. The information is scarce and the future is uncertain and unstable. On the other hand, the environment in which the subject interacts is not the same both diachronically (i.e., for the individual over time) and synchronously with the individuals with whom he or she interacts (Viale, 2022b). Ecological adaptation and cognitive success depend on the specific subjective *umwelt* (environment) of each individual (Uexkull, 2010). Each of us has a different *umwelt*, that is, we see and consider different aspects of the environment as relevant. The environment is not stable and fixed, but is perceived by the individual in a dynamic and unstable way in relation to the momentary perceptual, sensorial, emotional, motor, visceral, and cognitive characteristics of the individual. In other words, we see the environment differently at different times of our life and different in comparison to what others observe. The salience and attention to some aspects change within the same subject and between different subjects (Viale, 2022b). We have a peculiar view of the environment similar to the frog that does not see an insect if it is immobile, but can only see one in motion. The *umwelt* of the frog is made up only of moving objects. We can, however, coordinate our actions (speech and decisions) by the common characteristics we share.[23]

---

[22] Hubert Dreyfus (1972) was right when, in 1972, he criticized AI because of its disembodied dimension. Without the body, the mind is not able to work and to feel to be situated in the world ("dasein"): "Human beings are somehow already situated in such a way that what they need in order to cope with things is distributed around them where they need it, not packed away like a trunk full of objects, or even carefully indexed in a filing cabinet. This system of relations [...] makes it possible to discover objects when they are needed in our home or our world" (Dreyfus, 1972, p. 172).

[23] Current AI, on the other hand, lacks any *umwelt*. The environment of these machines is composed of the data they obtain to adapt to their parameters. But a deep neural network is not even aware that an environment exists. Alpha Zero can beat any human in chess or Go, but it does not know that it is playing a game called chess, or that there is a human opponent playing against it (Gigerenzer, 2022).

The *enactivist* part of *wide* and *strong* embodied cognition emphasizes the idea that perception is for action and that this action orientation shapes most cognitive processes (Gallagher, 2017, p. 40). The cognition is distributed across brain, body, and environment. How to explain this dynamical coupling? According to Van Gelder (1997) and Gallagher (2007), tools and methods of nonlinear dynamical systems can be used to capture the dynamical coupling between body and environment. Perception depends on sensory-motor skills and possibilities. Gallagher (2017) writes:

> Perception is a pragmatic exploratory activity mediated by movement or action and constrained by contingency relations between sensory and motor processes. One can think of this in terms of ecological psychology where one's perception of the environment includes information about one's own posture and movement, and one's own posture and movement will determine how one experiences the environment. (Gallagher, 2017, p. 41)

According to Husserl (1989), perception is guided by what he calls "I can." We see things in terms of what we can do with them and how we can reach and manipulate them. This also applies to social interaction and social cognition. When we see a person, our perception is based on what we can do with him or her, on the social context of rules and norms, and on the expectations of action caused by that person. Their possible actions represent affordances which leads us to act accordingly. If I play on a football team, it will be the position of my teammate on the field compared to that of the opponents and their known skills that influence how to approach the game and that guide my action of passing the ball. If I am in a political confrontation on TV, my opponent's facial expressions, postures, and gestures along with what I know about his or her previous performance in political debates will guide my style of both verbal and behavioral responses. Our physical engagement with another person in a socially defined environment includes sensory-motor and affective processes that may be online or offline based on past experiences. We inactively perceive the actions and emotional expressions of others as forms of intentionality that are meaningful and directed. Enactive perceptions of others means that we see their emotional expressions and contextualized actions as meaningful in terms of how we might respond to or interact with them (Gallagher, 2017, pp. 77–78). Social interaction involves a reciprocal dynamic, and enactive responses to the other's action, taking the action as an affordance for further action. Activation of mirror neurons that simulate the other's action may be preparatory for enactive responses rather than for a matching action (Newman-Norlund et al., 2007).

This characteristic of social affordances can be seen not only in a bilateral, but also in a multilateral way. In other words, the concept of "I can" can also become that of "*we can*." If in my engagement with others, my enactive response is shaped by my belonging to a social group, it will be calibrated by also referring to the action of the other members of the group. This seems obvious in any team game. In basketball, when a playmaker makes a pass to a teammate stimulated by their position on the pitch, his or her affordance is not only him or herself, but the set of positions of the other teammates in relation to those of the opposing team. In this sense, the sensory-motor stimulus is what "we can" derived from the positions and from the skills of the teammates compared to the opponents. This can be applied in any organization where organizational interaction involves a reciprocal dynamic, and enactive responses to the other's action, taking the action as an affordance for further action involving more members of organization. For example, when a technological prototype has to be created within an industrial company, problem enacting occurs based on the motor responses or the anticipation of the motor responses of the team members according to their specific skills to tackle the solution of the subproblems of the prototype to be created.

In an unorganized social situation such as a crowd in a square, the interaction does not take place on the basis of expectations of actions linked to organizational rules and knowledge of the skills of other people. When one is in a crowd of unknown people, the interaction can be bilateral in relation to the response to certain facial expressions, gestures, and postures. For example, an aggressive attitude makes us react defensively and with avoidance. On the other hand, a friendly attitude leads us to approach and reciprocate in a coherent way. Then, there are intermediate, semiorganized situations, such as being seated in a bar or restaurant in which one has expectations about the behavior of others according to rules codified by experience. The enacting interaction occurs in relation to the actions expected of others: toward a waiter we will act on the expectation of receiving the menu, ordering the dishes, and being served; one will expect a neighbor at the table to maintain a tone of voice that is not too loud, which one will reciprocate in kind; in the event that the neighbor shows postures and gestures of excessive curiosity toward us, we will react in the same way to stop an annoying invasion of our field.

596   R. Viale

7   Conclusion: Why Embodied Individualism?

When I introduced the term Methodological Cognitivism (MC) in the two volumes of 2012 and 2013, the goal was to mark a conceptual shift with respect to the prevailing theses of methodological individualism (MI). The MI was confined to an antipsychological approach; its aim was to explain the action on the basis of a priori rational principles; and its concept of mind was implicitly or explicitly against the identity theory. Contrary to the MI of authors like Popper (1960, 1966), Von Hayek (1973), and Watkins (1973) who deny the social sciences belonging to the sphere of natural sciences, reject the relevance of causal explanation models and propose an interactionist theory of mind (Popper and Eccles, 1977) the MC aims to overturn these philosophical premises. The explanation of social action must take place according to the constitutive causal model used in the natural sciences. Only this epistemological model can make the construction of the theory of social action something similar to what happens in the experimental research of the natural sciences. On the other hand, the same principle of rationality cannot be based on unfeasible and computationally intractable a priori assumptions, but as has been demonstrated by various authors such as Quine (1960, 1969, 1985), Simon (1981, 1982, 1986), Cohen (1981), Stich (1983, 1990), Goldman (1986, 1992, 1993, 1999), it can only be generated on a descriptive or reflexive basis (in the sense of reflective equilibrium à la Goodman, 1965). The normative dimension must not disappear, but must derive from the adaptive success of problem-solving and task-tackling. The concept of bounded and ecological rationality that has progressively become relevant in many disciplines of the human sciences, from microeconomics to organization theory to decision psychology (Viale, 2021) illustrates this change in a pragmatic and adaptive sense (Gigerenzer, 2021). The same interactionist theory of mind of Karl Popper (Popper & Eccles, 1977) is untenable today. The main confrontation in the philosophy of mind has shifted between the theses supporting the mind–body identity (Patricia Churchland, 1986; Paul Churchland, 1981, 1989) and those supporting property dualism (Davidson, 1970; Kim, 1993). The former proposes an eliminative program, which aims to create a unified conceptual language between mind and body, while the latter continues to support an irreducible difference between mental and physical properties. The MC embraces the first monistic and eliminative thesis as a research program under development through the discoveries of neuroscience.

The term MC was introduced in the late 90s when cognitivism was dominated by the information-processing psychology approach. The model of the mind was of the Cartesian type, and the analogy was that of the digital Turing machine. The mind was the space for symbol manipulation and amodal representations. By replacing the term Individualism with Cognitivism, I wanted to emphasize how the theory of social action should be based on the natural causal mechanisms of action. The MC program, also for its monist and eliminative philosophy of mind, was however open to incorporate within it the developments of neurocognitive sciences and embodied cognition. As we have seen in the previous sections a Cartesian model of mind (a mind in a vat) does not explain the role of the body and the environment in the generation of action. A dimension of wide embodied cognition of action incorporates both the causal role of the body and the causal variables of the extended, embedded, and enactive dimension of individual action. Ultimately, it is an *Embodied Individualism* that structurally includes the environmental and social dimension in individual action. The concept of social affordances and the individual enaction toward them demonstrates how there is a horizontal relationship and recursive interaction between individual action and the social environment. The action itself consists of this enactive coupling and the consequent feedback loop with the environmental affordances and the tasks they express. The embedded dimension of the action in the sense of the situational context that shapes it is another concept that expresses the founding social dimension of the action. The same person with the same objectives acts differently in relation to the change of the context and to the network of meanings that constitutes it. And conversely, different people in the same social context have different umwelt, that is, representations of the environment that leads them to act differently. Finally, the extended dimension of cognition, which is not limited only to what is contained in the individual's skull, but is present in all mnemonic supports and external computing devices of an artificial (the digital camera; the cellular phone; the files in our pc; the search algorithms; the expert systems; etc.) but also human type (the memory of the friend; the analysis of the consultant; the knowledge of the teacher; the institutional values contained in the documents; etc.).

Cognitivism has portrayed the mind that thinks and decides as if it were in a vat, separated from the body and the environment. The mind is "disembodied" from the body that carries it and "detached" from the environment in which it interacts. Thought takes place as a form of computation; every mental activity is performed by algorithms similar to the computer's machine language; cognition derives from computational procedures that are carried out on abstract symbolic structures. The new perspective introduced by neuroscience reveals instead an "embodied," "embedded," "enactive," and "extended" cognition: that is, a cognition integrated with the body through action and shaped by the social and physical environment with which the body interacts and where it is located. This bodily interaction with the environment shapes and models the same cognitive activity. The center of gravity of the social action is therefore no longer located in the computational and cognitive part, but it shifts to the pragmatic one, that is, to the possible actions that the body–social–environment interaction allows (Viale, 2011a, 2020, forthcoming).[24]

References

Agassi, J. (1973). Methodological individualism. In J. O'Neill (Ed.), *Modes of individualism and collectivism.* Heinemann.

Anderson, J. A., & Rosenfeld, E. (1988). *Neurocomputing: Foundations of research.* MIT Press.

Antiseri, D. (1996). *Trattato di Metodologia delle Scienze Sociali.* UTET.

Bhargava, R. (1992). *Individualism in social science: Forms and limits of a methodology.* Clarendon Press.

Boudon, R., & Viale, R. (2000). Reasons, cognition and society. *Mind & Society, 1,* 41–56.

Brooks, R. A. (1991). Intelligence without representation. In J. Haugeland (Ed.), *Mind design.* The MIT Press. (Brooks' paper was published in 1986).

Churchland, P. M. (1981). Eliminative materialism and the propositional attitudes. *Journal of Philosophy, 78,* 67–90.

Churchland, P. M. (1989). *A neurocomputational perspective. The nature of mind and the structure of science.* MIT Press.

Churchland, P. M. (1998). The neural representation of the social world. In P. A. Danielson (Ed.), *Modelling rationality, morality and evolution.* Oxford University Press.

Churchland, P. S. (1986). *Neurophilosophy, toward a unified science of the mind/brain.* MIT Press.

Churchland, P. S., & Sejnowski, T. (1992). *The computational brain.* MIT Press.

Clark, A. (2008). *Supersizing the mind: Reflections on embodiment, action, and cognitive extension.* Oxford University Press.

---

[24] I thank a reviewer for his/her remark on the enactive approach of Hayek: "Hayek liked Merleau-Ponty's book *The Structure of Behavior* because he considered it very similar to his book *The Sensory Order*. Hayek's book, whose first draft was written in the 20's, is regarded by scholars like Gerald Edelman, Joaquin Fuster, Jean Petitot and Barry Smith like the first proto-connectionist theory of mind." I do not agree that his theory of mind entails the validity of the interpretative approach of MI. The validity is endangered by Hayek's refusal of the embodied cognition explanation of social action.

Cohen, J. (1981). Can human irrationality be experimentally demonstrated. *The Behavioural and Brain Sciences, 4*, 317–370.

Damasio, A. (1994). *Descartes' error*. Avon.

Davidson, D. (1970). *Actions and events*. Clarendon Press, 1980.

Dilthey, W. (1989). *Introduction to the human sciences*. Princeton University Press.

Dreyfus, H. L. (1972). *What computers can't do: A critique of artificial reason*. Harper & Row.

Elster, J. (1983). *Explaining technical change*. Cambridge University Press.

Fodor, J. (1975). *The language of thought*. Harvard University Press.

Fodor, J. (1990). *A theory of content and others essays*. The MIT Press.

Gallagher, S. (2017). *Enactivist interventions: Rethinking the mind*. Oxford University Press.

Gallese, V., Mastrogiorgio, A., Petracca, E., & Viale, R. (2021). Embodied bounded rationality. In R. Viale (Ed.), *Routledge handbook on bounded rationality*. Routledge.

Gershon, M. D. (1999). *The second brain*. Perennial.

Gigerenzer, G. (2007). *Gut feeling*. Penguin.

Gigerenzer, G. (2021). What is bounded rationality? In R.Viale (Ed.), *Routledge handbook of bounded rationality*. Routledge.

Gigerenzer, G. (2022). *How to stay smart in a smart world*. The MIT Press.

Gigerenzer, G., & Gassamaier, W. (2011). Heuristic decision making. *The Annual Review of Psychology, 62*, 451–482.

Goldman, A. I. (1986). *Epistemology and cognition*. Harvard University Press.

Goldman, A. I. (1992). *Liasons: Philosophy meets the cognitive and social sciences*. MIT Press.

Goldman, A. I. (1993). *Philosophical application of cognitive science*. Westview.

Goldman, A. I. (1999). *Knowledge in a social world*. Clarendon.

Goldman, A., & de Vignemont, F. (2009). Is social cognition embodied? *Trends in Cognitive Sciences, 13*(4), 154–159.

Goodman, N. (1965). *Fact fiction and forecast*. The Bobbs-Merril.

Homans, G. C. (1970). The relevance of psychology to the explanation of social phenomena. In R. Borger & F. Cioffi (Eds.), *Explanation in the behavioural sciences*. Cambridge University Press.

Hurley, S. (1998). *Consciousness in action*. Harvard University Press.

Husserl, E. (1989). *Ideas pertaining to a pure phenomenology and to a phenomenological philosophy-second book: Studies in the phenomenology of constitution*. Kluwer Academic.

Kim, J. (1993). *Supervenience and mind*. Cambridge University Press.

Laudan, L. (1984). *Science and values: The aims of science and their role in scientific debate*. University of California Press.

Llinas, R. (2002). *I of the vortex: From neurons to self*. The MIT Press.

Marr, D. (1982). *Vision: A Computational Investigation into the Human Representation and Processing of Visual Information*. San Francisco: W. H. Freeman and Company. ISBN 0-7167-1284-9.

Marx, K., (1859). *A contribution to the critique of political economy* (M. Dobb, Ed.). Progress Publishers (S. W. Ryazanskaya, Trans.). Lawrence and Wishart (London), and International Publishers (New York) cooperated in the publication of the Progress Publishers edition.

Merleau-Ponty, M. (1962). *The phenomenology of perception.* The Humanities Press.

Mill, J. S. (1843 [2002]). *A system of logic ratiocinative and inductive* (1st ed.). University Press of the Pacific. ISBN 1-4102-0252-6.

Newman-Norlund, R. D., Noordzij, M. L., Mulenbroek, R. G. J., & Bekkering, H. (2007). Exploring the brain basis of joint attention: Co-ordination of actions, goals and intentions. *Social Neuroscience, 2*(1), 48–65.

Popper, K. (1960). *The poverty of historicism.* Routledge and Keegan Paul.

Popper, K. (1966). *The open society and its enemies* (5th ed.). Routledge.

Popper, K. R., & Eccles, J. C. (1977). *The self and its brain.* Springer International.

Putnam, H. (1975). *Mind, language, and reality.* CUP.

Quine, W. O. (1960). *Word and object.* Cambridge University Press.

Quine, W. O. (1969). *Epistemology naturalized: Ontological relativity and other essays.* Columbia University Press.

Quine, W. O. (1985). Epistemology naturalized. In H. Kornblith (Ed.), *Naturalizing epistemology.* Cambridge University Press.

Savage, L. J. (1954). *The foundations of statistics* John Wiley & Sons, Chapman & Hall.

Searle, J. (1983). *Why I am not a property dualist.* http://ist-socrates.berkeley.edu/~jsearle/132/PropertydualismFNL.doc

Simon, H. A. (1947). *Administrative behavior: A study of decision-making processes in administrative organization.* Macmillan.

Simon, H. (1981). *The sciences of artificial.* MIT Press.

Simon, H. A. (1982). *Models of bounded rationality. Volume 1: Economic analysis and public policy. Volume 2: Behavioural economics and business organization.* MIT Press.

Simon, H. A, (1986, October). Rationality in psychology and economics. *The Journal of Business, 59*(4), 209–224.

Spellman, B., & Schnall S. (2009). Embodied Rationality. *Queen's Law Journal, 117.*

Stich, S. (1983). *From folk psychology to cognitive science.* MIT Press.

Stich, S. (1990). *The fragmentation of reason.* MIT Press.

Thaler, R. (1994). Psychology and saving policies. *American Economic Review, 84*(2), 186–192.

Thaler, R. H., & Benartzi, S. (2004). Save more tomorrow™: Using behavioral economics to increase employee savings. *Journal of Political Economy, 112*(S1), S164–S187. https://doi.org/10.1086/380085

Tversky, A., & Kahneman, D. (1974). Judgment under uncertainty: Heuristics and biases. *Science, 185*(4157), 1124–1131.

von Uexkull, J. (2010). *A foray into the world of animals and humans.* University of Minnesota Press.

van Fraassen, B. (1980). *The scientific image.* Oxford University Press.

Van Gelder, T. (1997). Dynamics and cognition. In J. Haugeland (Ed.), *Mind design II* (pp. 439–448). The MIT Press.

Viale, R. (2011). Brain reading social action. *International Review of Economics, 58*, 319–336.

Viale, R. (2011b). Reasons and reasoning: What comes first? In R. Boudon, P. Demeulenaere, & R. Viale (Eds.), *L'Explication des normes sociales*. Presses Universitaires de France.

Viale, R. (2012). *Methodological cognitivism: Mind rationality and society*. Springer.

Viale, R. (2013). *Methodological cognitivism: Cognition, science and innovation*. Springer.

Viale, R. (2018). *Oltre il Nudge*. Il Mulino.

Viale, R. (2019). La razionalità limitata embodied alla base del cervello sociale ed economico. *Sistemi Intelligenti, 31*, 193–203.

Viale, R. (2020). Corpo e Razionalità. In G. Coricelli & D. Martelli (Eds.), *Neurofinanza*. Egea.

Viale, R. (Ed.). (2021). *Routledge handbook on bounded rationality*. Routledge.

Viale, R. (2022a). *Nudging*. The MIT Press.

Viale, R. (2022b). Artificial intelligence should meet natural stupidity. But it cannot! In R.Viale, M. Shabnam, U. Filotto, & B. Alemanni (Eds), *Artificial intelligence and financial behaviour*. Elgar.

Viale, R. (forthcoming). Enactive problem solving: An alternative to the limits of decision making. In G. Gigerenzer, R. Viale, & S. Mousavi (Eds.), *The companion to Herbert Simon*. Elgar.

von Hayek, F. A. (1973). From scientism and the study of society. In J. O'Neill (Ed.), *Modes of individualism and collectivism*. Heinemann.

Von Neumann, J., & Morgenstern, O. (1944). *Theory of games and economic bahavior*. Princeton University Press.

Watkins, J. (1973). Methodological individualism. In J. O'Neill (Ed.), *Modes of individualism and collectivism*. Heinemann.

Weber, M. (1978). *Economy and society*. University of California Press.

Weiss, D. J., & Shateau, J. (2021). The futility of decision-making research. *Studies in History and Philosophy of Science., 90*, 10–14.

Methodological Individualism and its Critics: A Roundtable Discussion

Methodological Individualism, Naive Reductionism, and Social Facts: A Discussion with Steven Lukes

Steven Lukes, Nathalie Bulle, and Francesco Di Iorio

Dr. Steven Michael Lukes, born in 1941, in the UK, is Professor of Sociology at New York University and a fellow of the British Academy. He is regarded as one of the most prominent social and political theorists of our time. The topics of his publications include Durkheim, power, morality, Marxism, and individualism in all its forms. Lukes' article "Methodological individualism reconsidered" (*The British Journal of Sociology*, 1968; republished as a book chapter in his edited volume *Essays in Social Theory*, Macmillan, 1977) and his monograph *Individualism* (Harper & Row, 1973; republished with a new introduction by ECPR Press in 2006) have been major influences on the debate about methodological individualism for over fifty years.

S. Lukes
Politics and Sociology, New York University, New York, NY, USA
e-mail: sl53@nyu.edu

N. Bulle
Sociology, French National Center for Scientific Research, Sorbonne University, Paris, France
e-mail: nathalie.bulle@cnrs.fr

F. Di Iorio (✉)
Department of Philosophy, Nankai University, Tianjin, China
e-mail: francedi.iorio@gmail.com

© The Author(s), under exclusive license to Springer Nature
Switzerland AG 2023
N. Bulle and F. Di Iorio (eds.), *The Palgrave Handbook of Methodological Individualism*,
https://doi.org/10.1007/978-3-031-41508-1_25

606   S. Lukes et al.

Lukes' approach is related to a conceptual framework and a style of reasoning that are typical of the philosophy of language and the logical-empiricist tradition. Because of this, his account of methodological individualism is based on assumptions that differ from the ones traditionally used by individualist social scientists. Unlike the latter, he conceives the individualist explanation as a linguistic problem linked to a specific kind of factual knowledge. According to Lukes, the doctrine of methodological individualism can be summarized as follows: "facts about society and social phenomena are to be explained solely in terms of facts about individuals" (1977 [1968], p. 178). By this, he means that, from the perspective of this doctrine, explanations should consist only of "predicates" about "individuals" and their properties and should not include predicates about social and institutional factors and their properties (ibid., p. 180). For example, if we adopt the canons of methodological individualism, Lukes contends, we are "compelled to talk about the tribesman but not the tribe, the bank-teller but not the bank" (ibid., p. 184). In his opinion, methodological individualism is "an exclusivist, prescriptive doctrine about what explanations are to look like" (Lukes, 2006 [1973], p. 99) and is centered on a "futile linguistic purism" (Lukes 1977 [1968], p. 184). According to Lukes, this doctrine must be regarded as clearly implausible. He recognizes that there are both more atomistic and less atomistic variants of methodological individualism. The latter (e.g., Popper's), unlike the former, consider the actor as influenced by the context and "allow 'situations' and 'interrelations between individuals' to enter into explanations" (ibid.., p. 186). However, these less atomistic variants are, in his view, only apparently more acceptable because: (i) it is difficult to see in which sense they are expressions of methodological individualism; and (ii) they cannot achieve the kind of linguistic purism they claim to adhere to.

In the introduction to the 2006 edition of *Individualism*, written more than thirty years after the book's first release, Lukes reframes and enlarges his earlier analysis in the light of more recent debates on methodological individualism. He argues that the approach he originally provided in 1973 must now be regarded as of limited utility because today three variants of methodological individualism may be distinguished and his old definition applies only to one of them. The two which fall beyond the scope of his earlier objections are: (i) "a reductionist doctrine, according to which holistic laws about social wholes must be reduced to individualist laws about individuals" that is inspired by Ernest Nagel's work and may be regarded as seriously problematic because this reduction seems to be impossible (Lukes, 2006, p. 5); and (ii) the view that causal claims about macro phenomena "must always be supplemented by accounts of causal mechanisms at the level of individuals," that is the view that macro-explanations need "microfoundations" (ibid., p. 6).

In Lukes' opinion, this third variant of methodological individualism developed by scholars such as Raymond Boudon, Jon Elster, and Daniel Little "has proved central to the debate over this doctrine since the 1970s" (ibid.) He maintains that it is much more defensible and useful than the two others because it "does not entail that all explanations must be exclusively in individualist terms" and "can allow for the socially structured shaping of individuals' desires, beliefs and dispositions" (ibid.). According to Lukes (ibid., p. 8), a shortcoming of "this third version of methodological individualism—the mere challenge to provide microfoundations for social phenomena"—is that it has "mainly been made in the name of one or another version of the rational choice perspective," which means with the assumption of an abstract model of action and thus, as highlighted by the Polish-American political scientist Adam Przeworski (1990, pp. 64–65) of "unidifferentiated, unchanging, and unrelated individuals." In Lukes' opinion, Przeworski (ibid.) summarizes the problem along lines that can be found in *Individualism*: while the critique of Marxism offered by methodological individualism is "salutary" because "history must have microfoundations," the theory of action cannot be abstract, but "must contain more contextual information."

Bulle and Di Iorio: From the standpoint of the history and practice of the social sciences, the interpretation of methodological individualism in terms of a linguistic exclusivism that rules out the possibility to develop explanations including predicates about social and institutional factors is difficult to understand and justify. This interpretation stems from the attempt to apply categories of the philosophy of language and logical positivism that do not belong to the individualist tradition to make sense of scholars like Menger's and Weber's thought in a way that looks uncharitable. These individualist social scientists never defined their methodology in terms of linguistic exclusivism, but used social predicates abundantly and conceived social phenomena as systems of interaction. Moreover, for them, individuals are not brute facts in the sense of logical empiricism, but theoretical constructs that depict situated actors, whose beliefs and actions can only be defined on the basis of social and cultural concepts. Methodological individualists like them attach crucial relevance to individuals not in the sense of linguistic exclusivism, but to challenge holism understood as the tendency to conceive social wholes in terms of naïve realism and actors as "puppets" that are remote-controlled by social and historical deterministic laws.

608   S. Lukes et al.

Could you clarify why you think otherwise and which evidence supports the claim that scholars like Menger and Weber are committed to a useless linguistic exclusivism?

Steven Lukes: I am no expert on Menger and accept your assertion that he was not an advocate of "linguistic exclusivism,...used social predicates abundantly and conceived social phenomena as systems of interaction." Menger's principal concern was, I take it, to reject the notion of a "national economy" advanced by the historical school of economists, and to see it rather as the outcome of interaction resulting from individuals' actions in pursuit of their respective material interests. Weber was similarly fully aware that macro institutions (bureaucracy, the state) and economic systems (capitalism) consist of typical modes of interaction among individuals, though he had a richer view of human interests and was less preoccupied than Menger was with refuting what you call "naïve realism." By this phrase, you mean the view that individual actors are remote-controlled "puppets." This might help explain the old puzzle of Weber's lack of interest in his contemporary Durkheim's methodological writings (Tiryakian, 1966), with its talk of social facts as "*sui generis*" (which Weber may well have dismissed as naïve realism as you characterize it, without bothering to inquire further into what Durkheim was seeking to explore, notably normative pressure and the power of symbolism). You are right to claim that neither Menger nor Weber was "committed to a useless linguistic exclusivism."

Bulle and Di Iorio: In your opinion, the multiple-realization problem, which was originally developed in the philosophy of mind, is a valid argument against methodological individualism. As applied to the criticism of this approach, the multiple-realization problem can be summarized as follows: since any social phenomenon "may supervene upon multiple individualist type descriptions" the reduction of a macro-level to the properties of the individuals (micro-level) "cannot be carried out" (Lukes, 2006, pp. 7–8). For example, a market system can be realized in various ways by distinct individuals who may be interrelated in different ways (Kincaid, 1986; Hansson Wahlberg, 2019). As a consequence, a market system is not reducible to the properties of concrete individuals, but is compatible with different micro-level or individual properties.

The validity of this criticism of methodological individualism depends on the validity of the assumption that this approach is tantamount to the view that social properties must be reduced to the properties of concrete individuals. A problem with this criticism is, to use Weber's words, the ideal-typical nature of individualist explanations. Individualist scholars do not account for macro-level phenomena like a market in terms of concrete individuals and their biographical features, but as pointed out by Schutz (2011, p. 86–92), they refer to ideal-typical "puppets" and typical forms of social interaction. Consider, for example, Weber's Calvinist entrepreneurs and his abstract model of a capitalist economy (Weber, 2013 [1905]). From the standpoint of methodological individualism, understanding a system of interaction like a market cannot be carried out in purely structural terms as some of its critics assume it can. According to this approach, focusing only on structural factors without giving any relevance to individual intentionality and human agency is mistaken because structural factors alone cannot account for the logic of social phenomena. However, as pointed out above, the fundamental information about the modes of action and social interaction is only taken into account in a typical form, without considering the particular individual causal histories at the actual origin of the macro-phenomenon under investigation.

Given the above, why do you regard methodological individualism as a form of naïve reductionism and the multiple-realization problem as a valid argument against this approach?

Steven Lukes: I am very sympathetic to the point you raise in asking this question. As it happens, I have recently published an essay about James Joyce's *Ulysses* (Lukes, 2022) in which I make exactly this point. I argue there that.

Individuals as viewed by the social sciences are always abstract individuals. They exist only in thought or as an idea and, being abstract, do not have a physical or concrete existence. The social scientific individual, therefore, is fictional, not real (351).

And that, by contrast.

In actual fiction, by contrast, as in novels and plays, the goal is not, I suggest, typically explanation but rather exploration. One concludes a process with closure, the other consists of a process of unfolding. Explanation yields understanding by resolving puzzles; exploration does so by pursuing questions that generate puzzles (352).

Moreover, I even make the same connection with Schutz, who, addressing the question: how is one to give an explanation of the social world, answers that doing so involves "replacing the human beings which the social scientist observes as an actor on the social stage by puppets created by himself, in other words, in constructing ideal types of actors." This requires imputing "motives in the mind of an imaginary actor" whose consciousness is "restricted in its contents only to all those elements necessary for the performance of the typical acts under consideration. These elements it contains completely but nothing beyond them" (Schutz, 1960, pp. 218, 220). Here, of course, the

610   S. Lukes et al.

"puppets" to which the early Schutz refers (later typifications) are not remote-controlled by putative collectivities but are rather manipulated by the social scientist for explanatory purposes.

Bulle and Di Iorio: you sharply distinguish Boudon's and Elster's micro-foundationalist approach from classical methodological individualism. This is because you regard these two scholars—along with Little, who, though unlike them, never endorsed methodological individualism—as the originators of a new and more acceptable variant of this orientation that is not committed to linguistic exclusivism. However, Boudon and Elster understood their approach as the implementation of the old methodological individualism rather than as an improved version of it. According to Boudon (1986, p. 55), the principle of methodological individualism, as applied in his empirical works, was "clearly indicated by the classical German sociological tradition (Weber and Simmel, for example)." In his opinion, it did not start with economists like Hayek in the twentieth century, but was much older and "had great influence on all the social sciences in Austria and Germany at the end of the nineteenth century" (ibid., p. 231, note 27). Elster (1985 ibid., pp. 3–4) considers his microfoundationalist approach as rooted in the eighteenth century "theories of history that saw it as the result of human action, not of human design" (see also 1989, p. 91ff.). Moreover, while he criticizes Weber's view that psychology is useless in the social sciences and partly disagrees with his theory of rationality, he does question whether Weber's sociology takes into account social conditioning and allows for reference to social predicates (Elster, 2000).

Could you clarify your view that Boudon's and Elster's microfoundationalist approach must be sharply distinguished from classical methodological individualism?

Steven Lukes: I see the force of your point and would not insist on distinguishing Boudon's and Elster's versions from the classical form of methodological individualism.

Bulle and Di Iorio: In *Methodological individualism reconsidered* you argue that methodological individualism is an explanatory theory, not an ontological one (Lukes, 1977, pp. 180–181). This is true. However, while some methodological individualists like Boudon avoid taking a clear ontological stand, others like Mises, Hayek, and Popper maintain (at least in some of their works) that, even if, as an explanatory principle, methodological individualism must not be confused with an ontological view, it can be regarded as underpinning a nominalist ontology. You strongly criticize this kind of ontology that you understand as the idea that all the words and predicates that do not refer to strictly individual properties are empty and nonsensical because "only individuals are real" (ibid., p. 180), while social factors and phenomena do not exist. The problem, you contend, is that these factors and phenomena exist and are explanatorily relevant.

It seems to us that the scholars mentioned above never deny this. They only argue that social factors and phenomena must not be interpreted in terms of naïve realism, that is in terms of *sui generis* entities that exist and function "independently of the individuals" (Hayek, 1948, p. 6; see also Popper, 1966, p. 421). They recognize the role of emergent properties and the necessity to study and analyze the structural features of society and their influence on individuals (see Hayek, 1967, p. 70; Popper, 1957, pp. 82, 76), but argue that these features, which are, in their opinion, often unintentional, must be conceived in terms of global properties resulting from the interaction of a collection of individuals rather than in terms of holistic entities (according to methodological individualism, the study of these global properties is carried out in terms of typically organized structures and typical situated actions). This view has nothing to do with the claim that all the words and predicates that do not refer to strictly individual properties are empty.

Your opposition to nominalism can be broken down into four arguments. Argument (i): in the social world "both individual and social phenomena" (for example, the procedure of a court) "are observable" (Lukes, 1977, p. 180). This is not denied by methodological individualists. According to them, both phenomena that are describable in terms of a vocabulary about individual properties (e.g., John running on a desert beach) and phenomena that are not (e.g., the French Revolution) are observable. Argument (ii): it is false "that individual phenomena are easy to understand, while social phenomena are not…: compare the procedure of the court with the motives of the criminal" (ibid.). Even in this case, we fail to see the difference between your view and that of methodological individualists. In the latter's opinion, the procedure of a court is tantamount to typical forms of interaction between individuals underpinning typical ways of thinking and acting and, as such, is not less understandable than the behavior of an isolated individual. Argument (iii): individuals do not "exist independently of…groups and institutions" because "as facts about social phenomena are contingent upon facts about individuals, the reverse is also true" (ibid.). We can only speak of soldiers because we can speak of armies. As clarified above, rejecting the holistic ontology does not prevent us from referring to semantically irreducible concepts and from using these concepts to make sense, for example, of the social roles of the individuals within a bureaucratic institution like an army. Even in this case, we fail to see any incompatibility between your view and the one of methodological individualists. Argument (iv): the idea that

612   **S. Lukes et al.**

"all social phenomena are fictional and all individual phenomena are factual" is absurd because it "would entail that all assertions about social phenomena are false or else neither true nor false" (ibid., p. 181). As pointed out above, methodological individualists do not consider social phenomena as fictional and do not assume that the assertions about these phenomena are false or meaningless.

Could you explain why you do not agree with our analysis?

Steven Lukes: It seems to me that maybe Durkheim and what we may broadly call the Durkheimian legacy, explored and expounded in Philip Smith's excellent book *Durkheim and After* (Smith, 2020), is the "elephant in the room" here. You offer a sharp contrast between your "naïve realism," where *sui generis* entities are claimed to exist and function "independently of the individuals," and the ways in which Mises, Hayek, and Popper recognize "the role of emergent properties" and "the structural features of society and their influence on individuals." You are correct in observing that they, like others in this and other traditions, conceptualize these as *emergent*—as the *outcome* of social interaction. Often the outcomes are unintended. I agree with Kenneth Arrow that "the *emergent* nature of social phenomena ...may be very far from the motives of the individual interactions" (Arrow, 1994, p. 3; see Schelling, 1978 for a variety of examples). I further agree with Arrow that "individual behavior is always mediated by social relations" and that these are "as much a part of the description of reality as is individual behavior" (5). In characterizing them as emergent, the authors you cite here acknowledge this. What they do not focus on and largely miss is the influence on individuals of structural features of society. This is what the Durkheimian perspective enables us to focus upon: namely, all the ways in which "social facts" are *experienced* by individuals as external, constraining, general, and independent of their individual manifestations, and understood as preexisting and outlasting individuals. This is how, for instance, moral norms and their weakening or absence (as in anomie) are experienced. It is, as anthropologists say, an *emic* account, as opposed to an *etic* account of how norms function to influence and shape both experience and behavior via networks of shared expectations. Durkheimian "*sui generis*" social phenomena do not exist "independently of the individuals" but exist within them and act upon them.

Bulle and Di Iorio: In your opinion, while the microfoundationalist variant of methodological individualism, which has been endorsed among others by Elster and Boudon, "has been of considerable value," it also possesses a shortcoming (Lukes, 2006, p. 8). This is because it has been developed "in the name of one or another version of the rational choice perspective," which means positing an abstract, invariable, and thus, partly

Methodological Individualism, Naive Reductionism ...

atomistic concept of the individual (ibid.) In agreement with Adam Przeworski (1990, pp. 64–65) you think that "the theory of individual action must contain more contextual information than the present paradigm of rational choice admits."

It is true that the rational choice model played a role in the history of methodological individualism, especially in economics, but this approach cannot be assimilated or reduced to the former. Boudon's individualist sociology is centered on a cognitive or ordinary rationality that is strictly related to a criticism of the limits inherent to the rational choice model (Boudon, 2001, 2009, 2013). According to him, this model can only explain a limited typology of actions, and fails to account for those dependent on positive beliefs (the descriptions of the world and its mechanisms) and normative beliefs (the endorsement of ethical standpoints). In Boudon's opinion, these two kinds of belief must be explained neither in illogical terms, nor in holistic ones, but result from a situated, non-utilitarian and non-instrumental rationality. Positive beliefs, Boudon contents (1994, pp. 235–246), are regarded as true or false independently of their utility and their social and economic consequences, but presuppose nevertheless a system of "good reasons" that are understandable taking into account the sociocultural background of the agent. Similarly, for him, normative beliefs are not accepted blindly or because of a sociocultural determinism, but result from an argumentative rationality that is at odds with the Cartesian model of rationality and the utilitarian theory of action, while matching well the assumptions of Perelman's and Toulmin's "New Rhetoric" (Boudon, 1995, pp. 194–222, 259–259).

As pointed out by Boudon and Bourricaud (1990, p. 14), methodological individualism assumes that the agent "operates in a framework of constraints determined both by socialization and by the structure of the situation." As a consequence, understanding his/her rationality means taking into account this framework and the way it affected his/her actions (ibid. pp. 287–289). In other words, methodological individualism assumes that "the idea of rationality must thus be seen as relative, that is to say as dependent upon the structure of situations. To be sure, it must also be seen as dependent upon the position and generally the characteristics of the actors" (p. 290).

Why do you consider all the recent variants of methodological individualism as committed to the rational choice model?

Steven Lukes: This question is largely about whether I have been fair to Boudon and Elster. I readily acknowledge that Boudon sought to supplement rational choice models with his account of 'argumentative rationality' and that Elster, having first embraced rational choice theory also later frankly

614   S. Lukes et al.

acknowledged its limitations while resolutely insisting on his adherence to methodological individualism. I do not insist that the latter is irretrievably committed to the rational choice model. I agree, once more, with Kenneth Arrow that "the individualist viewpoint is in principle compatible with bounded rationality, with violations of the rationality axioms, and with the biases in judgment characteristic of human beings" (Arrow, 1994, p. 4). And I continue to find considerable value in the insistence upon the need to find microfoundations for macro and meso explanations, as a guard against attempts to explain outcomes that misapply the language of human agency to disembodied collectivities (such as class or culture) or social relations (such as power relations).

References

Arrow, K. (1994). Methodological individualism and social knowledge. *American Economic Review, 84*(2), 1–9

Boudon, R. (1986). *Theories of social change*. Polity Press.

Boudon, R. (1994). *The art of self-persuasion*. Polity Press

Boudon, R. (1995). *Le juste et le vrai*. Fayard.

Boudon, R., and F. Bourricaud. (1990). *A critical dictionary of sociology*. University of Chicago Press.

Boudon, R. (2001). *The origins of value*. Transaction Publishers.

Boudon, R. (2009). *La rationalité*. Puf.

Boudon, R. (2013). *Sociology as science: An intellectual autobiography.* The Bardwell Press.

Elster, J. (1985). *Making sense of Marx*. Cambridge University Press

Elster, J. (1989). *Nuts and Bolts for the social sciences*. Cambridge University Press

Elster, J. (2000). Rationality, economy, and society. In S. Turner (Ed.), *The Cambridge companion to Weber*. Cambridge University Press.

Hansson Wahlberg. T. (2019). Why the social sciences are irreducible. *Synthese, 196*(12), 4961–4987.

Hayek, F. A. (1948). *Individualism and economic order*. University of Chicago Press.

Hayek, F. A. (1967). *Studies in philosophy, politics and economics*. University of Chicago Press.

Kincaid, H. (1986). "Reduction, Explanation, and Individualism". *Philosophy of Science, 53*(4), 492–513.

Lukes, S. (1977 [1968]). Methodological individualism reconsidered. In Lukes S (Ed.), *Essays in social theory*. Macmillan.

Lukes, S. (2006 [1973]). *Individualism.* Harper & ECPR Press.

Lukes, S. (2022). Joyce's Ulysses: Social science, fiction and reality. *Social Research, 89* (2), 351–360

Menger, C. (1985 [1883]). *Investigations into the method of the social sciences with special reference to economics*. New York University Press.

Popper, K. R. (1966). *The open society and its enemies, Vol. 2: Hegel and Marx*. Princeton University Press.

Popper, K. R. (1957). *The poverty of historicism*. Beacon Press.

Przeworski, A. (1990). Marxism and rational choice. In P. Birnbaum & J. Leca (Eds.), *Individualism: Theories and methods* (pp. 64–78) (J. Gaffney, Trans.). Clarendon.

Schelling, T. (1978). *Micromotives and macrobehavior*. Norton.

Schutz, A. (1960). The social world and the theory of social action. *Social Research, 27*(2), 203–221.

Schutz A. (2011). *Collected Papers V: Phenomenology and the social sciences*. Ed. L. Embree. Springer Science+Business Media.

Smith, P. (2020). *Durkheim and after the Durkheimian tradition, 1893–2020*. Polity

Tiryakian, E. A. (1966) A problem for the sociology of knowledge: The mutual unawareness of Emile Durkheim and Max Weber. *European Journal of Sociology, 7*(2), 330–336.

Weber, M. (2013 [1905]). *The protestant ethic and the spirit of capitalism*. Merchant Books

Methodological Individualism and Institutional Individualism: A Discussion with Joseph Agassi

Joseph Agassi, Nathalie Bulle, and Francesco Di Iorio

Dr. Joseph Agassi (born in Jerusalem on May 7, 1927) is Emeritus Professor at Tel-Aviv University, Israel, and York University, Canada. He is one of the most eminent Popperian epistemologists and has written extensively on a wide array of philosophical, logical, and historical issues. His contribution to the individualism-holism debate, which has captured the attention of both philosophers and social scientists over the past few decades, is centered on the idea that we need to reject not only holism, understood as the idea that social wholes "have distinct aims and interests of their own" (Agassi, 1960, p. 245), but also a large part of the individualist tradition. In his opinion, most of that tradition is more or less committed to psychologism, namely,

---

J. Agassi
Tel-Aviv University, Tel Aviv-Yafo, Israel

York University, Toronto, ON, Canada

N. Bulle
Sociology, French National Center for Scientific Research, Sorbonne University, Paris, France
e-mail: nathalie.bulle@cnrs.fr

F. Di Iorio (✉)
Department of Philosophy, Nankai University, Tianjin, China
e-mail: francedi.iorio@gmail.com

© The Author(s), under exclusive license to Springer Nature Switzerland AG 2023
N. Bulle and F. Di Iorio (eds.), *The Palgrave Handbook of Methodological Individualism*,
https://doi.org/10.1007/978-3-031-41508-1_26

618   J. Agassi et al.

the "theory that every social theory … is reducible to psychology; that every social explanation can be fully explained, in its turn, by a purely psychological explanation" (Agassi, 1987, p. 119). Focusing on the reasons for the untenability of psychologism, Agassi has highlighted the necessity of endorsing an anti-psychologist version of individualism developed by his mentor Karl Popper, which Agassi has called "institutionalism" or "institutional individualism" (Agassi, 1960, 1975, 1987). In his view, institutionalism is "Popper's great contribution to the philosophy of the social sciences" (Agassi, 1987, p. 119). A corollary of Agassi's analysis is that the term "methodological individualism" is not the most appropriate way to describe the proper method of the social sciences because it is too generic and confusing, referring indiscriminately to both the variants of individualism mentioned above. He considers the term "institutional individualism" to be a preferable option.

Two differences between psychologism and institutionalism should be emphasized. The first is that while, according to psychologism, only individuals exist, institutionalism claims "that certain social entities exist, and are of primary importance to the social sciences" (1987, p. 123), though they exist "not in the same sense in which people exist" (1987. p. 127). However, while institutionalism assumes that these entities—namely, institutions—exist, it denies that they have "(distinct) interests" (ibid.). According to this anti-holistic view, an institution "may have aims and interests only when people give it an aim, or act in accord with what they consider should be its interest" (ibid.). Yet, Agassi contends, both the individual and the institutional background exist, which means "that we cannot reduce psychology into sociology and we cannot reduce sociology into psychology" (ibid.). The second difference between psychologism and institutionalism is that the latter approach claims that the explanation of social phenomena assumes that "individuals are affected by social conditions, and in their turn affect them" (1987, p. 129). Social phenomena are produced not only by individual psychology, but also by the existing institutional structures (1987, p. 145). In Agassi's opinion, the problem with psychologistic explanations is that they do not give institutional factors any causal role and assume that the very analysis of institutional phenomena must be carried out in atomistic terms. He firmly rejects this atomistic view: "It is an error to assume that the only satisfactory explanation of institutions is by assumptions which say nothing about institutions" (1987, p. 146). In his view, arguing that institutions exist and produce social constraints means arguing that individual behavior takes places within a coordinated system of actions: the institutional framework. The framework of the "existing social co-ordinations" constitutes "an important factor in determining the rational or purposeful behavior" of the agent and

Methodological Individualism and Institutional ... **619**

its social consequences (ibid.). The "inter-personal means of co-ordination" that are represented by institutions "can be explained as … attitudes which are accepted conventionally or by agreement" (1987, p. 150). According to Agassi, his view "accords with the classical individualistic idea that social phenomena are but the interactions between individuals. Yet it conflicts with the classical individualistic-psychologistic idea that interaction depends on individuals' aims and material circumstances alone; rather it adds to these factors of interaction the existing inter-personal means of co-ordination as well as individuals' ability to use, reform, or abolish them, on their own decision and responsibility" (Agassi, 1987, p. 150).

Bulle and Di Iorio: Professor Agassi, you claim that institutional individualism—a term that is mentioned for the first time in an article that you published in *The British Journal of Sociology* in 1960—is an approach invented by your mentor Karl Popper. However, Popper never discussed the concept of institutional individualism. Following the members of the Austrian School of Economics, he always defined himself as a methodological individualist. Moreover, he never argued that the entire individualist tradition before him was committed to psychological reductionism. Could you discuss this point and try to summarize the contents of your personal conversations with Popper about methodological individualism, institutional individualism, and your interpretation of the pre-Popperian individualist tradition as mainly psychologistic?

Agassi: There was nothing to discuss: Popper's position was clearly stated in Chapter 14 of *The Open Society and Its Enemies* on the autonomy of sociology. His new argument against psychologism is stunning: psychologism entails the methodological myth of the social contract. Needless to say, we should avoid methodological myths. Popper never responded to my paper except once when I expressly asked for it. He said he liked the assertion that institutions have no aim of their own. This idea is found in his *The Open Society and Its Enemies*.

Bulle and Di Iorio. It seems to us that your analysis of the history of the individualist tradition, which leads you to brand most of this tradition as psychologistic, possesses some problematic features. The fictional and unrealistic models or situations outlined by mainstream mathematical economics and social contract theory can certainly be regarded as tendential forms of atomistic individualism that more or less deny the relevance of social structures. However, precisely because of their fictional assumptions, these approaches must not be confused with the explanatory logic of methodological individualism properly understood. For example, interpretative sociology has little to do with these oversimplistic constructs.

620   J. Agassi et al.

In your opinion, according to pre-Popperian individualism, only individuals exist, while institutions, understood as social coordinations accepted conventionally by individuals, do not exist as relatively independent entities. This is why you consider pre-Popperian individualism atomistic, that is, as a theory that denies that action takes place within a preexisting set of social structures that influence it.

However, it seems to us that well-known non-atomistic individualist sociologists and economists such as Menger, Weber, Simmel, Spencer, Mises, and Hayek never claimed that institutions do not have a real existence and are explanatorily irrelevant, but developed a more articulated and defensible position. This position can be summarized as follows: social factors and social institutions should not be conceived in terms of naïve realism. This means that they should be studied and taken into account in economic and sociological explanations, but they should not be interpreted in holistic terms as *sui generis* entities, that is as entities that "exist independently of the individuals which compose them" (Hayek, 1948, p. 6). Institutions should be regarded rather as abstract models that describe typical and systemic modes of interaction between concrete individuals. This is not a denial of the existence of the institutions and their influence in terms of social conditioning: it is rather a criticism of the misuse of collective concepts in the social sciences and the tendency to regard wholes as entities acting on their own as separate causes.

Take, for example, Menger's methodological reflections. He claims that social phenomena must always be explained in terms of individuals, but also affirms that action takes place within structures of interaction that affect it and are created by rules commonly accepted by the actors (1985 [1883], pp. 139–142). In his opinion, the methodology of the social sciences needs to focus on the systemic features of the social world: each part of the global system called society—each individual or each social subsystem (e.g., a firm, a university, a family)—he contended—"serves the normal function of the whole, conditions and influences it, and in turn is conditioned and influenced by it in its normal nature and its normal function" (ibid., p. 147).

While we admire Popper's social theory, it seems to us that what you call institutionalism is older than his work. In our opinion, Popper's theory of institutions is in line with the anti-holistic theory of the collective concepts developed by Menger and the other scholars mentioned above. For Popper (2002 [1957], p. 126), institutions are not *sui generis* entities, but "abstract objects." They "are theoretical constructions" or "models" that are "used to interpret our experience" (ibid.): "Even 'the war' or 'the army' – he pointed out – are abstract concepts, strange as this may sound to some. What is concrete is the many who are killed; or the men and women in uniform, etc." (ibid.). He pointed out that this anti-holistic view "does not assume the existence of collectives; if I say, for example, that we owe our reason to 'society', then I always mean that we owe it to certain concrete individuals – though perhaps to a considerable number of anonymous individuals – and to our intellectual intercourse with them" (Popper, 1966, p. 421).

Agassi: The question that you raise applies generally to all modern western social thinkers. Consider John Locke. He studied property, which is as much a social institution as you can find. Did he study it as an institution? Yes. Did he consider it as such? I suppose the answer is, no: why else would he present it as an extension of one's body? Did Weber and Mises assert that institutions exist? Did they deny that institutions exist? Twice, no. It seems you slightly misrepresent them. I may be in error, of course, but no one has corrected my assertion thus far: collectivists say only collectives exist; individualists say only individuals exist; Georg Simmel says both exist; Popper says, collectives have no aims of their own. Popper refers to Simmel once, regrettably dismissing him with little familiarity (since his epistemology is old-fashioned).

I have discussed briefly the contrast between ontological and methodological institutionalism: they are logically independent, since one is a proposal, the other is a statement.

Bulle and Di Iorio: From a methodological standpoint, individualism takes into account the relevance of social and intersubjective factors. Focusing on Popper's theory of World 3 helps us to understand this point. Your interpretation of Popper's methodology of the social sciences in terms of institutional individualism seems to be partly related to this theory. However, this point is not extensively discussed and carefully analyzed in your works. According to Popper (2003 [1977], 38), the relationship between mind and body has to be explained in terms of interaction between three worlds: (a) the physical world, or World 1;(b) the world of mental states, or World 2; and (c) the world of culture, or World 3. Authors such as Chalmers (1985) and Udhen (2001) argue that methodological individualism as traditionally understood is incompatible with Popper's three worlds theory. This is because, for Popper, World 3, which includes all cultural artifacts (e.g., scientific theories and institutions), must be considered real and somehow autonomous from the other two worlds. Moreover, in his opinion, World 3 is "irreducible" to either World 1 or World 2 and "exerts a causal influence upon" the World 2 of individual mental states (Popper [1977] 2003).

Agassi: All this is much ado about very little new. For example, the theory says that a symphony is not any of its performances, much less any of the scores of it (not even the initial score, since it is often lost). What is it, then? Or, what is a text? Carnap faced this question. He said it is ink spots. This is as silly as possible, since the ink spots may include errors, be translated, etc. Putnam said it is that or soundtracks. He forgets semaphore! How silly can one get? All Popper did is to avoid the current silly suggestions. A text is something abstract. This is all of the World 3 ideas.

Bulle and Di Iorio: We see no reason why Popper's theory of World 3 should be regarded as incompatible with methodological individualism. According to Popper (1978, p. 145), the objects of World 3 exist as "abstract objects." They are not merely mental, but objective knowledge pertaining to a public, intersubjective world. This requires that they are formulated "in some language" and therefore can be the "object of criticism" (1978, p. 159). For example, a scientific theory exists as such not because of subjective mental states, but precisely because it is publicly criticizable knowledge. In Popper's opinion, human beliefs, including empirical descriptions of the natural world scientific theories, are World 3 objects only if we can speak of their "truth" or "falsity" or of their "logical relationship" with other beliefs (1978, p. 160). If a mental state is not expressed in language, it cannot be discussed and criticized. As a consequence, it "is merely a World 2 process, it is merely a part of ourselves" (ibid.). It does not exist as objective knowledge. According to Popper, the relatively autonomous existence of World 3 objects also depends on the fact that sometimes their physical realization is impossible. For example, since the sequence of natural numbers is infinite, "there is no embodiment" of it (1978, p. 161). Popper also contends that, given that the objects of World 3 are expressed in an understandable language and can be publicly criticized, they can "cause" individuals "to think" (1978, p. 163).

Agassi: You say, "In Popper's opinion, human beliefs, including scientific theories empirical descriptions of the natural world, are…." This cannot be true. From the start Popper ignored beliefs as belonging to psychology and so logically after methodology, not prior to it.

Bulle and Di Iorio: We mean here the *contents* of beliefs. Their contents can be either true or false and thus can be regarded as belonging to objective knowledge, to World 3.

Let's focus on issues strictly related to the philosophy of the social sciences. It seems to us that the theory of World 3 presupposes methodological individualism and should not be regarded as incompatible with it for at least four reasons. First, since this theory defends the existence of objective knowledge, it is also supportive of universally shared fundamental rational mechanisms. The existence of these intercultural mechanisms is a basic assumption of interpretative sociology.

Second, Popper (2003, p. 43) stressed that his theory of World 3 must not be confused with the naive essentialist ontology rooted in Plato's thought that inspired the holistic views developed by Hegel and Marx: "Though Plato's world of intelligible objects corresponds in some ways to our World 3, it is in many respects very different. It consists of…. essences." While I stress the existence of World 3 objects, I do not think that essences exist…I am an opponent…of "essentialism." Thus, in my opinion, Plato's ideal essences play no significant role in World 3" (Popper, 2003). In Popper's view, Plato's essentialism is the presupposition of the holistic assumption that we can grasp the laws governing a society as a whole (Popper, 2002 [1957], p. 76), i.e., of the assumption that the true "significance" of an action is "determined by the whole" (Popper, 2002 [1957], p. 22), understood as "the structure of all social and historical events of an epoch" (ibid., p. 78).

Agassi: So much discussion of so simple a point: Popper opposes methodological essentialism but allows that some essences exist—of human products such as science—introduced as hypotheses, not as definitions. They dwell in World 3.

Bulle and Di Iorio: We have further reasons we wish to outline. The third concerns Popper's idea that World 3 has a causal role and is not committed to holism because, according to him, "World 3 theories and world 3 plans and programs of action must always be grasped or understood by a mind before they lead to human actions" (Popper, 1978, p. 164). This is at odds with the view that social factors determine human minds mechanically and unconsciously. For him, "world 2 acts as an intermediary between world 3 and world 1" (1978, p. 156). World 3 can affect the physical world only by means of human interpretations of it and according to individual goals and needs; that is, through a World 2 that Popper (2003, 36ff.) does not consider to be a simple epiphenomenon of World 3.

Fourth, the thesis that World 3 is "irreducible" to the World 2 of mental states can be regarded as incompatible with MI only if this approach is equated with the view that vocabulary about social concepts must be reduced to vocabulary about psychological properties. However, this equation seems to us impossible because historically MI is centered on explanations in terms of systemic phenomena and unintended consequences that are irreducible to mental properties.

Could you clarify the relationship between Popper's theory of World 3 and what you call institutional individualism?

Agassi: To discuss properly a theory, says R. G. Collingwod, one has to begin with the problem that it comes to solve. Quite unusually, Popper's World 3 theory comes out of the blue. He told me clearly that he held it for years but feared publishing it because of its bluntly metaphysical character. Indeed, in my memory, this event comes with Bar-Hillel expression of shock upon his encounter with it. What problem Popper viewed as the one that comes to solve? He never reported his answer to this question. Clearly, human language is an institution (unlike the languages of other animals). Here comes Popper's endorsement of the theory of levels of language of Karl Bühler. It also comes with no background and no questions. Perhaps, the omission is of something obvious.

My initial papers on Institutional Individualism offered a simple thesis: valuable social theory takes both individuals and institutions for granted, without necessarily any ontology.

Bulle and Di Iorio: During the last decades, the interpretation of methodological individualism in terms of psychologism or psychological reductionism has been defended by different scholars. Your view is centered on ideas that have also been expressed by Roy Bhaskar, Margaret Archer, and Lars Udehn. One of the most common accusations against MI is that, as an approach based on *Verstehen*, it neglects the influence of objective institutional constraints on action because these constraints exist independently of the actor's subjective standpoint. Legal constraints, for example, are not excused by the ignorance of law in the sense that they produce consequences independently of the individual's opinion about what she is free or not free to do. This interpretation of MI as an approach incapable of taking into account institutional constraints because it only focuses on the subjective standpoint of the actor seems to us incorrect because of two reasons that are strictly interconnected.

The first is that methodological individualists such as Weber, Simmel, Menger, and Mises never assumed that strictly subjective standpoints are the foundation of social life. In explaining the social world, they did not apply the interpretative approach—*Verstehen*—to subjective psychological content, but rather to shared meanings or collective beliefs, i.e. to typical ways of thinking and acting (Weber, 1978 [1922]). The second reason is that, according to MI, the interpretation of these shared reasons is strictly linked to the analysis of their consequences that are often unintended. This analysis allows for the explanation of social constraints in terms of objective and sometimes brutal results of these shared reasons (see Hayek, 1952, p. 34). This means that MI conceives social interactions as resulting from both the intentions of the individuals and the objective structures that are preconditions of these interactions. An exemplification of the incompatibility between MI and psychological reductionism is Weber's analysis of the caste system in India. Weber accounts for the existence and acceptance of this system, and its constraints, in terms of objective consequences resulting from shared normative beliefs that create the structure of the social positions and are in turn affected by it (Weber, 1946, pp. 396–415). The legitimacy of this structure depends on the relationship between its reality and the mental. According to Weber (1921/1962, p. 131), the caste system is "only a product of rational ethical thought [...] Only the wedding of this mental product with the real social order through the promises of rebirth gave this order the irresistible power over the thought and hope of the people moored in it."

Agassi: Lars Udehn has expressed full agreement with me. I ignore Roy Bhaskar as he wants certitude. Margaret Archer offers complex theories that are difficult for me to discuss for want of understanding the problems that engage her. For my view on Max Weber see my "Bye Weber," *Philosophy of the Social Sciences*, 21, 1991, 102–109; for my view on Ludwig von Mises see my review of his *Theory and History* in *Times Literary Supplement*, 16 May 1958.

Bulle and Di Iorio: Both your article on Weber and your article on Mises are interesting, but they do not discuss the interpretation of MI in terms of psychologism. Your review of Mises's (1957) *Theory and History* focuses on his criticisms of historicism and positivism, while your text on Weber's *Science as a Vocation* challenges the latter's claim that the social sciences can and should be value-free (Weber, 1917/1946). In your opinion, Weber did not understand that values change and are improved under the pressure of "empirical criticism," when some set of values entail conflicting recommendations. You argue that Weber did not mean by axiological rationality what is usually meant by this term today and also that axiological neutrality in Weber's sense is impossible. Do you regard science as necessarily committed to politics? Do you think that the empirical criticism of values mentioned above is indissolubly linked to politics?

Agassi: Popper has rightly viewed as the peak of his philosophy the unity of his theory of science and his theory of democracy. To repeat my claim, the combination of the negative theology of Maimonides with the theology of Spinoza ("*Deus sive Natura*") amounts to negative science, which is Popper's great contribution to the philosophy of science. He saw progress as correction and correction as the offspring of criticism. This is the Socratic view of both science and politics. The obstacle to the application of the Socratic method to science is Plato's view of science as certitude. All the contributions to the philosophy of science after Hume were made in effort to answer him. This was felt necessary since Plato's dismissal of the unproven was taken for granted. Kant even went so far as to recommend censorship against publishing hypotheses. All efforts to answer Hume followed Bacon's proposal to avoid hypotheses, especially bold ones. Popper said, science is bold hypotheses and (Socratic) efforts to refute them. Here Popper admitted

626   J. Agassi et al.

Hume's criticism as valid but denied his value system. In particular, he said Hume's great contribution was his criticism, yet he never asserted the (Socratic) view that criticism is valuable (as the fuel of scientific progress). The background to Popper's great contribution explains its greatness: the idea that some scientific errors are important was both taken for granted and vehemently rejected. Already Newton said (*Principia*, Bk. II) that Kepler's and Galileo's laws are false and (Bk. III) that they are true. Popper began with the recognition that Einstein has refuted Newton's theory. In Einstein's background was the effort to make electrodynamics fit Newtonian mechanics (obey Galileo's transformations); he decided to make mechanics fit electro-dynamics (obey Lorentz' transformation), thus applying the famous field energy equation ($E = mc^2$) to mechanics. He did that by showing that the Maclaurin development of $mc^2 - E = m_0c^2/(1 - (v/c)^2)^{1/2}$—is $E = m_0c^2 + 1/2.m_0v^2 + \ldots$ that replaces the one in the Newtonian framework that is $E = mv^2/2$. Here is Einstein's great contribution to methodology. He viewed Popper's contribution true as a matter of course but was appalled by his positivism. (In his "The Popper Legend" he says he never was a positivist, meaning, adherent to the doctrine of the Vienna Circle. He concealed his past traditional-style positivism in his 1959 English translation of his 1935 *Logik der Forschung*).

Bulle and Di Iorio: Let's come back to the topic of individualism. It is necessary to make a distinction between two different problems: the explanation of the reasons why institutions objectively exist and the explanation of how they causally influence decision-making. Let's focus on the second problem from a strictly sociological standpoint. It seems to us that you and the pre-Popperian methodological individualists share a similar view: i.e., the idea that social structures do not mechanically affect decision-making because their influence is mediated by the interpretative skills of the actor. In other words, social structures do not act as separate causes of social phenomena because action depends on the meaning it carries out for the individual. Institutional factors such as the dominant ethical values of a given society cannot mechanically cause any action because only the way these values are interpreted by the actor can affect his/her behavior.

The distinction between the two levels of analysis mentioned above—i.e., the objectivity of the institutions, on one side, and the individual interpreta-tions that cause the action, on the other side—is fundamental to correctly grasp the relationship between the social framework and action from the standpoint of MI (see Huff, 1975, pp. 73–74). To explain the objectivity of a political "institution" or "compulsory association" (Anstalt) like the state, Weber (1978, p. 277) argues that within it "the orientation" of human "action

Methodological Individualism and Institutional ...   627

to the rules is expected...because the individuals...are empirically 'obligated' to participate in the social action constituting the community and because the probability exists that, if necessary, their opposition (however mild in form) will be restrained through a 'coercive apparatus'." However, this does not mean that the members of this political community's typical behavior results mechanically from these constraints. Their actions are explainable, like those of the members of other institutional structures, in terms of typical rational evaluations that are "possible" only given the existing cultural "background" and "do not contain a single grain of "psychology" (Weber, 1978 p. 277 n96; original emphasis).

Do you agree that institutional individualism and pre-Popperian individualism share the view that social structures affect decision-making only through the meaning that the individuals attach to them? Is it correct to say that, for you, the difference between these two approaches mainly depends on the fact that, while the former recognizes the relative autonomy of the institutions and their causal role in the generation of social phenomena, the latter does not?

Agassi: To your question: no; I do not agree. Also, allow me to add (so as to prevent confusion), I do not disagree. This paragraph is very puzzling to me. Do you remember Kipling's Just So story about the invention of writing?

Bulle and Di Iorio: You maintain that institutions are "an important factor in determining the rational or purposeful behavior" of the agent and that institutions "can be explained as ... attitudes which are accepted conventionally or by agreement" (1987, p. 150). Your view is clearly anti-holistic. This is because, for you, the influence of the institutions on the individuals is neither direct nor mechanical. The conventional rules that create the systems of coordination that you call institutions play a normative role but do not impose deterministically any goal or aim.

Agassi: that is correct: institutions have no aim of their own. (This holds for whole societies too.)

Di Iorio and Bulle: In your opinion, what is the difference between pre-Popperian individualism and institutional individualism with regard to the study of social conditioning?

Agassi: Mill said, all references to institutions in an explanation should be eliminated for it to be satisfactory. Popper said, this leads to the methodological myth of the social contract. What can be simpler?

Bulle and Di Iorio: Different pre-Popperian individualists like Menger, for example, criticized the myth of the social contract and explained the emergence of institutions in terms of unintended consequences. Moreover, they did not provide an atomistic explanation of social action.

628   J. Agassi et al.

Agassi: Why did Menger insist that institutions are unintended when legislation is a common phenomenon?

Bulle and Di Iorio: You display a mixture of admiration for and dissatisfaction with Weber's approach that you consider the most refined form of psychologism (Agassi, 1987, pp. 142–143; 2017). In your opinion, one of the problems with this approach is that it is "applicable only to a narrow range of problems" (ibid., p. 142). This is because it "ties us too much to the typical" and, as a consequence, "leaves no room for sociologically significant yet untypical characteristics" (ibid., pp. 142–143). For example, "it does not enable us to explain satisfactorily effects of detailed characteristics of one prominent individual" (ibid.).

Agassi: Your expression "a mixture of admiration for and dissatisfaction with" is very disturbing to me. I express admiration and dissatisfaction with many ideas, including, say, Newton's metaphysics. Yet, there is no mixture here anymore than, say, my admiration for Einstein and my admiration for his theory of gravity.

Bulle and Di Iorio: You highlight that, since social change is often "untypical", Weber's individualism cannot account for it. Moreover, in your view, his approach does not allow room for intended institutional reform (Ibid., p. 143). This is also because it denies the existence of institutions. For Weber, you contend, the existence of institutions is simply inconceivable because of his psychologistic idea "that if institutions exist, they are things with independent aims, interests and destinies" (ibid.).

We do not agree with the criticisms that you level at Weber. First, the interpretation of the institutions as systems of interaction, which exist independently of the individual's mental contents must necessarily presuppose the use of typical concepts in Weber's sense, namely, the analysis of the typical ways of thinking and acting that produce these systems via their objective consequences. Moreover, the study of social change as understood by sociology must account for typical features such as, for example, Bolsheviks' political viewpoint that affected the communist revolution in Russia. It is unclear to us in which cases individual untypical characteristics can be relevant from the standpoint of the sociological analysis of the social structure, its constraints, and its historical change, while we recognize, like Weber, that untypical features can be relevant from the standpoint of other disciplines like psychology.

Max Weber's ideal–typical approach is strictly related to an explicit criticism of psychologism. It must be noted that he does not mean by this term exactly what you mean. For him, psychologism means that the explanation does not refer to the typical culturally and socially situated rationality

Methodological Individualism and Institutional ... 629

of the agents that produce the institutional forms of social behavior. In his opinion, to develop a sociological explanation, we do not "proceed to a 'psychological' analysis of the 'personality' by means of some odd mode of inquiry, but…to an 'objective' analysis of a given situation" (Weber, 1902/1974, p. 24). According to Weber (ibid.), "the specific character" of ideal–typical "constructions, their heuristic value, and the limits of their empirical applicability are based on the fact that they do not contain the least bit of 'psychology' in any sense of that term." Moreover, in his view, ideal–typical constructs do not play a normative role in explaining the average kinds of behavior of a given society. Weber clarifies that they are not to be compared to "laws of nature" because they do not describe reality, but "[serve]to facilitate an empirically valid interpretation by comparing given facts with a possibility of interpretation" (ibid.).

Our impression is that you interpret in too radical a manner Weber's ideal–typical approach. According to this approach, the pure models are universal only from a very abstract standpoint and can be applied to concrete historical phenomena to develop a comparative analysis.

Agassi: You express an opinion that differs from mine. That is fine. You expect me to comment on this expression. I do not know what for.

Bulle and Di Iorio: Can you give an example of individual untypical characteristics that can be relevant from the standpoint of the sociological analysis?

Agassi: Of course. (1) Typically, Churchill was straightforward. Untypically, his attitude to the Shoah was shifty. (2) Typically, the British revered royalty. In the twentieth century, they shifted their attitude untypically because they felt that the royal family was maltreating Diana. (3) Typically, Israel is racist; untypically it allows a few Vietnamese refugees settlement within its borders.

Bulle and Di-Iorio: Saying that the attitude of the British toward the royal family changed because they felt that the royal family was maltreating Diana means talking about the reasons the British had to change their attitude. Here you refer to a typical rational evaluation: reasons shared by millions of individuals. The consequences of Churchill's behavior can be regarded as sociologically relevant, but his behavior is per se not a sociological phenomenon. Weber's approach requires us to apply the concepts of typical rationality and typical action only to the study of intersubjective phenomena.

Agassi: I do not understand you. You asked for individual untypical characteristics. I gave you the example of Churchill and the example of the British public. Are these not examples of individual untypical characteristics?

Bulle and Di Iorio: In discussing your view that pre-Popperian methodological individualism is psychologistic you develop some criticisms of Weber's methodology of the social sciences, but do not analyze and discuss the approach of other major individualist scholars such as, for example, the members of the Austrian school of Economics: Menger, Mises, and Hayek. Popper was clearly influenced by these social scientists. What, in your opinion, is the relationship between his individualism and the individualism of these three Austrian thinkers?

Agassi: For the Austrians let me mention Mises. He did not care about the question, are institutions objective: he spoke of them. Unlike earlier economists such as Smith, he spoke of the market (and not only of trade). He recommended the return to the gold standard so as to limit the ability of authorities to intervene in the market. He did not care about institutions but about free trade.

Bulle and Di Iorio: do you consider Hayek a psychologist?

Agassi. I do not know. Why do you think he wrote his *The Sensory Order?*

Bulle and Di Iorio: Could you develop this point more, please?

Agassi: You ask me a question and I say I do not know the answer. Why should I develop this?

References

Agassi, J. (1958, May 16). Theory and history. *Times Literary Supplement.*

Agassi, J. (1960). Methodological individualism. *The British Journal of Sociology, 11*(3), 244–270. https://doi.org/10.2307/586749

Agassi, J. (1975). Institutional individualism. *The British Journal of Sociology, 26*(2), 144–155. https://doi.org/10.2307/589585

Agassi, J. (1987). Methodological individualism and institutional individualism. In J. Agassi & I. Jarvie (Eds.), *Rationality: The critical view* (pp. 119–150). Springer.

Agassi, J. (1991). Bye bye Weber. *Philosophy of the Social Sciences, 21*, 102–109.

Agassi, J. (2017). Methodological individualism. In B. S. Turner (Ed.), *Wiley Blackwell encyclopedia of social theory*. Wiley-Blackwell.

Hayek, F. (1944). Scientism and the study of society. Part III. *Economica, 11*(41), 27–39. https://doi.org/10.2307/2549942

Hayek, F. A. (1948). *Individualism and economic order*. University of Chicago Press.

Hayek, F. (1952). *The counter-revolution of science: Studies on the abuse of reason.* Liberty Press.

Huff, T. E. (1975). *Max weber and methodology of social science*. Routledge.

Menger, C. (1985 [1883]). *Investigations into the method of the social sciences with special reference to economics*. New York University Press.

Popper, K. (2002 [1957]). *The poverty of historicism*. Routledge.

Popper, K. (1966). *The open society and its enemies, Vol. 2: Hegel and Marx*. Princeton University Press.

Popper, K. (2003 [1977]). *The self and its brain: An argument for interactionism*. Routledge.

Popper, K. (1978, April 7). *Three worlds*, The Tanner Lecture of Human Values. Delivered at The University of Michigan.

Weber, M. (1917/1946). *Science as Vocation*. In From Max Weber, tr. and (Ed.), H. H. Gerth, and C. Wright Mills. Free press.

Weber, M. (1974 [1902]). Subjectivity and determinism. In A. Giddens (Ed.), *Positivism and sociology* (pp. 23–32). Heinemann Educational Books.

Weber, M. (1946). *From Max Weber: Essays in sociology* (H. H. Gerth & C. Wright Mills, Eds.). Oxford University Press.

Weber, M. (1962 [1921]). *The religion of India: The sociology of Hinduism and Buddhism* (H. H. Gerth & D. Martindale, Eds., and Trans.). The Free Press.

Weber, M. (1978 [1922]). *Economy and society*. University of California Press.

Weber, M. (2004 [1919]). Politics as a vocation. In O. David & T. Strong (Eds.), *The vocation lectures* (R. Livingstone, Trans.). Hackett Books.

Methodological Individualism and Methodological Localism: A Discussion with Daniel Little

Daniel Little, Nathalie Bulle, and Francesco Di Iorio

Daniel Little (born 1949) is chancellor emeritus and professor of philosophy at the University of Michigan-Dearborn as well as professor of sociology at the University of Michigan, Ann Arbor. He is one of the most influential living thinkers in the field of the philosophy and methodology of the social sciences and has written extensively on topics such as social explanation, Marx, the philosophy of history, organizational dysfunction, and the ethics of economic development. He has also provided a relevant contribution to the individualism-holism debate. Little (2012a, pp. 10, 12) rejects holism and praises an "an actor-centered approach to social explanation" on the ground that the methodological requirement of "microfoundations" for causal and structural claims is "a universal requirement on valid sociological research" (ibid., p. 12; see also Little, 2014). However, he partly questions the

---

D. Little
Philosophy and Sociology, University of Michigan-Dearborn and University of Michigan, Dearborn, Ann Arbor, MI, USA
e-mail: delittle@umich.edu

N. Bulle
French National Center for Scientific Research, Sorbonne University, Paris, France
e-mail: nathalie.bulle@cnrs.fr

F. Di Iorio (✉)
Department of Philosophy, Nankai University, Tianjin, China
e-mail: francedi.iorio@gmail.com

© The Author(s), under exclusive license to Springer Nature Switzerland AG 2023
N. Bulle and F. Di Iorio (eds.), *The Palgrave Handbook of Methodological Individualism*,
https://doi.org/10.1007/978-3-031-41508-1_27

634   D. Little et al.

assumptions of methodological individualism because, in his opinion, this approach is committed to "reductionism" (Little, 2016, p. 78) in the sense that it considers individuals to be "a-social" because it does not take into account structural and socio-culturally variable constraints on action (ibid.). According to Little, methodological individualism is the view that "social explanations must be couched in terms of the laws of individual psychology" (1991, p. 192). He suggests rejecting this approach in favor of "methodological localism," an orientation that combines microfoundationalism with anti-reductionism (Little, 2014, p. 55): "providing microfoundations for a social fact does not mean the same as reducing the social fact to a collection of purely individual facts" (Little, 2016, p. 79).

According to Little (1991, p. 183), methodological individualism consists of three related but distinct claims: an ontological thesis, "a thesis about the meaning of social concepts, and a thesis about explanation." In his opinion, the ontological thesis, which states that "social entities are nothing but ensembles of individuals," is trivially "true," while the two other claims are inadmissible (ibid., pp. 183–184). According to the meaning thesis, social concepts must be definable via a "meaning reduction," i.e., "in terms that refer only to individuals" (pp. 185–186). The problem with this claim is that this is impossible because social concepts cannot be defined using solely a strictly individualist vocabulary, i.e., without referring to "social institutions and social relations" (ibid., p. 185). The thesis about explanation contends that "all social facts and regularities must ultimately be explicable in terms of facts about individuals" (ibid., p. 186). Little (ibid., pp. 186–187) argues that it is equally mistaken given the causal role played by the "emergent properties" in social explanation.

For Little (2016, p. 139), while methodological individualism is incompatible with any kind of emergent causality, methodological localism is supportive of the idea of "weak emergence" as opposed to that of strong emergence because, as implied by the above, it "allows that the emergent factor is amenable to microreductive explanation." In his view, a major problem with methodological individualism is that, since "social phenomena supervene upon individual phenomena," but are not reducible "to individual-level concepts and explanations," methodological individualism fails to recognize the relative "autonomy of the social" from individual facts (ibid., p. 195; see also Little, 2016, pp. 118, 128). All "social facts are embodied in the states of mind and behavior of individuals…but…some social facts (institutions, social practices, systems of rules) have explanatory autonomy independent from any knowledge we might be able to provide about the particular ways in which these facts are embodied in individuals" (ibid., p. 128). In other words,

Methodological Individualism and Methodological ... **635**

ensembles "sometimes have system-level properties that exert causal powers with regard to their own constituents" (ibid., p. 143). In Little's opinion, the existence of these irreducible systemic effects is "a point in favor of a modest holism" (ibid.).

Within Little's theoretical framework, the "idea of relative explanatory autonomy" of the social is related to the claim that, while micro-foundation is in principle always available, it is not always necessary: "mid-level system properties are often sufficiently stable that we can pursue causal explanations at that level, without providing derivations of those explanations from some more fundamental level" (ibid., p. 143). For example, if we have empirical evidence that independently of the sociocultural context "a certain organizational structure for tax collection is prone to corruption of the ground-level tax agents" we can "use that feature as a cause of something else" without descending at the micro-level and account for the motivations of action (ibid., p. 145). This is because the "way an organization is structured makes a difference to its performance" and it "is a causal power all by itself" (p. 212). While methodological individualism is based on the assumption that "social causation proceeds always and exclusively through actions and interactions of individuals," methodological localism rejects it (p. 205). The latter approach challenges "the exclusive validity of one particular approach to social explanation, the reductionist approach associated with MI and Coleman's boat. Rather, social scientists can legitimately aggregate explanations that call upon meso-level causal linkages without needing to reduce these to derivations from facts about individuals" (p. 145). Little (p. 205) also thinks that "it is legitimate to postulate causal powers for structures whose effects are realized in other meso-level structures," that is a "meso-meso social causation" such as, for example, the following one: "decreasing social isolation causes rising intergroup hostility" (p. 214). In his opinion, this "meso-meso causal connection" is prohibited by methodological individualism (p. 209). The various versions of this approach "—microeconomics, analytical sociology, Elster's theories of explanation, and the model of Coleman's boat—presume that explanation needs to invoke the story of the micro-level events as part of the explanation" (208). On the contrary, methodological localism "requires that we be confident that...micro-level events exist and work to compose the meso level; but it does not require that the causal argument incorporates a reconstruction of the pathway through the individual level in order to have a satisfactory explanation. This account suggests an alternative diagram to Coleman's boat" (pp. 208–209). The causal powers of an organization "having to do with efficiency, effectiveness, and corruptibility can be disaggregated into the incentives and behaviors of typical individuals" (p. 212). However, "here is

636   D. Little et al.

the key point: we don't need to carry out this disaggregation when we want to invoke statements about the causal characteristics of organizations in explanations of more complex social processes" (p. 212). On this reading, there are not only stable micro-level mechanisms as postulated by methodological individualists like Peter Hedström and Thomas Schelling, but also "meso-level causal mechanisms" that are about meso-micro and meso-meso causal links (p. 214). According to Little (p. 213), while "meso-level social entities do indeed have causal powers that can legitimately be invoked in social explanations," it is preferable not to assume that macro concepts are the bearers of social powers. This is because it is hard to find macro-level regularities, i.e., stable macro-level features (p. 146). However, he also believes that "large social factors" can be regarded sometimes as "causes" (pp. 215–216).

In Little's view, another important difference between methodological individualism and methodological localism is about action and its presuppositions. The latter approach, unlike the former, insists on the inherent "social-ness" of the individual "who is both socially constituted and socially situated" (Little, 2014, p. 56). Actors are "embedded" within a set of local, space-variable, "social relations and institutions that create opportunities and costs for them" (ibid.). Moreover, their ways of thinking and acting "are themselves the products of a lifetime of local social experiences," namely, "the mechanisms of socialization" (ibid., pp. 56–57). Methodological localism opposes "much social science theorizing" because the latter "depends on an over-simple theory of the actor, often involving the Aristotelian ideal of means-end rationality" (ibid., p. 57). Micro-foundations are crucially important for explanation: Social phenomena "depend ultimately" on "actors whose actions and thoughts make them up," but actors must be conceived as socially embedded in the sense clarified above rather than in abstract and atomistic terms (ibid., p. 58). According to Little (ibid., p. 61), the social sciences need a more complex and "nuanced" model of action than the Cartesian, utilitarian, and instrumental one proposed by the rational choice theory. Instead of conceiving of the individual as always acting "on the basis of a calculation of costs and benefit" it is necessary to take into account the normative influence of the social environment as well as the fact that the presuppositions of action can be "improvisational," "habitual," and largely subconscious, that is only vaguely understood by the individual (ibid., 67–73).

Little: I would like to begin by thanking you, Professors Bulle and Di Iorio, for engaging with me on these important and difficult issues concerning the ontology of the social world. Your questions are probing and insightful, and you have made me realize that there are some areas of unclarity that have arisen in my own views on "individualism," microfoundations, and actor-centered sociology over the forty years that I have been writing about these ideas. Before turning to your specific questions, I would like to begin with a very brief outline of my current understanding of these ideas.

The most fundamental idea at work throughout these decades is what we now call "ontological individualism"—the idea that the social world depends on the actions, thoughts, mentalities, and interactions of individual actors. There is no social "stuff" that is independent of the actions, thoughts, and interactions of individual actors. Ontological individualism differs from methodological individualism because it does not presume that social explanations must proceed from facts about individuals to facts about the social world; it is not a "reductionist" or "generativist" doctrine about social explanation; instead, OI is agnostic about the direction and nature of causal powers and mechanisms at work in the social world (beyond the ontological fact that they are ultimately embodied in individual actors). My view of the role of microfoundations in social explanation has evolved significantly over the years. At the time of writing *The Scientific Marx* (1986), I took the view that claims about social-level properties and causal powers should be accompanied by some sort of account of the microfoundations of those properties and powers—the pathways through which actions and interactions among individuals lead to the postulated social facts. I came to believe in the 2000s that this requirement was too strict and failed to correspond to the practice of many convincing sociological and historical explanations; further, it ruled out by fiat the possibility that intermediate-level social entities (organizations, normative systems) might have stable causal powers of their own. I, therefore, relaxed my formulation of the microfoundations requirement to the idea that "the researcher must be confident that microfoundations exist, but does not need to provide them as part of an explanation." In this formulation, the requirement of microfoundations is equivalent to ontological individualism. The idea of emergence is plainly relevant to this discussion, since anti-individualist theorists sometimes maintain that "social facts are emergent" relative to individual-level facts. I have tried to distinguish sharply between weak emergence ("the properties of the ensemble are different from the properties of the components") and strong emergence ("the properties of the ensemble are independent from the properties of the components and cannot be derived from them"). I recognize that social facts, structures, and other social arrangements are "weakly emergent," but they are not strongly emergent. Finally, I endorse the idea introduced by Jerry Fodor (1974) about the special sciences concerning "relative explanatory autonomy": a researcher in biology, psychology, or sociology can investigate a range of phenomena at the supra-component level (supra-molecular, supra-neurophysiological, and

638   D. Little et al.

supra-individual) without needing to reduce the claims he or she advances to the underlying level. For the social sciences, this means that one can consistently maintain ontological individualism and the reality and durability of some social–causal properties (e.g., organizational tendencies, institutional logics, interactions of systems of norms).

In short, I don't regard my position as embracing "methodological individualism." I prefer the terms "ontological individualism," "actor-centered sociology," and "methodological localism." Methodological individualism is distinct from ontological individualism because it is reductionist, and because it stipulates that explanations should be generativist in the sense described above: they should proceed from an account of the circumstances and motivations of the actors to a derivation of the social outcome to be explained. Methodological localism asserts that we cannot understand individual actors without having some knowledge about their "social constitution"—the cognitive, affective, and normative frameworks they have absorbed through their histories of social development—and we cannot understand the actions of socially constituted individuals without having knowledge of their "social situation"—the specific constraints, incentives, disincentives, powers, resources, and behaviors of other actors within which they choose to act.

Now, let me turn to your specific questions.

Bulle and Di Iorio: Professor Little, the view that methodological individualism is flawed because it is a non-systemic approach that assumes that the social sciences should use solely a vocabulary referring to individual properties and provide exclusively explanations in terms of these properties is widespread. It seems to us that this reductionist interpretation of methodological individualism is unwarranted in the light of the history and practice of the social sciences. First, no methodological individualist conceives his/her approach in terms of a linguistic reductionism that prohibits the use of certain words and predicates, namely those that refer to non-individual properties. Second, many advocates of methodological individualism like Menger, Simmel, Popper, Hayek, Coleman, and Boudon highlight that systemic analysis is necessary and inevitable in the social sciences and that macro and meso structural properties affect the micro-level (sometimes these authors depict the relationship between macro and micro factors in terms of mutual influence or circular causality: consider, for example, Hayek's interpretation of the market in terms of a complex system and Popper's theory of the three worlds and their reciprocal influence). Third, there are no examples of empirical explanation provided by methodological individualists that do not involve the use of a vocabulary that refers to systemic social properties. Reductionist explanations in the sense of non-systemic explanations seem simply impossible to achieve.

Could you clarify why you regard methodological individualism as a non-systemic approach?

Little: I do not use the concept of "system" very often, so perhaps I can rephrase your question slightly and talk about social structures and meso-level social entities. My point about the limitations of strict methodological individualism is that individuals are always involved in "structural and institutional arrangements" (which you might refer to as "systemic"), and therefore we need to be able to invoke those involvements in explaining collective or social ensemble outcomes. This is not equivalent to what Hedström (2005) refers to as "structural individualism" because the analytical sociologists tend to look at the structural factors as exogenous and fixed; whereas I see them as interdependent with the actions and interactions of the individuals involved. This is the "socially embedded" part of methodological localism. Individuals are also "socially constituted" with ideas, mental frameworks, norms, practices, and habits that they have gained through the process of culturation and socialization. My basic point, then, is that the actors who make up a bank, a labor union, or a racial group are not pure purposive agents; rather, their current situation and their practical cognition are informed by antecedent social influences that have impacted them through proximate mechanisms (family, work environment, schooling, television, …). But in every case, the "social influences" of structures, ideologies, knowledge frameworks, etc., are conveyed through interactions with other socially constituted actors who make up the local school, labor union, activist social organization, etc.

I should also say that defining or refuting methodological individualism as a doctrine has not generally been a priority for me. In particular, I am not invested in the question of whether methodological individualism presupposes the idea of a "pre-social" individual (atomism), though Hobbes, Menger, and JS Mill seemed to make that assumption (Mill's version of the idea was that the science of psychology should allow us to infer the laws of sociology). Economists and game theorists come close to this view, in the sense that they define the agent's purposiveness in terms of a preference structure and an assessment of risks and benefits of various possible actions. And as Joseph Heath argues in his article on methodological individualism in the *Stanford Encyclopedia of Philosophy* (Heath, 2020), Popper himself was prone to a form of psychological reductionism in his arguments for methodological individualism. An important component of my own advocacy of an "actor-centered sociology" is emphasis on the fact that actors cannot be reduced to a short list of abstract and general characteristics (utility functions, satisficing rules, …). Instead, we need a rich theory of the actor to explain most real and complex social outcomes.

I agree with your point that social scientists who make use of the idea of methodological individualism today are not concerned with a *semantic* or terminological point—"use only terms definable in terms of individual psychological states." Rather, they are concerned with a claim about the logic of social *explanation*—"explain social outcomes as the aggregate consequence of individual actions." That said, the most insistent advocates of methodological individualism have often made quite stringent assumptions about what an individual-based explanation can postulate about the individual's mentality. Often the assumptions about the actor that are made by researchers committed to methodological individualism are extremely thin and abstract—narrow economic rationality, portable across all social and cultural contexts. This assumption is shared by economists, game theorists, and some rational-choice political scientists.

The more important point that distinguishes methodological individualism from other views of social science is a particular view about explanation—the idea that social outcomes must be explained as the aggregate outcome of individual-level actions. In this respect, methodological individualism is a variant of reductionism: the task of the social scientist is to demonstrate how social features can be "reduced" to facts about the individual actors who constitute them. Here, "A is reduced to B" means demonstrating how the properties of A can be fully explained by reference to the properties of things at the B level. This "bottom-up" model of explanation precludes a very large volume of excellent sociological research about meso-level entities—organizations, institutions, systems of norms, racial and gender systems, and the like. Methodological individualism serves to inhibit and discredit research done at this meso-level.

In the past decade or so, when I have treated microfoundations and ontological individualism, I have been most concerned to work out an idea about how the social world works, and how we should try to explain its dynamics and outcomes. This idea combines the thesis of ontological individualism and the idea of what I came to call "methodological localism." This view expresses the assumptions of an actor-centered approach to social explanation by suggesting that actors are *socially constituted* and *socially situated*. And second, I have argued for a pluralistic approach to social causation involving lateral and descending social causes (meso-meso causation) as well as ascending causes from the micro- to the macro-levels. This means rejecting an important implication of James Coleman's theory of sociology, expressed in his "boat" diagram, which is a view that has also become a defining feature of much analytical sociology: that explanations of the macro must always take the form of a deduction of the higher level fact from information about behavior at the micro level. This is the generativist model (Epstein, 2006) that underlies much of analytical sociology and the methodology of agent-based models. The explanatory maxim is: "Derive the macro from the micro." The idea of microfoundations is something like a "bridge" analysis, through which we link the macro-level properties of a social ensemble to the intentions and interactions of the individuals who constitute that ensemble. To provide microfoundations is to demonstrate how the actions of the individuals aggregate to the social property under study. As noted above in my preliminary remarks, my adherence to the requirement of the "strong microfoundations principle" has changed. In *The Scientific Marx* (1986), I held that social explanations needed to be grounded in microfoundational accounts of the ways in which individuals brought about the outcomes (perhaps unintentionally). In more recent years, I have come to recognize that this principle is too strong; instead, we need simply to be assured that a microfoundational account exists, without being obliged to provide that account. I like the analogy with materials science and the causal properties of metals: we do not need to deduce the properties of an alloy of steel from fundamental physics in order to have good causal explanations of collapsing bridges.

It is of course true that a narrow generativist account is entirely feasible for some specific kinds of social outcomes, and constitutes a perfectly legitimate social explanation. This is the ascending strut of Coleman's boat, and it underlies the rationale for agent-based models as a kind of social explanation. My critique of methodological individualism as a comprehensive doctrine is simply that I reject the view that **all** adequate social explanations must take this logical form (from lower level to derivation of higher-level properties), or tracing out the rising strut of Coleman's boat. Instead, I advocate for the idea of actor-centered conceptions of social structures within the theory of "ontological individualism."

Bulle and Di Iorio: Let's focus on a specific aspect of your reductionist interpretation of MI. You hold the individual/society relations in MI to be comparable to the reductionist approach of the mind in the mind-brain identity thesis. It seems to us that this comparison is unfair in relation to MI where individuals are immediately situated at a social level, without the distinction of two levels of manifestation of individual and social phenomena. As Gustav Ramström (2018, p. 372) notes in this regard, one cannot compare the relationship between mental phenomena and neural activity with that between social phenomena and the individual actions underlying it. Especially, in the social case, the relationship between individual actions and social phenomena is inferred from these very actions by the observer (i.e., a "riot" revolt) and does not involve empirically two different phenomena as in the neural/mental distinction.

Can you comment on this?

Little: I don't really think the analogy of "neurophysiological level/ mental level" is helpful for philosophy of social science, beyond noting a parallel between ontological individualism and physicalism. I prefer the analogy between the social world—composed of social actors—and metals— composed of fundamental particles. We don't need to provide a mathematical model of the micro-physical characteristics of the muons that make up particles in order to have a good and empirically informed material science. And likewise, we don't have to decompose the national United Auto Workers union from its headquarters in Detroit to its locals around the country to the social networks and communication pathways through which Detroit-based executives influence particular actions in Toledo local 000 in order to have a good sociology of the UAW worker in a Toledo engine plant.

Ramström's (2018) point seems to be that there is only a nominal difference between a macro-level description and a micro-level description of the same set of occurrences. Can we empirically or observationally distinguish between the social event—the unfolding and dynamics of a specific riot— from the actions, thoughts, and dispositions of the individuals who make it up? Are individuals analytically separable from riots in anything like the way that functioning neurons are analytically separable from the performance of mental arithmetic? His article seems to suggest that there is only a perspectival difference between a riot and a collection of riotous individual actors.

This view casts doubt on the reality of levels in the social world altogether. I considered this idea under the rubric of a "flat social ontology" (Little, 2016), but came to the conclusion that adequate social explanation requires analysis of causal properties of entities like value systems, organizations, institutions, labor unions, and social movements—each of which have their own constitution at the level of interactions of individual social actors. It is hard to dispute that social things like kinship systems, business firms, and armies have stable and knowable characteristics that can be studied empirically, and therefore we shouldn't adopt an ontology that excludes legitimate topics for empirical research. The relationship between a social description of a social event "mob attack on tax office" and a myriad of individual descriptions of actions in the same setting—"Alice forces a door," "Bob breaks a window," Charles shouts abuse at tax official"—doesn't seem as direct or transparent in

Methodological Individualism and Methodological ... **643**

many other instances of social dynamics. There is a reality to the riot over and above the actions of the various individuals; this is one of the very interesting things we can learn from McAdam, Tarrow, and Tilly, *Dynamics of Contention* (2001). Moreover, Ramström seems to be off the mark in judging that "micro-structural" properties are not relevant in the social sciences. if Alice, Bob, and Charles are involved in a resistance-oriented online chat group, whereas David, Edward, and Francis are not, this fact is pertinent to the unfolding of the tax riot. (This is part and parcel of the idea of methodological localism: the particular social interactions of the actors make a difference to the social outcome). Further, many social dynamics are the unintended and unperceived consequences of individual actions. The failure of a rail strike because of a lack of critical mass of participation is analytically separable from the individual calculations made by potential strikers who want the strike to succeed but who weigh private costs against public benefits and choose not to participate. It was a genuine discovery when Mancur Olson reconstructed the public goods/free-rider problem at the individual level—even though it was well known that collective actions often fail even when they serve the interests of the vast majority of a given group.

Bulle and Di Iorio: We would like to raise a point in defence of Ramström's argument, which we think underlines an essential problem. Maybe there is a misunderstanding. Ramström does not say that microstructural properties (relative to the individuals and their interactions) are not relevant for the social sciences, but that social phenomena do not have macro-phenomenal expressions and are based on the observation of micro circumstances. The problem is that there is a widespread confusion in the literature between the analytical approach of methodological individualism that considers the micro and macro levels to be relative theoretical constructs (in the sense that the "social" level refers to sets of individuals and their interactions and the "individual" level implies properties of social nature) and the empiricist approach of science that is couched in terms of observable levels of composition (which was developed, for example, by the neopositivists). The latter approach argues that, from an empirical standpoint, there are social properties that are clearly distinct from the microstructural properties.

Let's focus now on a new topic. In your opinion, methodological individualists fail to understand that, while a micro-foundation is in principle always available, it is not always necessary. You stress, for example, that since some kinds of organization may present system-level properties that are stable over different contexts and whose micro-level presuppositions are at least schematically already known and understood, we can invoke the causal

644   D. Little et al.

characteristics of these organizations in explanations of more complex social processes without implementing a micro-foundation.

It seems to us that your view that micro-foundation is not always necessary is compatible with methodological individualism. As argued by Raymond Boudon (1998a, p. 173), this approach is micro-foundationalist in the sense that it requires "explanations without black boxes." Since there is nothing left unexplained in the example above, the explanation seems to us individualist. In *The Foundations of Social Theory*, following Popper, Coleman (1990 p. 5), highlights that in his book "there is no implication that for a given purpose an explanation must be taken all the way to the individual level to be satisfactory. The criterion is instead pragmatic: the explanation is satisfactory if it is useful for the particular kinds of intervention for which it is intended. This criterion will ordinarily require an explanation that goes below the level of the system as a whole, but not necessarily one grounded in individual actions and orientations."

From the standpoint of the analysis of scientific practice, methodological individualists conform to the principle that it is useless descending to the micro-level if what happens at this level is already known or it is commonsensical. Their insistence on the role of micro-level mechanisms must be regarded more as a criticism of deterministic explanations in terms of holistic macro-factors than as the idea that it is always necessary to descend to the micro-level.

Can you provide some further comments on this topic?

Little: I'm a bit concerned that your liberalized defense of methodological individualism deprives the doctrine of its force as a guide to social research and explanation. It would appear that almost any empirically supported sociological investigation can be claimed as conforming to methodological individualism, properly construed. But I disagree with that, because I believe that MI is committed at its core to the idea that social explanation must proceed from micro to macro. And I would observe that there is a great volume of excellent sociological research that does not conform to that model of explanation. Could a view that systematically explained international politics purely based on a formal model of a "multi-polar world" be called "methodological individualist?" Could a Marxist theory of modes of production that derives the collapse of feudalism from the system-level properties of the feudal mode of production alone be called methodological individualist? My answer intuitively is "no." Both these claims reject completely the idea that explanation must proceed from the properties and interactions of the components to the characteristics of the ensemble or structure.

Now one might say that these are bad examples precisely because the explanations are completely divorced from the situation and actions of individual actors. So, consider a concrete example. Kathleen Thelen (2004) remarks upon the resilience of various meso-level political and economic institutions in different countries, including institutions providing for skilled labor training. And she attributes this resilience and persistence to features of the political institutions of the countries under study. This is a meso–meso causal claim. I believe that it is fairly straightforward to sketch out an indication of what the microfoundations of this claim might look like. But this is not at all a part of Thelen's research project. Rather, she remains at the level of meso-level social structures and forces to explain features of other meso-level structures. So, confidence in the availability of microfoundations is secure; but the argument surely is not one that methodological individualists would embrace.

None of these examples embodies the "explanatory order" commitment that I believe is key to the doctrine of methodological individualism: explain social outcomes based on the properties of the components of those outcomes.

Rather than finding ways of construing methodological individualism as being compatible with meso-meso explanations, I believe that we are better served by simply endorsing the view that the fundamental requirement is that of *ontological individualism*: all social properties are somehow created by the actions and interactions of the socially constituted, socially embedded individuals who make them up. So, at the most abstract level, we must affirm that "it must be scientifically possible to demonstrate how the properties of the individuals give rise to the social properties observed". But all this establishes is that "microfoundations must be in principle possible" (the "weak microfoundations thesis"). In order to be confident that this condition is satisfied in a prospective social explanation, we need to have some idea of how the microfoundations might work. But we are not required to provide them, and there is nothing deficient about an explanation that proceeds from those social properties (properly supported by empirical evidence) to the feature of another social arrangement we wish to explain. I accept that some social features can be explained in this micro- to macro-way, but I do not believe that all social features must be so explained. In particular, I hold that both descending and lateral causal explanations are possible (meso-micro, macro-meso, and meso-meso).

Bulle and Di Iorio: Let us try to clarify our view. We agree with Raymond Boudon, who regarded the concept of MI as defining the research practice of a large number of social scientists because they often explain human

646   D. Little et al.

phenomena as either intentionally or unintentionally resulting from understandable reasons (Boudon & Bourricaud, 1990, pp. 11–17). For Boudon, this practice was implemented from the beginning of the empirical social sciences and even scholars like Marx and Durkheim, who in some of their writings clearly endorsed a holistic methodology, often provided individualist explanations (ibid.). There are refined variants of this individualist explanatory logic as well as more simplistic and atomistic ones. In any case, it seems to us that the view that methodological individualism as understood by its theorists assumes that causality is only micro–macro is falsified by some historical evidence. Consider, for example, Mises' and Hayek's analysis of the way in which the price structure, which is an emergent effect that unintentionally results from countless bits of distributed information, affects the global structure of production (Hayek, 1948, 1973; Mises, 1981). This analysis is related to the idea that prices retroact on the micro-level, imposing constraints on agents' freedom of choice. In analyzing the relationship between the price structure and the structure of production Mises and Hayek point out that taking into account the detailed motivations of the agents whose actions produce this relationship is impossible. The fact that millions of people interacting in a market are involved in economic exchanges is understandable in terms of common sense: they typically share the willingness to make profits and satisfy their needs. However, the presuppositions of their actions cannot be known in detail by the economist: for example, the reason why a particular individual needs to buy a drill at time x on day y can only be known to this individual. The market dynamics are characterized by complexity, which means that they are based on the use of a heterogeneous distributed information that is related to particular circumstances of time and place. On the contrary, if the presuppositions of the actions that produce a particular phenomenon are homogeneous, do not quickly change and can be known in detail—think of Weber's historical analysis of the relationship between the stable reasons of the Calvinist entrepreneurs and the emergence of capitalism in Northern Europe—the phenomenon under investigation is simple. The study of market coordination in terms of complexity, which is related to the assumption that the presuppositions of the market are dynamic and unknown, is an implementation of a variant of methodological individualism, that, according to Hayek (1973), is as old as economic science. Because of market complexity, Mises' and Hayek's analysis of the relationship between the price structure and the structure of production does not focus on micro-level dynamics. These two scholars acknowledge that the market presupposes the typical, i.e., shared, and understandable willingness of individuals to be involved in economic exchanges, but their analysis is mainly

Methodological Individualism and Methodological ... 647

about the relationship between two emergent or macrosystemic phenomena. However, they consider their approach an implementation of methodological individualism because they do not assume that the economic agents are remote-controlled by holistic social factors that unconsciously and mechanically control their actions. In other words, these two Austrian economists reject the realist ontology of collective concepts and do not reduce economic dynamics to environmental determinants. In defining their methodology Mises and Hayek never argued that the social explanations mandatorily require a detailed and full understanding of micro-level interactions.

Little: This is a useful clarification of methodological individualism. I would paraphrase your point in terms of the idea of microfoundations. To say that "changes in the price structure [the price of natural gas relative to labor, let's say]" leads to "change in the production process [substitution of labor-intensive processes for natural-gas powered processes]" is a macro-macro causal claim. But this claim remains faithful to the premises of methodological individualism because it is straightforward to provide the microfoundations at the level of consumers and nature (price structure) and production managers (production process) that explain why change in price structure leads to change in production process. In a nutshell, rational production managers will minimize costs of production by substituting a lower price input for a higher price input in the production process. Therefore, the macro-entities have fully individualistic microfoundations.

Bulle and Di Iorio: Let's move to a new topic. In *Varieties of Social Explanation*, you link methodological individualism to three different, and interrelated, claims, **ontological** (social entities are logical compounds of individuals), **semantic** (social concepts refer only to individuals and their relations and behavior), *and* **epistemological** (higher level regularities are to be reduced to lower level regularities), all three of which imply your interpretation of MI as essentially justified by the ontological constituents of the social world. Moreover, you distinguish three major methodological approaches in the social sciences: causal, rational-intentional, *and interpretive*. Methodological individualism does not appear at this level of the great metatheoretical approaches. Its status is not clearly defined in this respect. Let's focus on this point.

Causal and interpretative dimension of MI. You associate the perspective developed by Max Weber with the interpretive approach. The close links between his interpretive approach and the tradition of methodological individualism are well known, but you do not discuss these links since MI for you represents something else. This is surprising because the tradition of MI involves interpretive sociology. Raymond Boudon, in order to account for

648   D. Little et al.

this, recalls in various places that Max Weber saw in "methodological individualism" the basic principle of what he called "comprehensive sociology," taking up Weber's letter to Robert Liefmann in 1920 (quoted in Roth, 1976, p. 306): "sociology...can only be pursued by taking as one's point of departure the actions of one or more (few or many) individuals, that is to say, with a strictly 'individualistic' method." These Weberian individuals are social actors integrated in institutions and social structures, let's not come back to that. The point is that Weberian comprehensive (or understanding) sociology is based on MI and that, as such, the great approaches you distinguish find through MI a joint realization: interpretation refers to the intentions, or even the motivations of social actors, and the latter is held to be the "causes" of actions in Weber. Of course, these causes involve situational analysis and do not engage any determinism. On this subject, you point out that "the mechanisms that link cause and effect are typically grounded in the meaningful, intentional behavior of individuals," which, as you know, is a principle derived directly from MI.

Rational-Intentional dimension of MI. Sometimes, you seem to equate MI with rational choice, especially the utilitarian model of action, and whereas you discuss classical examples of MI explanations such as those of Mancur Olson, Thomas Schelling, game theorists, etc., you do not mention this link in *Varieties*. The individualist tradition includes non-utilitarian, non-instrumental, and non-Cartesian theories of rationality. From the standpoint of MI, sometimes the individuals act on the basis of utilitarian and clear reasons, sometimes, they act on the basis of non-utilitarian reasons that, depending on the case, may or may not be vague and non-demonstrative in the sense of Perelman's *New Rhetoric*, but that are still the causes of action.

For us, MI does not represent a reductionist epistemological approach justified by the ontological constituents of the social, but participates in the three major explanatory approaches you consider, and especially has deep and extensive links with Weber's interpretive approach.

We wonder in the end whether your idea of MI might not represent a mere philosophical construct, discussed mainly by social science philosophers, but never really implemented. If not, do you have specific examples of social science work that falls under MI as you describe it?

Little: In the SEP article on methodological individualism mentioned above, Joseph Heath (2020) describes methodological individualism in these terms: "[MI] amounts to the claim that social phenomena must be explained by showing how they result from individual actions, which in turn must be explained through reference to the intentional states that motivate the individual actors." I believe this is the most common understanding of the doctrine of methodological individualism, and it is—in this formulation—a view of the nature of explanation and reduction. Social outcomes must be explained on the basis of facts about the intentional states of individual actors. The ontological thesis about individuals and social facts is not methodological individualism, but rather ontological individualism. The latter view leaves it open what social explanations should look like; all it requires is that the explanations we offer should be compatible with their being embodied in the actions, intentions, and interactions of individual actors. It is my view that the large theories in sociology offered by figures like Durkheim, Weber, Marx, and others are generally compatible with ontological individualism—even though they disagree about the nature of social explanation.

As a philosopher of social science, my primary concern is to focus on how best to "understand society" (which includes both causes and meanings), and how to avoid various apriori slips that lead to bad social research. This is the reason I think it is important to always keep in mind the idea of an "actor-centered social world" and an actor-centered sociology. It helps us avoid the error of reifying social structures (like modes of production or markets), and to recognize the inherent heterogeneity and plasticity of the social world. Because people make history, but within circumstances not of their own choosing, there will always be variations, path-dependencies, adaptations, unintended crises, and the like; and it is incumbent upon sociologists and historians to use their research methods to discover some of the particular pathways and factual/social circumstances that lead to one outcome or another. This approach to social and historical inquiry is inherently *pluralistic*, encompassing as it does meanings, cultural frames, institutional constraints, educational arrangements, and artifactual and environmental conditions. Neither Marx's theories of the economic structure, nor Weber's idea of meaning-seeking individuals, nor Durkheim's ideas about moral conscience can serve as the basis of general theories of social order and change—because inherently, there can be no such general theories.

Return for a moment to Kathleen Thelen's account of the stability and change of skilled-labor training institutions in Britain and Germany. Thelen's account takes political and economic factors in both countries as important causal influences on the nature of these institutions, without taking the effort to show how individual-level workers, politicians, and business owners played various n-person games in supporting or undermining the existing set of institutions. But her account is plainly compatible with an actor-centered understanding of the politics and institutional arrangements of both countries—as well as the contingency and path-dependence of the shape that those institutions eventually took.

650   D. Little et al.

A core assumption in strict MI is that individual actors have an orderly basis for action (rational preferences, habits, cultural practices, meanings, …), and that a good social explanation *must* take the form of a derivation of the explanandum from the aggregated actions of these individual actors. It is, in its purest form, the generativist paradigm (Epstein, 2006). As such, it is not inherently rational-choice, economistic, game-theoretic, or intepretivist. Whatever one's theory of the actor, narrow MI requires that we explain social outcomes as the aggregate (often unintended) outcome of actors carrying out their action framework in specified circumstances. So I agree with the thrust of your question: MI can be associated with interpretative approaches (Weber), rational-choice or economistic approaches, or even Marxist approaches to explaining the social world. My persistent source of disagreement has to do with the direction of causation that is postulated by MI: whereas methodological individualism postulates that causation proceeds from ensembles of purposive actors upward to higher level social structures, I maintain that causation flows in all directions—upward, downward, and laterally.

Bulle and Di Iorio: Regarding the causal role of social entities, it is interesting to evoke the sociology of Emile Durkheim, whom you (1991, *Varieties*) present as a "committed social holist," which you interpret to mean "a critic of methodological individualism." Durkheim does not, of course, refer to the MI approach, which was still not well known under this name, and seems to ignore Weberian sociology. Durkheim, however, openly opposes psychologism in *The Rules of Sociological Method*, as well as introspective methods in the social sciences. Durkheim's approach, opposed to psychologism and to any form of methodological reductionism, is not wholly incompatible with MI as understood in the sociological tradition that is the subject of this handbook. If we take Joseph Agassi's (1975, p. 145) definition of methodological holism (opposed to MI) as an approach in which individual ends and decisions are created by social forces, or Boudon's (2007, 46; 75) as an approach that explains individual behavior by forces that are external to individuals, then Durkheim cannot be called a holist in this sense, especially with regard to his non-doctrinal work. He holds individual motivations to be socially constrained and conditioned, but not in a deterministic sense, since in that case they could not be challenged. More precisely, recourse to the meaning of action for the actor in MI is not opposed to the idea that our actions are on the whole normatively performed by our social learning. In this respect, the question of meaning is not at once problematic, provided that it may become so when a problem arises for the social actor. Durkheim explains in different places that when faced with problems or contradictions, the mind awakens

Methodological Individualism and Methodological ... **651**

and puts into question received ideas. Moreover, according to Durkheim, the social environment that is the source of learning is composed, in addition to material objects, of all the products of the human mind, which Popper will define as World 3, and of people who represent the active factors of social transformations, which is compatible with MI. You even recognize that his analysis of the causes of suicide is "fully consistent with methodological individualism" Boudon (1998b), in his *Studies in Classical Sociology* notes that *Suicide* constitutes an application of the understanding methodology as defined by Weber (p. 119ff), with the implementation of a Simmel/Weber type of abstract psychology (cf. also Boudon 1989 on the analysis of correlations between suicide cycles and economic cycles).

Your recognition of the conformity of the Durkheimian analysis of suicide with MI seems to us to reveal a non-reductionist conception of MI that contradicts the way you formally define it. Can you clarify this point?

Little: You are right that my view of Durkheim's social ontology has shifted since 1991. I no longer regard Durkheim as a social holist, but rather as a sociologist who insists on the relative explanatory autonomy of the social world.

In fact, I take the view that Durkheim's theories are fully compatible with—in fact, affirmative of—the premises of ontological individualism, though not methodological individualism. As argued above, ontological individualism does not imply reductionism. Durkheim's supposed holism is actually an artifact of his stringent insistence on the separateness of sociology as a science. I believe it is plain in *Rules of Sociological Method* (1982) that he endorses the core principle of ontological individualism: "Yet since society comprises only individuals it seems in accordance with common sense that social life can have no other substratum than the individual consciousness. Otherwise it would seem suspended in the air, floating in the void" (Durkheim, 1982, p. 39). There is no fundamental ontological separation between the "social fact of French *politesse*" and the psychological realities of French individuals. The individuals are shaped by their formative immersion in these rules as instantiated by their elders, and in turn, go on to shape the behavior of others. Durkheim makes this clear in his comments about education: "Moreover, this definition of a social fact can be verified by examining an experience that is characteristic. It is sufficient to observe how children are brought up. If one views the facts as they are and indeed as they have always been, it is patently obvious that all education consists of a continual effort to impose upon the child ways of seeing, thinking and acting which he himself would not have arrived at spontaneously" (Durkheim, 1982, p. 53). This is precisely what is intended by the phrase "socially constituted," and the

652   D. Little et al.

individual-level mechanisms through which social consciousness is conveyed to the child are evident. (Note that Steven Lukes appears to agree with this assessment in his introduction to *Rules* (Lukes, 1982, p. 17).)

What Durkheim insists upon is about the "autonomy" of social facts. This is a claim about what we would now call "emergence"—that some properties of the social ensemble are distinct and separate from the properties of the individuals who make it up. But this view too is compatible with ontological individualism. It is uncontroversial, from an actor-centered perspective, that there are large historical or social forces that are for all intents and purposes beyond the control of any of the individuals whom they influence. The fact that a given population exists as a language community of German speakers or Chinese speakers has an effect on every child born into that population. The child's cognitive system is shaped by this social reality, quite independently from facts about the child's agency or individuality. The grammar of the local language is an autonomous social fact in this context—even though it is a fact that is embodied in the particular cognitive systems and actions of the countless individuals who constitute this community. This point is equally true when we turn to systems of attitudes, norms, or cognitive systems of thinking. This fact reflects the iterative nature of social processes: individuals incorporate local mores, they reproduce those mores in their own mental frameworks and actions, they sometimes create innovations in those mores, and the next generation absorbs the successor locally instantiated system of mores.

Bulle and Di Iorio: Now that the differences in our conceptions of MI are clear and that it is also clear that, for us, methodological localism and MI are more epistemologically related than you acknowledge, we would like to discuss what we consider the real differences between these metatheoretical approaches. The concerns in the first place are the question of the causal power of structures as opposed to their causal role which is fully put into consideration by MI. More precisely, for MI the influence of social structures on action exists, but it is indirect because it is mediated by the interpretative skills of the actors.

You write: "I believe that it is perfectly legitimate to attribute causal powers to meso-level social structures [at the level of groups and organizations] […] Political institutions exist - and they are embodied in the actions and states of officials, citizens, criminals, and opportunistic others. These institutions have real effects on individual behavior and on social processes and outcomes - but always mediated through the structured circumstances of agency of the myriad participants in these institutions and the affected society" (Little,

2012b, p. 139; p. 144; 2016). And you write that the theory of microfoundations warns against "magical thinking" in the social sciences, preventing us from considering that social entities can have causal powers and structures of their own.

Nevertheless, sometimes you seem to attribute a direct causal power of social structures to the incorporation, by individuals, of rules and norms through procedures of inculcation and enforcement, entailing that social actors are "brought to comply with the rules and norms (to some degree)" and that the causal power of structures can be treated without the mediation of individual actions and interpretations. This causal action may thus be qualified as "embodied," like the Bourdieusian habitus, which refers to dispositions of action underlying normatively regulated social actions that escape the possibilities of conscious control by social actors. This is consistent with your defence of a "new pragmatism" emphasizing habits and practice, and your frequent reference to the notion of "mentality," which suggests psychological or moral dispositions, habits of mind, rendering the conscious "meaning" of action without any real interpretative interest. To illustrate this "incorporation" of structures, let us borrow an example from Talcott Parsons' structuro-functionalism. A social role corresponds, writes Parsons, to an "internalized object of the actor's personality": "When a person is fully socialized with respect to the system of interaction, it is not as true to say that the role is something the actor 'has' or 'plays' as it is to say that it is something he is." The process of socialization tends in this way to make the needs of the social system defined in terms of roles, and the orientations of individual personalities defined in terms of motivations, coincide. One knows of course the numerous sociological theories having recourse to types of embodied structures, culturalisms, structuralisms, neo-Marxisms, etc. But you do not seem to be a supporter of this type of explanation based on dogmatic interpretative postulates.

Can you clarify in which sense individual habits of mind can be caused by social structures in a way that cannot be mediated in principle by the interpretative skills of individuals? Can you also provide some examples of social structures that cause the individual habits of mind in this mechanical way? You seem to think that sometimes the influence of social structures on individuals cannot be accounted for through an ideal–typical approach that regards human action as necessarily related to the (either implicit or explicit) understandable meaning that social actors attach to it. Is this correct?

Little: I'm not entirely sure I understand the question clearly. Let's take the subjective actor first. I do affirm that actors are "subjective"—that is, they are purposive, norm-sensitive, meaning-seeking, and affective making use of specific cognitive resources to arrive at a plan of action. Actors embody subjectivity and choice. If the question postulates that actors have embodied governing "scripts" from surrounding ideological/normative fields—"Protestant ethic," "bourgeois rationality," "patriarchal domination," …, and that their actions are not chosen but mechanically determined by these scripts, then I would demur. This is not to deny that individual mentalities are shaped and influenced by the ambient ideologies, value systems, and stylized schemes of action within which they develop; that is quite obviously true. What I deny is the idea of ideology as an "iron cage" from which the actor cannot escape. Like James Scott in *Domination and the Arts of Resistance: Hidden Transcripts* (1990), I would argue that the young person raised within Italian capitalism can nonetheless develop a critical perspective on property and domination. He or she is influenced by the dominant ideology; but other influences are also present, and the individual actor has a capacity to reflect critically about the assumptions about the social world that she will accept, question, or reject. And the question of how to either accept or reject those assumptions is itself an active choice—to doff the cap to the landlord, to make quiet fun of the landlord's uncouth behavior, or to engage in a bread riot. I find E.P. Thompson's *Making of the English Working Class* (1966) to be an exemplary approach to ideology and class.

Let me turn to your very good question about how "structures" influence personal identities and mentalities in ways not chosen or even recognized by the individual. Let's take the example of gender identity and the fashion industry. The fashion industry and commercial enterprises like department stores in the 1950s in America arrived at very specific standards of female and male fashion and attractiveness. Whenever girls and women would visit Macy's department store they would be immersed in examples, both blatant and subliminal, about what a woman should look like and how she should behave. The color of clothing, the body type to which clothing was best suited, the use of cosmetics—all of these standards of "being an attractive woman" were written into the experience of shopping. Further, there were specific causal influences leading to this genderized treatment of the public; marketing specialists had deliberate strategies for selling products and maximizing revenues that turned on marshalling these cues. This genderized experience was repeated on television and movies and in the behavior of other men and women, and had unchosen effects on the gender identity of girls and women. Similar examples could be given concerning racial identity, rural identity, or even criminal identity (as Diego Gambetta demonstrates in *Codes of the Underworld: How Criminals Communicate*). We could consider different examples taken from literature, from the *Odyssey* to the 1952 film

High Noon, that were influential in shaping Greek and American ideas of masculine courage.

Bulle and Di Iorio: Let's discuss more precisely the distinctions between MI and methodological localism with regard to the rationality of individuals. A fundamental difference is that you are not systematically concerned with the meaning of action for the social actor, which is made clear by your recourse to a new pragmatism, and to psychological principles that do not necessarily rely on the conscious activity of individuals and refer, for example, to habits and mentalities. You (2011) specify that "mechanisms through which social identities and mentalities are transmitted, transmuted, and maintained are varied; inculcation, imitation, and common circumstances are central among these," which are processes that do not engage the reflection of individuals. And, even if you write that "ultimately, all social phenomena are the result of agents acting for their own reasons," this recognition has for you an ontological and not a methodological value. In your presentation of methodological localism on your blog (https://understandingsociety.blogspot.com/2008/11/what-is-met hodological-localism.htmethodologicallocalism), a formula that summarizes methodological localism is "the 'molecule' of all social life is the socially constructed and socially situated individual, who lives, acts, and develops within a set of local social relationships, institutions, norms, and rules [...] the individual is formed by locally embodied social facts, and the social facts are in turn constituted by the current characteristics of the persons who make them up."

What is, in the end, the meaning of agency in methodological localism? Is the individual in question completely penetrated by the "social" (whatever that may be) or does he keep a minimum of autonomy, of distance from the social of which he is an active stakeholder? Or to say it in a more direct way: is the individual in question absorbed and formatted by the social which pre-exists him, or does he take part in the construction of the social which builds him? What is his capacity to distance himself from his "social construction?" This is a question about the relationship between agency and structure that has been much discussed, but which cannot be answered in a way that is methodologically vague. This indefiniteness in methodological localism seems to us to support the forms of methodological or causal holism that we mentioned earlier, but in a roundabout and unacknowledged way. Are we wrong?

Little: It is true that methodological localism is "indefinite," in the sense that it is not a specific empirical analysis. It is rather a mid-level ontological picture of the social world. It implies, among other things, that we make a

656   D. Little et al.

huge conceptual mistake when we make general statements such as: "Islam in America is a patriarchal force in the Muslim community," "Southerners are white-supremacist," or "university professors are liberals indoctrinating their students." These statements are all faulty because they assume homogeneity whereas heterogeneity is the rule. To know how Islam works in America, we need to consider a range of connected Muslim communities (Dearborn, Chapel Hill, Los Angeles, Denver), and study each as a connected network of believers, Imams, students, parents, etc., to know how racial attitudes have proliferated from the Jim Crow period to the present, we need to study specific locales, from Lowndes County to Asheville. What we will find in each case is heterogeneity, conveyed by specific local institutions, leaders, and neighborhood activists who have influenced, transformed, and transmitted a set of practices and values. And there will be variation across each of these kinds of groups across different regions and cities. So methodological localism is "definite" in a particular sense: it recommends that the researcher should study the local, community-level mechanisms and institutions through which a value system is conveyed and changed. And it is anti-holist in a specific sense as well: it casts doubt on the idea that there are "ruling" aggregate-level structures that determine local arrangements, beliefs, and behaviors. Rather, we must always work to trace out the pathways of influence that extend from "national" organizations to regional and local organizations, down to individual members of various communities.

My view of what we need in a "theory of the actor" is fundamentally richer than what is offered by rational choice theory, neoclassical economics, or analytical sociology. Here are some questions that I proposed that can help clarify what is needed in a theory of the actor; fundamentally, it means we need to consider meanings, emotions, loyalties, commitments, purposes, plans, thought processes, heuristics, modes of reasoning, knowledge frameworks, and learning … (Understanding Society 10/28/2011).

1. How does the actor represent the world of action—the physical and social environment? Here, we need a vocabulary of mental frameworks, representational schemes, stereotypes, and paradigms.
2. How do these schemes become actualized within the actor's mental system? This is the developmental and socialization question.
3. What motivates the actor? What sorts of things does the actor seek to accomplish through action?
4. Here too, there is a developmental question: how are these motives instilled in the actor through a social process of learning?

5. What mental forces lead to action? Here, we are considering things like deliberative processes, heuristic reasoning, emotional attachments, habits, and internally realized practices.
6. How do the results of action get incorporated into the actor's mental system? Here we are thinking about memory, representation of the meanings of outcomes, regret, satisfaction, or happiness.
7. How do the results of past experiences inform the mental processes leading to subsequent actions? Here, we are considering the ways that memory and emotional representations of the past may motivate different patterns of action in the future.

Different theories of the actor provide different answers to these questions. Most notable in this listing of questions is the attention that is given to the importance of cognitive and cultural aspects of the actor's frame, and the important degree to which these features are socially and historically specific.

Once again, thank you, Nathalie and Francesco, for the thoughtful analysis that you have offered of the domain of social ontology in which we all share an interest, and the very interesting and stimulating questions that you have formulated.

References

Agassi, J. (1975). Institutional individualism. *British Journal of Sociology, 26*(2), 144–155.

Boudon, R. (2007). *Essays on the general theory of rationality*. PUF.

Boudon, R. (1998a). Social mechanisms without black boxes. In P. Hedstrom & R. Swedberg (Eds.), *Social mechanisms: An analytical approach to social theory*. Cambridge University Press.

Boudon, R. (1998b). Should we still read Durkheim's Rules after a hundred years? Raymond Boudon interviewed by Massimo Borlandi. In R. Boudon (Ed.), *Études sur les sociologues classiques* (pp. 137–163). PUF.

Boudon, R. (1989). Subjective rationality and the explanation of social behavior. *Rationality and Society, 1*(2), 173–196.

Boudon, R., & Bourricaud, F. (1990). *A critical dictionary of sociology*. University of Chicago Press.

Coleman, J. S. (1990). *Foundations of social theory*. Harvard University Press.

Durkheim. (1938 [1895]). *Rules of sociological method*. Free Press.

Gambetta, D. (2011). *Codes of the underworld: How criminals communicate*. Princeton University Press.

Hayek, F. A. (1948). *Individualism and economic order*. University of Chicago Press.

Hayek, F. A. (1973). *Law, legislation and liberty, Vol. 1: Rules and order*. University of Chicago Press.

Little, D. (2016). *New directions in the philosophy of social science*. Rowman & Littlefield International.

Little, D. (2014). Actor-centered sociology and the new pragmatism. In J. Zahle & F. Collin (Eds.), *Rethinking the individualism-holism debate* (pp. 55–76). Springer.

Little, D. (2012a). Analytical sociology and the rest of sociology. *Sociologica* (ISSN 1971-8853), Fascicolo 1, gennaio-aprile 2012. https://doi.org/10.2383/36894

Little, D. (2012b). Explanatory autonomy and Coleman's Boat. *Theoria. An International Journal for Theory, History and Foundations of Science, 27*(2), 137–151.

Little, D. (1991). *Varieties of social explanation. An introduction to the philosophy of social science*. Westview Press.

Mises, L. (1981). *Socialism: An economic and sociological analysis*. Liberty Found.

Ramström, G. (2018). Coleman's boat revisited—Causal sequences and the micro-macro link. *Sociological Theory, 36*(4), 368–391.

Roth, G. (1976). History and sociology in the work of Max Weber. *The British Journal of Sociology, 27*, 306–318.

Additional References

Durkheim, É. (1982). *The rules of sociological method*. The Free Press.

Epstein, J. M. (2006). *Generative social science: Studies in agent-based computational modeling. Princeton studies in complexity*. Princeton University Press.

Fodor, J. (1974). Special sciences and the disunity of science as a working hypothesis. *Synthese, 28*(2), 97–115.

Heath, J. (2020). Methodological individualism. In E. N. Zalta (Ed.), *Stanford encyclopedia of philosophy* (Summer 2020 ed.). https://plato.stanford.edu/archives/sum2020/entries/methodological-individualism

Hedström, P. (2005). *Dissecting the social: On the principles of analytical sociology*. Cambridge University Press.

Lukes, S. (1982). Introduction. In *The rules of sociological method*, Emile Durkheim. The Free Press.

McAdam, D., Sidney, G. T., & Tilly, C. (2001). *Dynamics of contention. Cambridge studies in contentious politics*. Cambridge University Press.

Scott, J. C. (1990). *Domination and the arts of resistance: Hidden transcripts*. Yale University Press.

Thelen, K. A. (2004). *How institutions evolve: The political economy of skills in Germany, Britain, the United States, and Japan*. Cambridge University Press.

Thompson, E. P. (1966). *The making of the English working class*. Vintage Books.

Methodological Individualism and Critical Realism: Questions for Margaret Archer

Nathalie Bulle and Francesco Di Iorio

Margaret Archer was originally slated to answer the following questions and engage in a debate on the interpretation of MI within the framework of critical realism. However, due to personal reasons, she was ultimately unable to do so.

Margaret Archer (born January 20, 1943 in Grenoside) is an Emeritus Professor of Sociology at the University of Warwick, England, and formely professor at l'Ecole Polytechnique Fédérale de Lausanne, Switzerland. She is one of the most prominent contemporary English theorists in sociology, best known for her "morphogenetic approach," which represents a methodological development for the social sciences of British philosopher Roy Bhaskar's critical realism.

Margaret Archer began her career in the 1970s with comparative research on educational systems with the aim of providing "a theoretical account of macroscopic patterns of change in terms of the structural and cultural factors which produce and sustain them" (Archer, 1979, p. 613). This sociological approach was later formalized in her morphogenetic theory which proposes to explain social phenomena as the result of the actions of individuals and

---

N. Bulle (✉)
French National Center for Scientific Research, Sorbonne University, Paris, France
e-mail: nathalie.bulle@cnrs.fr

F. Di Iorio
Department of Philosophy, Nankai University, Tianjin, China

© The Author(s), under exclusive license to Springer Nature
Switzerland AG 2023
N. Bulle and F. Di Iorio (eds.), *The Palgrave Handbook of Methodological Individualism*,
https://doi.org/10.1007/978-3-031-41508-1_28

659

groups in structurally and culturally predetermined situations (Archer 1982). In this framework, Archer postulates the existence of an efficient causal power of both structures and actions: Structures exert their own influence on actions, by enabling and constraining them, while sociocultural actions and interactions influence the development of social and cultural structures.

Q1 Bulle Di-Iorio: In order to understand your approach to the relationship between structure and agency, it is necessary to introduce the philosophical background of critical realism. Let us recall that the name of this philosophical approach goes back to the Canadian-American philosopher, Roy Wood Sellars (father of the American Philosopher Wilfrid Sellars) in his 1922 work, *Evolutionary Naturalism.* As Sellars (1922, p. 20) explains, "it is realism because it maintains that the human mind can build up knowledge about extramental realities; and it is critical realism because it holds that these realities cannot be presented to an immediate awareness, as naive realism inclines to assert." Critical realism thus developed in the United States at the beginning of the twentieth century (see Blake et al., 1941) although its connection to contemporary critical realism is rarely discussed (see Groff, 2008). Bhaskar himself defines his critical realism in opposition to two main traditions in the philosophy of science, classical Humean empiricism, and Kantian transcendental idealism. The objects of knowledge refer neither to empirical phenomena, nor to "transitive" structures, that is artificial, human constructs, but to "intransitive" structures that exist independently of their identification. Such intransitive structures involve an explanatory move from manifest phenomena to the structures that produce them (Bhaskar, 1979/2015, p. 13).

It is the conception of an ontologically stratified reality in levels of emergence that accounts for the formation of causally efficient structures at different levels of reality, typically involving a "layered model" of the real. Such a model was developed in the work of the emergentist theorists in the early twentieth century (see for example Mclaughlin, 1992). This work assumes that entities, understood as real structures of the world, have been hierarchically ordered into increasingly complex levels in the course of the evolutionary process, and, at each level, represent new causal powers: "Emergent" wholes or agents "are constituted at different levels of complexity and organization" and, as such, are irreducible to their "atomistic" components (Bhaskar, 2013, p. 114). On this basis, you defend the efficient causality of social conditioning, on the one hand, and of individuals in the conduct of their lives, on the other. Moreover, like Bhaskar, you develop a critique of both methodological individualism and methodological collectivism.

Methodological Individualism and Critical Realism ... **661**

It seems important to us here to have your point of view on the specific position of critical realism vis-à-vis emergentism, especially in relation to the critique of downward causation. Also, how would you describe the position of critical realism in the landscape of contemporary analytic philosophy?

Q2 Bulle-Di Iorio: The first thing that strikes the methodological individualists is their agreement with the critical realists on the role of human agency—even if they prefer to talk about the "social actor"—in explaining social phenomena. They cannot but find themselves very close to the critical realists when the latter explain that human action is characterized by the phenomenon of intentionality and that the causal effect of social forms is mediated through it. In your morphogenetic theory too, the effects of social and cultural structures are always mediated by the individual actors themselves. You insist on the importance of mental reflexivity (the absence of which you note in Pierre Bourdieu's sociology) through the notion of "internal conversation," according to which "reflexive deliberation constitutes the mediating process between structure and agency" (Archer, 2003, p. 130). The causal role of social and cultural structures described as the "results of past actions (…) deposited in the form of current situations" (Archer, 1995, p. 201) thus always involves the reflection of social actors: "These structural and cultural properties only become causally efficacious in relation to human projects in society" (Archer, 2003, p. 132). Proponents of MI would not deny the validity of these ideas. However, critical realists strongly reject MI on the basis of an account of its assumptions that all methodological individualists would find caricatural and unacceptable. This problem will be the subject of our next questions.

Bhaskar describes MI as a "naive" approach that is mistaken because of its reductionism, arguing that MI's conception of agency perpetuates a myth of (logical or historical) creation (Bhaskar, 1979/2015, p. 36). In this regard, Bhaskar evokes the parallel drawn by John Watkins (1952a, 1952b, 1957) between the individualist approach in the social sciences and the mechanism approach in physics (Bhaskar, 1979/2015, p. 26, note 5). However, in the 1952 text cited by Bhaskar, Watkins' evocation of physical constructs illustrates the analytical approach "of resolution and re-composition" (Watkins, 1952b, p. 188). The parallel between MI and Newtonian physics is thus established simply on the epistemological basis of the analytical method of decomposition into basic units (e.g., social actions), which are not assumed to exist in a state independent of the whole in which they participate. Watkins (1952a, p. 43), for example, argues for basing social action on "typical, socially significant dispositions." Watkins has not always been a well-informed and effective advocate of MI, but in his 1957 article also cited by Bhaskar, he (1957, p. 111) criticizes structural determinism rather than cultural conditioning: "What the methodological individualist denies is that an individual is ever frustrated, manipulated or destroyed or borne along by irreducible sociological or historical laws" (Watkins, 1957, p. 115, note 1).

According to you, MI sees individual actions and social or cultural structures as always temporally covariant, so that no structures would preexist actions (Archer, 1995, ch. 3). Its central assumption, in your view, is that "every aspect of the social context is the result of contemporary behavior, whatever its origin," so that "the 'present' (wherever it is situated historically) is always cut off from the future and the past" (Archer, 1979, p. 17). You also refer to Watkins in this regard, especially because he argues (1957, p. 107) that an individual's social environment is constituted by "other people" and that there is no social tendency that could not be changed if the individuals concerned both wanted to change it and had the appropriate information to do so. You conclude that social contexts in MI are essentially dependent on the social actors who participate in them, but you write elsewhere that "conditional influences may be agentially evaded, endorsed, repudiated or contravened" (Archer, 2003, p. 131), which is reminiscent of Watkins' argument in this regard.

Moreover, you subscribe to what you call "situational individualism," associated with Popper's approach, and your conception of culture, which includes all the elements that underlie human understanding and belong to the propositional register of society (Archer, 1988), is very close to Popper's World 3. In particular, you support the idea that the objects of World 3, which refer to thought contents that are distinct from thought processes, are part of the individuals' situational data. This ensures the possibility of a comprehensive (interpretative) sociology to which partisans of MI subscribe. Popper called himself a methodological individualist, and Watkins was an advocate of Popper's World 3 (see Watkins, 1974).

On what grounds, then, do you think the proponents of MI fail to consider the anteriority of social structures? And what kind of social tendency do you have in mind that cannot be changed by individual actions?

Q3 Bulle-Di Iorio: We think you are also unfair to Friedrich Hayek whom you quote in your critique of MI, in the first part of your 1995 book *Realist Social Theory*. You claim that the defects of MI and its explanatory program derive from empiricism, and you see two main reasons for this: the empirical evidence of the existence of individuals and the equally empiricist rejection of the hypostasis of collectives. You think so because Hayek writes that "no collective term ever designates definite things in the sense of stable collections of sense attributes" (Archer, 1995, p. 35).

In his three articles on *Scientism and the Study of Society*, Hayek (1942, 1943, 1944) defends the distinction between the methods of the natural and social sciences on the basis of the latter's reference to internal, subjective elements that involve understanding. On the contrary, he criticizes (metaphorically and perhaps provocatively) the positivists' empiricism, which leads them to understand collectives as things with a direct causal role. His critique of empiricist realism is such that he evokes "the empiricist prejudice" that identifies existence with empirical objectivity (Hayek, 1943, p. 62). He sees these collectives as existing and generally influencing on bases that cannot be directly observed because they depend constitutively on the interpretations of individuals ("the concepts which guide individuals in their actions" Hayek, 1942, p. 286). Hayek explains that in order to grasp these wholes that we cannot observe directly, we can only resort to hypotheses about the structures of the relationships connecting the data of experience that these wholes imply. Their existence depends on the validity of these hypotheses and can only be tested indirectly: "The terms for collectives which we all readily use do not designate definite things in the sense of stable collections of sense attributes which we recognize as alike by inspection; they refer rather to certain structures of relationships between some of the many things we can observe within given spatial and temporal limits and which we select because we think we can discern connections between them" (Hayek, 1943, pp. 43–45). Hayek's view is thus centered on critique of positivist empiricism. In his 1952 book *The Sensory Order*, he defends against empiricism in particular the view that all knowledge is based on subjective abstract constructs. Hence, the peculiarity of the subjectivist approach of the social sciences, which deal "with the phenomena of individual minds, or mental phenomena, and not directly with material phenomena" (Hayek, 1942, p. 279).

Could you then explain your criticism of Hayek's position?

Q4 Bulle-Di Iorio: We agree with Anthony King (1999) who wonders whether you are not attacking what is ultimately a "straw man," since for the interpretative tradition you seem to want to dismiss, the social and historical contexts in which individuals are situated are essential to understanding their reasons and practices. In particular, King notes that you barely cite the strands you identify with MI, except for the symbolic interactionists and the neo-phenomenological school, which include in particular the interactionist and ethnomethodological strands of the interpretive tradition (Archer, 1995, p. 60, 84). These strands do not summarize MI or interpretive sociology, as your other references to Watkins or Hayek show, nor do they present the absurdities that you ascribe to them. As Maurice Natanson (1968, p. 221) evocatively writes of Alfred Schütz's phenomenological approach to social reality, "of course, the social world does not spring magically into being with my birth or yours; it is historically grounded and bears the marks and signs of the activity of our ancestors, most remarkable of all the typifying medium of language." Schütz (1932/1967) does not deny that the various forms of the social world from social collectives, such as the "state," the "press" or the "economy," cultural objects, such as sign systems and all artifacts in the broad sense, to all kinds of social institutions, are not born with contemporaneous individual social actors. Rather, like critical realists, he asserts that these social forms have an influence based on individual conscious experience.

Could you elaborate on which sociologists, theories, or specific analyzes you have in mind when you suggest that methodological individualists assume that no social or cultural structures preexist social actions?

Q5 Bulle-Di Iorio: We think that the proximity of MI and CR on the above-mentioned points, and their common criticism of empirical realism, reveals a misconception of MI on the part of CR advocates. MI implies a scientific conception of causality (i.e., based on theoretical assumptions that refer indirectly to observation), that does not oppose CR's notion of efficient causality (according to which real causes refer to intrinsic powers or tendencies of real things underlying observational regularities), but proposes a provisional and indirect approach to causality involving theoretical models (cf. Ekström, 1992; Hamlin, 2003). MI's theoretical stance defines parts as abstract constructs that may refer to real entities and properties, but are not considered independent of the wholes they constitute. Conversely, the critical realists' interpretation of the "parts" in MI as representing a lower ontological level assumes such independence from structures, so that the structures MI refers to, according to your interpretation, should always be composed and created by contemporaneous actors. This leads you to wonder whether the individualistic approach is not contradictory when, in explaining the interaction between several persons, it implies some prior cultural formation. What you see as nonsense is the result of a profound difference in perspective. This is what we would like to address in what follows.

The layered representation of reality and the associated metaphysics of emergence lead critical realists to interpret MI's approach as referring to a level below the social, so that methodological individualists would have to endorse emergence (Archer, 1995, pp. 40–41) in order to accept socialization. But MI's perspective is in a sense the reverse: Individuals are situated at the social level from the outset and (theoretical) analysis leads to individualization, that is assumptions about individuals' situations and reasons for acting. In other words, social actors are from the outset members of social and cultural groups and individualist analysis leads us to account for their action according to these social and cultural participations, on the basis of the principle of understanding, which appeals to the idea of subjective rationality.

Let us take a classic example of an explanation in terms of MI (and interpretative tradition): that of the links between the development of capitalism and the Protestant ethic, which Max Weber (1904–1905/1992) developed to oppose to the determinism of economic structures in Karl Marx. The causal role of religious beliefs—along with other situational factors such as the legal system, the division of labor, etc.,—is obviously not inherent in the behavior of individuals in the present, but is part of the historical factors that explain the goals of economic success in terms of religious values: The work ethic and the Calvinist belief in predestination, represent cultural and social legacies.

If our diagnosis is correct, you assume something quite different, based on the emergentist ontology of CR, which underlies the misunderstandings mentioned above. You assume that since MI supports a conception of the causal power of individuals but not of social and cultural structures, according to MI individuals must preexist structures, but no structures can preexist the individuals. Taking the correct methodological perspective of MI, individuals, and structures are relative constructs, not levels of emergence. Structures can have independence from actual individuals (Popper's World 3 objects do) even if this independence does not mean that structures have causal power of their own, but only a causal role through individual situations.

Do you agree with this diagnosis, if not could you explain your point of view?

Q6 Bulle -Di Iorio: MI is an axiologically neutral methodological approach in the sense that it does not make value judgments about the relative importance of individuals versus society, since individuals and society represent relative constructs for the explanation. For its part, by speaking of structural tendencies (other than metaphorically), CR hypothesizes an interaction, ontologically speaking, of causal powers, so that two logics would be at work concurrently, and possibly in opposite directions.

For example, as you know, Raymond Boudon's thesis in *Inequality of Opportunity* is a paradigmatic example of an MI explanation that demonstrates why, in a society at a given time, the social opportunities of individuals according to their level of education depend on the overall distribution of social status in the society. The causal role of structures is defined not only at the level of individuals (students' cultural capital explains differences in achievement and students' social position explains difference in decision-making contexts), but also at a collective level that characterizes the state of society at a given moment. In this regard, Boudon advocates the use of systems analysis, especially because it allows structural constraints to be taken into account in understanding social processes (cf. Boudon, 1973, 1974). Nevertheless, Cynthia Hamlin (2003, p. 47) suggests revising the Boudonian explanation through the CR approach by writing: "There is a (structural) tendency for low status individuals to be low achievers at school, but this tendency is not always actualised because the causal powers of certain individuals (high IQ, verbal achievement, etc.) may neutralise the causal powers of the social structure, thus leading to a transformation of that structure." This interpretation involves a form of interaction between structural and individual causal powers, which represents an unnecessary, even ad hoc, assumption and seems to contradict the idea, which you share, that the causal role of social and cultural structures is mediated by the reflection of individuals.

Rom Harré (under whose supervision Bhaskar wrote his Ph.D thesis in philosophy), who initiated a realist conception of causal powers in the philosophy of science and developed a generative view of explanation that is in fact largely shared by MI (Harré & Madden, 1973), holds that causal powers involve "powerful particulars" and that, in the social world, the only entities endowed with causal power are persons. In this regard, Harré and Varela criticize Bhaskar's notion of causal power of structures which, by treating social structures as things, commits "the fallacy of 'collectivism,' that is, the fallacy of reifying a property of a group of social actors into an entity" (Harré & Varela, 1996, p. 314). Harré (2002) explains that if, for example, a rule were to be considered to have causal power, even the people involved in its creation could not escape it.

What is your response to Harré? Also, how would you explain this aspect of CR's ontology and its possible links to political positioning, knowing that contemporary CR bears the mark of Marx and gives rise to sophisticated dialectical materialism (Groff, 2009)?

References

Archer, M. (1979). *Social origins of educational systems*. Sage.

Archer, M. (1982). Morphogenesis versus structuration. *British Journal of Sociology, 33*, 455–483.

Archer, M. (1988). *Culture and agency: The place of culture in social theory.* Cambridge University Press.

Archer, M. (1995). *Realist social theory: The morphogenetic approach.* Cambridge University Press.

Archer, M. (2003). *Structure, agency and the internal conversation.* Cambridge University Press.

Bhaskar, R. (1979/2015). *The possibility of naturalism. A philosophical critique of the contemporary human sciences.* Routledge.

Bhaskar, R. (2013). *A realist theory of science.* Routledge.

Blake, D., Lovejoy, A. O., Pratt, J. B., Rogers, A. K., Santayana, G., Sellars, R. W., & Strong, C. A. (1941). *Essays in critical realism.* Peter Smith.

Boudon, R. (1973). *L'inégalité des chances.* Armand Colin.

Boudon R. (1974). The sociology of inequalities in a dead end? In margin of the book of Christopher Jenks: Inequality. *Analyse & Prévision, XVII*, 83–95.

Ekström, M. (1992). Causal explanation of social action: The contribution of Max Weber and of critical realism to a generative view of causal explanation in social science. *Acta Sociologica, 35*(2), 107–122.

Groff, R. (2008). *Situating critical realism philosophically.* http://www.pages.drexel.edu/~pa34/GROFF.pdf

Groff, R. (2009). Special issue on causal powers. *Journal of Critical Realism, 9*(3), 267–386.

Hamlin, C. L. (2003). *Beyond relativism. Raymond Boudon, cognitive rationality and critical realism.* Taylor and Francis.

Harré, R., & Madden, E. H. (1973). Natural powers and powerful natures. *Philosophy, 48*(185), 209–230.

Harré, R., & Varela, C. R. (1996). Conflicting varieties of realism: Causal powers and the problems of social structure. *Journal for the Theory of Social Behaviour, 26*, 313–325.

Harré, R. (2002). Social reality and the myth of social structure. *European Journal of Social Theory, 5*(1), 111–123.

Hayek, F. (1942). Scientism and the study of society. *Economica, 9*(35), 267–291.

Hayek, F. (1943). Scientism and the study of society. Part II. *Economica, 10*(37), 34–63.

Hayek, F. (1944). Scientism and the study of society. Part III. *Economica, 11*(41), 27–39.

King, A. (1999). Against structure: A critique of morphogenetic social theory. *Sociological Review, 47*, 199–227.

Mclaughlin, B. P. (1992). The rise and fall of British emergentism. In A. Beckermann, H. Flohr, & J. Kim (Eds.), *Emergence or reduction? Essays on the prospects of nonreductive physicalism* (pp. 49–93). Walter de Gruyter.

Natanson, M. (1968). Alfred Schutz on social reality and social science. *Social Research, 35*(2), 217–244.

Popper, K. (1978, April 7). Three worlds, *The Tanner lecture of human values*. Delivered at The University of Michigan.

Schütz, A. (1932/1967). *The phenomenology of the social world*. Northwestern University Press.

Sellars, R. W. (1922). *Evolutionary naturalism*. Open Court.

Watkins, J. W. N. (1952a). Ideal types and historical explanation. *The British Journal for the Philosophy of Science, 3*(9), 22–43.

Watkins, J. W. N. (1952b). The principle of MI. *The British Journal for the Philosophy of Science, 3*, 186–189.

Watkins, J. W. N. (1957). Historical explanation in the social sciences. *British Journal for the Philosophy of Science, 8*, 104–117.

Watkins, J. W. N. (1974). The unity of Popper's thought. In P. A. Schilpp (Ed.), *The philosophy of Karl Popper* (pp. 371–412). Open Court.

Weber, M. (1904–1905/1992). *The protestant ethic and the spirit of capitalism*. Routledge.

Author Index

A

Abbruzzese, S. 140, 141, 146
Acquaviva S.S. 146
Afsaruddin, A. 328
Agassi, J. 79, 425, 427, 428, 448–450, 466, 477, 551, 577, 617–619, 621–623, 625, 627–630, 650
Ahdieh, R. 78
Ajayi, A. 412
Albert, G. 532
Albert, H. 339
Alchian, A.A. 93
Alexander, J.C. 531
Allen, W. 414
Alpert, H. 461, 462
Althusser, L. 430, 459
Anderson, B.E. 412
Anderson, J.A. 577, 581
Anscombe, G.E.M. 60, 61
Antiseri, D. 429, 432, 433, 437, 447, 456, 457, 576
Archer, M.S. 3, 15, 22, 23, 425, 624, 625, 659–663, 665
Arjomand, A. 337

Arnsperger, C. 188
Aron, R. 132
Arrow, K. 17, 78, 383–386, 501–504, 506, 507, 510, 511, 514, 518, 612, 614
Ash, T.G. 237
Auffarth, C. 340
Axelrod, R. 154
Aydinonat, E.E. 272

B

Baert, P. 557
Baier, A.C. 12
Bainbridge, W.S. 254
Bales, R. 4, 112, 116
Barth, F. 301, 304–306, 314
Bastiat, C.F. 274, 275, 277, 278, 280
Basu, K. 78
Bates, R.H. 119
Bauer, P.T. 287, 288
Bearman, P. 430, 466
Becker, G.S. 39, 78, 82, 87–89, 92
Becker, S.O. 533

© The Editor(s) (if applicable) and The Author(s), under exclusive
license to Springer Nature Switzerland AG 2023
N. Bulle and F. Di Iorio (eds.), *The Palgrave Handbook of Methodological Individualism*,
https://doi.org/10.1007/978-3-031-41508-1

Author Index

Beck, U. 190
Behne, T. 23
Beidelman, T.O. 68
Bellah, R.N. 167
Bell, D. 162
Bendix, R. 457
Bénichou, M. 166
Berger, P.L. 131, 146, 221
Berkowitz, S.D. 156
Bertrand, E. 84
Beshara, R.K. 412
Bessis, F. 519
Bhargava, R. 576, 577, 580
Bhaskar, R. 425, 427, 624, 625, 659–661, 666
Bissoondath, N. 167
Bizer, K. 286
Blumenberg, H. 335, 336
Blumer, H. 552, 554–560, 565, 568
Boehm, C. 43, 48
Boettke, P.J. 277, 282, 291, 433, 456, 457
Böhm-Bawerk, E. von 457
Boldrin, M. 284
Bolton, P. 507
Bonilla-Silva, E. 412
Borch, C. 67
Borlandi, M. 464
Bošković, A. 304
Bouchard, G. 169
Boudon, R. 31, 44, 45, 79, 107, 109, 118, 119, 127, 128, 136, 137, 139–141, 145, 153, 159, 160, 164, 170, 171, 177, 190, 204, 206–211, 213–216, 218–222, 351–353, 356, 368, 428–430, 432, 433, 436, 437, 439, 447, 455, 457–466, 527, 530, 546, 574, 576, 607, 610, 612, 613, 638, 644–646, 650, 651, 665, 666
Bouglé, C. 153, 155
Bourdieu, P. 351–353, 459, 460, 464, 466, 534, 661

Bourricaud, F. 136, 160, 430, 433, 457, 462, 463, 613, 646
Bouvier, A. 162, 434, 435, 437, 438, 451, 457
Bowles, S. 39, 154, 515
Boyd, R. 40
Boyer, P. 40, 340
Boylan, R. 414
Bratman, M.E. 10–13, 20–22, 24
Braudel, F. 205
Breen, R. 353, 355
Brehm, H.N. 416
Bridel, P. 187
Brisset, N. 184
Bronner, G. 154, 428, 430, 433
Brooks, R.A. 353, 592
Brossard, B. 552, 557
Brown, W. 508
Brüggemann, J. 286
Bryant, W.W. 412
Buban, S.L. 559, 563
Buchanan, J. 207–209, 291
Bulle, N. 7, 9, 152, 262, 265, 429, 431, 433, 434, 439, 447, 450, 527, 546, 568, 608, 610, 612, 619, 621–630, 636, 638, 641, 643, 645, 647, 650, 652, 655, 662, 663, 665
Bunge, M. 7, 435–438, 476–478
Burkert, W. 326
Burman, P. 485, 486
Burns, S. 290
Burns, T.R. 559, 563
Buskens, V. 434, 435
Butler, J. 409
Bylund, P.L. 278, 290

C

Cahuc, P. 517
Calabresi, G. 78, 86, 87, 90, 91, 93–96
Caldwell, B. 447
Caldwell, P.C. 54, 55

Author Index **671**

Call, J. 23
Callon, M. 185
Camerer C.F. 381, 515
Campagnolo, G. 432
Cancian, F. 115
Candela, R.A. 433
Cantor, G. 476
Card, D. 279
Carnap, R. 74, 107, 621
Carnoy, M. 356
Carpenter, M. 23
Cartwright, N. 104, 517
Chaserant, C. 519
Chen S.-H. 424, 434, 435, 437
Cherkaoui, M. 429, 433
Chmielewski, A.K. 527
Chomsky, N. 482
Churchland, P.M. 582, 596
Churchland, P.S. 577, 581, 582, 596
Clark, S. 354
Clarke, A. 561
Coase, R.H. 78, 82–86, 90, 91, 93, 94
Cohen, A.K. 112
Cohen, G.A. 263
Cohen, J. 574, 596
Cohen, M. 504
Colander, D. 513, 515
Cole, J.H. 284
Coleman, J.S. 32, 33, 104, 106, 119, 162, 177, 195, 216, 227, 235, 355, 374–378, 380, 383, 384, 386, 388, 390, 395, 432, 435, 447, 448, 455, 456, 462, 463, 465, 526, 531, 532, 534, 536, 551, 635, 638, 640, 641, 644
Collin, F. 418, 450
Commons, J.R. 82
Comte, A. 31, 110, 132, 138, 139
Conein, B. 179
Connell, R. 416
Cook, D. 326, 335
Coser, L.A. 115

Côté, J.-F. 552, 558, 559
Couch, C.J. 559, 563
Coyne, C.J. 282, 286, 288–291
Cronin, A.K. 342, 344
Crosetto, P. 286
Crozier, M. 204
Cuchet, G. 146

D

Dahme, H.-J. 331
Damasio, A. 40, 591
D'Amico, D.J. 291
Damir-Geilsdorf, S. 335
Dardot, P. 508
Davidson, D. 16, 62, 586, 596
Davis, A.K. 112
Davis, J.C. 118, 119
d'Avray, D.L. 256, 262, 263, 268
Dawkins, R. 325
Deacon, T. 154
Deaton, A. 517
Debreu, G. 500, 501
De Bruin, B. 191
De Certeau, M. 186
Degenne, A. 153–156
Delsol, C. 146
Demeulenaere, P. 31, 38, 44, 184, 187, 425, 430, 433, 434, 508
Demo, D.H. 412
Demsetz, H. 87, 93
DeNardo, J. 243
Descartes, R. 585
Descombes, V. 29, 449, 519
Dewatripont, M. 507
Diallo, Y. 298
Diekmann, A. 377–379, 381, 383, 395
Dietrich, D. 412
Di Iorio, F. 7, 9, 154, 172, 424, 426, 428, 429, 433–438, 440, 447, 450, 466, 490, 568, 607, 608, 610, 612, 619, 621–630,

636, 638, 641, 643, 645, 647, 650, 652, 655, 662–665
Dilthey, W. 577
Di Nuoscio, E. 431, 433, 434, 437
Djankov, S. 288
Donahoe, B. 310
Donati, P. 3, 15, 22, 23, 144
Dostaler, G. 514
Downs, A. 90, 236
Dray, W. 104
Dreitzel, H.P. 108
Dressler, J. 528
Dreyfus H.L. 593
Dubreuil, L. 169
Dumont, L. 484
Dumouchel, P. 434, 435
Dunbar, R. 157
Dupuy, J.P. 434, 435
Durkheim, É. 3–5, 34, 35, 37, 39, 46, 47, 66, 67, 73, 110, 128, 130–132, 138, 139, 143, 144, 151, 159, 203, 204, 214, 215, 262, 263, 425, 428, 462–464, 466, 493, 528, 605, 608, 612, 646, 649–652
Duru-Bellat, M. 354
Dyer, P. 487

Easterly, W. 288
Eberlein, G. 105
Eccles, J.C. 596
Edwards, W. 584
Eickelman, D.F. 329
Eidson, J.R. 310
Einar, L. 281
Eisenberg, R.S. 284
Eisenstadt, S.N. 111, 117
Elder-Vass, D. 431
Electronic Freedom Foundation 285
Elias, N. 40, 205
Ellwood, C. 67

Elster, J. 45, 424, 433, 447, 448, 451, 454, 455, 460, 465, 466, 477, 576, 577, 607, 610, 612, 613, 635
Engels, F. 484
Epstein, B. 435–439, 450, 479, 480
Epstein, J. 641, 650
Epstein, R. 85
Erikson, R. 352
Esser, H. 106–108, 119, 331, 332, 526, 532–534, 536, 543
Etzioni, A. 109
Evans-Pritchard, E.E. 304

F

Fagiolo, G. 515
Faia, M.A. 114
Faure, M.G. 94
Favereau, O. 510, 517–520
Fawzi, I. 329, 343, 344
Fehr, E. 40
Ferguson, A. 466
Festinger, L. 534
Fetzer, J. 104
Feyissa, D. 310
Fielding, N. 334
Fine, G.A. 551, 555, 557, 563, 568
Fiske, A.P. 44
Fodor, J.A. 580, 586, 637
Førland, T.E. 491
Forsé, M. 156, 171
Foster, J.B. 416
Foucault, M. 268, 460, 508, 510
Fouda, Y. 334
Freeman, R.E. 193
Freud, S. 66–68
Freyer, H. 74
Friedman, I. 341
Friedman, M. 517
Frobert, L. 459
Fuest, V. 310
Fullbrook, E. 519

G

Gadamer, H.-G. 536
Gagner, N. 412
Gallagher, S. 590, 591, 594
Gallese, V. 590
Galt, A. 103, 109
Gambetta, D. 532, 540–542,
    544–546, 654
Gamson, W. 236
Garner, R.A. 109
Gassamaier, W. 589
Gauchet, M. 131, 146
Geertz, C. 336
Géhin, E. 428, 430, 433
Geoffard, P.-Y. 517
Gershon, M.D. 591
Gërxhani, K. 375
Gibbard, A. 195
Gibson, W. 551
Giddens, A. 475
Giesen, B. 113
Gigerenzer, G. 307, 515, 589–591,
    593, 596
Gilbert, M. 15, 18–20, 24, 25, 449
Gilbert, M.A. 409
Gillman, L. 410
Gintis, H. 154
Girard, R. 326
Gluckman, M. 337
Godin, M. 527
Goldman, A.I. 575, 590, 591, 596
Gold, N. 449
Goldthorpe, J.H. 107, 353, 355,
    375
Goodman, N. 289, 596
Graf, F.W. 534, 536
Granovetter, M. 46, 119, 158, 159
Greenblatt, S. 339, 341
Greene, J. 48
Greif, A. 119
Greshoff, R. 532
Greve, J. 23, 107, 531
Grier-Reed, T. 412
Groff, R. 660, 666

Gross, N. 558
Guessous, M. 112
Guo, C.B. 356
Gurr, T. 229, 234

H

Habermas, J. 250, 333
Habicht, C. 340
Hacking, I. 501
Hackney Jr. 96
Hadhazy, A. 195
Hagen, E.E. 118, 119
Hägerström, A. 54
Hahn, F. 501, 502, 506, 518
Haidt, J. 35, 41, 43, 47
Halévy, E. 459
Hall, A.R. 282, 290
Hallett, T. 559, 563
Hall, P.M. 559, 563, 566, 568
Hann, C. 311
Harberger, A.C. 81
Harding, S. 410
Harnay, S. 78
Harper, C.L. 109
Harré, R. 666
Harris, D. 414
Harris, S. 325
Hart, H.L.A. 69, 70
Hartog, J. 354, 356, 357
Hatchuel, A. 518
Hatemi, P.K. 41
Hayek, F. 79, 277, 278, 427–429,
    432–437, 439, 447, 448, 451,
    455–459, 462, 465, 466, 514,
    515, 598, 610–612, 620, 624,
    630, 638, 646, 647, 662–664
Hazlitt, H. 290
Heacock, R.T. 242
Heath, J. 7, 249, 250, 457, 525,
    639, 648
Hechter, M. 301
Hedström, P. 32, 430, 466, 636, 639

Author Index

Hegel, G.W.F. 74, 110, 403, 404, 407, 452, 623
Hegghammer, T. 326, 327
Heiner, R. 189
Heine, S.J. 40
Heinich, N. 169
Heller, M.A. 284
Helly, D. 169
Hempel, C. 104, 107, 114, 375, 540
Hengel, M. 339
Henrich, J. 40, 44
Herfeld, C. 436, 438
Hernes, G. 118, 119
Hervieu-Léger, D. 146
Hield, W. 114
Higgs, M.A. 274
Hinde, R. 228
Hindriks, J. 527
Hintz, R.A. 559, 563
Hirschman, A.O. 187, 213, 215
Hitchens, C. 325
Hobbes, T. 64, 112, 194, 325, 585, 639
Hodgson, G.M. 5, 7, 8, 79, 448, 454, 457, 478, 479
Homans, G.C. 31, 114–117, 424, 578
Horsley, R.A. 339, 340
Horstmann, A. 299
Howitt, P. 515
Huebner, D.R. 552, 558, 559
Huff, T. 9, 626
Hughes, M. 412
Hume, D. 45, 625, 626
Hummell, H.J. 376
Hunt, S. 340
Hurley, S. 586
Hurst, T.E. 412
Hurwicz, L. 503, 504, 518
Husserl, E. 428, 594

Iggers, G. 491

Ikeda, S. 277, 278
Inkeles, A. 204
Irvine, L. 565
Isambert, F.-A. 141
Israel, J. 484
Izadi, E. 564

Jackson, M. 352
Jaffray, J.Y. 504
Jäger, W. 109, 110
Jarvie, I.C. 433
Jepperson, R. 6, 7, 531, 532
Jerryson, M. 333
Jetten, J. 354
Joas, H. 330, 333, 552, 558–560
Johansson, I. 15, 16
Joksch, H.C. 281
Jones, C. 414
Jones, S.G. 342
Joppke, C. 167
Juergensmeyer, M. 333, 339, 341

Kahneman, D. 43, 153, 381, 514, 515, 583, 585
Kalberg, S. 253
Kalman, L. 94
Kampen, J. 339
Kant, E. 71, 625
Kantorowicz, E. 55
Kapitan, T. 330, 342
Kaplow, L. 96
Karpin, M. 341
Katouzian, H. 337
Katz, E. 217, 218
Kaye, H.L. 68
Kazemi, F. 337
Keith, V.M. 412
Kelsen, H. 54–74
Kiecolt, K.J. 412
Kim, J. 23, 596

Author Index   **675**

Kincaid, H. 30, 430, 431, 433, 440, 450–454, 460, 464–466, 475, 482–484, 608
King, A. 427, 428, 663
King, A.M. 412
King, M.L.J. 232, 241
Kinsella, S.N. 284
Kippenberg, H.G. 255, 334, 337, 341, 344
Kirman, A. 515
Kirzner, I. 277, 278, 285
Kitts, M. 333
Kivisto, P. 554
Klandermans, B. 238
Klein, D. 180
Knack, S. 288
Knörr, J. 299
Kohl, C. 299
Kornaï, J. 502
Kranton, R.E. 47, 519
Kreps, D. 505
Krueger, A.B. 279
Kuhn, T. 111, 466, 467
Kumar, C. 484
Kunkel, J.H. 118
Kutsch, T. 103, 108, 118, 119
Kutz, C. 10, 23
Kymlicka, W. 166

Lacey, N. 69
Lacroix, A. 188
Ladson-Billings, G. 413
Lambert, K.J. 289
Lamy, E. 191, 197
Landes, W.M. 90–92
Langlois, S. 165
Laudan, L. 575
Lauer, R.H. 103, 109
Laurent, A. 433–435
Laval, C. 508
Lavoie, D. 277
Lawrence, B. 342, 343

Lawrence, S.K. 412
Lawson, T. 425, 427, 519
Lazarsfeld, P. 218
Leach, E. 304
Le Bon, G. 66–68, 229–232, 234
Lee, H. 412
Lemasson, P. 518
Lemieux, V. 156, 163
Lengermann, P.M. 553
Lesourne, J. 515
Levine, A. 484
Levine, D.K. 284
Lévi-Strauss, C. 153
Levy, D. 84
Lewis, H.S. 304, 306
Libicki, M.C. 342
Liefmann, R. 648
Lightfoot, J. 413
Li, H. 357
Lilla, M. 168
Lindenberg, S. 117, 118, 375, 376
Lipset, S.M. 214
List, C. 383, 385, 450, 452, 454, 478
Littlechild, S. 456, 457
Little, D. 104, 431, 435, 450, 607, 610, 633–636, 642, 652
Littrell, B. 559
Liu, Y. 351, 356–359, 369
Livet, P. 179
Llinas, R. 587
Lloyd, C. 107
Loewenstein, G. 515
Lohlker, R. 329, 334
Lohse, S. 5, 7
Loyalka, P. 357, 358
Lübben, I. 329, 343, 344
Luckmann, T. 144
Lüders, M. 326, 327, 342
Luhmann, N. 110, 263
Lukes, S. 7, 8, 424, 430, 431, 440, 475, 605–613, 652

676 Author Index

M

MacEoin, D. 329
Machiavelli, N. 545
Macy, M.W. 654
Maines, D.R. 551, 558, 559
Maine, Sir H. 110, 315
Majone, G. 188
Malinvaud, E. 500
Mandelbaum, M. 109, 110, 430, 431, 475
Manent, P. 132, 146
Mannheim, K. 70, 213
Manzo, G. 430, 434
March, J.G. 533
Marciano, A. 78, 83, 91, 92, 94, 507
Marginson, S. 352, 356, 357
Marion, J.-L. 147
Markakis, J. 298
Marr, D. 586
Martin, D. 130
Martindale, D. 108
Marx, K. 110, 403–405, 407, 425, 459, 460, 484, 499, 576, 623, 633, 646, 649, 665, 666
Maryanski, A. 49, 114
Mastrogiorgio, A. 590
Maurer, A. 104, 119, 120
Mauss, M. 44, 138
Mayr, E. 109
McAdam, D. 241, 643
McCarthy, J. 236
McClelland, D.C. 118
McDermott, T. 327
McDougall, W. 66
McElreath. R. 40
McGinty, P.J.W. 552, 558, 559, 563, 566
McGraw Donner, F. 334
Mclaughlin, B.P. 660
McLeish, J. 108, 118, 119
McLennan, J.F. 138
McLewis, C. 414
McPhail, C. 552, 557

Mead, G.H. 3, 552, 556, 558–560, 563, 568
Medema, S.G. 78, 82–85, 93, 94
Mehrling, P. 515
Meijer, R. 329, 343
Meijers, A.W.M 15, 21, 22
Menger, C. 79, 271, 425, 432–434, 447, 450, 454, 455, 457–459, 499, 576, 607, 608, 620, 624, 627, 628, 630, 638, 639
Menzel, H. 217
Mercier, H. 45
Mercuro, N. 93
Merleau-Ponty, M. 592, 598
Merton, R.K. 273, 274, 331, 433
Mesure, S. 159
Meub, L. 286
Meyer, J.W. 6, 7, 531, 532
Michels, R. 210
Miebach, B. 104
Miller, D.E. 559, 563
Miller, K. 13, 14
Mills, C.W. 222
Miron, J.A. 282
Mises, L. von 271, 272, 275–279, 426, 434, 436–438, 447, 458, 461, 576, 610, 612, 620, 621, 624, 625, 630, 646, 647
Mitrović, B. 491
Moll, H. 23
Mongin, P. 503
Montalvo, J.G. 288
Moore, W.E. 113, 117
Morgenstern, O. 504, 573
Morin, J.-M. 137, 159
Mouzelis, N.P. 117
Muench, R. 528
Münch, R. 531, 535
Munk, M.D. 354
Munshi, I. 416
Murray, H.A. 21

Author Index   **677**

N

Nadeau, R. 437
Nagel, E. 108, 114, 606
Nagel, T. 41
Napoletano, M. 515
Neck, R. 79
Needham, J. 537
Nee, V. 119, 361, 381
Nemo, P. 434, 435
Nesbitt, T.M. 281
Neumark, D. 280
Nieswand, B. 317
Nik-Khah, E. 80
Nisbet, R.A. 110, 111, 115
Norenzayan, A. 40
Nutter, G.W. 81

O

Oberschall, A. 226, 230, 240
Ogburn, W.F. 108
Olen, P. 74
Oppenheim, P. 104, 540
Opp, K.-D. 383
Opwis, F. 329, 344
Orey, B.D. 412
Orléan, A. 505, 515
Ostrom, E. 226
Ourghi, M. 329, 337, 343
Ozment, S. 226

P

Padioleau, J.-G. 181
Page, S.E. 107
Paino, M. 414
Pareto, V. 133, 183, 203, 383, 385,
    459, 466
Parodi, M. 171
Parsons, T. 4, 21, 70, 110, 112–117,
    204, 330, 331, 333, 558, 653
Passeron, J.-C. 353, 459, 464
Peart, S. 84
Pellow, D.N. 416

Peltzman, S. 280, 281
Petitot, J. 423, 434, 435, 439, 598
Petracca, E. 590
Petroni, A.M. 430
Pettit, P. 17–19, 430, 431, 449, 450
Phan, D. 433, 450
Pharo, P. 162
Picavet, E. 195, 428, 430
Piketty, T. 39
Pinker, S. 40
Pink, J. 344
Piore, M. 508, 515
Polinsky, A.M. 88
Pope, A.T. 281, 340
Popkin, S. 208
Popper, K. 7, 74, 79, 103–107, 115,
    165, 189, 403–410, 415, 416,
    418, 419, 427, 429, 432, 433,
    436–438, 448–450, 452, 457,
    460, 461, 519, 525, 576, 577,
    596, 606, 610–612, 618–627,
    630, 638, 639, 644, 651, 662,
    665
Portier, P. 147
Posner, R.A. 78, 80, 83, 90–96
Poulantzas, N. 459, 460
Powell, B. 283
Pratten, S. 84
Pribram, K. 437
Priest, G. 81
Pritchard, D. 191
Prou, C. 451, 466
Przeworski, A. 607, 613
Putnam, H. 586

Q

Quine, W.O. 574, 596

R

Rabin, M. 340
Radcliffe-Brown, A.R. 304
Radkau, J. 457

678 Author Index

Rahimi, B. 336
Raines, H. 232
Ramadan, T. 329
Ramström, G. 641–643
Randall, S.C. 103
Raub, W. 117, 120, 374, 375, 377, 381, 426, 434, 435
Reardon, S.F. 527
Redford, A. 283
Renzulli, L.A. 414
Rexroat, C. 552, 557
Reynal-Querol, M. 288
Reynaud, B. 179
Riesman, D. 204
Riker, W.H. 301, 385
Risjord, M. 12
Ritchie, K. 475, 476
Ritzer, G. 104
Robbins, L. 271, 279
Robertson, L.S. 281
Roby, Y. 155
Romer, P. 513
Roosens, E.E. 309
Rosenthal, J.-L. 119
Ross, E.A. 67
Rössel, J. 118
Rothbard, M.N. 277
Roth, G. 648
Roth, P.A. 62
Rottenberg, S. 87
Rousseau, J.-J. 207
Roventini, A. 515
Ruben, D.-H. 431, 476
Rucht, D. 235
Ruiz-Junco, N. 551–553, 557
Rüpke, J. 327
Rutherford, M. 79, 92, 93

S

Safner, R. 285
Sahlins, M.D. 110
Salanié, B. 507
Sale, K. 233

Salvatore, A. 329
Sanderson, S K. 109–111
Sandler, T. 377
Savage, L.J. 501, 503, 504, 510, 584
Sawyer, R.K. 430, 431
Schacht, J. 334
Schäfer, P. 338
Scheff, T.J. 105
Schelling, T. 612, 636, 648
Schelling, T.C. 10, 389–391, 393
Scheuch, E.K. 108
Schimank, U. 105, 532
Schluchter, W. 332, 534, 536
Schmid, H-B. 3, 13, 15, 20–23
Schmid, M. 104, 106–108, 115, 119, 120
Schnabel, A. 531
Schnall, S. 590
Schneider, L. 110, 119
Schulze, R. 329, 330
Schumpeter, J.A. 184, 448–450, 453–460, 462–465, 485
Schutz, A. 609
Schützeichel, R. 531
Schweikard, D. 3, 13, 18
Scott, J.C. 654
Searle, J.R. 13–16, 20–24, 62, 586
Segre, S. 554, 556
Sejnowski, T. 577, 581
Sellars, W. 449
Selten, R. 515
Sen, A.K. 38, 195, 383, 385
Service, E.R. 110
Shalin, D. 552, 558–560
Shavell, S. 88, 96, 507
Sherif, M. 47
Shils, E. 4
Shirley, P. 280
Shongolo, A.A. 298, 320, 321
Shvarts, A. 545
Simmel, G. 66, 79, 425, 432, 436, 477, 552–555, 610, 620, 621, 624, 638, 651
Simon, H. 77, 80

Simons, A. 309
Sizgorich, T. 340
Smelser, N.J. 112, 115, 117, 531
Smith, A. 14, 110, 170, 228, 272, 434, 466, 502
Smith, A.D. 110
Smith, B. 598
Smith, D. 409
Smith, D.T. 412
Snijders T.A.B. 119, 381
Snow, D.A. 552, 557
Sober, E. 484
Sombart, W. 213, 214
Sommer, A.U. 330
Song Y. 357
Sorokin, P. 235, 464
Spellman, B. 590
Spencer, H. 110
Sperber, D. 45
Spiekermann, K. 450, 452, 454, 478
Spinoza, B. 194, 625
Sprinzak, E. 341
Stark, R. 254
Stark, W. 205
Staub, E. 231
Steinberg, G. 335
Stern, J. 326, 327
Steward, J.H. 110
Stich, S. 596
Stigler, G.J. 80–83, 85, 86, 88, 94
Stiglitz, J.E. 284, 313
Stoellger, P. 335
Stoutland, F. 12
Strasser, H. 103
Strauss, A.L. 559–561
Sugden, R. 183, 354, 356, 357, 448, 449
Summers, L. 513
Sunstein, C.R. 515, 583
Sutton, F.X. 112
Svensson, J. 288

T

Tarrit, F. 484
Tate, A.S. 414
Tavory, I. 557, 568
Taylor, C. 166, 519
Thaler, R.H. 515, 583
Thatcher, M. 482
Thelen, K.A. 645, 649
The New York Times 233, 289
Thévenon, O. 519
Thompson, E.P. 654
Thomsen, J.P. 354
Thornton, M. 282
Thucydides 206, 207
Tilly, C. 234, 235, 643
Tirole, J. 188, 376, 507
Tiryakian, E.A. 608
Titus, P. 305, 306
Toboso, F. 79
Tocqueville, A. de 132–137, 141, 143, 144, 160, 161, 211, 212, 220
Tollison, R.D. 281
Tomasello, M. 23, 48
Travisano, R.V. 559
Trieu, M. 412
Tuboly, A.T. 74
Tucker, A. 491
Tullock, G. 207–209
Tuomela, R. 13, 14, 450
Turner, J. 114
Turner, J.C. 519
Turner, J.H. 43, 49, 108
Turner S. 62, 68, 74
Turner, V. 337
Tversky. A. 153, 381, 514, 515, 583, 585
Tylor, E.B. 138, 146
Tyrell, H. 536

U

Udehn, L. 5, 82, 425–429, 461, 490, 493, 525, 624, 625

680 Author Index

Uexkull, J. von 593
Ullmann-Margalit, E. 117
Uzquiano, G. 476, 481

V

Vago, S. 109
Vaihinger, H. 74
Van Assen, M.A.L.M. 434, 435
van de Rijt, A. 391–394
van Fraassen, B. 578, 580
Van Gelder, T. 592, 594
Van Horn, R. 78, 80
Varese, F. 545
Velleman, J.D. 12
Vermes, G. 339
Vertocec, S. 166
Viale, R. 574, 575, 577, 578, 582, 584, 590, 593, 596, 598
Vignemont, de F. 590, 591
Viner, J. 291
Voget, F.W. 111
vom Lehn, D. 551
Von Neumann, J. 504, 573, 584
Voss, T. 117, 120, 374, 377, 381

W

Walliser, B. 451, 466, 515
Walter, C. 181, 191
Wang, P. 545
Warren, R.L. 109
Watkins, J.W.N. 4, 5, 7, 403, 407, 448–450, 452, 454, 458, 466, 480, 576, 596, 661, 662, 664
Watson, E.E. 298
Weber, M. 4, 7–9, 32, 34–37, 39, 41, 45–47, 53, 54, 60, 63, 64, 79, 129, 130, 132, 139, 141–144, 147, 162, 214, 219, 227, 249–270, 331, 332, 425–430, 435, 447, 448, 450, 452, 455, 457–460, 462, 463,
526, 528, 530–537, 542, 544–546, 552–556, 576, 577, 607–610, 620, 621, 624–630, 646, 649, 651, 665
Wei, J. 357
Weiss, D.J. 584
Weldes, J. 475, 484
Wellman, B. 156
Wessendorf, S. 166, 167
Weymann, A. 109
White, E.G. 86
White, L.A. 110
Whyte, W. 204
Wieczorek, O. 528
Williamson, O.E. 93
Winch, P. 449, 460
Winckelmann, J. 255
Windelband, W. 330, 331
Winter, E. 166, 169
Wippler, R. 31, 104, 375, 426
Wiswede, G. 103, 108, 118, 119
Wittek, R. 119, 381
Wittgenstein, L. 16, 179, 449
Woessmann, L. 533
Wright, E.O. 484
Wrong, D.H. 557

Y

Yaish, M. 352
Ylikoski, P. 33, 450, 454, 455
Young, J. 298

Z

Zahle, J. 6, 418, 431, 450–452, 454, 460, 464–466, 475, 483, 484
Zald, M. 236
Zaman, M.Q. 329
Zenker, O. 310
Zerubavel, E. 553, 557, 565
Zwiebel, J. 282
Zylberberg, A. 517

Subject Index

A

Abstract psychology 214, 220, 651

Adaptive 110, 228, 231, 238, 241, 245, 308, 586, 596

Adequacy 22, 32, 184, 376, 532–534

Affect 35, 38, 42, 43, 77, 79, 94, 108, 167, 168, 259, 272, 278, 310, 318, 353, 355, 359, 363, 365, 375, 377, 380, 385, 395, 434, 438, 440, 486, 556, 565, 566, 618, 620, 626, 627, 638, 646

Affective 530, 567, 588, 589, 591, 594, 638, 653

Agency 16–18, 20, 21, 23, 71, 73, 85, 235, 251, 268, 269, 288, 297, 299–304, 306, 308, 310, 311, 318, 321, 404, 411, 544, 552, 556, 563, 565, 567, 609, 614, 652, 655, 660, 661

Agent-based 374–376, 391, 641

Aggregation 18, 88, 92, 117, 163, 171, 183, 207, 228, 229, 238, 386, 434, 500, 530, 534–536, 538–540, 579, 581, 582

Altruism 146, 187

Analytical sociology 423, 425, 466, 635, 641, 656

Analytic philosophy 60, 430, 439, 447, 449, 454, 661

Anomie 215, 612

Anthropological 70, 130, 187

Anthropology 130, 134, 136, 153, 298, 304, 306

Atomism 7, 18, 24, 423, 577, 578, 639

Atomistic 7, 9, 20, 136, 424, 425, 439, 606, 613, 618–620, 627, 636, 646, 660

Austrian School of Economics 432, 438, 619, 630

Axiological 127, 159, 163, 179, 223, 625

B

Behaviorism 515

Behaviorist 515

© The Editor(s) (if applicable) and The Author(s), under exclusive
license to Springer Nature Switzerland AG 2023
N. Bulle and F. Di Iorio (eds.), *The Palgrave Handbook of Methodological Individualism*,
https://doi.org/10.1007/978-3-031-41508-1

682   Subject Index

Belief 5, 8, 10, 14, 15, 17–20, 25, 36, 37, 43, 45, 46, 62, 63, 104, 128–131, 133, 134, 136–141, 143–148, 151, 159, 171, 177, 179, 185–187, 189–192, 197, 213, 219, 220, 225–227, 230, 234–236, 251, 257, 262, 267, 269, 331–333, 375, 380, 381, 383, 418, 426–429, 493, 531, 532, 574, 582, 607, 613, 622, 624, 656, 665
Biology 40, 109, 307, 411, 637
Bounded rationality 514, 515, 573, 574, 589, 590, 614

C

Capitalism 32, 39, 82, 226, 266, 268, 415, 416, 435, 457–459, 463, 465, 484, 487, 489, 531–534, 536–539, 544, 546, 608, 646, 654, 665
Causal 6, 8, 9, 23, 32, 54, 71–73, 107, 115, 116, 152, 163, 172, 178, 231, 250, 258, 273, 426, 437, 439, 492, 526, 527, 530, 532–534, 541, 575–581, 596, 597, 607, 609, 618, 621, 623, 627, 633–637, 641–643, 645, 647, 649, 650, 652–655, 660, 661, 663, 665, 666
Causal explanation 579, 596, 635, 641, 645
Causality 5, 6, 9, 32, 70, 71, 73, 107, 162, 163, 172, 273, 333, 434, 435, 575, 634, 638, 646, 660, 664
Causation 6, 228, 268, 279, 435, 437, 534, 635, 640, 650, 661
Cause 87, 135, 166, 194, 221, 226, 228, 230, 232, 245, 252, 253, 255, 269, 280, 285, 325, 326, 342, 364, 425, 430, 433, 448, 482, 532, 537, 538, 542, 622, 626, 635, 648, 653
Chance 44, 129, 222, 241, 275, 282, 365–368, 387, 395, 528
Class struggle 258
Cognitive 47, 131, 137, 138, 140, 148, 152–156, 159–161, 163, 171, 191, 203, 299, 310, 330, 466, 514, 534, 573, 575, 578, 580–594, 598, 613, 638, 652, 654, 657
Cognitive bias 94, 153, 458
Cognitive psychology 40, 153, 584
Cognitivism 581, 584, 586–588, 590, 596, 597
Cognitivist 586
Collective action 117, 154, 155, 178–180, 196, 198, 210, 214, 226, 227, 229, 232, 234–236, 238–246, 643
Collective belief 171, 220, 428, 429, 624
Collective concept 53, 54, 56, 61–64, 165, 430, 455, 456, 458–460, 462, 620, 647
Collective consciousness 34, 37, 66, 73, 462
Collective construct 250
Collective entities 8, 60, 61, 63, 438, 449, 455, 457–460, 463, 465, 518
Collective good 210, 211, 226, 228, 235, 244, 245, 377–380, 395
Collective intentionality 3, 4, 10–15, 20, 21, 23, 25
Collective motivation 251, 257
Collective phenomenon 35, 138, 540
Collective properties 63
Collective representation 29–31, 33–39, 42, 43, 47–49, 463
Collective whole 438, 461
Collectivism 18, 451, 466, 666
Collectivist 16, 406, 451, 621, 660

Communism 254, 456, 458

Communist 230, 234, 237, 242, 456, 484, 628

Communitarian 138, 167, 169, 170

Community 36, 37, 55, 57, 66, 90, 127–131, 136–138, 140, 142–144, 147, 148, 151, 158, 161, 163, 164, 166, 170, 188, 196, 218, 260, 264, 265, 267, 268, 303, 313, 327–331, 333, 334, 339–341, 344, 410, 508, 516, 535–538, 540, 566, 627, 652, 656

Complex 5, 6, 10, 24, 42, 44, 45, 47, 57, 58, 61, 70, 96, 109, 155, 178, 185, 195, 198, 203, 207, 228, 286, 290, 291, 356, 357, 386, 395, 419, 433, 435, 518, 538, 539, 555, 567, 592, 625, 636, 638, 640, 644, 660

Complexity 6, 30, 46, 47, 110, 180, 182, 194–196, 198, 213, 309, 356, 436, 507, 574, 576, 646, 660

Comprehensive sociology 427, 648

Computational 374, 391, 514, 519, 581, 585–587, 589–593, 597, 598

Concept 3, 4, 7, 8, 11–13, 15, 16, 20–22, 24, 31, 41, 45, 53–56, 58–64, 66, 69, 74, 86, 107, 108, 110, 115, 129, 138, 141, 142, 153, 155, 165, 166, 168, 169, 189, 227, 253–255, 257, 258, 291, 305, 327–330, 334, 343, 377, 410, 423, 424, 427–431, 433, 434, 436, 437, 449, 455–460, 462, 463, 465, 476, 486, 520, 526, 533, 554, 558, 561, 563, 567, 568, 574, 582, 585, 590, 591, 595–597, 607, 611, 613, 619, 620, 623, 628, 629, 634, 636, 639, 645, 647, 663

Conceptual 7, 58, 74, 109, 114, 249, 254, 280, 299, 393, 403–406, 408, 410, 448, 456, 457, 459, 461, 533, 554, 556, 563, 568, 577, 584, 596, 606, 612, 656

Conflict 30, 47, 48, 57, 60, 70, 114, 116, 119, 165, 171, 231, 234, 287, 298, 299, 310, 311, 331–333, 341, 356, 405, 415, 541, 566

Consciousness 15, 19, 21, 22, 34, 37, 66, 71, 73, 110, 269, 308, 409, 412, 413, 462, 560, 562, 609, 651, 652

Consequentialist 43, 159, 255

Cooperation 14, 24, 48, 154, 156, 206, 315, 566

Coordination 46, 119, 154, 158, 164, 167, 179, 183, 242, 434, 498, 501–504, 506, 507, 509, 510, 512–516, 518–520, 627, 646

Critical realism 425, 427, 659–661

Critical realist 661, 664

Cultural 32, 33, 40, 41, 74, 79, 82, 88, 113, 127, 128, 130, 131, 133, 144, 146, 151, 152, 163–168, 170, 171, 183, 185, 234, 245, 259, 260, 267, 286, 299, 304, 307, 319, 331, 351–356, 360, 363–365, 367–369, 393, 423, 426, 428, 429, 509, 527, 533, 536, 537, 539, 545, 555, 607, 621, 627, 640, 649, 650, 657, 659–662, 664–666

Cultural capital 351–354, 527, 666

Culturalism 127, 169, 171, 653

Culturalist 430

Culture 40, 41, 146, 151–153, 163, 164, 166, 170, 231, 234, 243–245, 259, 301, 305, 316, 319, 321, 355, 426, 431, 451,

684 Subject Index

487, 491, 532, 533, 546, 562, 564, 614, 621, 662, 664

D

Decision-making 44, 92, 178–182, 184, 185, 192, 208, 209, 227, 228, 307, 308, 353, 355, 357, 361, 383–385, 391, 488, 574, 582–587, 626, 627, 666
Deliberation 46, 307, 661
Democracy 132, 137, 143, 230, 233, 234, 237, 238, 245, 288, 625
Democratic 132, 134, 135, 137, 168, 171, 208, 240, 245, 384, 417
Determinism 127, 128, 130, 190, 353, 425, 433, 613, 648, 662, 665
Determinist 136, 439
Dualism 7, 15, 54, 70, 588, 596

E

Economic behavior 33, 38, 39, 90, 509, 582
Economic order 435
Economics 38, 40, 77–88, 90, 91, 93, 95–97, 153, 209, 227, 271, 273, 279, 287, 290, 304, 312, 313, 315, 317, 318, 332, 376, 382, 383, 385, 386, 393, 430, 438, 447, 456, 473, 482, 485, 487, 490, 492, 497–501, 504–508, 511, 513, 514, 516, 517, 519, 520, 573, 582, 584, 613, 619, 656
Economist 38, 77, 78, 80–83, 85–87, 94–96, 210, 288, 384, 427, 447, 448, 454, 458, 460, 465, 497, 499, 505–509, 516, 517, 583, 608, 639, 640, 646, 647

Education 134, 146, 164, 188, 197, 351–356, 360, 363, 367, 369, 413, 414, 483, 528, 533, 651, 659, 666
Egoism 214, 215, 307
Emergence 3, 85, 108, 117, 152, 154–156, 166, 167, 171, 182, 183, 188, 194, 195, 206, 259, 285, 291, 301, 326, 342, 414, 429, 435, 520, 531, 538, 541, 544, 627, 634, 637, 646, 652, 660, 664, 665
Emergentism 437, 661
Emergentist 435, 660, 665
Emergent properties 434, 611, 612, 634
Emotion 45, 71, 73, 231, 307, 309, 565–567, 581, 588, 590
Emotional 45, 131, 145, 148, 307–311, 366, 582, 588, 591, 593, 594, 657
Empirical 49, 58, 61, 81, 84, 116, 132, 139, 160, 177, 178, 182, 186, 272, 273, 279–282, 284, 285, 300, 308, 312, 333, 355, 375, 377, 381, 383, 385, 386, 393, 394, 408, 414, 438, 439, 457, 502, 503, 506, 517, 562, 564, 566, 568, 574, 575, 577–579, 584, 586, 610, 622, 625, 629, 635, 638, 642, 643, 645, 646, 655, 660, 662–664
Empiricism 516, 607, 660, 662, 663
Empiricist 606, 643, 662, 663
Epistemological 30, 49, 112, 163, 190, 191, 423, 425, 451, 454, 461, 463, 504, 505, 516, 540, 575, 580, 584, 596, 647, 648, 661
Epistemology 407, 410, 411, 417, 418, 574, 621
Equality 169, 170, 234, 305
Equilibrium 112–118, 180, 189, 378–380, 383, 391–393, 395,

424, 498, 500–504, 506, 507, 512, 513, 518, 519, 592, 596
Ethnic 36, 155, 160, 163–165, 167–169, 171, 226, 231, 245, 288, 299, 301, 302, 304, 306, 314, 321, 412, 413, 417, 452
Ethnicity 37, 152, 164, 167, 168, 297, 302, 321, 412, 413, 415
Evolutionary 39, 41, 43, 48, 109–111, 307, 498, 514
Evolutionism 110–113, 116, 514, 515
Explanation 4–10, 33, 41, 48, 54, 58, 61–63, 68, 79, 85, 103–109, 111, 112, 114, 115, 118–120, 129, 131, 133, 136, 147, 152–154, 156, 159, 160, 171, 177, 178, 181, 183, 186, 188, 197, 205, 215, 216, 222, 229, 235, 250–252, 257–259, 265, 307–309, 326, 337, 374, 375, 380, 381, 407, 413, 418, 424, 427, 428, 431–435, 438, 439, 449–455, 460, 463, 464, 477, 478, 480, 489, 491, 500, 525–528, 530–535, 540, 542, 544–546, 575, 577–582, 587, 590, 596, 606–609, 614, 618, 620, 623, 624, 626–628, 633–642, 644–650, 653, 665, 666
Explanatory 6, 61, 68, 105–110, 114–116, 118–120, 132, 177, 183, 188, 189, 418, 426, 437, 526, 529, 530, 533, 567, 577, 579, 610, 619, 634, 635, 637, 641, 645, 646, 648, 651, 660, 662
Explanatory model 7, 104–106, 119, 120, 578

F

Feminist 409–413

Freedom 110, 117, 134, 161, 167, 170, 183, 193, 198, 234, 236, 237, 244, 245, 277, 285, 290, 310, 329, 406, 407, 425, 426, 429, 434, 435, 646
Free will 253, 331
Functional 112, 114, 115, 117, 119, 130, 144, 218, 265, 587
Functionalism 92, 111, 117, 118, 127, 262, 265, 557, 558
Functionalist 109, 112, 114–117, 119, 120, 265, 267, 580, 581

G

Game theory 96, 206, 207, 395, 507, 519
General equilibrium 498, 500–504, 506, 512, 518
Good reason 137, 139, 144, 145, 147, 148, 154, 168, 177, 179, 191, 213, 215, 429, 540, 613
Group Agency 18

H

Habit 134, 135, 141, 145, 259
Habitus 171, 353, 430, 464, 534, 653
Hermeneutic 178, 179, 182, 184, 192, 193, 198, 327, 532
Heuristic 228, 374, 584, 629, 657
Historical 39–41, 78, 84, 109, 132, 138, 161, 162, 195, 257, 262, 278, 279, 284, 300, 302, 303, 317, 319, 321, 330, 331, 337, 404, 405, 407, 413, 416, 417, 439, 454, 466, 467, 490, 491, 493, 512, 525, 531, 539, 540, 558, 607, 608, 617, 623, 628, 629, 637, 646, 649, 652, 661–663, 665
Historicism 403–408, 625
Historicist 403–407, 410, 577

686   Subject Index

History 39, 48, 80, 83, 93, 109,
111, 153, 165, 166, 171, 205,
251, 254, 257, 258, 266,
268–270, 273, 280, 285, 289,
303, 304, 306, 311–314,
317–319, 321, 326–328, 330,
331, 333, 360, 386, 404–407,
412, 414, 425, 435, 439, 463,
475, 491, 493, 499, 553, 565,
574, 607, 610, 613, 619, 633,
638, 649
Holism 4, 7, 9, 22, 29, 30, 34, 49,
79, 81, 82, 403, 404,
406–409, 412, 414, 415,
417–419, 425, 426, 428, 430,
433, 448–454, 463, 464, 466,
473, 474, 477, 482, 487, 489,
491, 493, 499, 607, 617, 623,
633, 635, 650, 651, 655
Holistic 4, 6, 7, 22, 110, 411, 418,
419, 424–426, 430, 433, 439,
451, 453, 459, 460, 484, 487,
498, 499, 576, 606, 611, 613,
620, 623, 644, 646, 647
Homo economicus 38, 508, 510, 515,
518
Human Action 71, 107, 165, 171,
255, 271, 272, 275, 280, 290,
427, 428, 433, 438, 440, 466,
560, 568, 575, 591, 610, 623,
653, 661
Humean 660
Hypostasis 455, 662
Hypostatized 34, 438, 460

Ideal 133, 168, 170, 208, 220, 227,
234, 235, 340, 459, 546, 585,
623, 636
Ideal type 250, 252, 257, 258, 265,
268, 609
Ideal typical 159, 580, 608, 609,
628, 629, 653

Identity(ies) 47, 48, 57, 62, 63, 66,
67, 138, 140, 145, 151–153,
155, 160, 163, 165–169, 172,
196, 235, 297–300, 302–306,
309–319, 321, 353–356, 359,
364–366, 368, 409–414, 417,
418, 518, 519, 555, 556, 580,
581, 596, 641, 654, 655
Ideology 14, 55, 57, 68, 304, 415,
419, 487, 529, 654
Immigrant 44, 158, 160–169, 171,
234, 302, 318, 354, 412
Immigration 44, 151, 152,
159–166, 169, 171, 172
Income inequality 529
Inequality 45, 166–168, 185, 351,
352, 365, 368, 369, 394, 412,
413, 416, 483, 484, 527, 528,
558, 561, 568
Institution 5, 84, 157, 265, 281,
329, 352, 353, 356–359,
361–363, 366, 367, 369, 476,
500, 501, 526, 535, 611, 618,
621, 624, 626
Institutional 6, 18, 36, 40, 79, 81,
83, 84, 88, 93, 96, 117, 152,
157, 179, 184, 190–197, 204,
209, 210, 227, 234, 268, 291,
353, 390, 391, 424, 426–429,
437, 438, 453, 475, 481, 500,
507, 509, 512, 531, 537, 539,
597, 606, 607, 618, 624,
626–629, 638, 639, 649
Institutional analysis 291
Institutional individualism 79, 427,
428, 618, 619, 621, 623, 624,
627
Institutionalized 10, 46, 67, 68, 157,
197, 239, 287, 367, 369, 461,
526, 538, 540, 543
Instrumental 64, 87, 140, 148, 159,
163, 184, 189, 231, 232, 250,
251, 254, 256, 257, 265, 267,

268, 270, 310, 311, 315, 355, 368, 636

Instrumentalism 517

Instrumentalist 516

Interaction 4, 7, 46, 65–67, 108, 118, 119, 154, 181, 183, 186, 192, 195, 203, 204, 206, 207, 212, 222, 235, 245, 300, 311, 477, 490, 510, 535, 538, 553–555, 559–563, 566–568, 578, 580, 588–592, 594, 595, 597, 598, 607–609, 611, 612, 619–621, 628, 653, 665

Interactionism 551–559, 561, 568

Interactionist 551–561, 564, 565, 568, 596, 663

Interdependence 204, 208, 277, 378, 507, 517

Interpretative 4, 39, 131, 139, 181, 182, 196, 433, 436, 519, 532, 619, 622, 624, 626, 647, 650, 652, 653, 662, 663, 665

Interpretive 333, 428, 429, 433, 519, 553–555, 560, 647, 648, 663

Invisible Hand 14, 272, 312, 433, 434, 502

Irrational 130, 139, 141, 210, 219, 232, 252, 254, 298, 307, 308, 369, 384, 405, 408, 417

Irrationality 139, 252–254, 298, 585

K

Knowledge 12, 19–21, 49, 67, 80, 93, 107, 112, 115, 131, 138, 139, 146, 159, 163, 186, 191, 273, 275, 277, 286, 288, 291, 300, 311, 318, 319, 321, 327, 330, 331, 362, 381, 404, 410, 411, 416, 417, 448, 534, 540, 542, 545, 553, 565, 574, 575, 577–579, 582, 586, 595, 597,

606, 622, 634, 638, 639, 656, 660, 663

L

Law 31, 53–59, 61–65, 67, 69–74, 77–83, 85–97, 104, 115, 116, 132, 134, 183, 194, 204, 209, 210, 226, 229–232, 234, 240, 243, 260, 268, 275, 279, 280, 284, 286, 329, 332, 334, 338, 339, 343, 360, 364, 365, 405, 426, 431, 434, 452, 458, 488, 506, 507, 510, 520, 534, 537, 579, 582, 583, 624

Liberal 94, 132, 133, 135, 169, 170, 194, 241, 473, 482, 485, 487, 488, 492, 508

Liberalism 170, 254, 418, 488, 508, 510, 515

Logical 19, 49, 104, 106, 108, 109, 112, 115, 118, 133, 142, 203, 204, 253, 254, 268, 274, 280, 448, 458, 466, 475, 484, 492, 505, 517, 535, 544, 580, 585, 606, 607, 617, 622, 641, 647, 661

M

Macroeconomic 498–500, 512, 513

Macrosociological 152, 162–164, 166, 172, 212, 217, 464, 552, 557

Market 6, 82–85, 91, 93, 94, 133, 180, 184, 185, 188, 189, 195, 204, 210, 211, 272, 276–279, 282, 284, 287–289, 312, 356, 357, 361, 362, 369, 382, 383, 429, 434, 487, 488, 498, 500–504, 506–510, 512–515, 518, 519, 538, 541, 545, 567, 579, 588, 608, 609, 630, 638, 646

688 Subject Index

Marxism 605, 607
Marxist 180, 404, 450, 460, 498, 499, 644, 650
Materialism 134, 136, 484, 666
Materialist 257
Meaning 6, 8, 9, 11, 18, 32, 33, 42, 45, 48, 56, 57, 59, 61–63, 108, 141, 153, 155, 158, 159, 164, 182, 188, 194, 222, 251, 259, 266, 274, 326, 328, 330–332, 335, 337, 339, 395, 424, 426, 428, 431, 447, 448, 452, 454, 461, 463, 498, 516, 527, 532–536, 554, 555, 559, 560, 565, 566, 589, 626, 627, 634, 649, 650, 653, 655
Meaningful 4, 8, 32, 63, 119, 143, 144, 183, 195, 227, 249, 287, 404, 566, 594, 648
Means-ends 133, 271, 330
Mechanism 32, 41, 48, 107, 112, 113, 116, 119, 153, 156, 178, 183, 184, 187, 195, 197, 211, 290, 308, 355, 379, 390, 464, 513, 576, 578, 580–582, 589, 590, 597, 607, 613, 622, 636, 637, 639, 644, 648, 652, 655, 656
Mental 4, 10, 15, 16, 21, 67, 71, 107, 153, 229, 230, 260, 308, 314, 425, 426, 428, 429, 461, 476, 478, 483, 486, 488, 490, 528, 577, 578, 580, 581, 585, 586, 588, 589, 596, 597, 621–623, 625, 628, 639, 641, 642, 652, 656, 657, 661, 663
Metaphysical 40, 41, 53, 55, 56, 58, 143, 219, 336, 435, 436, 438, 623
Metaphysics 60, 335, 628, 664
Methodology 7, 56, 79, 82, 86, 93, 118, 132, 139, 148, 214, 226, 235, 359, 407, 408, 410, 411, 418, 419, 439, 479, 499, 525,

532, 573, 575, 576, 581, 590, 607, 620–622, 626, 630, 633, 641, 646, 647, 651
Microeconomic theory 81, 596
Microfoundations 118, 452–454, 463–465, 544, 546, 607, 614, 633, 634, 636, 637, 640, 641, 645, 647, 653
Micro-macro 374, 382, 388, 646
Microsociological 152, 162, 163, 166, 172, 449, 464, 551–553, 557, 568
Mind 11, 15, 17, 18, 24, 31, 41, 45, 66, 67, 74, 89, 110, 114, 137, 190, 228, 229, 254, 257, 265, 271, 272, 297, 321, 329, 408, 426, 456, 459, 467, 566, 575–578, 580–582, 585–589, 592, 593, 596–598, 609, 621, 623, 634, 641, 649–651, 653, 660, 664
Model 20, 29, 38, 39, 44, 65, 73, 87–90, 92, 104, 107–109, 112, 114, 115, 117–120, 152, 154, 155, 161, 165, 167, 169, 171, 178–182, 184–191, 196, 197, 204, 214, 217, 231, 238, 250, 258, 281, 284, 298, 303, 307, 312, 331–335, 337–341, 344, 355, 373–377, 379, 381, 382, 387–391, 393–396, 439, 461, 485, 486, 499, 501, 502, 504, 510, 512–515, 531–534, 536, 537, 543, 546, 578, 579, 581–586, 588–590, 596–598, 607, 609, 613, 614, 619, 620, 629, 635, 636, 640–642, 644, 648, 660, 664
Modernization 118, 337
Monism 71
Moral 36, 43, 66, 110, 115, 128, 132, 134–138, 140, 144, 146, 147, 223, 231, 234, 235, 260, 265, 311, 316, 328, 333, 403,

406, 407, 500, 509, 510, 589, 612, 649, 653

Morality 55, 65, 145, 226, 406, 506, 510, 605

Moral norm 612

Moral sentiments 231

Motivation 38, 79, 82, 106, 114, 128, 181, 182, 185–187, 189, 190, 192, 226, 250–254, 257, 262, 265, 269, 270, 278, 304, 308, 312, 327, 430, 456, 463, 473, 474, 480, 482, 490, 527, 528, 530, 542, 543, 546, 635, 638, 646, 648, 650, 653

Motive 145, 265, 267, 581

Multi-agent 515

Multiculturalism 152

Multiple-realization 431, 608, 609

Naturalism 190

Neo-Marxism 127, 653

Network 6, 16, 147, 155–159, 163, 190, 233, 362, 368, 374, 387–389, 396, 510, 563, 593, 597, 656

Nominalism 435–438, 611

Nomological 107, 108, 114, 115, 579

Nomologico-deductive 106, 114, 540, 545, 546

Norm 35, 44, 48, 49, 60, 63–65, 177–182, 184, 187, 190, 196, 197, 355, 533, 540, 582, 583, 653

Normative 8, 39, 54, 55, 58, 60–65, 68–70, 72–74, 88, 89, 91, 171, 177–184, 186, 189, 190, 192, 194, 195, 198, 228, 231, 238, 245, 383, 412, 500, 509, 510, 574, 582, 584–586, 596, 608, 613, 624, 627, 629, 636–638, 654

Objectivity 170, 192, 404, 410, 626, 663

Ontological 5–8, 11, 15, 17, 20, 22, 23, 32, 34, 48, 436–439, 451, 453, 461, 463, 474, 479, 480, 484, 485, 487, 488, 491, 492, 519, 538, 542, 576–578, 610, 621, 634, 637, 638, 640–642, 645, 647–649, 651, 652, 655, 664

Ontological individualism 5, 6, 437, 474, 479, 480, 487, 491, 576, 578, 637, 638, 640–642, 645, 649, 651, 652

Ontology 7, 427, 435, 436, 491, 586, 610, 611, 622, 624, 636, 642, 647, 651, 657, 665, 666

Organicism 450

Organization 6, 138, 156–158, 177, 178, 184–186, 188, 189, 191, 194–196, 204, 206–208, 210, 211, 219, 226, 227, 236, 241, 245, 321, 343, 359, 438, 506, 508, 510, 514, 529, 531, 559, 563, 564, 568, 595, 596, 635, 639, 643, 660

Organizational 6, 179, 187, 189, 194–196, 198, 207, 208, 509, 531, 559–561, 563, 564, 595, 633, 635, 638

Paradox 17, 55, 228, 237, 384, 385, 405, 497, 498, 503

Pareto-optimality 383

Passions 133, 135, 145, 187, 228, 267, 361, 365

Personality 129, 143, 165, 204, 232, 249, 261, 414, 415, 629, 653

Phenomenology 251, 254, 261

Philosophy 3, 40, 54, 55, 74, 169, 178, 218, 219, 330, 383, 404,

690   Subject Index

406, 423, 439, 440, 447–450,
466, 490–493, 507, 515, 553,
558, 574, 575, 577, 580, 585,
586, 592, 596, 597, 606–608,
618, 622, 625, 633, 642, 660,
666

Philosophy of mind 40, 466, 577,
580, 586, 596, 597, 608

Physicalism 642

Political economy 185, 188, 227

Political individualism 455, 485, 487

Political science 385, 386, 393, 447,
584

Positivism 59, 69, 70, 607, 625, 626

Positivist 546, 553, 626, 663

Pragmatism 552, 555–557, 568

Praxeology 271

Prediction 111, 178, 181, 184,
186–190, 381, 393, 408, 520,
577

Pre-social 7, 639

Problem-solving 104, 106, 111, 534,
552, 560, 562, 564, 566, 567,
586, 590, 596

Psychic 32, 577

Psychical 34, 37

Psychological 29–31, 33, 37, 39, 40,
42, 45, 48, 68, 107, 108, 153,
215, 231, 253, 254, 262, 278,
307, 308, 424–426, 429,
431–434, 440, 462, 464, 480,
490, 561, 568, 576–581, 585,
590, 591, 618, 619, 623, 624,
629, 639, 640, 651, 653, 655

Psychological individualism 462, 464

Psychologism 424, 427, 490, 573,
576–578, 580, 617–619, 624,
625, 628, 650

Psychology 7, 34, 39, 40, 42, 43,
45, 49, 62, 67, 68, 214, 215,
253, 317, 393, 473, 490, 492,
493, 519, 574, 575, 577, 578,
582–586, 594, 596, 610, 618,
622, 627–629, 634, 637, 639

R

Radical 15, 16, 85, 114, 153, 164,
234, 316, 326, 405, 408, 453,
454, 498, 514, 629

Radicalization 243, 327

Rational 9, 17, 30, 37, 42, 45–49,
64, 72, 82, 86, 88, 90, 92–94,
104, 116, 118, 129, 134–136,
139, 141, 159, 207, 227, 229,
231, 250, 252, 256, 259,
268–270, 298, 300, 307–309,
312, 374, 375, 379, 381–384,
386, 393, 404, 408, 418, 500,
504, 508–510, 518, 525, 526,
532, 533, 536–539, 542, 583,
584, 589, 596, 607, 612–614,
618, 622, 625, 627, 629, 636,
640, 647, 648, 650, 656

Rational choice 46, 82, 227, 250,
254, 255, 268, 269, 298,
307–309, 353, 355, 356, 360,
375, 381–383, 386, 504, 538,
607, 612, 613, 648

Rational Choice Theory (RCT) 137,
249, 250, 254, 255, 270, 308,
374, 381–383, 386, 499, 501,
503, 507, 512, 525, 532, 533,
636, 656

Rationalism 143, 534

Rationalist 189, 439

Rationality 30, 33, 42–46, 48, 49,
96, 134, 137, 140, 142, 152,
154, 159, 160, 171, 189, 190,
203, 215, 227, 228, 251,
254–257, 267, 268, 307, 308,
353, 379, 381–383, 386, 395,
405, 498–508, 510–515, 518,
519, 533, 573–575, 582, 584,
586, 590, 596, 610, 613, 614,
625, 628, 629, 636, 640, 648,
654, 655, 665

Rationalization 129, 132, 141–143,
148

Subject Index

Realism 62, 457, 459, 516, 607, 608, 611, 612, 620, 660, 661, 663, 664

Realist 647, 661, 666

Reason 8–10, 18, 34, 38, 39, 41, 42, 44–46, 58–61, 79, 88, 92, 110, 115, 117–119, 129–137, 139–141, 144–148, 152, 154, 159, 160, 163, 164, 177–179, 181–185, 188, 189, 194, 196–198, 206, 209, 211–213, 219–221, 223, 227, 252, 259, 265, 269, 272, 276, 277, 291, 301, 310, 311, 325, 331, 342, 344, 352, 381, 385–387, 389, 394, 406, 411, 415, 429, 431–435, 437, 460, 465, 474, 475, 481, 483, 484, 490, 491, 507, 520, 527, 528, 530, 573, 574, 576, 577, 582, 585, 588, 618, 621–624, 626, 629, 646, 648, 649, 655, 659, 662, 663, 665

Reasoning 30, 38, 43–50, 57, 58, 67, 68, 95, 131, 187, 274, 275, 307, 308, 377, 382, 384, 386, 389, 393, 395, 396, 476, 489, 574, 581, 582, 585, 589, 590, 606, 656, 657

Reduction 9, 17, 21, 23, 141, 242, 282, 430, 431, 473, 490, 492, 493, 576, 580, 606, 608, 634, 649

Reductionism 9, 21, 37, 39, 40, 104, 136, 153, 154, 423–426, 428–435, 439, 440, 552, 557, 568, 609, 619, 624, 634, 638–640, 650, 651, 661

Reductionist 10, 11, 13–15, 79, 108, 172, 418, 423, 424, 430, 436, 439, 440, 527, 552, 568, 579, 580, 606, 635, 637–639

Relativist 467

Religion 127, 131–144, 146–148, 152, 155, 160, 163, 167, 168, 219, 251, 254–257, 259–261, 263, 267, 268, 270, 297, 298, 302, 303, 320, 321, 325–329, 333, 336, 339, 344, 534

Religious 72, 127–135, 137–148, 156, 163, 164, 186, 187, 219, 225, 226, 234, 237, 245, 251–259, 262, 264, 265, 267, 299, 301–303, 306, 320, 321, 325, 326, 328, 330, 332–337, 339–342, 408, 429, 431, 436, 438, 457, 461, 462, 528, 529, 531, 532, 535–539, 665

Rules 6, 8, 32, 38, 56, 62, 65, 69, 73, 77, 78, 80–82, 84, 86–89, 91–93, 95–97, 117, 129, 133, 155, 171, 178–184, 189, 218, 222, 256, 272, 291, 316, 321, 344, 369, 375, 376, 378, 380, 384–386, 404, 410, 425, 427, 440, 465, 488, 506, 507, 509, 510, 539, 579, 581, 594, 595, 607, 620, 627, 634, 640, 651, 653, 655

S

Science 3, 6, 30, 40, 41, 104, 108, 127, 136, 153, 222, 227, 273, 290, 330, 331, 356, 357, 373, 374, 385, 386, 403, 404, 406–411, 415–419, 423, 427, 429–433, 439, 447–454, 457, 460, 461, 465, 466, 473, 474, 477, 482, 487, 489–493, 514, 573, 576, 582, 584, 590, 596, 597, 607, 609, 610, 618, 620–622, 625, 630, 633, 636–638, 643, 646, 650, 653, 659, 663

Scientific 71, 96, 103, 106, 108, 110, 120, 132, 139, 158, 192,

Subject Index

193, 273, 285, 330, 374, 408, 410, 411, 415, 417–419, 423, 450, 478, 497, 501, 503, 505, 506, 509, 510, 516, 530, 532, 575, 581, 582, 609, 621, 622, 626, 644, 664
Self-interest 44, 72, 134, 171, 182, 187, 189, 227, 272, 508
Situational analysis 648
Social change 103, 106, 108, 109, 111–114, 116, 118–120, 130, 204, 210, 222, 298, 303, 408, 491, 628
Social control 113, 114, 229, 237, 239, 241, 245, 277, 290
Social forms 155, 337, 553, 661, 664
Social influence 131, 392–395, 639
Socialization 117, 138, 171, 555, 613, 636, 639, 653, 656, 665
Socialized 300, 653
Social movement 155, 210, 236, 238, 245, 404, 461, 478, 642
Social need 113, 265, 418, 462, 620, 636
Social norm 30, 33, 38–44, 46, 48, 71, 72, 154, 178, 228
Social structures 5, 9, 35, 106, 112, 118, 120, 152, 299, 301, 318, 354, 427, 432, 451, 455, 459, 464, 465, 526, 527, 619, 620, 626, 627, 639, 641, 645, 648–650, 652, 653, 662, 666
Social value 455, 456
Solidarity 36, 37, 42, 44, 46, 48, 151, 167, 196, 229, 231, 235, 241, 245, 299
Spontaneous order 183, 291, 429
Structural functionalism 127, 653
Structural reasons 424
Structure 6, 13, 15, 23, 24, 31, 39, 81, 106, 112, 113, 115, 119, 130, 152, 154, 156, 157, 171, 194, 195, 203, 205, 206,

216–218, 236, 245, 251, 258, 259, 271, 283, 297, 299, 300, 304, 313, 318, 321, 330, 331, 378, 380, 388, 413, 425, 427, 429, 430, 436, 438, 458, 459, 483, 498, 505, 527–529, 546, 563–565, 576, 577, 580, 581, 584, 586, 588, 589, 613, 623–625, 628, 635, 639, 644, 646, 647, 649, 655, 660, 661, 666
Subjectivism 438, 456, 458
Subjectivity 654
Suicide 159, 203, 204, 214, 215, 234, 451, 453, 454, 462, 528, 529, 651
Supraindividual 485
System 4, 31, 33, 39, 55, 56, 58–60, 62–65, 68–70, 72–74, 78, 82, 84–86, 90–92, 96, 110, 112–116, 118, 132, 144, 154, 166, 188, 189, 192, 195, 204, 206, 207, 212, 217–219, 222, 267, 272, 281, 282, 288, 291, 304, 311, 393, 404, 408, 410, 411, 413, 414, 429, 432–435, 457, 458, 484–488, 501, 507, 510, 529, 532, 542, 562, 564, 565, 586, 587, 591–593, 608, 609, 613, 618, 620, 624–626, 635, 638, 639, 643, 644, 652, 653, 656, 657, 665
Systemic 7, 30, 79, 107, 152, 172, 272, 423–425, 431, 432, 434–437, 439, 440, 463, 561, 620, 623, 635, 638, 639

T

Three-worlds 621, 638
Totalitarian 236, 406, 418
Totalitarianism 277, 407, 408
Tradition 3, 29, 30, 33, 67, 111, 127, 129–134, 137, 139, 148,

170, 180, 190, 220, 221, 227, 264, 320, 333, 340, 405, 425, 427, 431, 447–450, 452, 454, 455, 457, 458, 463, 465, 466, 478, 484, 491, 498–501, 504, 506, 517, 518, 530, 532, 535, 545, 546, 552–554, 556, 559, 561, 563, 578, 606, 607, 610, 617, 619, 647, 648, 650, 663, 665

U

Unconscious 30, 37, 42, 43, 45, 49, 67, 363, 368, 536
Understanding 23, 24, 34, 44, 61, 73, 78, 114–116, 120, 137, 141, 144, 147, 148, 177–179, 183, 186, 197, 214, 222, 232, 234, 253, 254, 256, 269, 270, 273, 275, 280, 291, 326, 328, 341, 342, 344, 419, 429, 434, 449, 460, 474, 475, 477, 479, 481, 485–489, 493, 508, 525, 526, 530, 532, 534, 540, 544, 554, 556, 560, 563–565, 567, 582, 586, 591, 609, 613, 625, 637, 647–649, 651, 656, 662, 663, 665, 666
Understanding explanation 183, 458
Understanding sociology 137, 214
Unexpected 128, 130, 132, 167, 195, 197, 228, 327, 393, 560, 561

Unintended 9, 10, 164, 167, 204, 207, 208, 265, 272–275, 277–280, 282–287, 289–291, 373, 380, 433, 466, 488, 490, 493, 557, 612, 623, 624, 627, 628, 643, 649, 650
Universalism 168–170
Utilitarian 134, 159, 222, 613, 636, 648
Utilitarian individualism 502
Utilitarianism 91

V

Value 66, 91, 96, 104, 144, 153, 164, 170, 182, 186, 195, 227, 241, 245, 251, 254–257, 268, 273, 288, 298, 302, 305, 306, 312–318, 330–332, 344, 355–357, 394, 404, 419, 434, 438, 455, 456, 474, 510, 525, 530, 533, 566, 579, 589, 612, 614, 625, 626, 629, 642, 654–656, 665
Value-rational 255, 530
Value rationality 254–257
Verstehen 227, 249, 428, 429, 458, 532, 554, 624

W

World 3 621–623, 651, 662, 665